CW00496247

R. Hassler F. Mundinger T. Riechert

Stereotaxis in Parkinson Syndrome

Clinical-Anatomical Contributions to Its Pathophysiology

With an Atlas of the Basal Ganglia in Parkinsonism by R. Hassler

Foreword by E. A. Spiegel

With 163 Figures in 235 Separate Illustrations

Springer-Verlag Berlin Heidelberg New York 1979

Professor Dr. med. Rolf Hassler
Direktor der Neurobiologischen Abteilung des Max-Planck-Instituts für Hirnforschung,
Deutschordenstraße 46
D-6000 Frankfurt/Main 71, Fed. Rep. of Germany

Professor Dr. med. Fritz Mundinger
Ärztlicher Direktor der Abteilung Stereotaxie und Neuronuclearmedizin der Abteilungsgruppe Neurochirurgische Universitätsklinik, Albert-Ludwigs-Universität Freiburg, Hugstetterstraße 55
D-7800 Freiburg i.Br., Fed. Rep. of Germany

Professor Dr. med. Traugott Riechert
em. Direktor der Neurochirurgischen Universitätsklinik Freiburg i.Br.
Sonnhalde 10
D-7800 Freiburg i.Br., Fed. Rep. of Germany

ISBN-13: 978-3-642-66523-3 e-ISBN-13: 978-3-642-66521-9
DOI: 10.1007/978-3-642-66521-9

Library of Congress Cataloging in Publication Data. Hassler, Rolf, 1914–. Stereotaxis in Parkinson syndrome with an Atlas of the Basal Ganglia in Parkinsonism. Includes bibliographical references and index. 1. Parkinsonism, Pathophysiology – Surgery. 2. Stereotaxis. 3. Brain – Functional Localization. I. Mundinger, Fritz, 1924– joint author. II. Riechert, Traugott, 1905– joint author. III. Title. [DNLM: 1. Stereotaxis technics. 2. Parkinsonism. WL359 H355s] RD594.H38 616.8′33 76-30573

Softcover reprint of the hardcover 1st edition 1979

Reproduction of the figures: Gustav Dreher GmbH, Stuttgart

Monophoto typesetting, printing, and bookbinding: Universitätsdruckerei H. Stürtz AG, Würzburg

2123/3130-543210

Foreword

Despite the amazing progress made by the stereotactic technique, particularly regarding the localization of the target, despite the extreme caution,which stereotactic neurosurgeons apply at every step of the procedures, despite the routine roentgenologic and physiologic controls (depth EEG, electric stimulation) preceding the production of a lesion, there remains a certain degree of uncertainty regarding the position, shape and extent of the lesion as well as of the electrode track and also regarding unintended lesions in the vicinity. The final answer to these questions depends on the anatomical control.

Thus we must be grateful to the authors, who performed 3700 subcortical stereotactic interventions, that they present a careful comparison of the clinical and of the pathologic-anatomic findings, in the cases whose brains could be studied. They demonstrate what degree of exactness can be expected from the stereotactic technique. They also present a careful outline of the morphology, physiology, biochemistry and pathology of the structures that are affected in Parkinson patients. In addition this monograph permits one to familiarize oneself with the methods of stimulation and destruction practiced by the authors. The relevant literature is conscientiously quoted. A study of this monograph will benefit not only the stereotactic neurosurgeon, but also neurophysiologists, pathologists and practicing neurologists.

Philadelphia, Pennsylvania

E.A. Spiegel

Preface

This book is essentially the outcome of the joint work of three men over a period of years: a brain researcher and neurologist, a skilled neurosurgeon who was thoroughly conversant with Kleist's concepts of cortical localization of psychic and somatic functions, and a young neurosurgeon.

In scientifically backward and impoverished post-war Germany there was much discussion in neuropsychiatric circles of the merits and demerits of leucotomy and prefrontal lobotomy, introduced in 1935 by Egas Moniz. At this time the eminent psychiatrist and physician Kurt Beringer assembled a team of neuroscientists to develop new therapeutic methods in psychiatry and neurology for therapy-resistant patients facing permanent disability. Traugott Riechert was called to Freiburg University to become Professor of Neurosurgery and Rolf Hassler was sponsored to become Docent there, Beringer having recognized the significance for his project of Hassler's research on the functional Organization of the human thalamus. Fritz Mundinger also became involved in the project. E.A. Spiegel and H. Wycis had designed, after the Horsley-Clarke instrument (1908), a stereotaxic apparatus for use in human patients and they performed the first operations placing lesions in the dorsomedial thalamic nucleus (SPIEGEL and WYCIS, 1947, 1948). Riechert and Hassler were to explore the possibilities of treating extrapyramidal disorders by inactivating small circumscribed subcortical structures instead of destroying extensive areas in the motor cortex.

Riechert took on the task of developing the surgical instruments and with the physicist M. Wolff designed a targeting device with a phantom ring. With this instrument any subcortical structure could be reached from any point on the surface of the skull. Hassler devised a method for determining the specific target points in individual patients which avoided the use of average values (HASSLER and RIECHERT, 1954). The results enabled Riechert and Hassler to perform stereotaxic coagulations, first in the medial and then in motor thalamic nuclei. In 1950, they used the stereotaxis technique for the surgical treatment of athetotic hyperkinesia. HORSLEY (1909); ANSCHÜTZ (1910); PAYR (1921); BUCY and CASE (1937) and others had suggested that this could be influenced favourably by extirpations in the contralateral motor and premotor cortex. The harmful side-effects of ablating these cortical areas could probably be avoided by inactivating the thalamic nuclei that project specifically to the motor and premotor cortex, as suggested by Hassler. The results proved that coagulation of the target structures markedly reduced or even abolished the athetotic movements, without impairing active movements.

The positive results of stereotaxic coagulation in these motor thalamic nuclei, and earlier reports of relief of parkinsonian tremor by ablation of area 4 (BUCY, 1938; BUCY and CASE, 1939) or by interrupting the lateral pyramidal tract at the cervical level ('pyramidotomy'; PUTNAM, 1940) encouraged Hassler and Riechert in 1951 to destroy stereotaxically the thalamic targets in a patient with postencephalitic parkinsonism. Immediately after coagulation the tremor in the contralateral limbs as well as the muscular rigidity disappeared completely. Richard Jung and his associates, especially R. von Baumgarten, made it possible to record EEG potentials during stimulation and coagulation. This task was taken over by Wilhelm Umbach and subsequently by J.A. Ganglberger, who some years later elucidated the physiological background of the stimulation and coagulation effects by analysing the EEG and EMG recordings.

In the years 1953/1954 the success of inactivating the thalamic nucleus V.o.p for tremor and the V.o.a nucleus or the pallidum for rigor ushered in the world-wide application of stereotaxic therapy in Parkinson's disease. In 1955 Mundinger and Riechert improved the technique by using a modified pure sine-wave high-frequency coagulator (originally designed by Wyss in 1945), because it appeared that the irregular high temperature peaks generated by the spark-gap diathermy instrument were causing harmful side-effects. Two years later they further refined the procedure by controlling the temperature with a thermoprobe at the tip of the wire electrode.

A further advance in tremor treatment was the additional inactivation with a string electrode of the dentato-thalamic fibers caudally before they enter the V.o.p nucleus. These improvements reduced the incidence of complications and 2000 operations had been performed by 1962.

The stereotaxic inactivation of specific thalamic projection nuclei was based on the hypothesis of a functional antagonism between descending nigral and thalamo-corticospinal systems. The therapeutic effect of the operation seemed to validate this, but pathoanatomical corroboration was still lacking. It was, therefore, imperative to study histologically the brains of patients who had died from intercurrent diseases or from complications of the operation. Although the mortality rate was only about 0.8% we were able to examine serial sections of brains from 17 patients with 23 operations.

The scope of this book is to elucidate the clinico-anatomical correlations in parkinsonism between the pathophysiology and the mechanisms underlying the therapeutic effect of stereotaxic surgery. Recent advances in drug therapy, e.g., in the form of L-dopa treatment, have greatly reduced the number of patients with parkinsonism who need stereotaxic surgery. Nevertheless we consider that the book will be of importance for further research in this field because it explains the pathophysiological basis of parkinsonian symptoms, including the physiology of the interaction of the neurotransmitters involved.

The plan to write this book was conceived in the late fifties, long before the bulk of material contained in it had become available. The processing of the material from autopsy for histological study and the examination of the serial sections took longer than anticipated, and several sections had to be rewritten as new results emerged. The writing suffered further delay owing to the many commitments of the authors to their clinical work and scientific studies.

We wish to express our special thanks to the translator, Dr. S.L. Last, London, and to Mrs. P.W. Fried of Springer-Verlag. Thanks are also due to Dr. J.F. Christ of the Max-Planck-Institut für Hirnforschung who helped in many ways in the completion of the text. We are indebted to our photography team for the excellent photomicrographs and to Mr. Kampe for his perseverance and precision in lettering and preparing the illustrations for printing. The final manuscript was the product of many secretaries at the Neurochirurgische Klinik in Freiburg (especially Ms. Heinicke) and the Max-Planck-Institut für Hirnforschung in Frankfurt (especially Mrs. Trojan). We would like also to thank the staff of Springer-Verlag for their expert advice and help.

September, 1979

R. Hassler
F. Mundinger
T. Riechert

Contents

I. Introduction

During the last 25 years, stereotaxic operations have achieved general recognition as one of the most successful treatments for the Parkinson syndrome. Their outcome depends on circumscribed surgical destruction of certain nuclei and tracts in the diencephalon. These areas can be reached by aiming instruments built for this purpose and by a special method of localization (SPIEGEL *et al.*, 1947; TALAIRACH *et al.*, 1949; LEKSELL, 1949; RIECHERT and WOLFF, 1950, 1951 a, b; GUIOT and BRION, 1952; NARABAYSHI and OKUMA, 1953; HASSLER and RIECHERT, 1954a; RIECHERT and MUNDINGER, 1956). During these operations, stimulation produces neurologic and psychological effects that are linked to certain diencephalic structures. Thus far, it has been possible to confirm in only a few cases at autopsy whether the structures aimed at were, in fact, reached by the stimulating electrode and the area of coagulation (SPIEGEL and WYCIS, 1962; MARK *et al.*, 1963; KRAYENBÜHL *et al.*, 1964; MACCHI *et al.*, 1964; HASSLER *et al.*, 1965, 1969, 1970; PAGNI *et al.*, 1965; MARKHAM *et al.*, 1966; SMITH, 1967; HANIEK and MALONEY, 1969; HARTMANN-VON MONAKOW, 1972).

This monograph therefore has four aims: (1) to relate post-mortem findings to the corresponding particular effects of stimulation and of functional loss as well as to the clinical result; (2) to deduce functions of certain cerebral systems; (3) to decide on the best therapeutic steps and their targets; (4) to demonstrate in a series of frontal sections the diencephalic structures from a Parkinson patient. In addition, certain complications that have been observed in these cases and their relationship to the anatomical substrate will be described.

II. Basis of the Parkinson Syndrome: Morphology, Physiology, Biochemistry, and Pathology

The Parkinson syndrome is the most important and common extrapyramidal disease. According to international estimates, it constitutes 0.3–0.5% of all diseases. The term "extrapyramidal" was coined in 1898 by PRUS in reference to some animal experiments. Following STARLINGER (1897), he caused generalized convulsions in the dog by cortical stimulation, even after severing both pyramids in the medulla oblongata. After ANTON (1896) had described status marmoratus in double athetosis, the concept of the extrapyramidal system was specially formulated by C. VOGT (1911) and C. and O. VOGT (1920). It was taken from clinical pathology and was used to denote motor disorders that were due not to a lesion of the pyramidal tracts but to damage to another motor system. To this negative characterization of the system, VOGT (1911) and WILSON (1912) added positive findings observed in the nuclei of the basal ganglia. VOGT (1911) described a "striate (or extrapyramidal) system" in the narrower sense, which consisted of the following structures: (1) the striatum with its two parts, the putamen and the caudate nucleus, (2) the pallidum, (3) the nucleus subthalamicus (corpus Luysii), (4) the nucleus ruber, and (5) the substantia nigra (Figs. 1, 2, 4, 6). The effects of this system on motor activity travel by efferent pathways, which were not identified for many years. In C. and O. VOGT's (1920) concept this system had no connection with cortical areas, and this concept led the experimental investigation of the structures in other directions for a long time. Today, even after many new investigations, the efferent pathways of the extrapyramidal system are known only in part. It can be regarded as established that the extrapyramidal system predominantly affects the peripheral motor neurons via the pyramidal tract (HASSLER, 1949a, c, 1953b) and other corticospinal systems, as shown in Figure 2.

1. Afferent Inputs of the Extrapyramidal Motor Nuclei

A motor system without regulation by the receptors of muscles and joints is unthinkable; it could never induce a useful movement. This observation

---

Fig. 1. Schematic drawing of afferent and efferent pathways of extrapyramidal motor system (*in color*). Basal ganglia and thalamus are exposed through rectangular window in middle of left hemisphere. Specific projections from discrete thalamic nuclei to discrete cortical fields of somatosensory region are drawn in. Projection from somatosensory thalamic nucleus V.c.p (here *c.p*) to area 2 in postcentral gyrus (*blue*) represents joint sensibility [*c.a* (*blue*) means V.c.a]. Projection from thalamic nucleus V.im (here *im*) to area 3a in bottom of central sulcus (*green*) is the central representation of Ia muscle spindle afferents. These somatosensory cortical fields send out descending pathways through internal capsule to substantia nigra (*Ni*) and pons (*Po*) (*green*). Thalamic nucleus V.o.p (here *o.p*) is supplied by dentato-thalamic fibers and projects immediately to area 4γ (all *violet*), in anterior wall of central sulcus, where the thickest corticospinal fibers originate. They descend through internal capsule and cerebral peduncle with collaterals to substantia nigra and pontine gray matter and proceed, crossing in pyramids of bulb into contralateral funiculus of spinal cord, to facilitate also directly the motoneurons (*violet*). Anterior part of V.o nucleus (here *o.a*) receives its specific input from internal segment of pallidum (*red*), which is under direct control of putamen (*Put*) and caudate nucleus (*Cd*) (both *green*). Nucleus V.o.a projects directly to cortex in front of area 4γ, especially to area 6aα (*red*); this area projects downward to spinal cord through direct and indirect cortical-spinal connections. The most anterior thalamic nucleus, lateropolaris (here *po*), receives specific input from anterior part of lateral segment of pallidum (*red*), which is also under the primarily inhibitory influence of caudate nucleus and putamen. Afferent pathway to both (mainly to the outer) segments of the pallidum starts from intralaminar thalamic nuclei (*i.La*). L.po, roughly identical with VA nucleus, projects specifically to supplementary motor area (*6spl*). – The motor control apparatus of putamen and caudate nucleus consists partly of corticostriatal fibers, omitted here. Caudate nucleus and putamen regulate activity at outer segment of pallidum (*Pa.e*), of nerve cells of substantia nigra (*Ni*), and of pontine gray matter (*Po*). Substantia nigra and

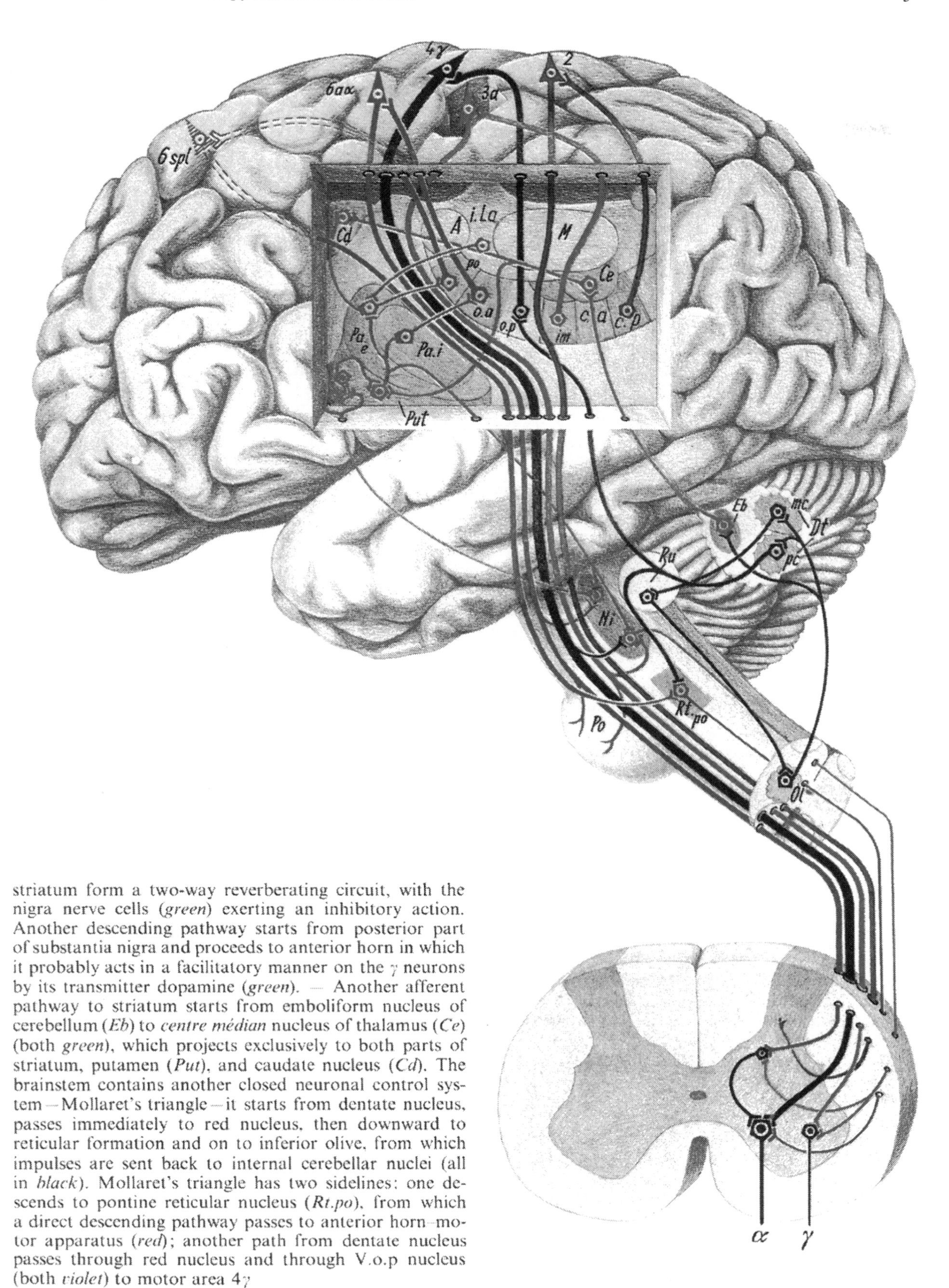

striatum form a two-way reverberating circuit, with the nigra nerve cells (*green*) exerting an inhibitory action. Another descending pathway starts from posterior part of substantia nigra and proceeds to anterior horn in which it probably acts in a facilitatory manner on the γ neurons by its transmitter dopamine (*green*). — Another afferent pathway to striatum starts from emboliform nucleus of cerebellum (*Eb*) to *centre médian* nucleus of thalamus (*Ce*) (both *green*), which projects exclusively to both parts of striatum, putamen (*Put*), and caudate nucleus (*Cd*). The brainstem contains another closed neuronal control system — Mollaret's triangle — it starts from dentate nucleus, passes immediately to red nucleus, then downward to reticular formation and on to inferior olive, from which impulses are sent back to internal cerebellar nuclei (all in *black*). Mollaret's triangle has two sidelines: one descends to pontine reticular nucleus (*Rt.po*), from which a direct descending pathway passes to anterior horn–motor apparatus (*red*); another path from dentate nucleus passes through red nucleus and through V.o.p nucleus (both *violet*) to motor area 4γ

is valid also for the extrapyramidal system in the narrower sense. Sensory inputs come from the vestibular nerve and certainly also from muscle receptors. Internal afferent influences come from nuclei and areas of the extrapyramidal system, which will be described (Figs. 1, 2, 4).

1.1. Putamen and Caudatum (= Striatum)

The *putamen* and *caudatum*, the two highest centers, receive their afferent input mainly from the cortical areas, the Nigra and the *centre médian* Figs. 1, 4. The magnocellular part (Ce.mc) of this nucleus projects onto the caudatum, and the parvocellular part (Ce.pc) projects onto the putamen[1] (C. and O. VOGT, 1941; HASSLER, 1949c). The *centre médian* belongs to the nonspecific projection system of the thalamus. It is, therefore, a thalamic nucleus without direct connections to the cortex. There are two sources of its afferent input: (1) the reticular formation of the midbrain through short nonbundled fibers (HASSLER, 1964) and (2) the brachium conjunctivum (Figs. 1, 2, 6). From findings in man it is known that its cerebellar input originates in the nucleus emboliformis of the cerebellum (HASSLER, 1950a), which was recently confirmed in cats by the Nauta-Gygax method (HASSLER, 1969, 1972). It follows that the two highest centers of the extrapyramidal system, the putamen and caudatum, are linked to the cerebellum, the central organ for motor coordination. The indirect afferent input from the cerebellum is of much greater importance for the coordination effect of the putamen and caudatum than is the nonspecific input from the reticular formation and the Nigra.

[1] Recently proved by experimental bouton degeneration (HASSLER, 1975; CHUNG *et al.*, 1977; HASSLER *et al.*, 1978) the transmitter seems to be acetylcholine (WAGNER *et al.*, 1975).

1.2. Pallidum

The *pallidum* has a histologic structure that merits the name "reticular formation" more than does the part of the brainstem. Its netlike structure is an expression of a specially marked convergence of stimuli and of an intensive linking of the individual neuronal elements. Afferent input from the sensory system, i.e., the lemniscus medialis, comes through the internal capsule to the pallidum and possibly even from the spinothalamic system for protopathic sensations. Convergence of sensory inputs into the pallidum has been established by electrophysiologic investigations of individual cells of the pallidum (SEGUNDO and MACHNE, 1956). In addition, the cells of the pallidum can be stimulated from the acoustic and visual systems. Their pathways were unknown for a long time, particularly since no anatomical evidence could be produced for a direct connection between the reticular formation and the pallidum. This afferent input goes through the nuclei of the unspecific projection system of the thalamus, to which the intralaminar nuclei (Figs. 2, 6), the nucleus limitans, and the previously mentioned *centre médian* belong. These thalamic nuclei, also called truncothalamic, do not project directly to the cortex but

Fig. 2. Scheme of afferent pathways of pyramidal and extrapyramidal motor systems and their connections to cortical areas in frontal, central, and parietal regions. Most of these thalamocortical connections are two-way connections and run in parallel. Superficial dorsal nucleus (*D.sf* or *LD*) (*blue*) receives afferents from pallidum internum; it projects to upper parietal lobe, especially to area 7. Input to area 5 from dorsal nuclei and corticostriate fibers are not drawn in. Posterior column input for joint sensibility through medial lemniscus runs to V.c.p thalamic nucleus, which projects to area 2 in the postcentral gyrus. Cervicothalamic tract after passage of lateral column terminates in thalamic V.c.a nucleus (*blue*), which projects to area 1. Spinothalamic tract for pain sensation (*blue*) terminates among other structures in thalamic nucleus V.c.pc projecting to area 3b (*blue*). Ascending pathway (*red*) for muscle spindle excitations starting in CLARK's column enters intermediate ventral nucleus (*V.im.e*) (*green*), which projects to area 3a in depth of central sulcus. Thalamic nucleus V.o.p, supplied by dentato-thalamic fibers, projects to area 4γ of precentral cortex. Rostrally adjacent area 6aα (*red*) is supplied by projections from V.o.a, which receives its input from inner segment of pallidum (*Pa.i*) with afferents from outer segment of pallidum (*Pa.e*) and from putamen and caudate nucleus. Vestibulothalamic pathway terminates in inner part of nucleus ventro-intermedius (*V.im.i*), which projects to area 6aδ of premotor cortex. Interstitial nucleus of Cajal (*Ist*) responsible for rotatory movements to same side is connected with inner part of oral ventral nuclei (*V.o.i*), which projects to area 6aβ. Magnocellular part of lateropolar nucleus (*L.po.mc*) of thalamus (*red*), itself supplied by afferents from pallidum externum, projects to supplementary motor area (*6 supl*). Finally, rostral part of V.o.i projects to area 8—frontal oculomotor area. The often-mentioned pallidum is under control of putamen and caudate nucleus (*green*), respectively. Both form reverberating circuits with distinct cell groups of substantia nigra. Two parts of striatum, caudate

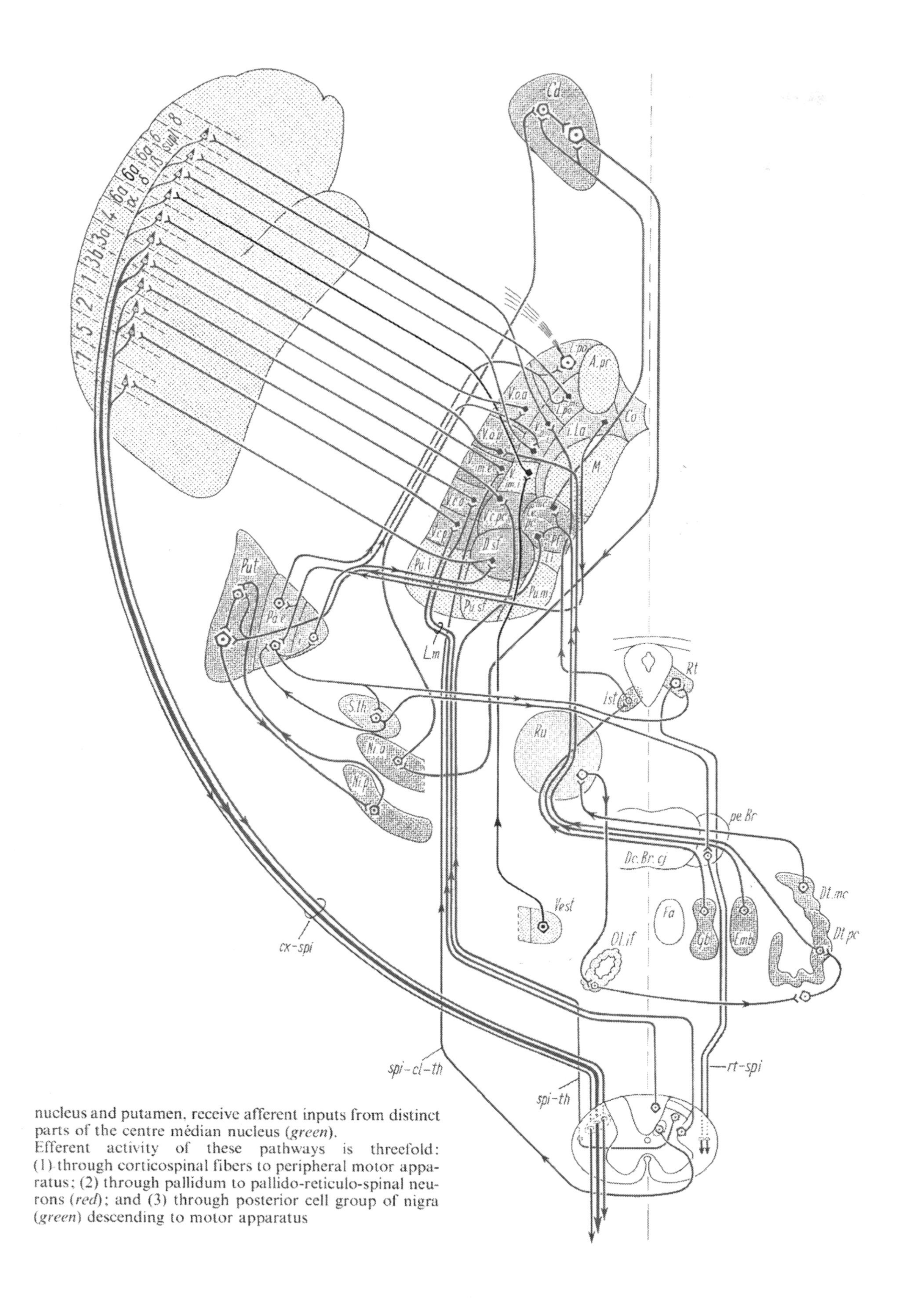

nucleus and putamen, receive afferent inputs from distinct parts of the centre médian nucleus (*green*).
Efferent activity of these pathways is threefold: (1) through corticospinal fibers to peripheral motor apparatus; (2) through pallidum to pallido-reticulo-spinal neurons (*red*); and (3) through posterior cell group of nigra (*green*) descending to motor apparatus

have direct fibers going to the external part of the pallidum. The afferent pathways to the intralaminar nuclei run from the reticular formation of the midbrain (Fig. 6) through short pathways and from the spinothalamic tract (MEHLER *et al.*, 1960; BOWSHER, 1957; HASSLER, 1960) and the ascending vestibular system by long pathways, e.g., through the midbrain interstitial nucleus of Cajal.

1.3. Nucleus Subthalamicus

The nerve cells of the *nucleus subthalamicus* (corpus Luysii) degenerate after destruction of the pallidum externum, since they are reciprocally connected (Fig. 2). Statements in the literature regarding the afferent input into the subthalamic nucleus are contradictory; in particular, it is not certain that it is linked to the sensory systems. There are no connections with the substantia nigra. Different parts of the subthalamic nucleus probably receive different groups of fibers from the mass of fibers of the pedunculus cerebri (MEYER and BECK, 1954; MEYER *et al.*, 1947). Stripes through the subthalamic nucleus showing reduction of cells, demyelination, and gliosis can be seen with lesions in the internal capsule. Instances of this are shown in Figures 77a and 133b.

1.4. Nucleus Ruber

The *nucleus ruber* or red nucleus obtains its afferent input mainly from the cerebellum. The whole mass of fibers of the brachium conjunctivum, the upper stalk of the cerebellum, enters the red nucleus. The majority of these fibers pass through the red nucleus. Only the fibers originating in the magnocellular part of the dentate nucleus terminate here (Figs. 1, 2) in human material (HASSLER, 1950a), while the experimental material points to the interpositus nucleus as the origin of rubral afferents. In addition, the red nucleus receives afferent input from various cortical areas of doubtful significance. Connections with other extrapyramidal nuclei are minimal, if they exist at all. There are no fibers linking the red nucleus to the substantia nigra. One has to be especially cautious in transferring some results from animal experiments regarding the red nucleus to man, since some portions of the red nucleus develop to form its main part only in the higher primates and in man. In the cat, on the other hand, the bulk is formed from different parts; from these, direct pathways descend to the motor neurons in the rubrospinal tract.

1.5. Substantia Nigra

As far as is known, the *substantia nigra* receives no afferent input from the sensory system, nor has it any direct connections with the reticular formation. Two large systems of afferent impulses to the substantia nigra are known: (1) from the putamen and caudatum and (2) from various cortical areas. As regards the strionigral fibers, it has been found that circumscribed parts of the striatum are linked to only one group of cells in the substantia nigra in both directions. Other areas of the putamen and caudatum have connections with other cell groups (ROSEGAY, 1944; HASSLER, 1950c, 1953b; SZABO, 1962, 1967; USUNOFF *et al.*, 1974). The same specificity is found for the fiber connections of the corticonigral systems originating in the frontal lobe, the central areas, and the parietal and temporal lobes (MURATOFF, 1893; DÉJERINE, 1901; MINKOWSKI, 1924; BECK, 1950; HASSLER, 1950c). It follows that each group of cells in the substantia nigra has both a circumscribed cortical input and an input from a distinct part of the striatum. If both afferent input systems to a cell group in the substantia nigra are interrupted, only the cells in this group atrophy and eventually disappear altogether.

This brief and incomplete survey shows that every nucleus of the extrapyramidal system receives afferent input from several sources, which are able to regulate its activity.

2. Connections Within the Extrapyramidal System and the Functional Significance of Individual Nuclei

2.1. Putamen and Caudatum (=Striatum)

The *putamen* and *caudatum*, being the highest nuclei of the extrapyramidal system, act on other extrapyramidal nuclei through two tracts: (1) the striopallidal fibers and (2) the strionigral connections. Because the disappearance of cells in the putamen and caudatum is the most important pathologic finding in Huntington's chorea, the striatum was considered, following C. and O. Vogt (1920), to have an inhibitory function. This could not be confirmed in experimental animals with small lesions (Wilson, 1914). Mettler *et al.* (1939) inactivated larger areas in the putamen and caudatum of monkeys, which led to signs of motor disinhibition, particularly with regard to running movements. If the lesion of the caudatum was one-sided, the movements were directed toward the damaged side. Comparable phenomena had already been observed in the last century by Magendie (1841) and Schiff (1858) as well as by Lussana (Lussana and Lemoigne, 1871; Lussana, 1886) and Nothnagel (1873, 1889) using crude techniques. As a result, the nucleus lentiformis was called the *nodus cursorius*. Only Kennard (1944) additionally saw involuntary movements of head and limbs in the first few days after setting a lesion in the putamen. These locomotor movements as related to putamen and caudatum have been demonstrated with certainty and with accurate localization within the nuclei (Fig. 3a–f) by our own experiments on cats (Dieckmann and Hassler, 1968; Hassler and Dieckmann, 1968). Stimulating the caudatum usually leads to running movements to the contralateral side (Akert and An-

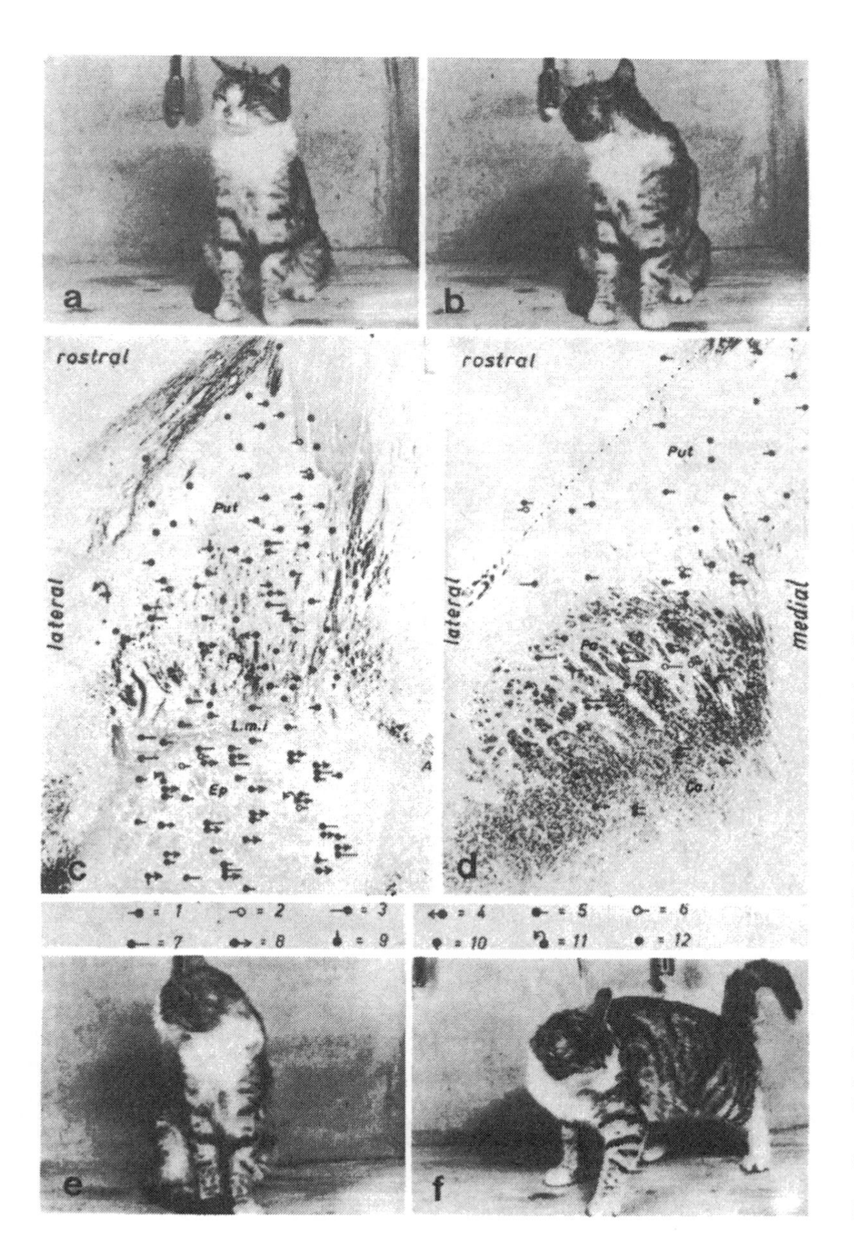

Fig. 3. (a–e) 30 cps stimulation in cat putamen (*Put*) without general anesthesia causes ipsiversive turning of head. [Compare (b) and (e) to head posture before stimulation in (a).] Stimulation of dorsal margin of pallidum (*Pa*) and of adjacent internal capsule (*Ca.i*) elicits contraversive turning of head (f). *Ep*: entopeduncular nucleus; *L.m.i*: lámella pallidi interna. Marchi staining. × 19. (c, d) Diagrams of horizontal sections at two different levels of basal ganglia (1.5 mm apart), showing sites of stimulating points producing direction-specific movements. Contraversive turning and circling are produced mainly by stimulation in entopeduncular nucleus (*Ep*) and pallidum (*Pa*). Ipsiversive movements are obtainable primarily from putamen (*Put*). Explanation of signatures—1: ipsiversive turning of head; 2: ipsiversive turning of eyes; 3: ipsiversive turning of body; 4: ipsiversive circling; 5: adversive turning of head; 6: adversive turning of eyes; 7: contraversive turning of body; 8: adversive circling; 9: raising of head; 10: lowering of head; 11: head rotation around anterior-posterior axis of head; 12: negative effect. (According to Hassler and Dieckmann, 1968)

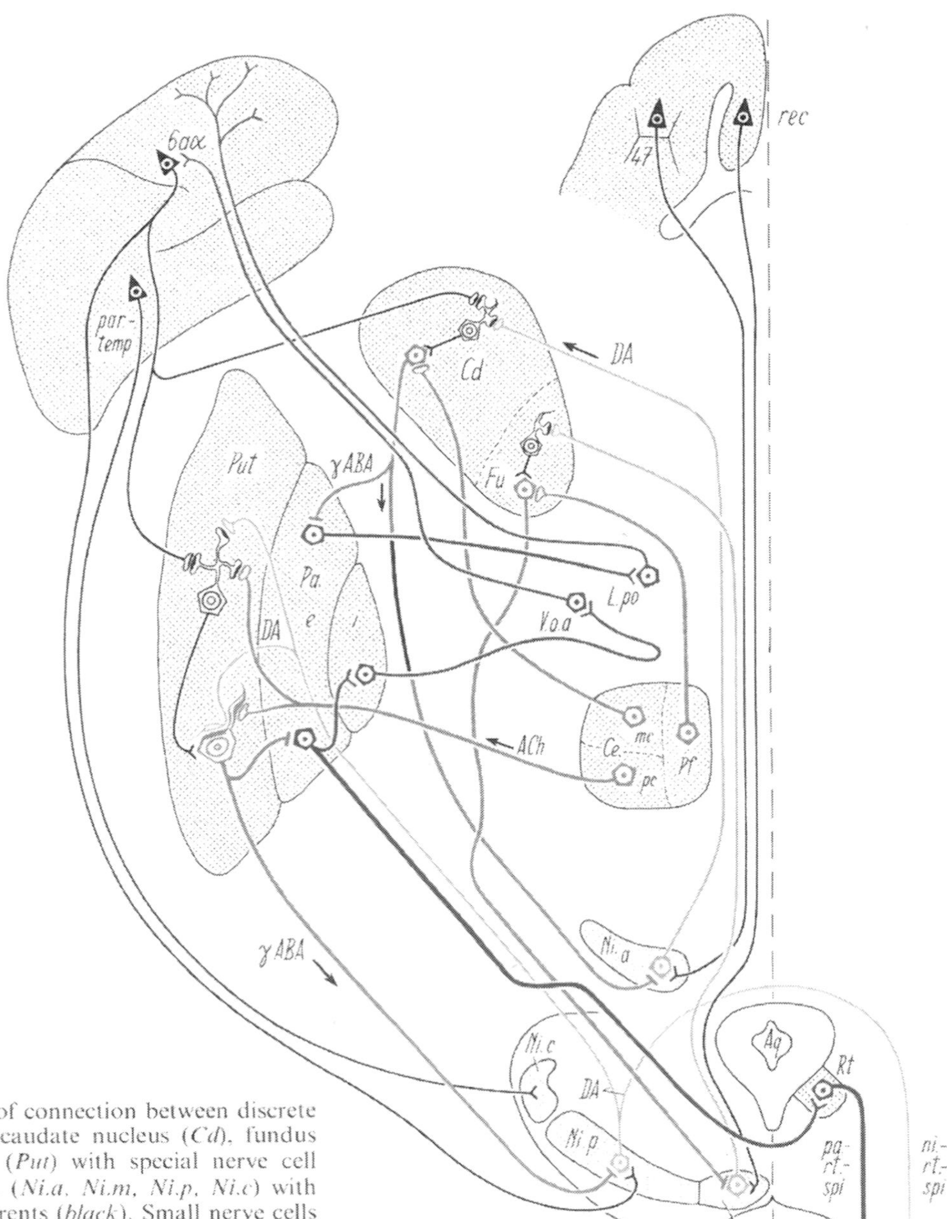

Fig. 4. Schematic drawing of connection between discrete parts of striatum such as caudate nucleus (*Cd*), fundus striati (*Fu*), and putamen (*Put*) with special nerve cell groups of substantia nigra (*Ni.a*, *Ni.m*, *Ni.p*, *Ni.c*) with their probable cortical afferents (*black*). Small nerve cells of striatum (*black*) receive impulses from substantia nigra (*orange*) with transmitter dopamine (*DA*) by axospinous synapses. Same nigrostriatal pathways influence large nerve cells by a *bouton en passant*. Putamen receives nigral input from posterior main section of nigra (*Ni.p*), caudate nucleus from anterior main section (*Ni.a*), and fundus from medial cell group (*Ni.m*). At the same time, small and large striatal nerve cells receive a special type of bouton from centre médian nucleus of thalamus, probably with acetylcholine as transmitter. On small striatal cells there are axospinous synapses and on large, axodendritic synapses (according to Hassler and Chung, 1976 and Chung *et al.*, 1976/1977).
Small striatal nerve cells receive third input from different parts of cerebral cortex, shown in detail only in putamen and caudate nucleus with glutamic acid (*Glu*) as transmitter (Kim *et al.*, 1977).
Fundus striati receives second afferent input from parafascicular nucleus of thalamus. Efferent mechanism of strionigral circuits goes mainly through pallidum (*red*); there, direct pathway from pallidum externum to rostral pole of thalamus (*L.po*), main ascending pathway through pallidum internum, and pallidothalamic fibers passing Forel's fields H_2, H, H_1 terminate in thalamic nucleus (*V.o.a*). This nucleus projects to area 6aα of premotor cortex. Nucleus L.po projects diffusely to transition zones of parietal, prefrontal and premotor cortex. Third efferent pathway descends from Pa.e to midbrain reticular formation (*Rt*) to spinal cord (*pa.-rt.-spi*). Another efferent pathway seems to branch off from posterior main section of nigra (*Ni.p*) and descends, after crossing midline in midbrain, toward spinal cord (*ni.-rt.-spi*)

DERSSON, 1951) during which muscular coordination in the limbs is completely preserved, as is the coordination of the limbs themselves. Stimulating the putamen clearly has the opposite effect upon turning movements, since the animal's circling movements are directed to the side of the stimulation (Fig. 3a–f). This indicates an inhibitory effect of the putamen on the functionally subordinate pallidum. The pallidum is the hub of the systems of the central nervous system for movements (Fig. 3f) to the opposite side (adversive movements) (MONTANELLI and HASSLER, 1964). This inhibitory function was demonstrated by AKERT and ANDERSSON (1951) by stimulating according to the method of HESS (1941, 1954). They described a syndrome of inactivation that was later interpreted by HESS as partial sleep (1954). During stimulation of the caudatum, spontaneous movements were slowed down, reduced, and eventually stopped (delayed inhibition). A similar effect can be obtained from the putamen; however, it depends on the frequency. Higher frequencies of stimulation lead to an arrest reaction, i.e., immediate interruption of all spontaneous or other movements. This response is evidence of the inhibitory function of the putamen and also of the caudatum (DIECKMANN and HASSLER, 1968). This inhibition affects by GABA-ergic synapses the pallidum, whose excitatory symptoms can be suppressed by stimulation of the putamen. Monosynaptic inhibitory function (YOSHIDA and PRECHT, 1971) can be ascribed to the connections of caudatum with the anterior substantia nigra, and probably of putamen with the posterior main part, the stimulation of which nevertheless results in contraversive turning (Fig. 26). It is the opposite effect of putamen stimulation (WINKELMÜLLER, 1972).

2.2. Pallidum

The *pallidum* receives a large number of fibers from the putamen and caudatum in addition to the afferent input described previously. The activity of the pallidum influences the function of the motor system in two different ways. The first efferent pathway leads through the ansa lenticularis to the subthalamic nucleus and various motor regions of the mesencephalon, including part of the reticular formation. The second pathway, by which the pallidum can influence other systems, starts at the pallidum internum, passes through the internal capsule at the level of the subthalamus (Fig. 5a, b), and goes via the fasciculus thalamicus (Forel's field H_1) to the anterior part of the oral ventral

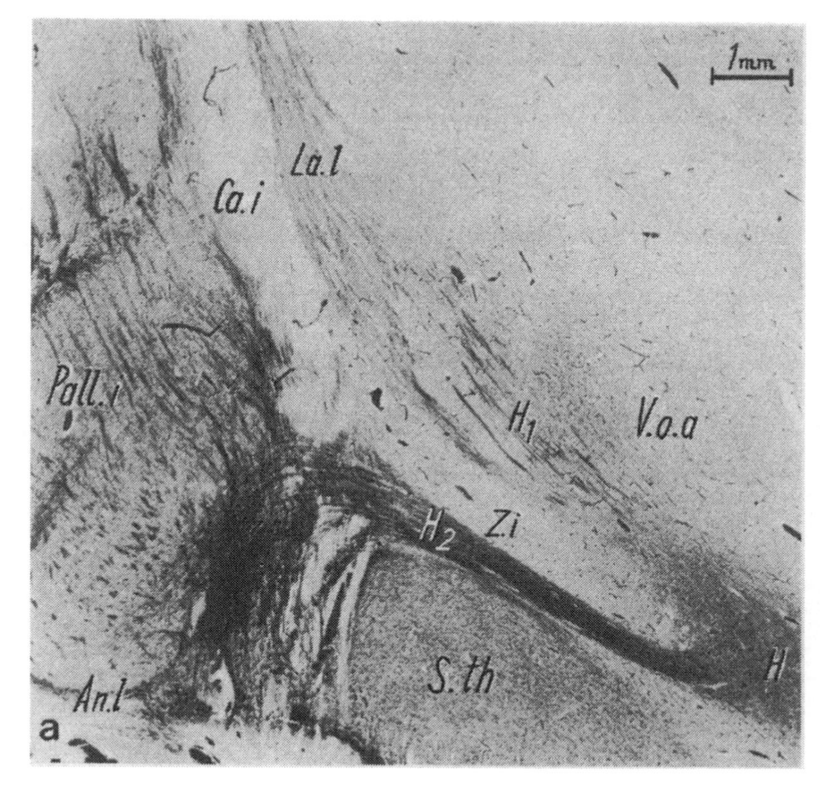

Fig. 5. (a) Transverse section through anterior thalamus of 8-month human fetus. Internal capsule and most of thalamus are not myelinated in this figure, so they appear white; pallidar neurons, still myelinated, are prominent. Efferent pallidar fibers, starting from ansa lenticularis (*An.l*) and from internal segment of pallidum (*Pall.i*), converge on medial border of pallidum. They curve upward medially, cross internal capsule, and form H_2 bundle above subthalamic nucleus (*S.th*). Lenticular fascicle, H_2, runs medially and curves upward in Forel's field H. Its fiber bundles are distributed in region of thalamic nucleus V.o.a. The terminals of these pallidothalamic fibers are marked by H_1. These fibers and some fiber bundles in lamella lateralis thalami (*La.l*) are myelinated in this section. (After HASSLER, 1949a)

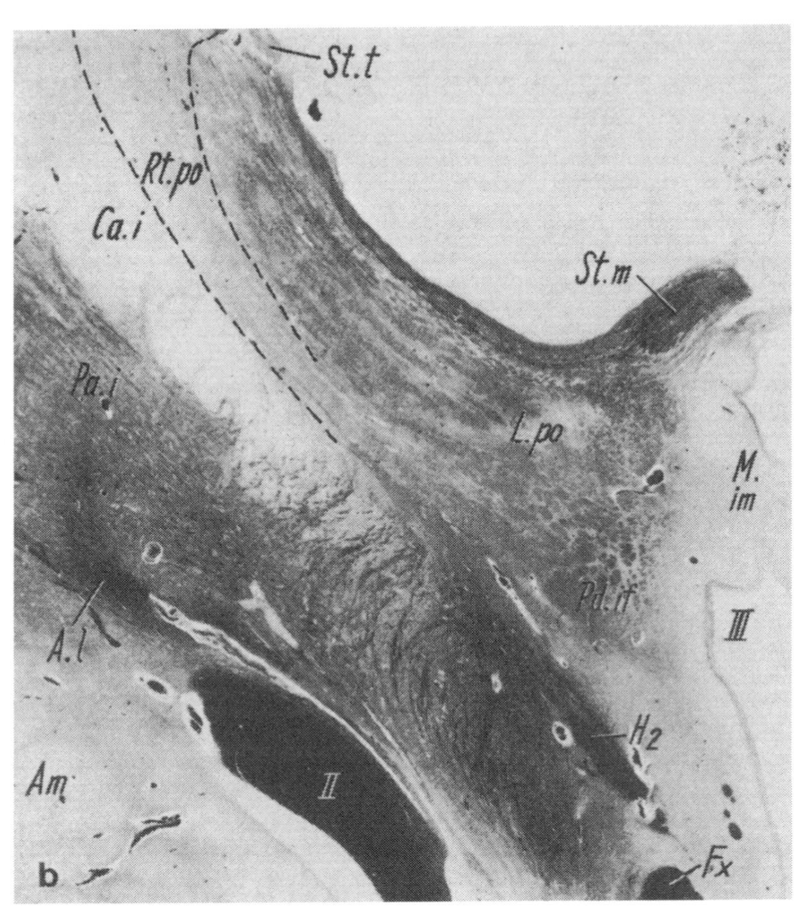

Fig. 5. (b) Cross section through anterior pole of left thalamus in a case of striatal apoplexy that shows complete degeneration of all fibers of internal capsule; thus, internal capsule (*Ca.i*) appears white in fiber staining. It is passed by many single fibers originating in pallidum (*Pa.i*) and passing through internal capsule to anterior pole of thalamus (*L.po*) where they disappear in fiber network. These are direct pallido-anterior-thalamic fibers. The fiber bundle H_2 of Forel, which conducts fibers from pallidum to V.o.a nucleus, and other efferent fibers are not degenerated. Inferior thalamic peduncle (*Pd.if*) is almost completely preserved because it is supplied by efferent pallidar fibers and by other fibers from basal nucleus, which reach the base of thalamus below internal capsule

---

Fig. 6. Color schematic drawing of non-specific thalamo-cortical system and of the subcortical efferents of pallidar system. From the periaqueductal reticular formation (*Rt*) an activating afferent connection (*yellow*) arises to the non-specific projection nuclei of the thalamus (*yellow*). The intralaminar (*i.La*) and the commissuralis (*Co*) nuclei project to the outer segment of the pallidum (*Pa.e*), the centre median (*Ce*) projects to the caudate nucleus (*Cd*) and Putamen (*Put*). The striatum (*green*) is inhibitory in relation to start reaction and to increased awake activity. Descending pathway to anterior and to posterior sections of substantia nigra (*Ni.a; Ni.p*) arises from caudate nucleus as well as from putamen. These descending strionigral neurons have γABA as inhibiting transmitter. The efferent pathways from both sections of substantia nigra project to caudate nucleus but another efferent pathway of posterior section of nigra descends to motor apparatus, crosses midline, and passes through reticular formation (*ni.-rt.-spi*). The transmitter seems to be dopamine (*DA*). From posterior section of substantia nigra, also an ascending dopaminergic pathway runs to putamen where it exerts an inhibitory action. Stimulation effect of putamen is ipsiversive movement marked by a circle with a line to left side. Caudate and putamen neurons exert an inhibitory influence via γABA on neurons of pallidum, external and internal segments as well. One efferent pathway of pallidum runs through H_2 bundle (*red*) and terminates in subthalamic nucleus (*S.th*). Another crosses midline in anterior midbrain and terminates on contralateral reticular substance around aqueduct (*red*). Stimulation effect of this pallido-subthalamo-reticulo-spinal (*pa.-rt.-spi*) pathway is contraversive movement marked by small circle with arrow to the right. Main connections from pallidum pass through internal segment and terminate in two cortico-dependent nuclei of thalamus, *V.o.a* and lateropolaris (*L.po*). These nuclei project to area 6aα in premotor cortex or to area 6aβ and supplementary area (*6 supl*) on medial aspect of hemisphere; its multiple red lines spread to almost all cortical fields. Another efferent pathway of pallidum supplies middle part of medial nucleus and anterior nucleus of thalamus (not drawn in this Figure). Both project to cortex, namely to prefrontal

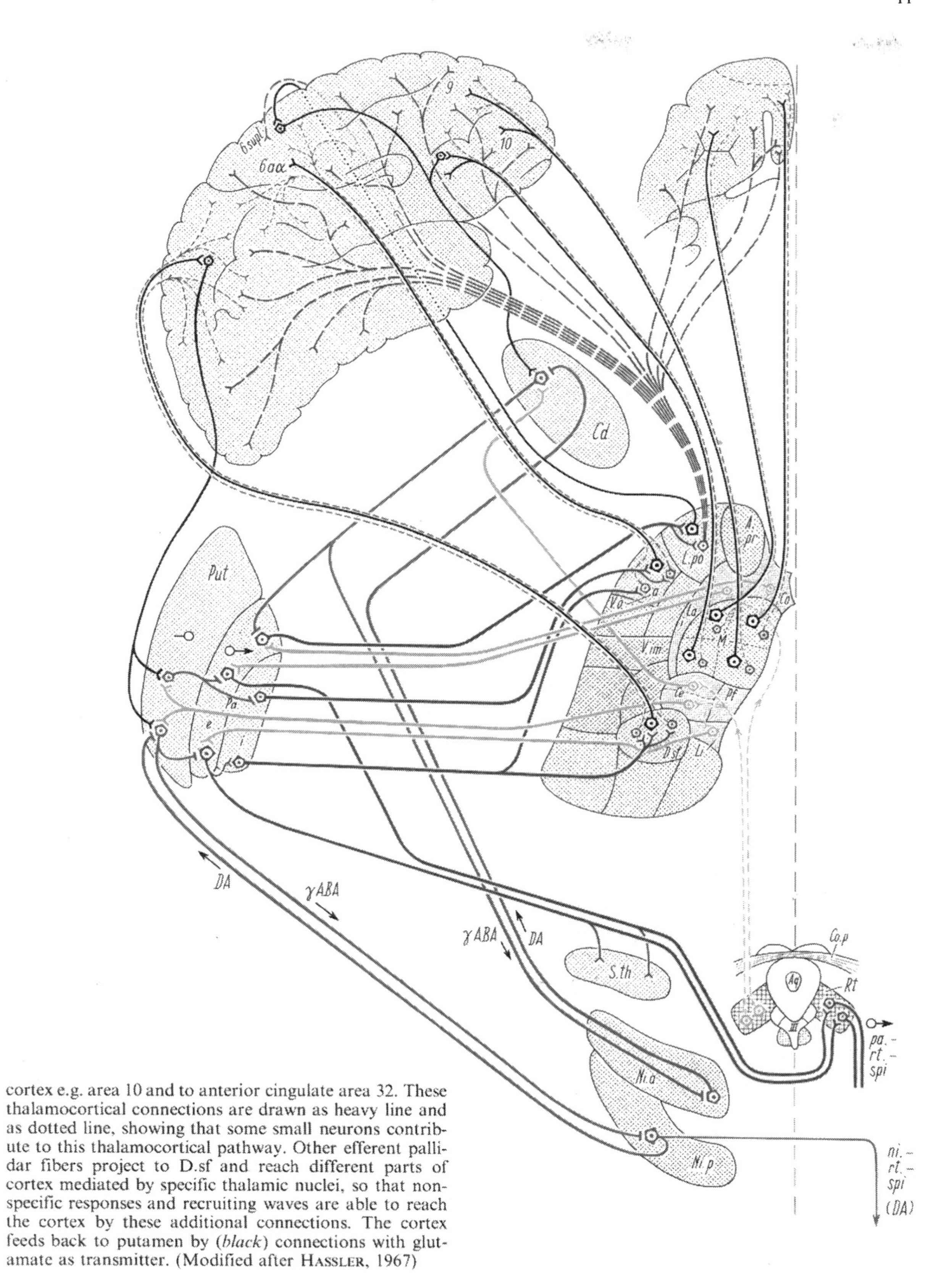

cortex e.g. area 10 and to anterior cingulate area 32. These thalamocortical connections are drawn as heavy line and as dotted line, showing that some small neurons contribute to this thalamocortical pathway. Other efferent pallidar fibers project to D.sf and reach different parts of cortex mediated by specific thalamic nuclei, so that nonspecific responses and recruiting waves are able to reach the cortex by these additional connections. The cortex feeds back to putamen by (*black*) connections with glutamate as transmitter. (Modified after HASSLER, 1967)

nucleus (V.o.a, or rostral part of VL) of the thalamus. This nucleus has an efferent projection to the motor cortex. This efferent pathway of the pallidum conducts excitatory impulses to area 6aα of the motor cortex; it influences the peripheral motor apparatus via the premotor cortex and the pyramidal fibers originating in it. Hence one can formulate the following paradox: The pyramidal tract is an important efferent pathway of the extrapyramidal system. Fibers from the pallidum (Fig. 56) reach the rostral pole of the thalamus (*L.po*), the Ventralis anterior (VA) of ARONSON and PAPEZ (1934), which is of great electrophysiologic interest. According to DIECKMANN and SASAKI (1970), low-frequency stimulation of the pallidum causes typical recruiting responses in field potentials in area 6aβ of the cat; this is a so-called unspecific influence of pallidum-stimulation on the cerebral cortex.

Until 1920 the Parkinson syndrome was regarded as due to disappearance of neurons from the pallidum (status desintegrationis of C. and O. VOGT, 1920). Since TRÉTIAKOFF'S (1919) thesis, Parkinsonism without etiologic distinction was attributed to nigra cell loss. The anatomical examination of further cases of postencephalitic Parkinsonism showed that in all of them cells had disappeared from the substantia nigra. For some years this caused workers (e.g. SPATZ, 1927; 1935) to assume that only in paralysis agitans were the Parkinson signs due to a lesion of the pallidum while in postencephalitic Parkinsonism they were due to loss of cells in the substantia nigra. Since the post-mortem investigations of Parkinson syndromes of different etiologies in serial sections have revealed constant cell loss, not in the pallidum, but in the nigra (HASSLER, 1938), the localization of all kinds of Parkinsonism in the nigra has been accepted. Only for cases of Parkinsonism following carbon monoxide poisoning is it still claimed (ZIMMERMANN, 1949) that the lesion of the pallidum is the only basis. Such a cystic defect without Parkinson

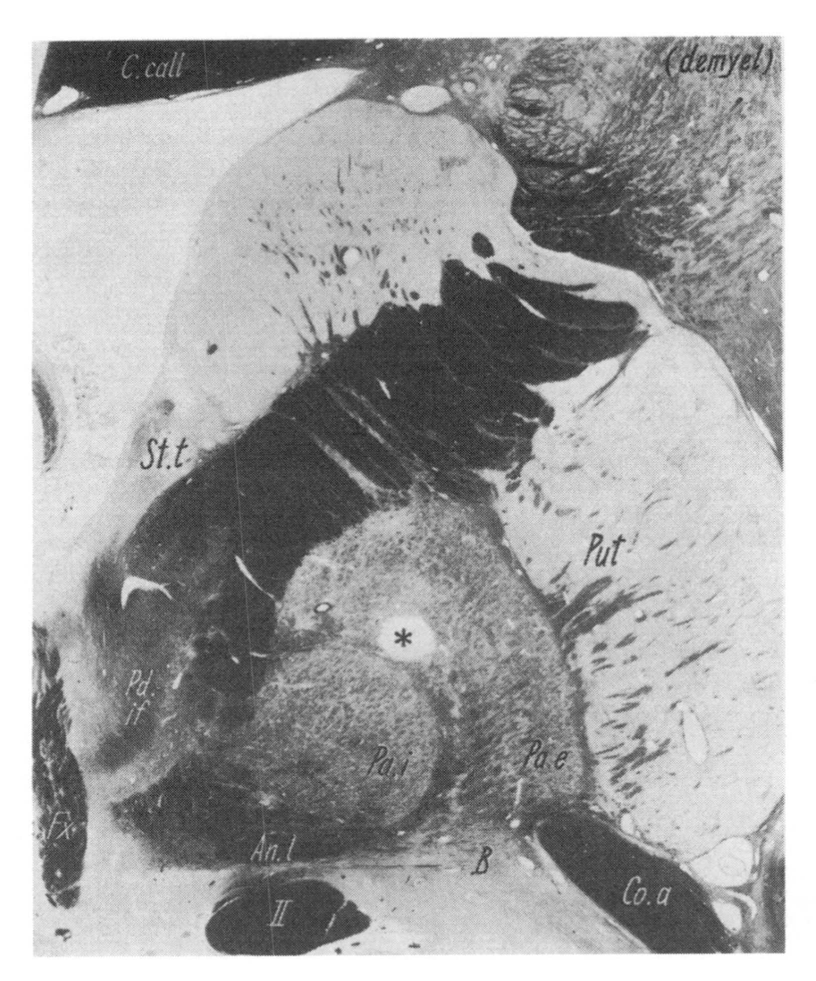

Fig. 7. Transverse section of bassal ganglia in a case of CO intoxication. Patient had survived for many years a deep coma due to CO intoxication without any sign of Parkinsonism. There is small cystic defect (*), surrounded by demyelination, near Lamella pallidi interna in outer segment of pallidum (*Pa.e*). Even strong demyelination (*demyel.*) in white matter of hemisphere, in contrast to corpus callosum, did not produce clinical signs. (Observation of W. KRÜCKE, courtesy of Prof. KRÜCKE)

signs is demonstrated in Figure 7. Some forms of athetosis are also said to be due to lesions of the pallidum, i.e., athetosis in cases of progressive atrophy of the pallidum of VAN BOGAERT (1946), and cases of status dysmyelinisatus of C. and O. VOGT (1920) following kernicterus. The extension of status marmoratus from the putamen to the exterior part of the pallidum or the transneuronal degeneration of pallidar cells in status marmoratus can be the cause of athetosic disorders of movement. In animal experiments, bilateral destruction of the pallidum has as a rule not led to motor signs. Only F.A. and C.C. METTLER (1942) reported inactivity with cataleptic postures in monkeys after bilateral pallidum destruction. In especially large lesions of the pallidum that involved neighboring structures, KENNARD (1949) sometimes observed hypertonus and action tremor. WILSON (1914) found no effect on stimulating the pallidum, while METTLER *et al.* (1939) achieved phasic movements of the limbs and a plastic muscle tone by the same means. Stimulation of both parts of the pallidum using the technique of HESS (1948, 1954) in the nonanesthetized cat regularly caused turning movements to the opposite side (Fig. 3c, d, f). Their speed could be increased by increasing the frequency and intensity of the stimulation (MONTANELLI and HASSLER, 1964).

Stimulating the pallidum internum in man during stereotaxic brain operations at higher frequencies leads to conjugate eye movements to the contralateral side (Fig. 104). They take the place of turning the head, which is impossible because the head is fixed (HASSLER, 1957b; HASSLER *et al.*, 1960; HASSLER and RIECHERT, 1961). They are usually combined with widening of the palpebral fissure and pupils and sometimes with excitement (Fig. 105). To this extent experiences on stimulating the pallidum in man coincide with those in animal experiments. The pallidum is a system of neurons that governs adversive movements and turning to the opposite side (Fig. 6).

Destruction of the pallidum internum in man has been performed almost exclusively in cases of hyperkinesis and the Parkinson syndrome (Fig. 98). The immediate result is a reduction of muscle tone in the contralateral limbs, evidently in direct proportion to the volume of pallidum that has been destroyed. In the pallidum there is a somatotopic arrangement for head, arm, and leg from rostral to caudal (see Fig. 29) (HASSLER, 1961, 1975; MUNDINGER and POTTHOFF, 1960).

From the nucleus subthalamicus, which is directly linked to the external part of the pallidum, descending tracts go into the brainstem, though their exact endings are insufficiently known. Fibers run to the symmetrical structure in the other hemisphere through the commissura hypothalamica posterior. In experiments on immobilized cats, stimulation of the nucleus subthalamicus led only to mydriasis and opening of the eyes; when the cats could move freely they reacted by turning toward the other side (HESS, 1954). Inactivating this area leads to a mirrorlike turning to the side of the lesion. The results of the stimulation are due to activation of efferent fibers from the pallidum (HASSLER 1956b) in the fasciculus lenticularis (Forel's field H_2) and probably not to activation of the neurons of the nucleus subthalamicus itself. In animal experiments, inactivating the nucleus subthalamicus leads to rhythmic movements toward the same side, comparable to the turning movements (MELLA, 1924; WALLER, 1940). These movements are probably due to additional damage to the efferent pathways from the pallidum in bundle H_2. Older descriptions (LAFORA and GLUECK, 1911; D'ABUNDO, 1925) of hyperkinesia following lesions of the subthalamic nucleus in carnivores were confirmed by CARPENTER *et al.* (1950) in monkeys. These authors describe the hyperkinesia as choreoid hyperkinesia. Since JAKOB (1923), localized lesions of the subthalamic nucleus have been regarded as causing contralateral hemiballismus. In a large number of post-mortem observations, the link between a lesion in the nucleus subthalamicus and contralateral hemiballismus has been confirmed (MARTIN, 1927; MARTIN and ALCOCK, 1934; VON SÁNTHA, 1928; BALTHASAR, 1930; TITECA and VAN BOGAERT, 1946; WHITTIER, 1947; KRAYENBÜHL *et al.*, 1964; HOPF *et al.*, 1968).

2.3. Nucleus Ruber

The *nucleus ruber* or red nucleus obtains its major afferent input from the dentate nucleus. Large bundles of fibers coming from the deep cerebellar nuclei run through it and terminate in the thalamus. Impulses from the red nucleus use two pathways (Fig. 2): (1) that from the magnocellular red nucleus forms the rubrospinal tract, which in man is rudimentary as is the magnocellular part from which it comes; (2) the major part of the neurons in the red nucleus pass into the central tegmental tract, as demonstrated by WEISSCHEDEL (1937). Most of its fibers terminate in the inferior olive.

Since the inferior olives project exclusively to the cerebellum, the nucleus forms a feedback system to the cerebellum (HASSLER, 1950a). GUILLAIN and MOLLARET (1931) showed that circumscribed lesions of this triangle — red nucleus, inferior olive, dentate nucleus — cause myorhythmias or myoclonus (Fig. 1). In animal experiments lesions of the red nucleus that are restricted to the nucleus itself cause a coarse ataxia with slow intention tremor and myoclonus. The hypertonus, which had previously been ascribed to the red nucleus, was always present if lesions in the neighborhood of the red nucleus involved the substantia nigra as well. In waking and unrestrained cats stimulation of the rostral part of the red nucleus or of the rubrospinal fibers caused upward movements of the head and anterior part of the trunk (HESS and WEISSCHEDEL, 1949; HASSLER, 1960b, 1966a). Inactivating these fibers does not lead to complementary negative effects with lowering of the head. However, this result of stimulation is related to the magnocellular part, which is sharply reduced in man.

Experience in human pathology with lesions of the red nucleus in the BENEDIKT and WEBER syndromes shows action myoclonus, coarse tremor on the opposite side, severe lack of synergia, and occasionally athetosic movements (see HASSLER *et al.*, 1975).

2.4. Substantia Nigra

The substantia nigra consists of a posterior and an anterior main section (HASSLER, 1937); both of them are divided into a reticular zone and a compacta zone (SPATZ, 1922). The latter is composed of many constantly arranged nerve cell groups which begin to accumulate melanin pigment in the human at the age of 7 years. In the many species of mammals investigated so far, even in primates, no evidence exists that the substantia nigra receives an afferent input from any kind of sensory system. The single possible exception is the nucleus tractus peduncularis transversus, which forms a paramedian ventral expansion of the substantia nigra, and is well developed in the prosimians such as *Tarsius spectrum*. That this latter nucleus incorporates crossed optic fibers has been ascertained by experiments involving nerve cell degeneration following an interruption of one optic

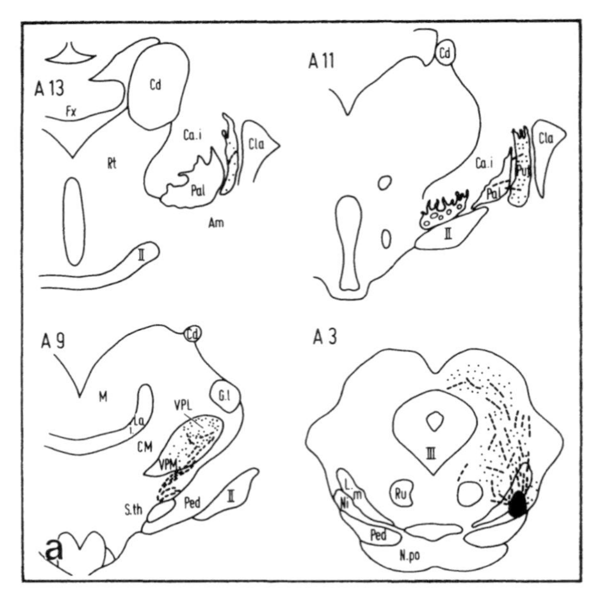

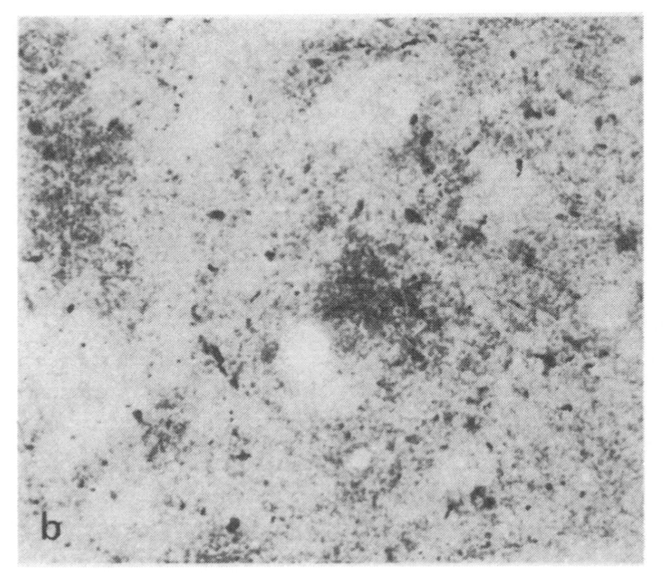

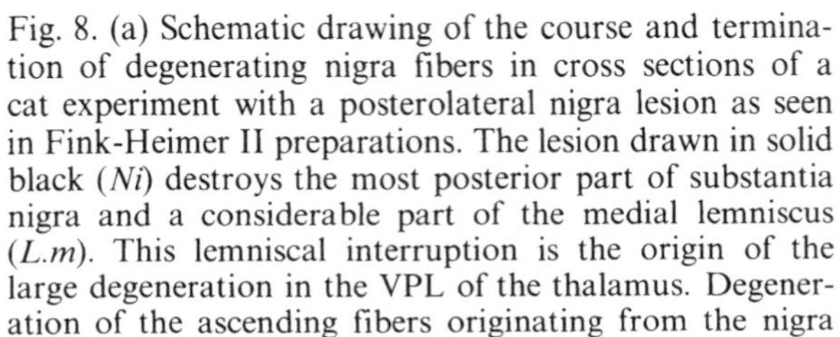

Fig. 8. (a) Schematic drawing of the course and termination of degenerating nigra fibers in cross sections of a cat experiment with a posterolateral nigra lesion as seen in Fink-Heimer II preparations. The lesion drawn in solid black (*Ni*) destroys the most posterior part of substantia nigra and a considerable part of the medial lemniscus (*L.m*). This lemniscal interruption is the origin of the large degeneration in the VPL of the thalamus. Degeneration of the ascending fibers originating from the nigra is restricted to the putamen, whereas the caudate nucleus (*Cd*) shows no degeneration (USUNOFF *et al.*, 1976).
(b) The Ebbeson-Heimer preparation of the same experiment shows in the dorsal putamen highest density of terminal degeneration following destruction of the caudolateral substantia nigra, as presented in Figures 8a (Ebbeson-Heimer preparation), initial magnif. ×600). (After USUNOFF *et al.*, 1976)

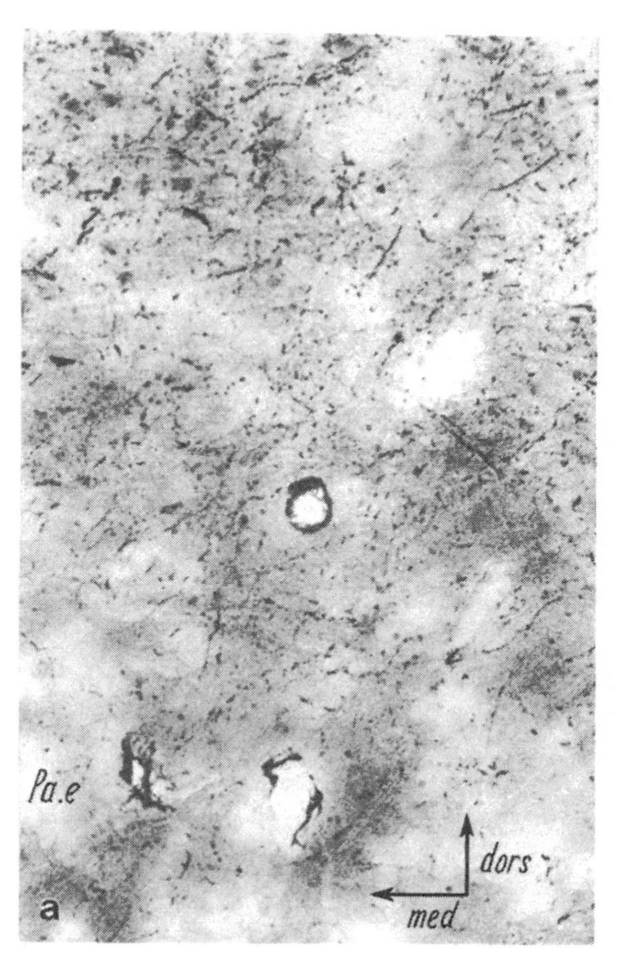

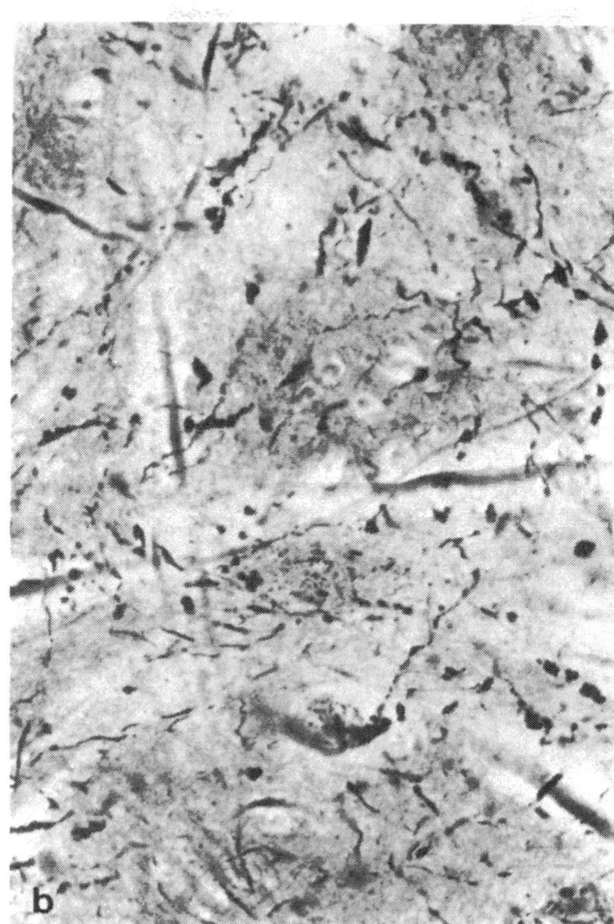

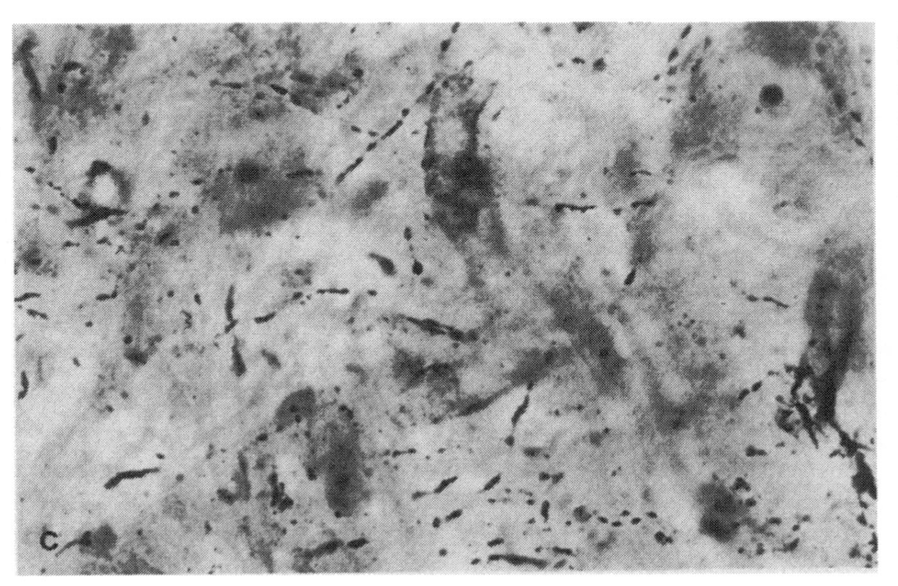

Fig. 9. (a) After coagulation in cat caudate nucleus degenerating fine fibers of caudatal origin are found in cat's pallidum. Their density decreases from dorsal (upper part of the figure) to ventral (lower part of the figure). Ebbeson-Heimer. ×714.

(b) After caudate nucleus coagulation very dense degenerated fibers appear in rostral substantia nigra pars reticulata. Wiitanen method. ×1,786.

(c) After caudate nucleus lesion there are many degenerated fibers between nerve cells of zona compacta of cat's substantia nigra. Fink-Heimer method I. ×1,500. (After Usunoff *et al.*, 1974)

nerve by HASSLER (1966c); it is, however, not definitely established whether this nucleus actually belongs to the substantia nigra.

The main afferent input to all parts of the substantia nigra comes from the striatum. The three large groupings of nigra compacta nerve cells, the Ni.a (Nigra compacta anterior), the Ni.m (Nigra compacta medialis), and the Ni.p (Nigra compacta posterior) each receive a bulk of afferent fibers from a special part of the striatum. According to the detailed descriptions by HASSLER (1950c), SZABÓ (1962), USUNOFF *et al.* (1974), and HASSLER *et al.* (1975) the caudate nucleus projects to the Ni.a, the fundus striati of BROCKHAUS (1942) to the Ni.m, and the putamen to the Ni.p, which comprises by far the largest amount of melanin-pigmented nigra nerve cells.

Only since the pioneer studies of LLAMAS (1969) and MOORE *et al.* (1970/71), followed by a more detailed study by USUNOFF *et al.* (1976), has it been known that almost all nigra cell groups possess a reciprocal efferent pathway (back) to the same part of the striatum from which they receive projections. Thus, at least three reverberating circuits (Fig. 4) are formed (1) between the caudate nucleus and Ni.a, (2) the fundus striati and Ni.m, and (3) the putamen and Ni.p (Fig. 8a, b) without connections to the caudate nucleus. Only the lateral cell groups of the posterior main section of the nigra (Ni.p) send efferent fibers downward (Fig. 8a) to the bulbar reticular formation after crossing the midline in the commissure of the colliculi (HASSLER, 1966; USUNOFF *et al.*, 1976); whereas, RINVIK *et al.* (1976) have given new evidence for the existence of descending nigrotectal and nigroreticular fibers.

It is surprising that the terminal degeneration of the connections from the parts of the striatum to the substantia nigra is, for the most part, accumulated in the rostral reticular zone of the nigra (Fig. 9b) and less around the compacta nerve cells (Fig. 9c), although there is only a restricted number of nerve cells in the reticular zone. The electron-microscopic picture provides an explanation by showing that the large dendrites of nigra compacta cells are extended mainly in the reticular zone; therefore, the dense accumulation of degenerated boutons of the striato-nigral pathway actually forms connections from the striatum to the nigra compacta nerve cells.

Connections from motor cortical areas to the anterior main section of the substantia nigra have been noted since the work of MURATOFF (1893) many times in human pathologic material and by RINVIK (1966) in the cat. The most detailed statement of HASSLER (1950c) that many distinct cortical areas are connected with discrete nerve cell groups of the substantia nigra, which receives at the same time fiber connections from discrete parts of the striatum, requires confirmation by other methods. For the description of the nigrocortical tract by MOLINA and REINOSO-SUAREZ (1965) only the retrograde chromatolysis method has been used; consequently, its existence is not yet firmly established.

2.4.1. Synapses of the Striato-Nigral Circuit and Their Transmitters. In the last 15 years intensive biochemical, histochemical, and electron-microscopic studies of the nigral and striatal neurons have been made. For a long time the melanin content of the nigra cells has been thought to indicate the presence of catecholamines in the nigra, since the breakdown of catecholamines can lead to the formation of melanin. The high concentration of catecholamines in the caudate nucleus was first demonstrated by BERTLER and ROSENGREN (1959a, b) and BERTLER *et al.* (1960). The content of catecholamines in the nigra of man was first determined by EHRINGER and HORNYKIEWICZ (1960) in post-mortem material at least six hours after death. In spite of this delay they found in each gram of fresh substance 0.4 mg of dopamine (DA) and 0.5 mg of serotonin (5-HT), but less than 0.1 mg of noradrenalin (NA). In animal experiments that allow determination of the concentration of catecholamines without post-mortem enzymatic destruction, the monoamine content of the caudatum is about ten times higher for DA, for 5-HT, and for NA (SOURKES and POIRIER, 1966). Application of the histochemical method for demonstration of DA and NA by green fluorescence according to FALCK and HILLARP (FALCK, 1962; FALCK *et al.*, 1962; FALCK and OWMAN, 1965) showed a high degree of diffuse fluorescence (Fig. 10d) in the neuropil of the caudatum, i.e., the tissue between the nerve cells. In the substantia nigra, however, the green fluorescence of DA is restricted to the cytoplasm of the nerve cells (Fig. 10a) while the neuropil does not show any fluorescence (BAK *et al.*, 1969). The green fluorescence of the cytoplasm of nigra nerve cells indicates the presence of either DA or NA. Since the latter is nearly lacking in substantia nigra, according to biochemical studies, the green fluorescence of the nigra cells is due to the presence of DA.

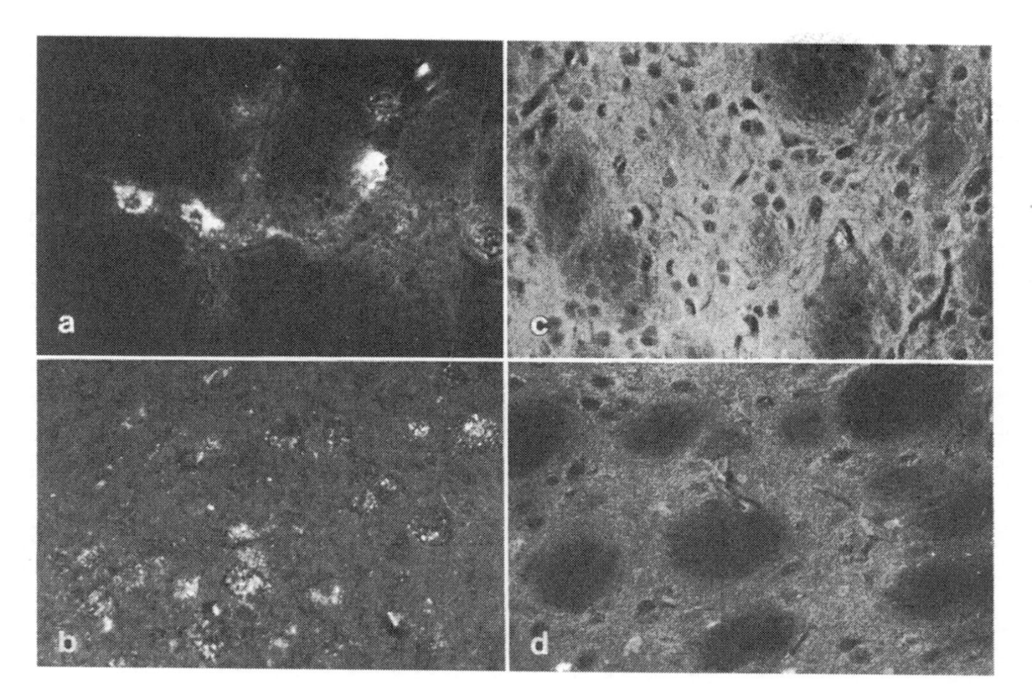

Fig. 10. (a) Fluorescence microphotogram of nerve cells of rat substantia nigra with intracytoplasmic green fluorescence. Some fibers in this section also have faint green fluorescence.
(b) After 35 mg/kg oxypertine, cytoplasmic green fluorescence is reduced and has become granular; it has disappeared in the single fibers.
(c) Control fluorescence microphotogram of rat striatum shows strong green fluorescence of neuropil, typical for dopamine and noradrenaline. The small nerve cells are free of fluorescence.
(d) Fluorescence microphotogram of rat striatum 2 h after 70 mg/kg oxypertine. Green fluorescence of neuropil is strongly diminished due to heavily reduced content of dopamine and noradrenaline. Dark ovals show location of efferent striatal fiber bundles. (After HASSLER *et al.*, 1970)

Following interruption of the connections between substantia nigra and the caudate nucleus the fluorescence is, in the caudate nucleus, much reduced due to the decreased concentrations of DA and NA. The same diminished concentration of DA can be obtained by application of oxypertine (35 mg/kg); then the intracytoplasmic green fluorescence is reduced and has become granular (Fig. 10a, b). The diffuse green fluorescence of the striatum indicating catecholamine concentration is strongly diminished after 70 mg/kg oxypertine (Fig. 10c, d) (HASSLER *et al.*, 1970).

These observations correspond to the biochemical findings of BERTLER *et al.* (1960) that a reduction of catecholamines in the caudatum and putamen was obtained following interruption of their connections with the substantia nigra (SOURKES and POIRIER, 1966). Also in baboon a hemitransection made between the substantia nigra on the one side and the striatum on the other side is followed by a sharp decrease of DA in the putamen and caudatum above the hemitransection; whereas below in the nigra the DA concentration remains almost unchanged (Fig. 11, HASSLER, 1974). Therefore, it must be concluded that the nigra neurons projecting to the striatum are dopaminergic while the striatal neurons projecting downward to the nigra contain another transmitter. The nature of the ascending dopaminergic projection that leaves the substantia nigra on the dorsomedial border nevertheless remains unclear because it gives a strong acetylcholinesterase reaction (OLIVIER *et al.*, 1970) like the peripheral sympathetic (noradrenergic) fibers.

The biogenic monoamines including DA in the central and peripheral neurons are stored for the most part in bound form in special vesicles. This has been unequivocally demonstrated in the peripheral sympathetic fibers and in central DA-ergic neurons. Those vesicles storing catecholamines, and especially DA, can be recognized by the fact that the vesicle contains an electron-dense core of 0.1 μm diameter (Fig. 12a). By treatment with reserpine (depleting all catecholamines) such dense

Effect of premesencephalic hemitransection on concentration of GABA and dopamine in striatum, substantia nigra, and pallidum in baboon (10–14 days p. op)

		Dopamine		GABA		n
Nc. caudatus	unop. side	8.05 ± 1.40	−66%	3.40 ± 0.54	+20%	4
	op. side	2.76 ± 0.51		4.08 ± 0.39		4
Putamen	unop. side	9.52 ± 1.50	−90%	3.99 ± 0.80	− 4%	4
	op. side	0.93 ± 0.17		3.83 ± 0.90		4
Sub. nigra	unop. side	1.76 ± 0.32	−17,5%	8.49 ± 0.99	−66%	4
	op. side	1.45 ± 0.10		2.89 ± 0.97		4
Pallidum	unop. side	1.40 ± 0.09	−20%	8.48 ± 0.62	− 6%	4
	op. side	1.13 ± 0.20		7.96 ± 0.81		4

Fig. 11. After interruption of the connections between striatum and substantia nigra in baboon a significant drop of GABA in substantia nigra occurs below the hemitransection, whereas the 20% increase of GABA in caudate nucleus is not significant. Above the hemitransection in the caudate nucleus and putamen a large drop of dopamine occurs (66% and 90% respectively), whereas the dopamine concentration below in substantia nigra is not significantly changed. (After HASSLER, 1974)

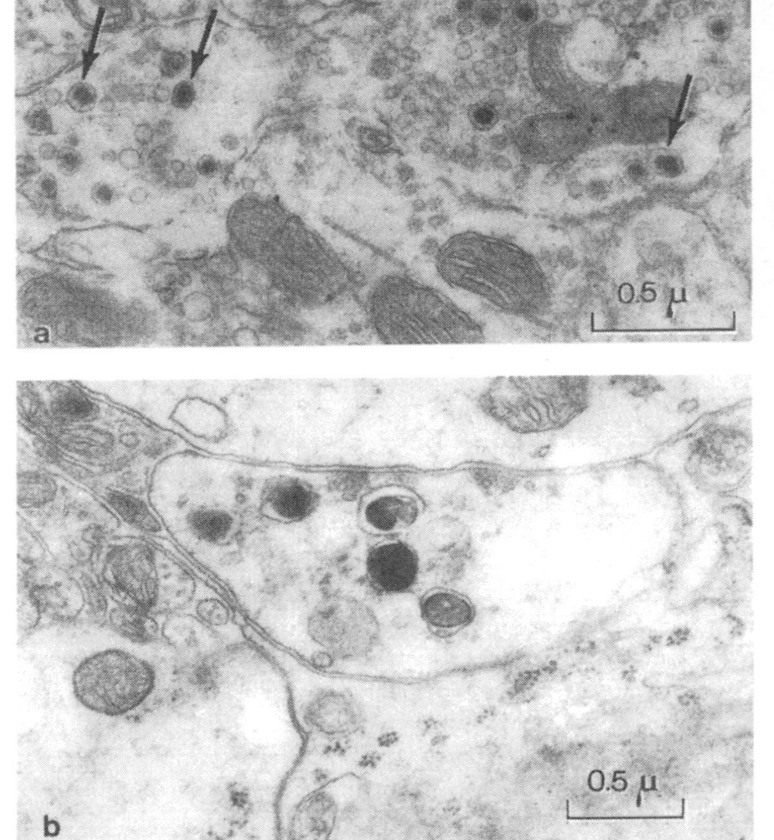

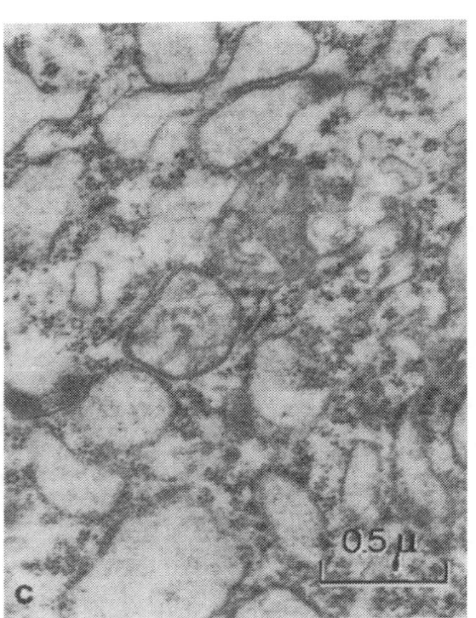

Fig. 12. (a) Nerve terminals of rat substantia nigra of dense core vesicle type, probably axon collateral of nigra cell. In contrast to small empty vesicles, large ones of 0.1 µm diameter contain dark core, believed to be a store of monoamines. (After HASSLER *et al.*, 1970.)
(b) After 3 days administration of iproniacide and L-dopa, granulated vesicles in synapse-type of substantia nigra are strongly increased and its dense core is enlarged. This seems to be expression of increased storage of catecholamines in the dense-core vesicles. Notice dense accumulation of small empty vesicles twice near axodendritic contact. (After BAK and HASSLER, 1967.)
(c) Five days after daily treatment with reserpine (5 mg/kg), granulated vesicles are completely depleted, and some seem to be dissolved with membrane damage. Notice swelling of cisterns of endoplasmic reticulum. (After HASSLER and BAK, 1966)

cores of vesicles can be emptied (PELLEGRINO DE IRALDI, 1965; HASSLER and BAK, 1966; BAK, 1967), at least in sympathetic fibers of the pineal gland and in nigra boutons (Fig. 12c), and enlarged by administration of monoamine oxidase inhibitor and L-dopa, the precursor of DA, in the axon collaterals of nigra cells (Fig. 12b) (BAK and HASSLER, 1967). This enlargement of the dense cores cannot be observed in the caudatum where dense core vesicles are very rare (BAK, 1967). They only occur *en masse* within preterminals (HASSLER and CHUNG, 1976).

By means of light- and electron-microscopic autoradiography it was possible to show (PARIZEK *et al.*, 1971) that ^{3}H-labeled DA and ^{3}H-labeled NA administered to the brain ventricles accumulate in the dendrites and perikarya of nigra nerve cells (Fig. 13a, b), while ^{3}H-labeled serotonin (^{3}H-5-HT) remains absent in these perikarya and dendrites. It is, however, accumulated in some kinds of synapses contacting the large dendrites (Fig. 13c, d), mainly these containing dense-core vesicles. The nigra neurons take up specifically catecholamines (without differentiation of NA and DA), and some synapses in the nigra take up indolamines. This pattern of uptake in the nigra makes it very credible that DA is the transmitter formed by nigra neurons.

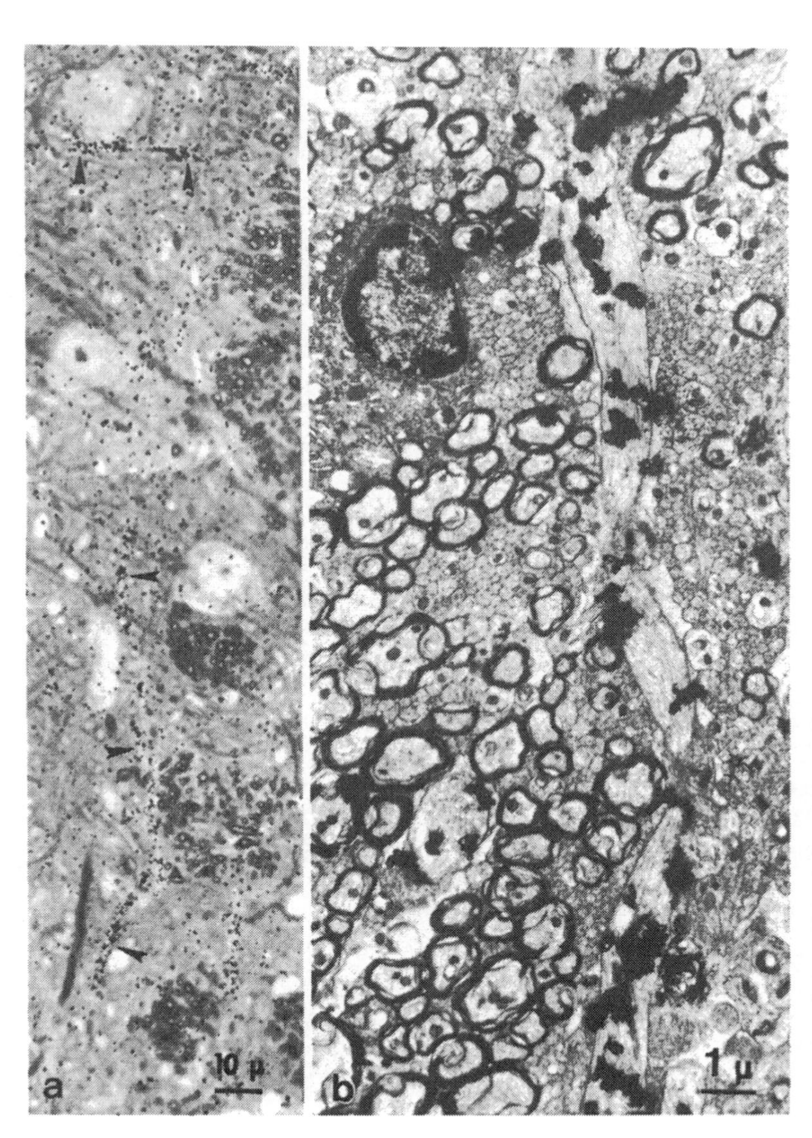

Fig. 13. (a) Light-microscopic autoradiogram of substantia nigra after administration of ^{3}H-dopamine into brain ventricles of rat. An accumulation of silver grains due to radioactive dopamine deposit is seen, also in perikarya of nigra cells, and in long rows (*arrow*) that correspond to dendrites of nigra nerve cells. (After PARIZEK *et al.*, 1971.)
(b) Electron-microscopic autoradiogram of rat substantia nigra shows longitudinally-sectioned dendrite of nigra cell. This dendrite is occupied by many silver grains from ^{3}H-dopamine, administered in brain ventricles. Labeled dopamine is accumulated in nerve cells and in their dendrites. (After PARIZEK *et al.*, 1971)

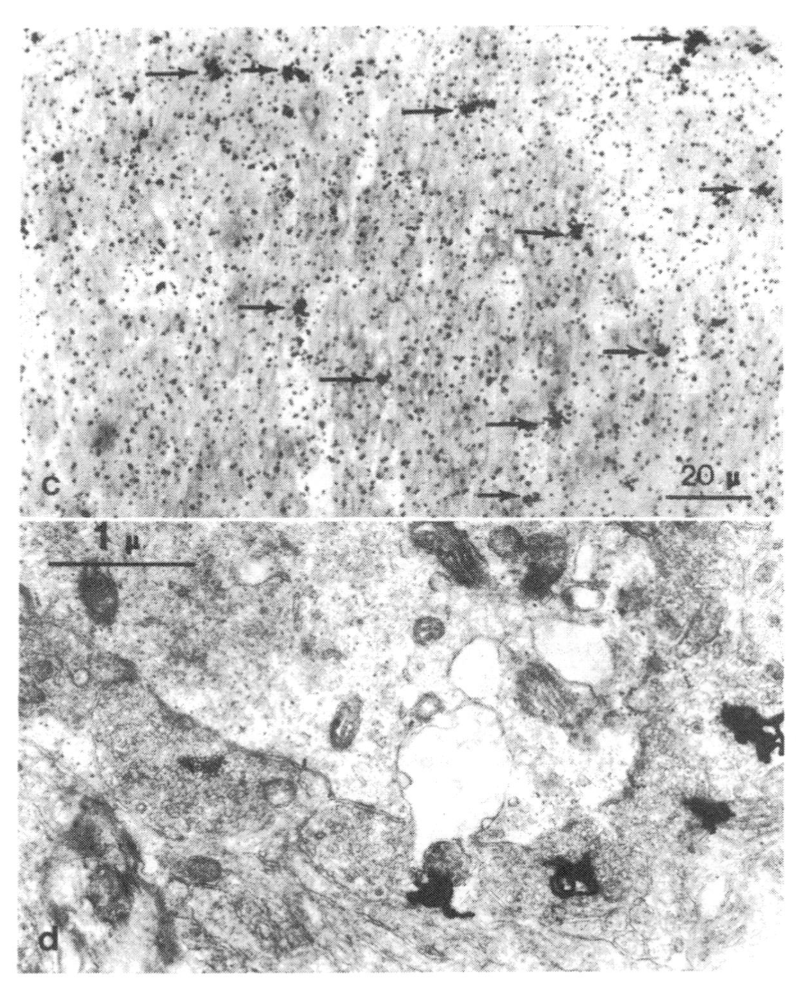

Fig. 13. (c) Light-microscopic autoradiogram of rat substantia nigra after intraventricular application of ^{3}H-serotonin. Serotonin is accumulated (*arrow*) in small amounts in neuropil of substantia nigra. It is absent in perikarya and in dendrites of substantia nigra.
(d) Electron-microscopic autoradiogram of rat substantia nigra after intraventricular administration of ^{3}H-serotonin. Labeled particles of serotonin are accumulated in synapses on a large dendrite only in substantia nigra. Dense-core vesicles appear in 3 of these synapses, which suggests that some synapses with dense-core vesicles in substantia nigra contain serotonin as transmitter. (After PARIZEK *et al.*, 1971)

The descending strio-nigral neurons terminate in the nigra mainly in one form of bouton, the pleomorphic type (I) (Fig. 14a, b), which comprises almost 40 to 50% of all boutons in the nigra (HAJDU *et al.*, 1973). This has been demonstrated in experiments showing a dark degeneration of this type I bouton after an interruption between the nigra and the striatum (RINVIK and GROFOVÁ, 1970) (Fig. 14e, f), in which the original shape and size of the pleomorphic vesicles can still be observed (HAJDU *et al.*, 1973). After the same hemitransection between striatum and substantia nigra is made, a drop of γ-aminobutyric acid (GABA) below the transection in the nigra occurs, as shown in Figure 11, however, not in the striatum above the lesion. Therefore, it seems reasonable to conclude that the (descending) strio-nigral neurons contain GABA as a transmitter and conduct it downward toward the nigra. This can be strongly confirmed by the autoradiographic observations showing that after intranigral injection of ^{3}H-GABA the remaining labeled material is accumulated (Fig. 15) around the perikarya and the dendrites of nigra cells (BAK *et al.*, 1975). This suggests that the pleomorphic boutons on the nigra cells and dendrites are GABA-ergic.

Knowledge of the origin of the other boutons (II, III, IV) in the nigra is insufficient; however, it is known that the small, round vesicled boutons (type III) are perhaps the bouton type of the cortico-nigral fibers because some of them degenerate after lesions of the internal capsule. Some of the type IV boutons, which contain many dense-core vesicles, are probably axon collaterals of efferent DA-ergic nigra cells, while others are probably preterminals containing serotonin as transmitter.

The *ascending nigro-striatal* projections terminate in the caudate nucleus and putamen via two types of synapses: the axospinous type I, contacting small striatal neurons (Fig. 16a) and the axo-

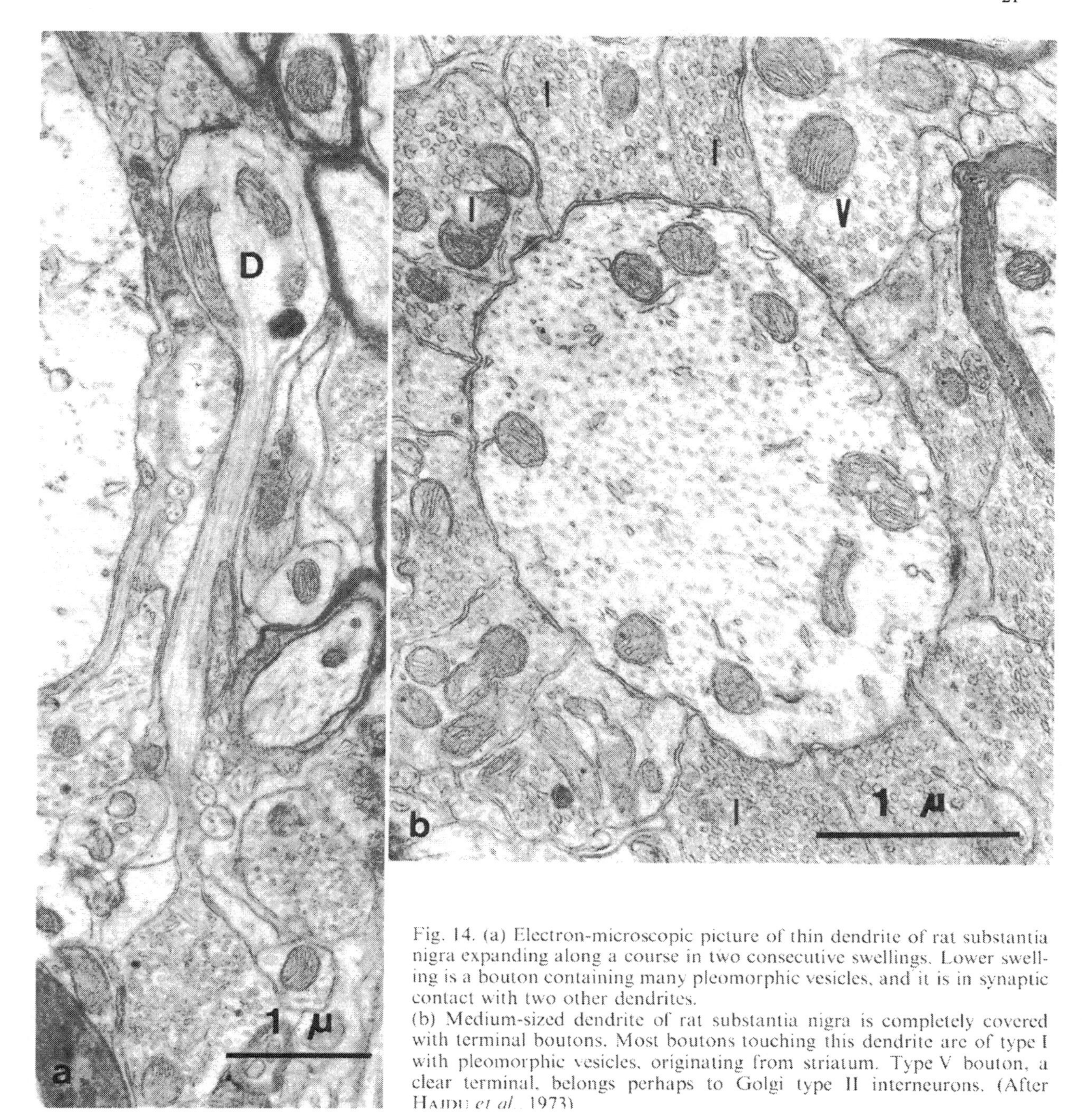

Fig. 14. (a) Electron-microscopic picture of thin dendrite of rat substantia nigra expanding along a course in two consecutive swellings. Lower swelling is a bouton containing many pleomorphic vesicles, and it is in synaptic contact with two other dendrites.
(b) Medium-sized dendrite of rat substantia nigra is completely covered with terminal boutons. Most boutons touching this dendrite are of type I with pleomorphic vesicles, originating from striatum. Type V bouton, a clear terminal, belongs perhaps to Golgi type II interneurons. (After Hajdu *et al.*, 1973)

dendritic type II contacting *en passant* the large striatal neuron (Fig. 16b). This has been determined by degeneration experiments or by hemitransection made in front of the substantia nigra; in these cases only type I (Fig. 17b) and type II terminals degenerate (Bak *et al.*, 1975; Hassler *et al.*, 1975; Chung *et al.*, 1977). Both types of synapses are very probably dopaminergic because after the hemitransection mentioned above the DA content of the striatum drops (Fig. 11c). In all parts of the striatum there are at least seven additional distinct types of synapses (Hassler and Chung, 1976) that have an origin other than the substantia nigra: two degenerate after decortication (Fig. 17c) (Bak *et al.*, 1975; Hassler *et al.*, 1977) and two degenerate following coagulation in the centre médian nucleus of the thalamus (Hassler, 1975; Chung *et al.*, 1977).

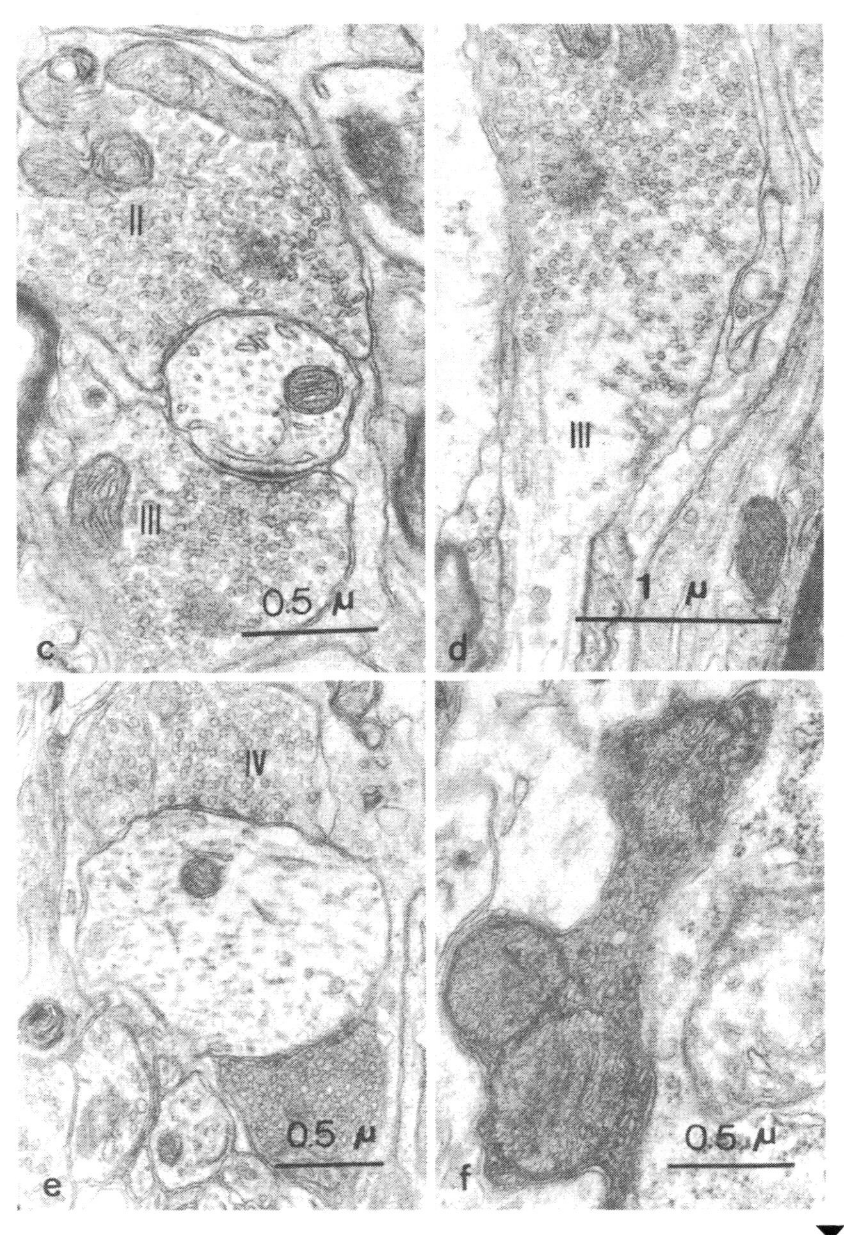

Fig. 14. (c) Axodendritic synapses of rat substantia nigra filled with elongated vesicles (type *II*) and below, another bouton with small round vesicles (type *III*).
(d) In rat substantia nigra, an axon ascending from below this loses its myelin sheath and expands in a vesicle containing preterminal bouton type III. Bouton type III forms axosomatic contact.
(e) Two days after hemitransection between striatum and substantia nigra, a degenerated axodendritic bouton appears on lower side of picture. Matrix between pleomorphic vesicles is much darker than normal. Non-degenerated bouton with large round vesicles (type *IV*) is seen on the opposite side of dendrite.
(f) Two days after interruption between striatum and substantia nigra in rat, an axosomatic synapse in substantia nigra shows dark degeneration in which shape and size of pleomorphic vesicles are still recognizable. (After HAJDU *et al.*, 1973, and HASSLER *et al.*, 1974)

Fig. 15a and b. Caption see opposite page

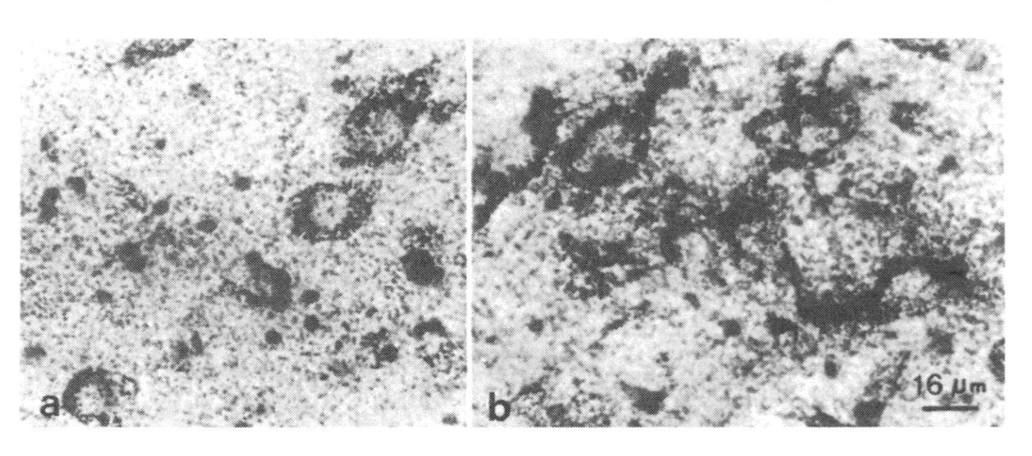

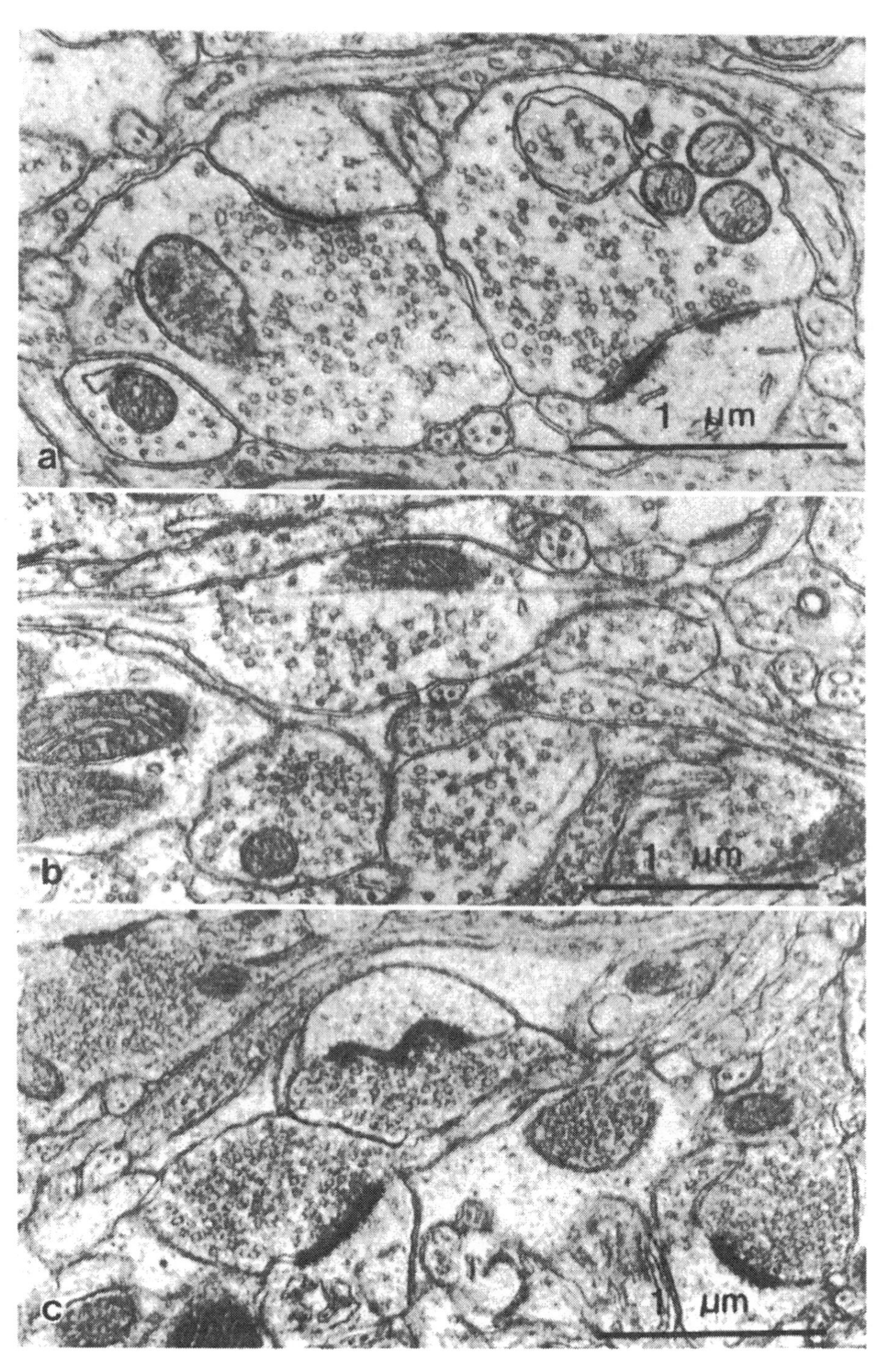

Fig. 16. (a) Electronmicrographs of 2 terminal boutons type I of nigrostriatal fibers in caudate nucleus. They form a thickened contact with small dendritic spines of the small striatal nerve cells. The loosely arranged apparently empty vesicles contain dopamine.
(b) Striatum of rat. Two longitudinally sectioned *en passant* boutons (type *II*). Enlargement of unmyelinated fiber contains loosely arranged synaptic vesicles of 45 nm diameter. The same type of bouton is seen in cross section through left part of (c).
(c) Several small boutons (type *III*) with densely packed, relatively small synaptic vesicles. The boutons contact two dendritic spines with prominent postsynaptic thickenings. They probably represent cortico-striatal terminals. (After BAK *et al.* 1975)

◀ Fig. 15. (a) Light-microscopic autoradiogram prepared from animals after intranigral injection of ^{3}H-GABA. They demonstrate accumulation of silver grains around nigral perikarya and, to some extent, around primary dendrites. These neurons are some distance from injection focus. (After HASSLER *et al.*, 1974)
(b) Light-microscopic autoradiograms taken from region adjacent to injection of ^{3}H-GABA. Surfaces of perikarya and proximal dendrites are heavily labeled. (After BAK *et al.*, 1975)

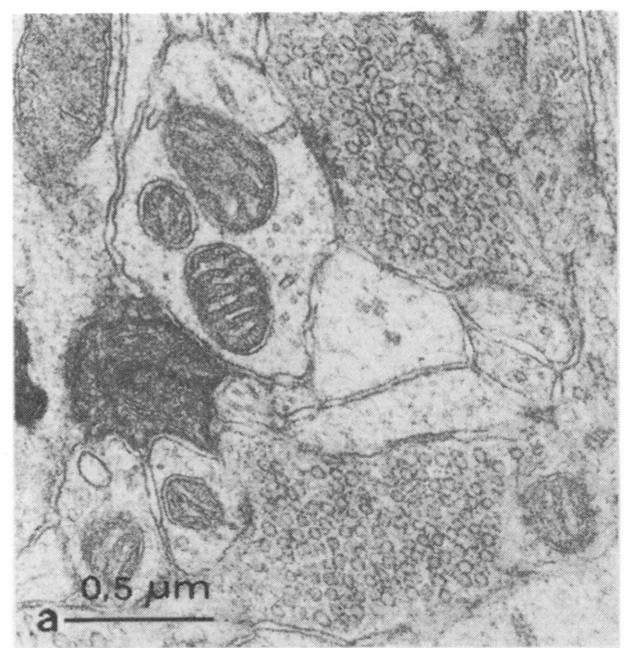

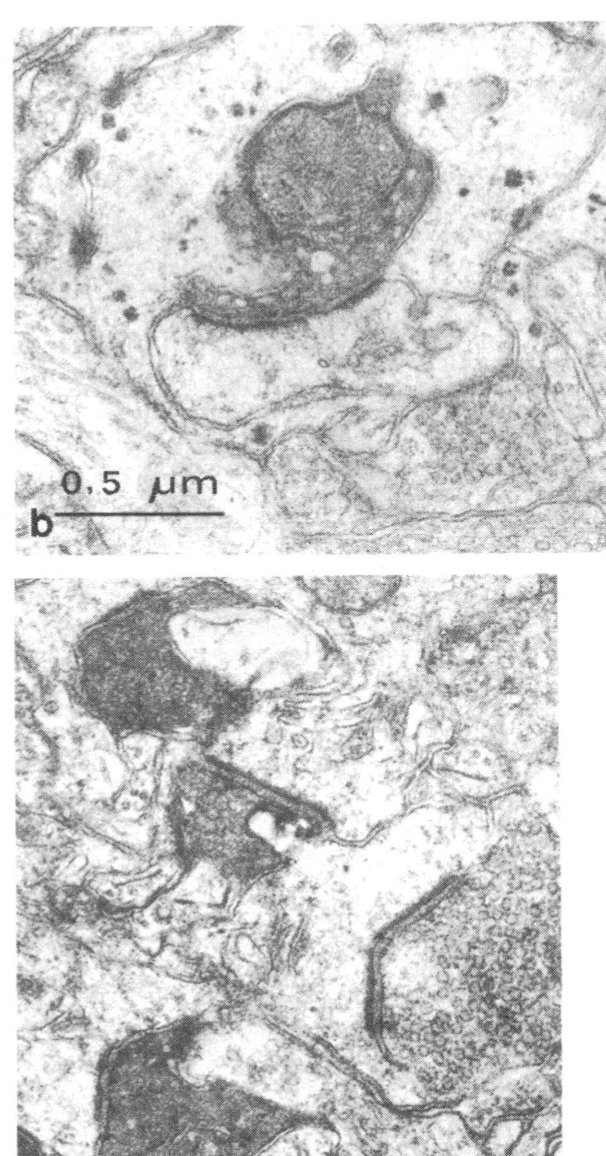

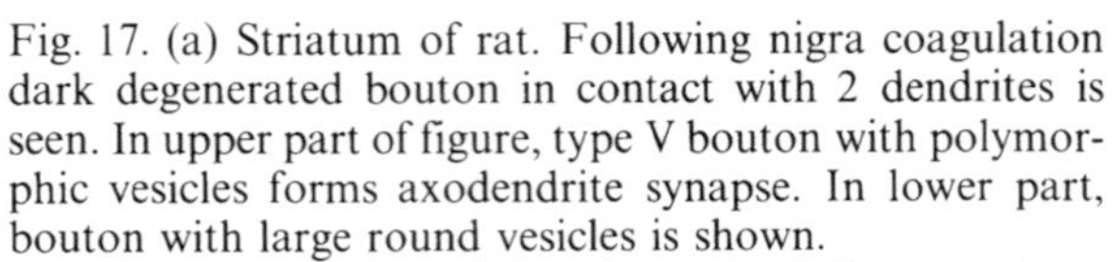

Fig. 17. (a) Striatum of rat. Following nigra coagulation dark degenerated bouton in contact with 2 dendrites is seen. In upper part of figure, type V bouton with polymorphic vesicles forms axodendrite synapse. In lower part, bouton with large round vesicles is shown.
(b) After interruption of the nigrostriatal fibers a nigrostriatal bouton (type *I*) still thickened in contact to a striatal spine has undergone a dark degeneration and shrinking, losing the clear synaptic vesicles. It is enclosed in a translucent astroglia process.
(c) Three degenerating boutons of type III in dorsolateral part of caudate nucleus 4 days after extirpation of sensorimotor cortex in cat. (After BAK *et al.*, 1975)

The nigro-striatal circuit enters and passes through the network of the striatal internuncial cell apparatus where its activity is modified by and integrated with the cortical and centrothalamic impulses. The resulting messages of the small striatal interneurons are transmitted to the large striatal neurons by cholinergic boutons containing large round vesicles. Their activity can be further checked and modified by all three striatal afferent neurons (nigra, centre médian, cortical) before they send the efferent messages to the pallidum or back to the substantia nigra and transmit them by GABA-ergic terminals (KIM *et al.*, 1971; BAK *et al.*, 1975; HASSLER *et al.*, 1975.

2.4.2. Mode of Action of the Anticholinergic Drugs Against Parkinson Symptoms. The therapeutic effectiveness of atropine, scopolamine, and most synthetic drugs used to treat the Parkinson syndrome, all being anticholinergic, favor a cholinergic mechanism of the rigidity- and tremor-producing system. Moreover, investigation of the acetylcholine (ACh) content of the striatum in rats treated by harmaline has shown that the acetylcholine content parallels the rise and subsequent course of the serotonin content (Fig. 18). The experimental animals showed flexor rigidity, akinesia, and a tremorlike restlessness. The mode of behavior is unchanged if, in similar experiments

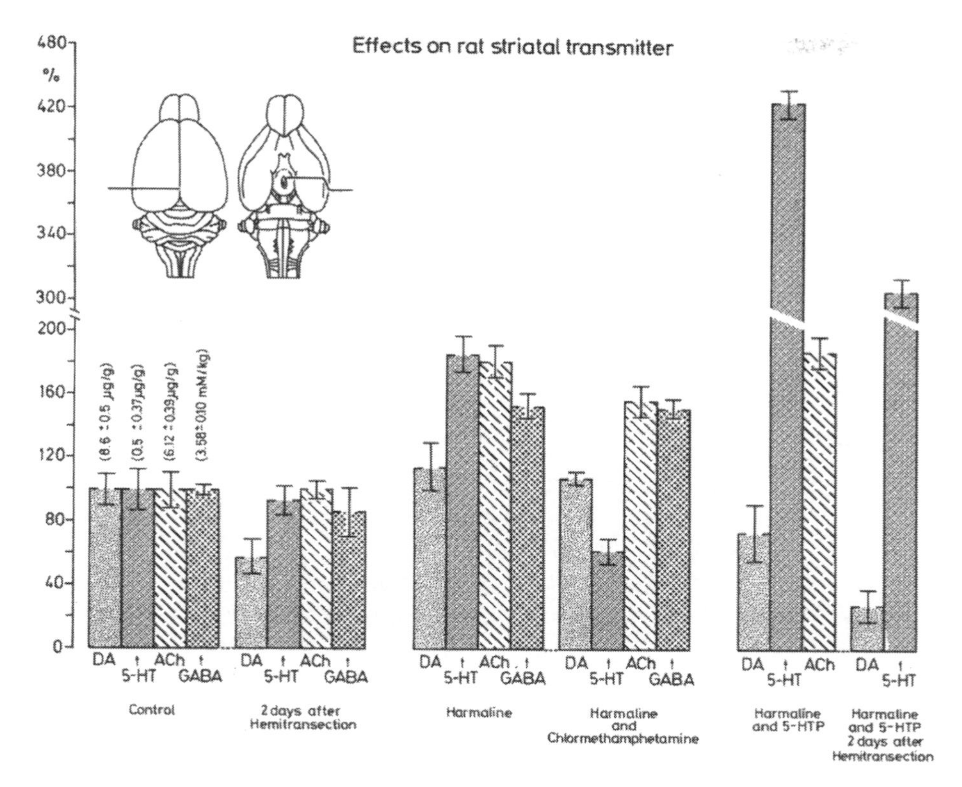

Fig. 18. Diagram of striatal transmitters under hemitransection and administration of harmaline. Two days after hemitransection between basal ganglia and substantia nigra, shown in insert, the concentrations of dopamine (*DA*), serotonin (*5-HT*), acetylcholine (*ACh*), and GABA are altered—mainly those of DA and of GABA. Third block of columns: Three hours after harmaline (10 mg/kg), the concentrations of 5-HT, ACh, and GABA are increased. If harmaline is combined with chlormethaphetamine, the increase in GABA is unchanged. ACh increases 160%, whereas 5-HT is reduced to 60% in the striatum. – The fifth block shows the effect of harmaline combined with 5-hydroxytryptophan (*5-HTP*). This results in an additional increase of 5-HT, whereas DA is reduced. The same combination of harmaline and 5-HTP given 2 days after hemitransection reduces increase of striatal 5-HT to 320% and lowers concentration of dopamine to 25%. (After Hassler and Bak, 1969, and Kim *et al.*, 1970)

with harmaline, the rise of serotonin in the striatum is suppressed by chlormetamphetamine (Fig. 18), while the acetylcholine in the striatum is allowed to rise (Kim *et al.*, 1970). Therefore, the extrapyramidal motor signs following administration of harmaline run parallel to the acetylcholine, and not to the serotonin, content of the striatum. There is, however, in the striatum an antagonism between ACh-concentration and DA-concentration insofar as the increase of DA-concentration in the striatum leads to hypotonia and an increase in the speed of movements, while the rise of ACh- or 5-HT-concentration produces a decrease in the speed of movements. DA/ACh–5-HT is the quotient for accelerations and hypotonia against the anti-action of slowing down and rigidity by ACh–5-HT.

Consequently, it can justifiably be concluded that the action of the anticholinergic drugs takes place, for the most part, in the striatum, even in the case of their therapeutic action against Parkinson symptoms. This is strongly supported by a human case of hemiparkinsonism due to bilateral extensive degeneration of nigra cells by demyelinating foci (see Fig. 142), because the manifestation of Parkinson symptoms has been suppressed in the other limbs by severe unilateral perinatal damage to the putamen (status marmoratus: see Figs. 140 and 141). The extensive reduction of putamen tissue did not allow the neurochemical basis of increased ACh concentration to develop, so that instead of rigidity a severe action myoclonus with hypotonia developed on this side. Consequently, the anticholinergic therapy relieved only those Parkinson symptoms contralateral to the normal putamen and could not exert any effect on the action myoclonus of the other limbs due to the deficit of putamen tissue (Hassler *et al.*, 1975b).

Among the nine types of synapses in the striatum, the intrinsic synapses (HASSLER *et al.*, 1977) (containing large, spherical vessicles), which connect the small "spiny" with the large efferent neurons, are cholinergic. The axospinous and axodendritic synapses (containing small, round vesicles), which originate from the center median nucleus (HASSLER, 1975), are probably also cholinergic. Whether the type of synapse in the substantia nigra (also containing small round vesicles: Fig. 14a, b; HAJDU *et al.*, 1973) is cholinergic must remain undecided.

2.4.3. Efferent Pathways of the Substantia Nigra. The electron-microscopic findings support the earlier anatomical assertion that every cell in the substantia nigra has a double input, namely, from a cortical area as well as from an area in the striatum (see Fig. 4). It has been postulated that this strionigral circuit must have an output.

ROSEGAY'S (1944) finding of degenerated cells in the substantia nigra following interruption of the connections from cortex and from striatum to the nigra should be interpreted as a retrograde as well as anterograde (transsynaptic) degeneration of nigra cells. The myelogenetic (HASSLER, 1966) and Marchi studies have shown that the efferent pathways of the substantia nigra cross over in the midbrain under the colliculus superior and then descend on the other side. As mentioned above, this pathway might be confirmed by scrutinizing the efferent fibers ascending to the commissura colliculorum (USUNOFF *et al.*, 1976) by experimental degeneration studies made after lesions of the posterior lateral nigral cell group.[2]

[2] The efferent nigro-fugal descending fibers were described by HASSLER in 1950 (1950c) and their presence was supported by myelogenetic studies in 1966 HASSLER (1972a) HASSLER and WAGNER (1971/1975) reported manifold physiologic evidence as did YORK (1972, 1973). Since the morphologic findings concerning nigrotectal and nigro-reticular fibers reported by RINVIK *et al.* (1976) additional descriptions of descending, efferent nigral fibers have appeared.

2.4.4. Neuropathology of the Substantia Nigra. A neuropathologic involvement of the substantia nigra is found in many inflammatory diseases, for instance, in lyssa, encephalitis, thyphoid fever, typhus, and in poliomyelitis. Destruction of cells has also been observed in these diseases. However, involvement of the substantia nigra in late sequelae of such acute types of encephalitis has rarely been observed. Acute stages already showing signs of the Parkinson syndrome are found only in encephalitis epidemica or lethargica, which first appeared in 1916 (VON ECONOMO, 1917; ACHARD, 1921). These acute inflammatory Parkinson syndromes first supported the anatomical localization of the Parkinson syndrome by histologic investigation (Fig. 21b). Previously, BLOCQ and MARINESCO (1893) and BRISSAUD (1895) had already related the Parkinson syndrome to the destruction of the substantia nigra in cases of tuberculoma (as in Fig. 19a, b) and other circumscribed processes. However, his view was rejected in favor of that of localization of the Parkinson syndrome in the pallidum, which was stressed by numerous authors, e.g., LEWY, from 1913 until 1938. FOIX (1921), GOLDSTEIN (1922), LUKSCH and SPATZ (1923) found destruction of nerve cells in the substantia nigra to be the decisive anatomical finding in chronic cases of Parkinsonism in which signs appeared months or years after the acute encephalitis. It was only after these findings that the thesis by TRÉTIAKOFF was taken notice of, although in 1919 he had already described the regular disappearance of cells in the substantia nigra of brains of patients with Parkinson's disease. His material must cer-

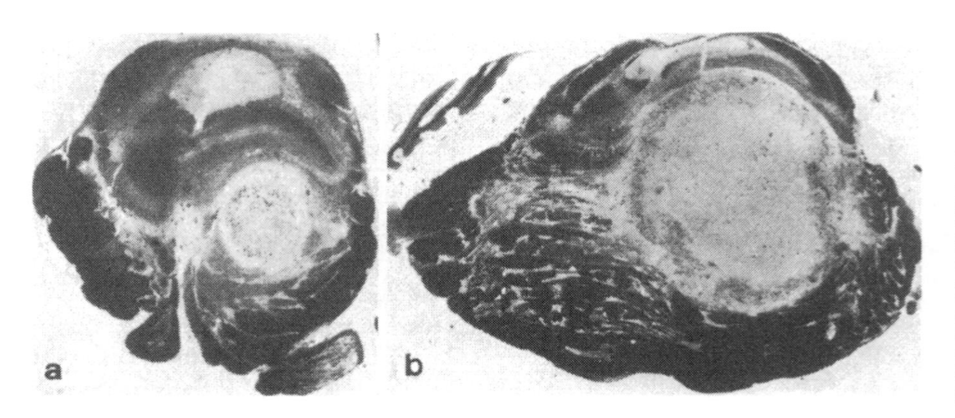

Fig. 19a and b. In a human patient, tuberculoma of midbrain has destroyed right substantia nigra. It was followed by contralateral Parkinson syndrome. This tuberculoma, however, also expanded into tegmentum of pons (b). (After HASSLER, 1953c)

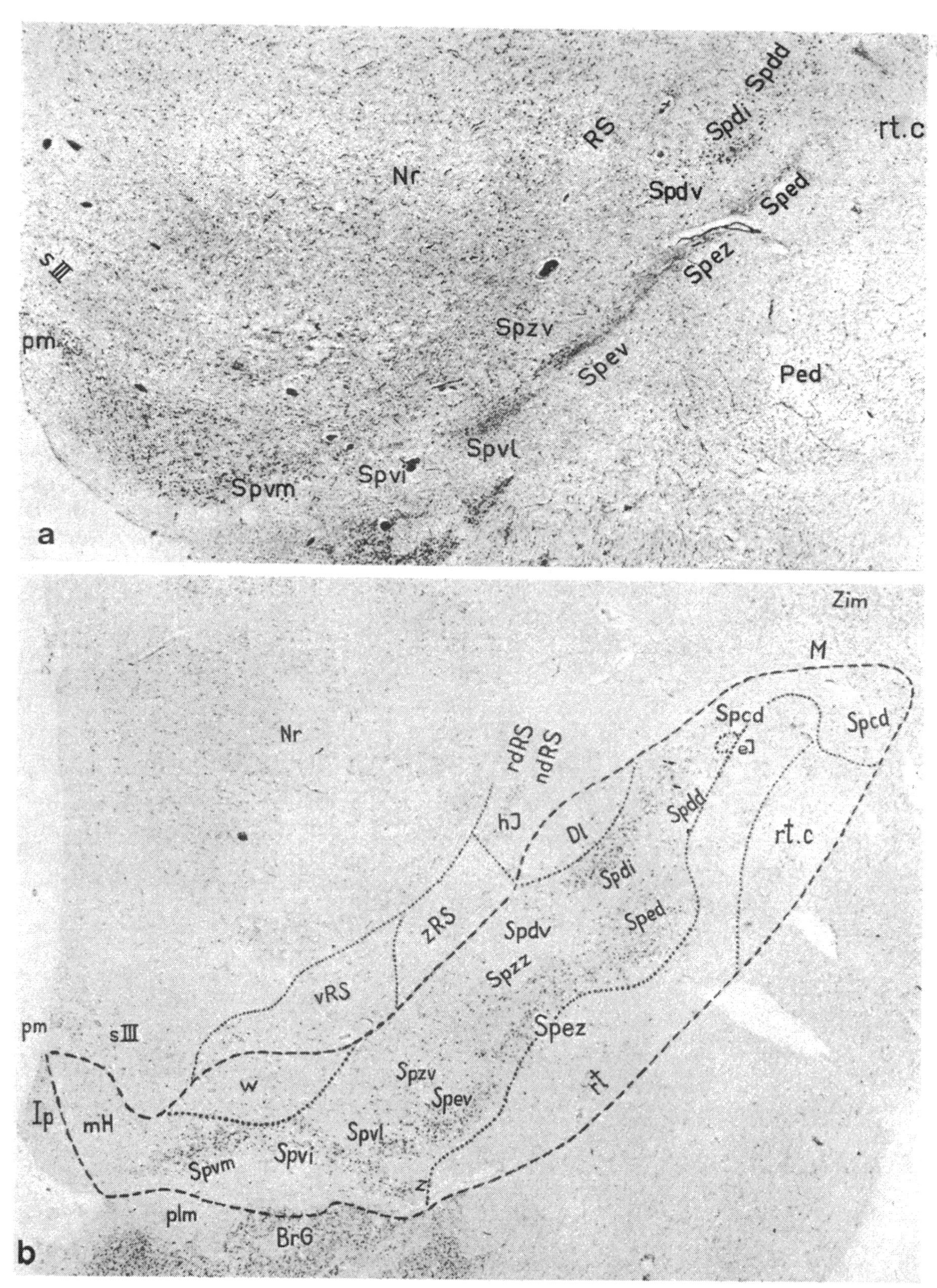

Fig. 20. (a) Cell loss in posterior main section of substantia nigra in hereditary case of paralysis agitans. (Hassler, 1938) (b) For comparison, normal arrangement of cell groups of substantia nigra compacta in posterior main section. (Hassler, 1937)

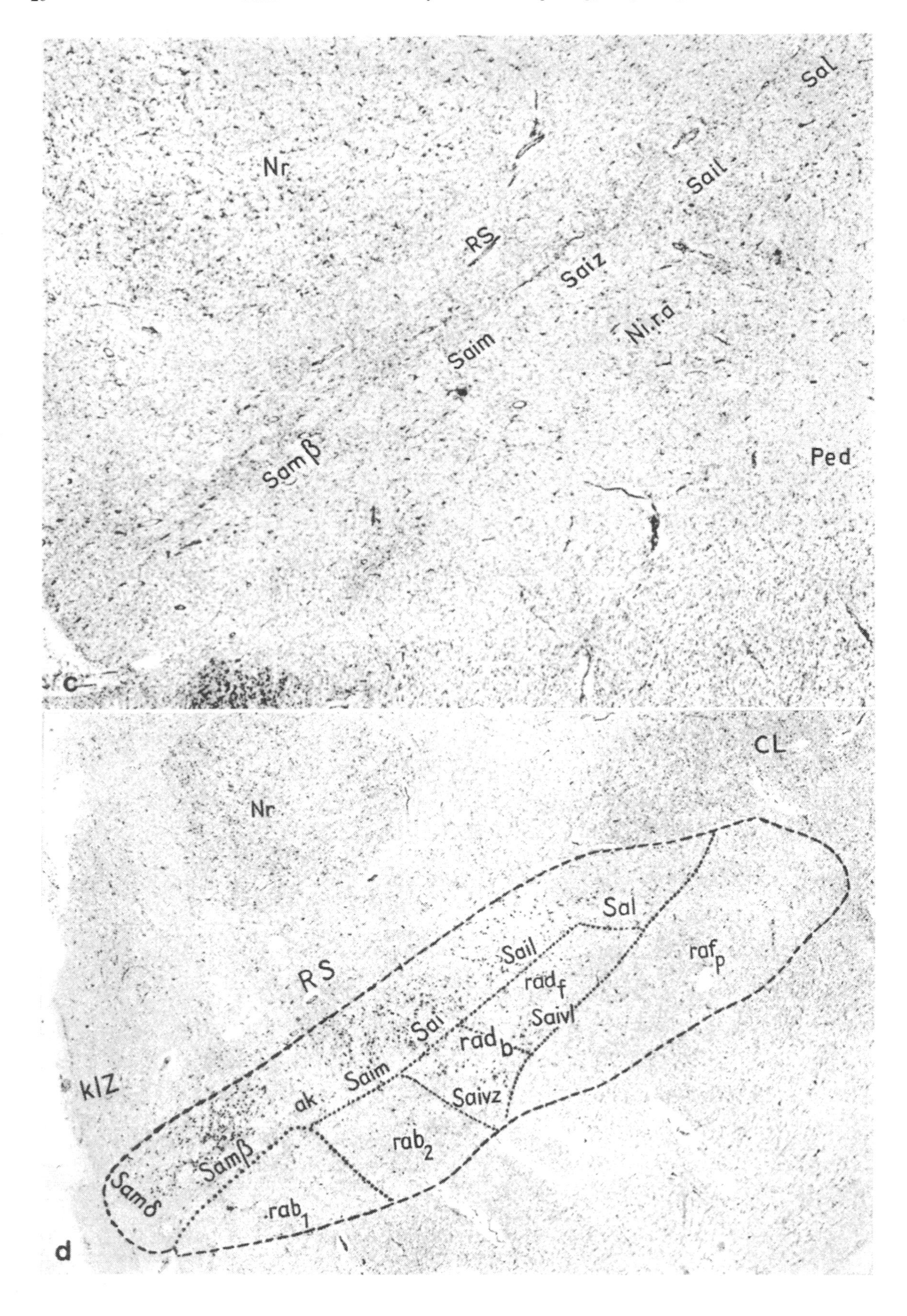
Sal
Nr
Sail
RS
Saiz
Ni.r.a
Saim
Sam β
Ped
c
CL
Nr
Sal
Sail
raf p
RS
rad f
Saivl
Sai
rad b
kIZ
Saim
ak
Saivz
rab 2
Sam β
Sam δ
rab 1
d

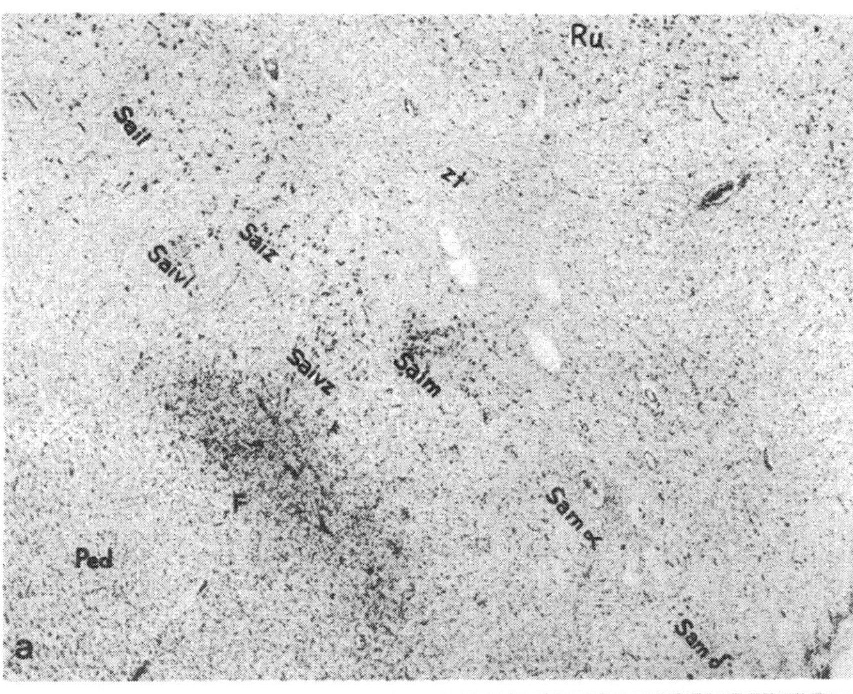

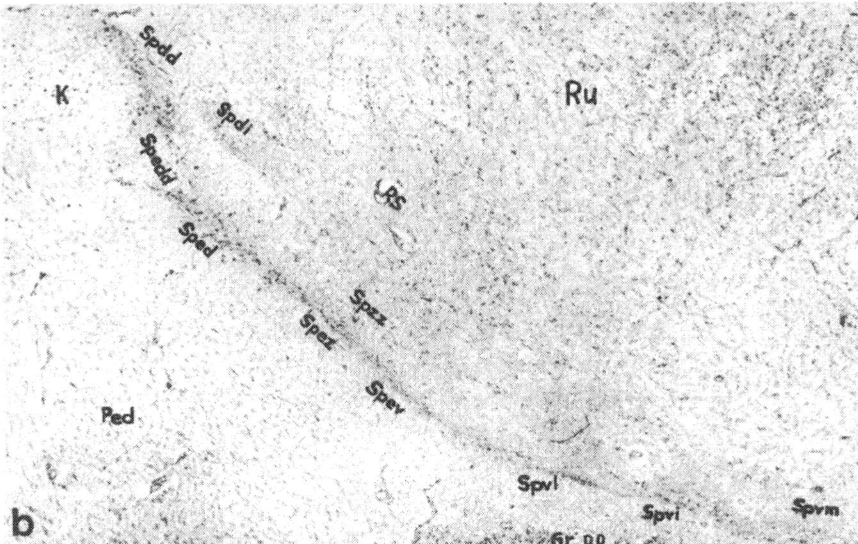

Fig. 21. (a) Anterior main section of substantia nigra in a case of arteriosclerotic Parkinsonism. Reticular zone contains focus of glia cell proliferation and detritus below melanin-pigmented nerve cell groups. This incomplete softening due to arteriosclerotic changes in midbrain vessels is followed by slow degeneration of black pigmented nerve cells, which does not imitate the pattern of idiopathic Parkinsonism. (After HASSLER, 1938)
(b) In a case of postencephalitic Parkinsonism, all lateral cell groups have lost nerve cells with compensatory gliosis, but the most dorsal groups are better preserved in contrast to the idiopathic cases. (After HASSLER, 1938)

◀ Fig. 20. (c) Severe nerve cell loss in cell groups of anterior main section in a case of postencephalitic Parkinsonism.
(d) For comparison, normal arrangement of cell groups of substantia nigra in anterior main section. Notice that pattern of cell loss in hereditary as in idiopathic cases involves ventral and lateral cell groups, except for most medial cell group, which is always preserved. Even in a case of degenerative hereditary paralysis agitans, cell loss in ventral and lateral cell groups is replaced by dense glia proliferation. The cases of postencephalitic Parkinsonism have no specific pattern of nerve cell disintegration; furthermore, every cell group of melanin-pigmented nerve cells can be involved. Sometimes glia proliferation in these old postencephalitic cases is less than in more acute degenerative cases. (After HASSLER, 1938)

tainly have contained some postencephalitic cases in addition to a few cases of paralysis agitans, although TRÉTIAKOFF (1919) made no distinction as to etiology. Most workers now agree that there is disappearance of nerve cells in every case of Parkinson syndrome, whether postencephalitic or idiopathic. One of us demonstrated in 1938 (HASSLER, 1938) that this destruction shows a predilection for certain groups of nerve cells in cases of idiopathic paralysis agitans, while in postencephalitic cases the destruction follows no rule and also attacks the medial cell groups, which are spared in paralysis agitans (Fig. 22). The results of this serially investigated material were further supported by BEHEIM-SCHWARZBACH (1956) and confirmed by GREENFIELD and BOSANQUET (1953) and by HALLERVORDEN (1957). Reexamination, especially of the cases in the VOGT collection, showed that the reported changes in the pallidum were really vascular lesions. Similar lesions or even more severe damage can be found in non-Parkinsonian patients of the same age. Similarly, senile cell atrophies in the pallidum, found in cases of Parkinsonism in the higher age groups, also occur in older people without Parkinsonian signs. The vascular lesions in the putamen, the so-called lacunae of PIERRE MARIE, were already regarded as having no functional significance when first described. Later they were regarded by DENNY-BROWN (1962) as the basis of the Parkinson syndrome; he pointed to the allegedly negative result of the attempts to produce Parkinson-like signs in animal experiments by damaging the substantia nigra. The findings of POIRIER (1960) have now made this view obsolete (Fig. 25). Since the idiopathic paralysis agitans has a late manifestation – age of almost 50 years – the inveterate supposition can still be found very often that it is due to an arteriosclerosis of brain vessels, the more so as the measurement of cerebral blood flow in Parkinson patients shows

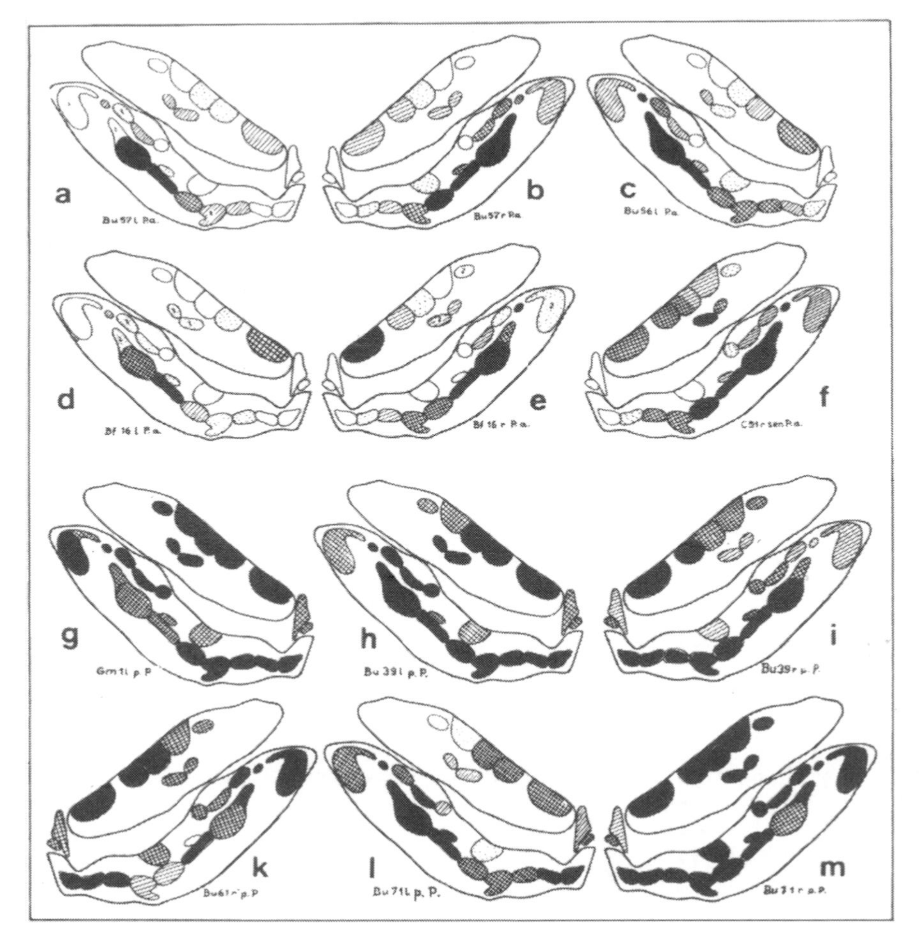

Fig. 22a–m. Diagram of involvement of different nerve cell groups of substantia nigra in cases of idiopathic or hereditary and of postencephalitic Parkinsonism ($p.P$ = g–m). Diagram shows regular pattern of cell group involvement only in cases of idiopathic and hereditary Parkinsonism ($P.a$ = a–f). Lateral and ventral cell groups always are most severely affected by deterioration process. Most medial cell group in posterior main section and lateral cell group of anterior main section of substantia nigra are spared. (After HASSLER, 1938)

an increase in vascular resistance (SCHMIDT, 1963). The neuropathologic reinvestigation of Parkinson syndromes of different etiologies has shown (HASSLER, 1938) that there really exist some cases of arteriosclerotic Parkinsonism with a parenchymal necrosis or softening in the substantia nigra (Fig. 21a), however, in considerably less than 10% of all Parkinson cases (HASSLER, 1938).

2.4.5. Anatomical Differential Diagnosis of the Parkinson Syndrome. The anatomical changes in the substantia nigra in postencephalitic and genuine Parkinsonism differ not only with regard to distribution of the cell loss but also in the changes occurring in the nerve cells. ALZHEIMER (1911) changes of the fibrils in postencephalitic Parkinsonism were first described by FENYES (1932) and HALLERVORDEN (1933). They were confirmed in the Vogt serial material (Fig. 23e, f) by HASSLER in (1938, 1955b), and by GREENFIELD and BOSANQUET in 1953. Even when cell destruction is far advanced, one can observe this typical finding in serial sections in every postencephalitic case of Parkinsonism, whereas it is always absent in cases of idiopathic or senile paralysis agitans. In a tuberculoma of the substantia nigra (Fig. 19a, b) HASSLER (1953c) found in the neighboring nerve cells a cytopathologic process without the formation of tangles of neurofibrils.

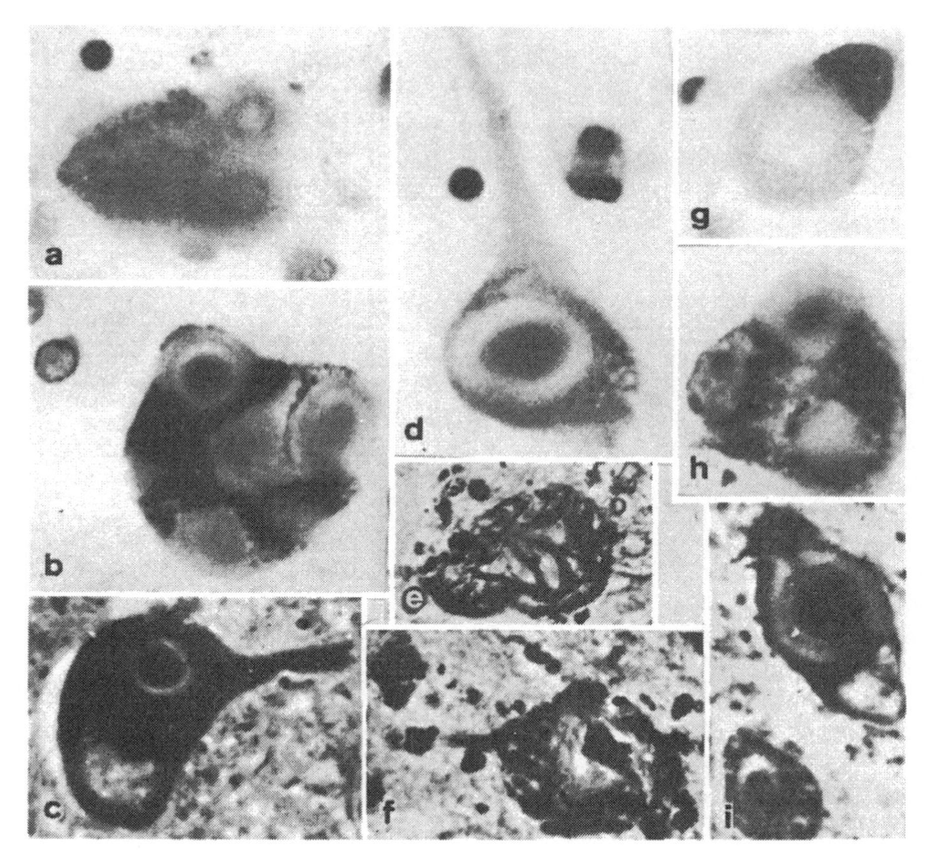

Fig. 23a–i. Cytopathologic pictures of substantia nigra and locus caeruleus nerve cells in idiopathic and postencephalitic Parkinsonism.
(a) Two cytoplasmic spherical laminated inclusion bodies in nigra nerve cell in a case of familial paralysis agitans. Nissl stain. ×900.
(b) Large nerve cell of locus caeruleus of same case is inflated by many spherical and sharply laminated cytoplasmic bodies with metachromasy. Nissl stain. ×900.
(c) Intraplasmic inclusion bodies in a case of idiopathic paralysis agitans in nigra nerve cell. Inclusion body is argyrophilic. Bielschowsky stain. ×800.
(d) Another large inclusion body like a drop of colloid in nerve cell of locus caeruleus of familial case of paralysis agitans. Nissl stain. ×900.
(e, f) Neurofibrillar tangles of Alzheimer in nigra nerve cells in case of postencephalitic Parkinsonism. Pile of melanin is preserved in middle of cell (f). Bielschowsky stain. ×800.
(g) Nigra cells in hereditary case of paralysis agitans with large spherical inclusion body in cytoplasm with faint lamination that shifts nucleus to the surface. Nissl stain. ×900.
(h) Nerve cell of locus caeruleus in case of familial Parkinsonism: cytoplasm contains at least two vacuoles, one with an inclusion body. Nissl stain. ×900.
(i) Intraplasmic body in a case of idiopathic paralysis agitans in nigra nerve cell. Layered inclusion body is argyrophilic. Bielschowsky stain. ×800. (After HASSLER, 1938, 1955)

On the other hand, in paralysis agitans argyrophilic substances are seen forming spheroid layered inclusion bodies in the cytoplasm (Fig. 23a, c, g, i) of the remaining nerve cells of the substantia nigra. Despite their much simpler form, they are reminiscent of the comparable cytoplasmic inclusions (Fig. 23b, d, h) first described in the locus caeruleus by LEWY (1923), which have since then been called Lewy bodies. They do not occur in the cell groups of the substantia nigra that are spared in paralysis agitans. They are therefore frequently overlooked in neuropathologic examinations if one does not look for them in the affected and already-thinned-out cell groups of nigra. Here the inclusions can be found in every case in serial sections. It is, therefore, possible to make a differential diagnosis between idiopathic and postencephalitic Parkinsonism also on the basis of the cytopathology. These argyrophilic layers of protoplasmic inclusions (Fig. 23c, i) are also found regularly in the cells of the substantia nigra in cases of paralysis agitans with proved heredity (HASSLER, 1938).

2.4.6. Localization of the Various Signs in Parkinsonism. The three main motor signs of Parkinson syndrome are probably due to the disappearance of cells in the substantia nigra. The vegetative signs of Parkinson syndrome, however, evidently do not depend on damage to the substantia nigra just as they do not occur in animal experiments following the setting of lesions in the substantia nigra. In cases of Parkinsonism with vegetative signs, whether of the postencephalitic or the idiopathic kind, changes are regularly found in two nuclei of the brainstem: the locus caeruleus (Fig. 23d) and the dorsal nucleus of the vagus (FOIX and NICOLESCO, 1925). In adults, the nerve cells of both nuclei contain a black melanin pigment, just like the substantia nigra. Both nuclei always show loss (Fig. 37a, b) of nerve cells in Parkinson brains (HASSLER, 1938; BEHEIM, 1954). In the nonpostencephalitic cases, this loss is regularly found in combination with the inclusion of layered argyrophilic cylinders in the cytoplasm of some remaining cells. Thus far, it has not been ascertained whether such vegetative signs as hypersalivation, lacrimation, hyperhidrosis, and sialorrhea are associated with lesions of these two nuclei.

Another constant finding in cases of genuine Parkinsonism is that of changes in the substantia innominata, or the basal nucleus. This nucleus shows a normal senile involution (Fig. 24b), which starts about the 60th year of life. This involution is much increased (Fig. 24c) in cases of genuine Parkinsonism, in which it comes earlier than usual, leading to loss of many cells (HASSLER, 1938, 1955, 1956; VON BUTTLAR-BRENTANO, 1955). It may be that the functional significance of the involution of this basal nucleus is that it causes the normal signs of mental aging, which appear much earlier and more markedly in Parkinsonism; this may apply, for instance, to bradyphrenia.

2.4.7. Experimental Lesions in Substantia Nigra. In the early investigations, destruction of the substantia nigra in dogs and cats was not followed by changes of movements or of muscle tone. Only D'ABUNDO (1925) observed choreiform movements and later rigidity in newborn cats following destruction of the substantia nigra, but extensive lesions in the immediate neighborhood were present in these cases.

In animal experiments tremor at rest was first produced by BISHOP *et al.* (1948) and PETERSON *et al.* (1949) by small lesions dorsomedial to the substantia nigra area adjacent to the red nucleus. The resting tremor appeared exclusively on the contralateral side. Such coagulations dorsomedial or medial to the substantia nigra, are generally used to produce Parkinson-like tremor in monkeys (POIRIER *et al.*, 1966). The explanation that these lesions interrupt the dorsomedially accumulated efferent pathways of the substantia nigra was long doubted until POIRIER (1960) was able to demonstrate a severe cellular atrophy in the substantia nigra of macaca monkeys (Fig. 25) several months after such a lesion. In experiments on monkeys it has now been ascertained that, following damage to the neuronal system of the substantia nigra, resting and postural tremor can be produced.

Fig. 24a–d. Transverse section of basal ganglia, showing sublenticular part of basal nucleus (*B.sl*) below pallidum in Nissl stain, taken (a) from normal brain, (b) from brain of neurologically unaffected person of 84 years of age, and (c) from a case of hereditary paralysis agitans of 66 years of age. Basal nucleus of Parkinson patient has undergone strong involution with rarefication of nerve cells and compensatory glia proliferation. Insert (d) shows Lewy's glassy nerve cell alteration, by which basal nucleus nerve cells deteriorate. The normal senile involution of these nerve cells, as shown in (b) is precocious and enhanced in Paralysis agitans. (After HASSLER, 1938)

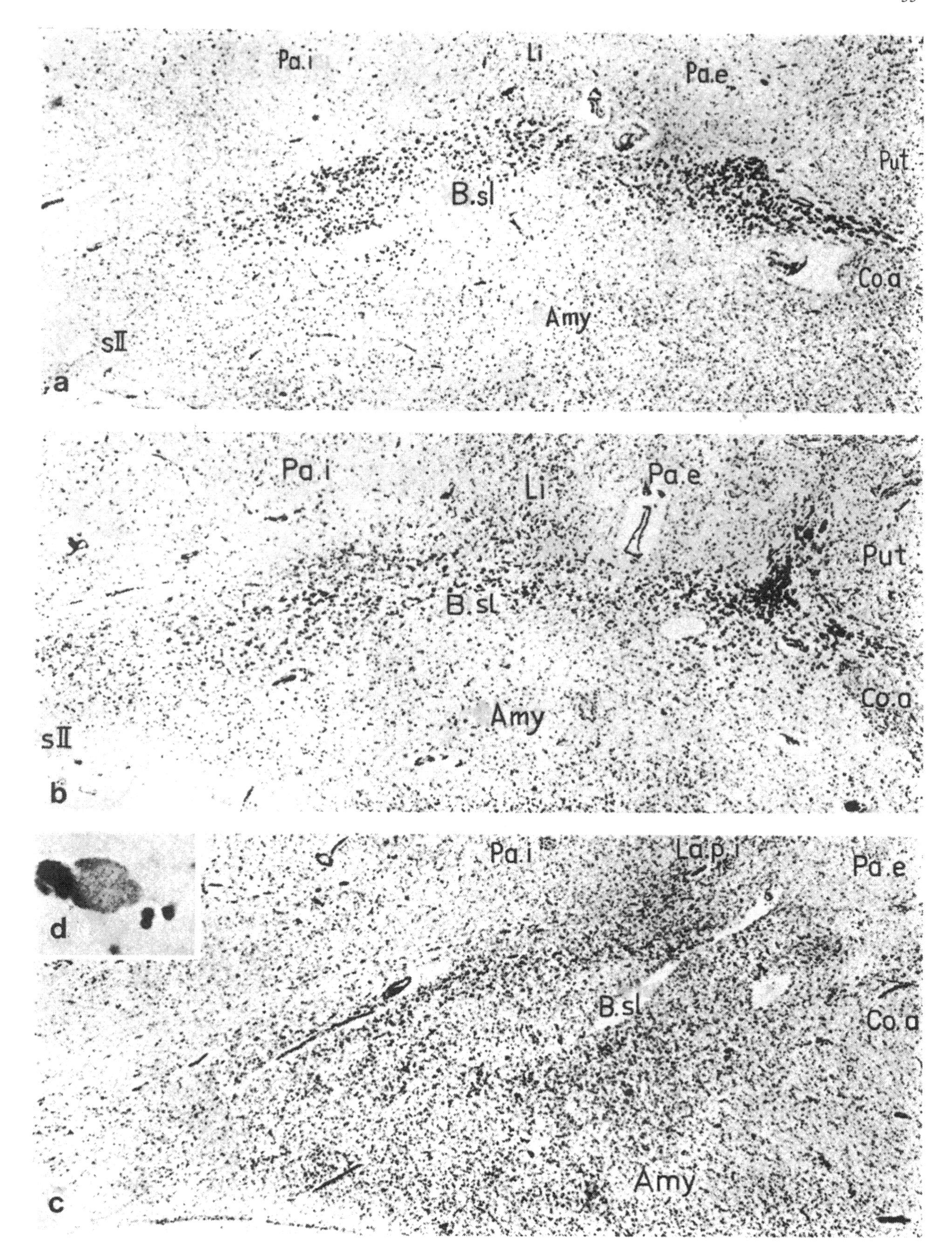
Pa.i
Li
Pa.e
Put
B.sl
Co.a
Amy
sII
a
Pa.i
Li
Pa.e
Put
B.sL
Co.a
Amy
sII
b
Pa.i
La.p.i
Pa.e
B.sl
Co.a
Amy
d
c

However, it is strange that thus far no true rigidity has thereby been produced in experimental animals. As yet, it has not been possible to decide whether nerve cell loss in substantia nigra produces this sign only in the higher primates and in man. One of the authors (HASSLER) observed in one chimpanzee an increased resistance to passive movement of the contralateral arm after this small lesion had been made dorsomedial to the substantia nigra (unpublished finding).

The third main motor sign in man, akinesia, has been observed in experimental animals in which either the efferent fibers of the substantia nigra or the substantia nigra itself have been damaged. Here it was found that the opposite limbs were used to a reduced extent. Since the work of BAILEY and DAVIS (1943), it has been known that severe states of akinesia with aphonia, or fading of expressive sounds, are produced following coagulation in the gray matter surrounding the aqueduct (substantia grisea periaquaeductalis). These findings were extended by FERNANDEZ DE MOLINA and HUNSPERGER (1962) and by MELZACK *et al.* (1958). In experimental lesions of the substantia nigra we have never observed the vegetative signs seen in Parkinsonism.

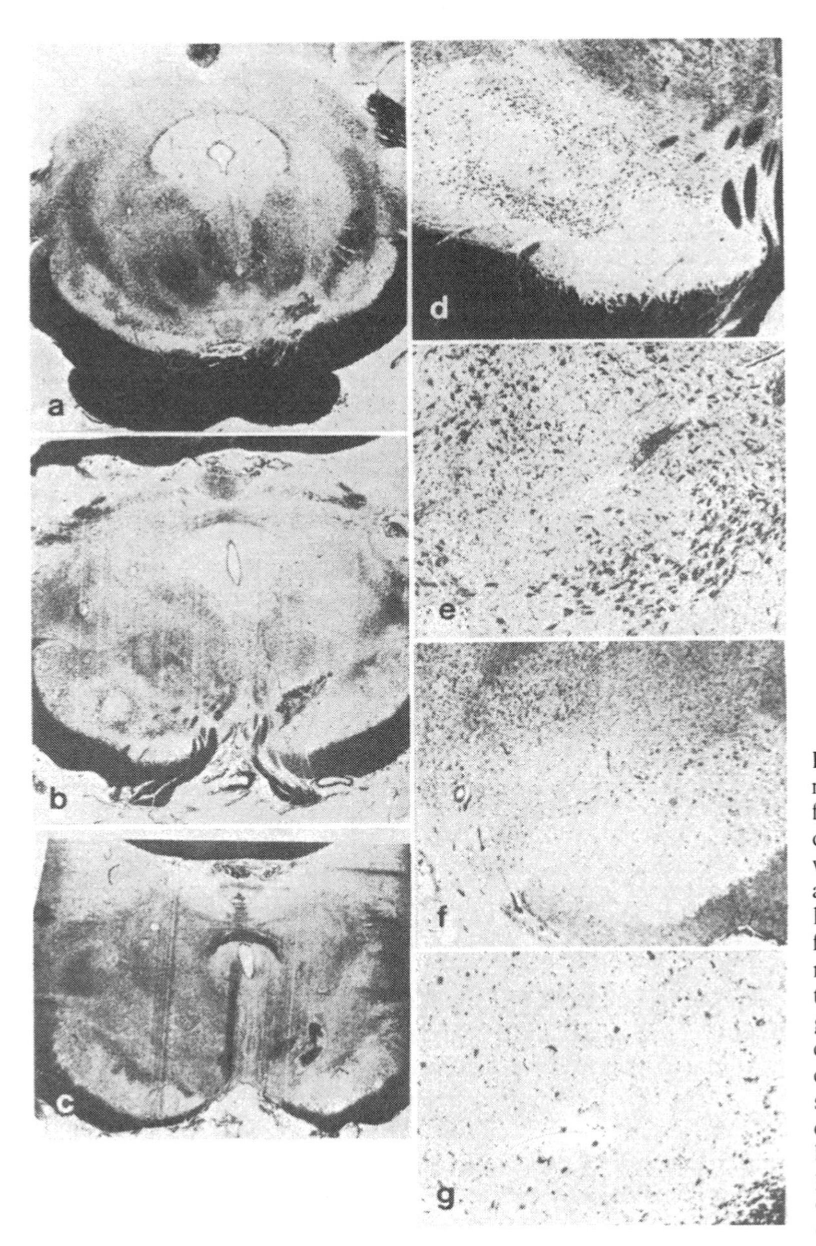

Fig. 25a–g. Transverse sections through midbrain of *Macaca mulatta* in three different levels (a–c), showing rostral-caudal extent of a unilateral lesion, which was followed by hypotonia, hypokinesia, and sustained postural tremor of contralateral limbs, sections (a), (b), and (c) in fiber staining according to Laidlaw's modification of Klüver technique. Partial figures (d) and (e) are microphotographs of same brain, showing myeloarchitecture and cytoarchitecture of pars compacta of substantia nigra on normal side. Partial figures (f) and (g) show pars compacta of substantia nigra on side of lesion with nerve cell loss of posterior part of substantia nigra after coagulation of efferent pathways on dorsomedial side of nigra. (After POIRIER, 1960)

2.4.8. Stimulation of the Substantia Nigra. Stimulation experiments of the substantia nigra with old techniques (now obsolete) resulted in rhythmic chewing and swallowing movements (VON BECHTEREW, 1909–1911; JURMAN, 1900; VON ECONOMO, 1902). These results have been confirmed and elaborated by WINKELMÜLLER (1972), using Hess's method (Fig. 26). Very few stimulation experiments have been performed in the last 30 years to elucidate the function of the substantia nigra. METTLER *et al.* (1939) found that, in cats, stimulating the substantia nigra with higher frequencies increased the extensor tonus even if the motor cortex had been destroyed. This type of stimulation reduces the extent of phasic movements, even when they are produced by simultaneous stimulation of the cortex. If the substantia nigra is stimulated by a modified Hess technique in cats that are not restrained and not anesthetized, an acceleration of spontaneous movements in the oral areas and in the limbs is seen. The medial part of the substantia nigra is connected with the oral area and the head (chewing, licking, swallowing, side-to-side glancing) (Fig. 26), while the lateral part is related to the limbs (WINKELMÜLLER, 1972; YORK, 1973). Further experiments are needed to show whether the frequently seen circling movements toward the other side (Fig. 26d, f) are excitation effects of substantia nigra itself or of adjacent structures like the cerebral peduncle.

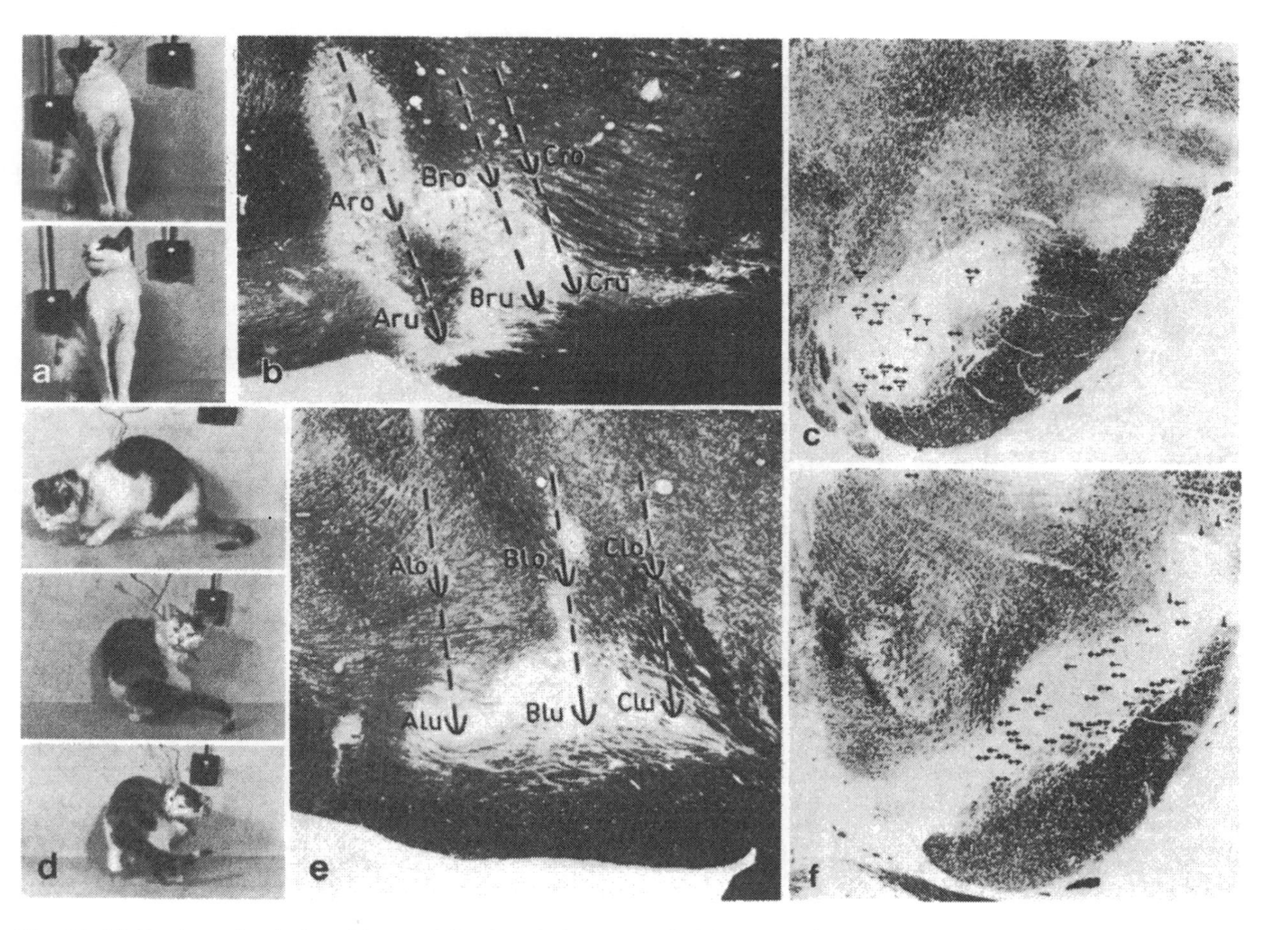

Fig. 26. (a) During stimulation (30 cps; 0.3 mA) of electrode position Aru, located in most anterior part of substantia nigra (b), according to a modified Hess technique, a quick glancing from side to side is obtained. During stimulation (30 cps; 0,4 mA) of electrode position Blu (e) the drowsy unanesthetized cat (d_1) immediately becomes alert (d_2) and performs a fast circling to contralateral side, combined with pupillary dilatation (d_3). (c, f) Diagrams of anterior and posterior main sections of substantia nigra show stimulation points that elicited contraversive circling (*arrows to the right*). These stimulation points are concentrated in the posterior lateral part including the reticulate zone on border of cerebral peduncle. The stimulation points of quick glancing from side to side are situated in the anterior part of substantia nigra (c). ↔ = side-to-side glancing (after WINKELMÜLLER, 1972)

The acceleration of spontaneous movements by stimulating the substantia nigra suggests a dependence of the peripheral innervation of the muscle spindles on the substantia nigra. During anesthesis, stimulation of the caudal half of the substantia nigra (and up to the caudal two-thirds) and of the dorsal side (Fig. 27a and insert) regularly results in an acceleration of the activity of individual γ fibers (spindle motor). Following coagulation of the area stimulated, this effect can no longer be produced. On the other hand, stimulation in the anterior third of the substantia nigra and at its anterior edge results in inhibition of the γ activity (WAGNER, 1964; WAGNER and KALMRING, 1968; HASSLER, 1972a; HASSLER and WAGNER, 1975). Inhibition that is similar but not as marked results from stimulation in the pallidum and nucleus entopeduncularis of the cat (which corresponds to the pallidum internum of man). The γ activity is greatly increased as soon as a decerebrating cut severs the substantia nigra from its rostral afferent input (Fig. 27b).

There is unequivocal physiologic evidence that this nigra action of the γ-neurons is not conducted upward to the striatum but directly downward to the motor apparatus, perhaps over one intercalated

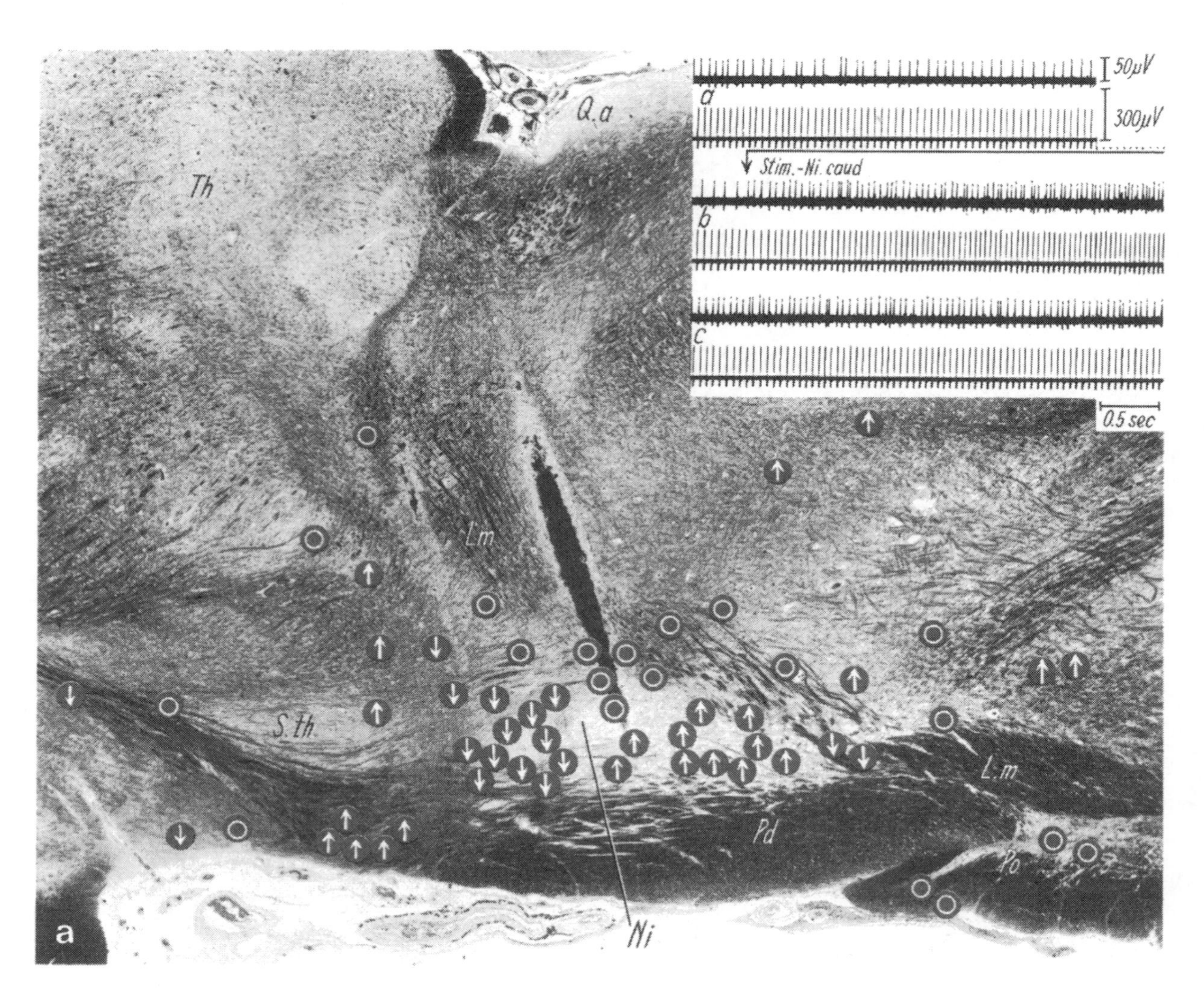

Fig. 27. (a) In sagittal section of cat brain, effects of circumscribed midbrain stimulation on activity of single γ fibers in anterior root L7 are registered. Anterior half of substantia nigra (*Ni*) is filled with arrows pointing downward, indicating that γ activity is inhibited by stimulation. Posterior half of substantia nigra contains arrows pointing upward, indicating that single γ fiber activity is facilitated by low frequency stimulation. Circles indicate no stimulation effect on γ activity. Insert: single γ fiber activity and single Ia fiber activity (below) are recorded before (a) during (b) and after (c) stimulation of posterior part of substantia nigra in cat. *L.m* medial lemniscus; *Pd* cerebral peduncle; *S.th* subthalamic nucleus. (After HASSLER, 1972, based on Wagner's experiments)

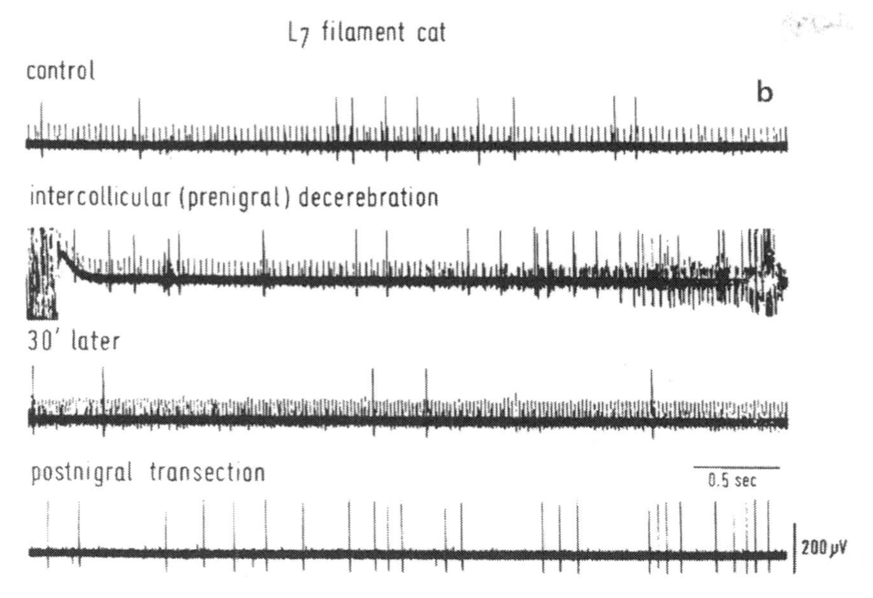

Fig. 27. (b) Recordings from a thin filament of 7th lumbar anterior root show the regular spike activity of a single γ fiber (60 μV) and irregular activity of a smaller γ fiber: a few α spikes (200 μV) are among them. Following classical intercollicular decerebration, which separates the substantia nigra from all forebrain structures, the γ and α discharges are irregularly increased.
Thirty minutes later the 60 μV γ spikes reach the double frequency, the smaller γ spikes are accelerated, and the α discharges are diminished: γ rigidity. Following a postnigral decerebration the higher γ spikes are abolished and the α discharges predominate. (WAGNER'S recording in HASSLER, 1972a)

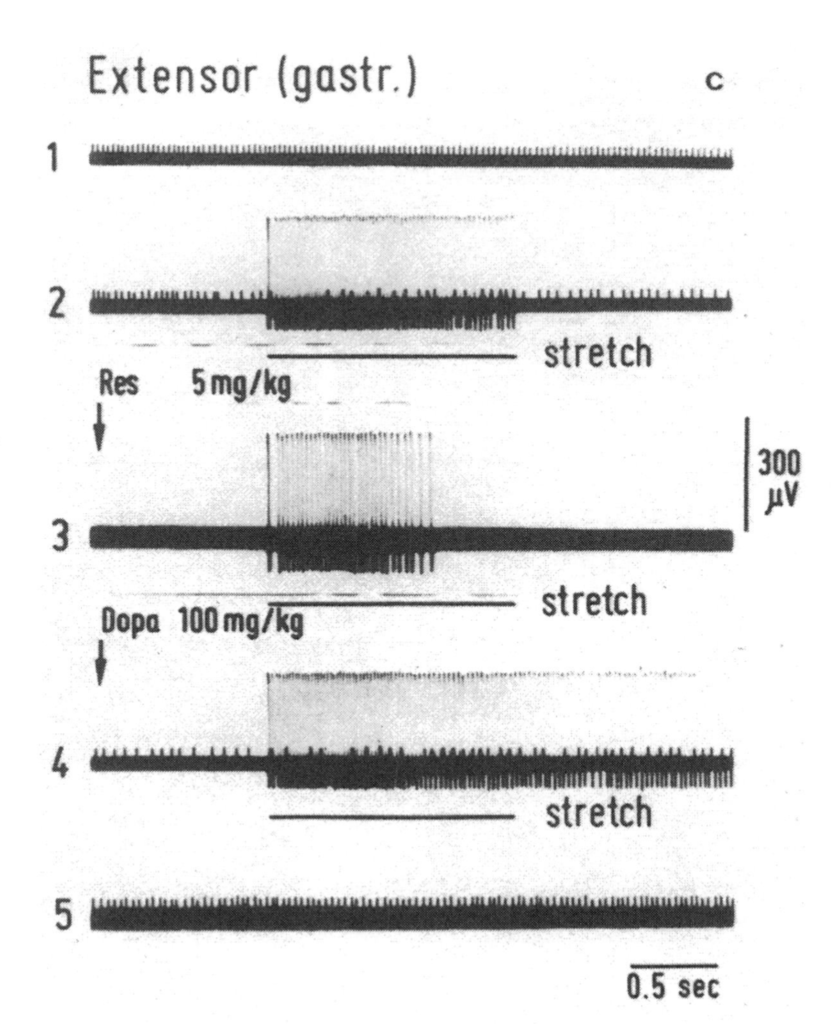

Fig. 27. (c) Recording from a thin anterior root filament conducting to gastrocnemius muscle of cat.
1. Control: very regular γ discharge (35 μV)
2. During muscle stretch decreased γ activity and a burst of α spikes last as long as the muscle stretch
3. Following 5 mg/kg reserpine the γ activity is completely suppressed: the α burst in response to muscle stretch is much shorter than the stretch
4. 20 min after 100 mg/kg L-dopa a slow single γ fiber activity reappears. During muscle stretch the γ activity is slowed down because of the α burst, which outlasts the stretch for many seconds
5. 40 min after 100 mg/kg L-dopa regular single γ fiber activity at higher frequency than control. (After WAGNER, DUPELJ, HASSLER, 1971)

synapse. If a postnigral transection is made through the midbrain (Fig. 27b), all spontaneous γ activity is lost, further evidence for the fact that the γ-activating influence originates in the substantia nigra. This is true, however, only for the fast γ neurons (Fig. 27b), the reinforcement apparatus for fast muscle actions, and the monosynaptic reflex mechanism.

The same activation and acceleration of γ impulses in the anterior root of cat can be elicited by systemic application of L-dopa (Fig. 27c), if dopamine (DA) stores have been depleted hours before by reserpine (WAGNER *et al.*, 1971). Therefore the descending excitatory effect of the posterior two-thirds of the substantia nigra seems to be transmitted to the γ neurons by DA, which is in addition to the usually assumed ascending action of DA to the striatum. The acute depletion of DA by reserpine abolished the spontaneous activity recorded on single γ fibers (Fig. 27c), and produced acute akinesia, rigidity, and tremor-like movements—the complete picture of a Parkinson syndrome. In the

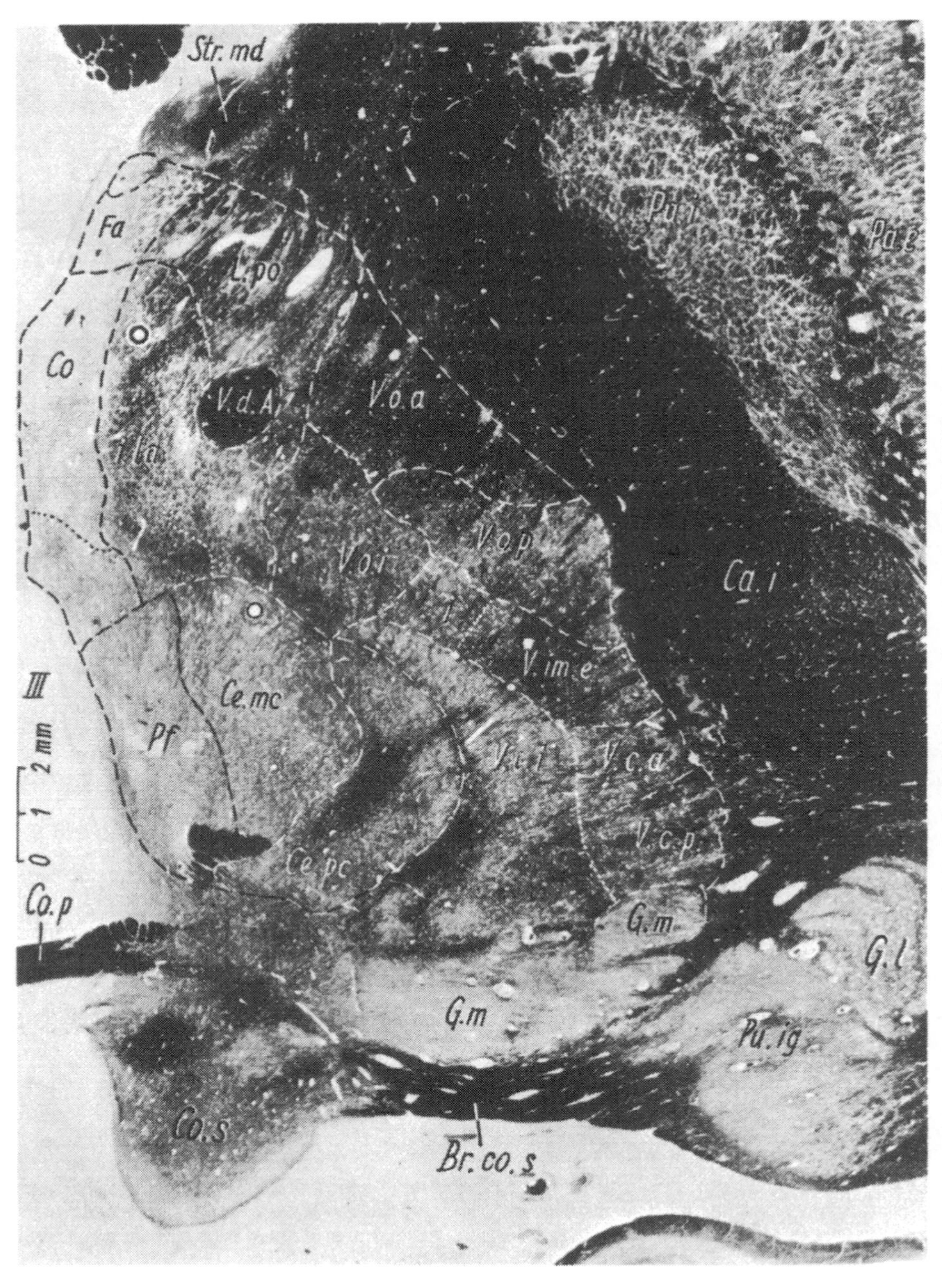

Fig. 28. Horizontal section through ventral nuclei of human thalamus in fiber staining. Adjacent to internal capsule starting in front at genu of internal capsule, ventral nuclei are located one behind the other from rostral to caudal: lateropolar nucleus (*L.po*) supplied by ext. pallidar fibers, corresponding to VA; anterior part of ventro-oral nucleus (*V.o.a*), supplied by pallidar fibers (H_1); posterior part of ventro-oral nucleus (*V.o.p*) supplied by dentato-thalamic fibers [*V.o.p.* and *V.o.a* form the ventral part of *VL*]; intermediate ventral nucleus (*V.im.e*), supplied by I a-fibers; anterior part of ventrocaudal nucleus (*V.c.a*) supplied by cervico-thalamic fibers, corresponding to anterior part of VPL; posterior part of ventrocaudal nucleus (*V.c.p*) supplied by lemniscus medialis, corresponding to posterior part of VPL. Behind this follow the subnuclei of medial geniculate body (*G.m*). ×3.5

same way, monosynaptic reflex activity is governed by DA concentration. The long-lasting α activity of the cat gastrocnemius circuit elicited by rapid muscle stretch is considerably shortened after application of reserpine, which abolishes the underlying γ activity by depleting its exciting transmitter DA. After a subsequent overdose of DA, the γ activity reappears and is increased in frequency. The same muscle stretch of gastrocnemius then elicits a prolonged reflectory α activity. DA is the transmitter of the contraction-reinforcing servomechanism.

These findings suggest that the substantia nigra has a positive influence on the dynamic innervation of muscle spindles and hence on the facilitation of the monosynaptic stretch reflex, which plays a part in rapid movements. This central dopaminergic excitation of γ activity is normally under the inhibitory influence that putamen and caudate nucleus exert (by GABA) on the substantia nigra.

2.5. Other Neuronal Systems Linked to the Extrapyramidal System

Up to now we have discussed the nuclei of the extrapyramidal system in the narrower sense. They have several additional connections with other nuclei and with the cerebral cortex (see Figs. 1, 2, 4). Although they conduct impulses that originate in the extrapyramidal nuclei, lesions in these additional neuronal systems do not as a rule lead to extrapyramidal disorders. In the past some of these nuclei and systems of fibers were regarded as part of the extrapyramidal motor system in the broader sense (Spatz, 1927). It is significant that there are many cortical areas that have a close functional connection with the extrapyramidal nuclei (Fig. 2). Certain ventral thalamic nuclei (see Fig. 28) are dealt with here because the purpose of these investigations was (1) to interpret the pathophysiologic mechanisms of extrapyramidal signs, and (2) to

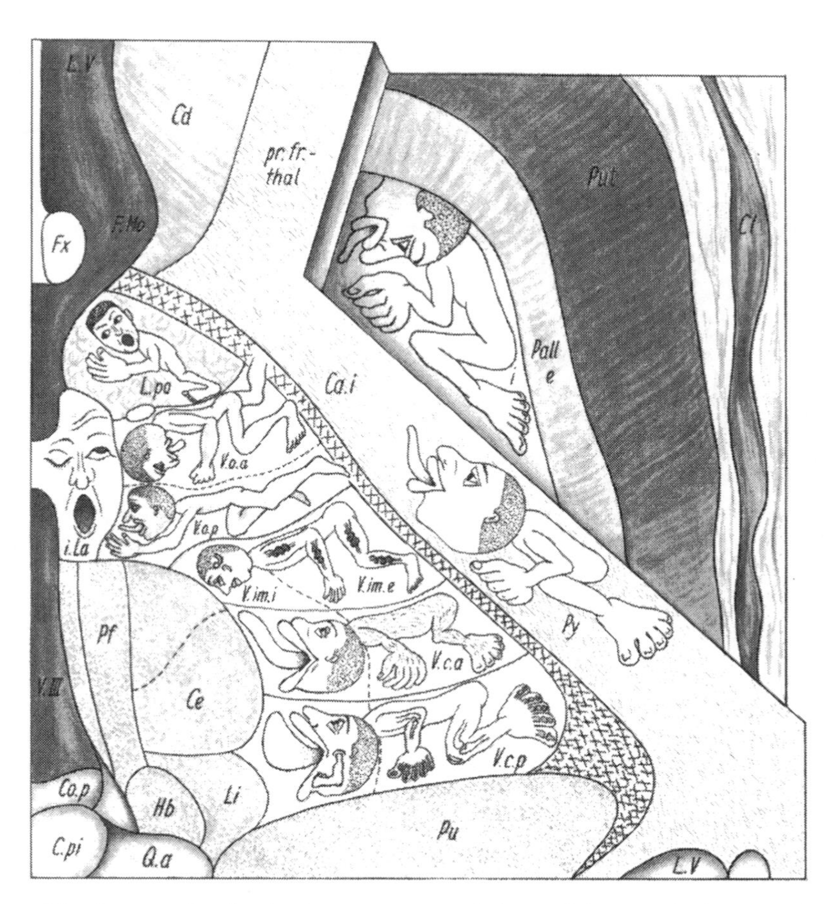

Fig. 29. The homunculi represent the somatotopic arrangement of the 6 functionally different ventral relay nuclei of the thalamus, in a representative horizontal section. The relative size of the body parts of the homunculi indicates the size of the representation of each body region in the respective thalamic nucleus. From back to front follow, first the homunculus of the joint sensibility in V.c.p, the homunculus of the sensory hair sensibility in V.c.a, the homunculus of the muscle spindles length and tension sensation in the V.im.e and medially of the organ of equilibrium in V.im.i, then the sprinter homunculus of the fast muscle action induced by cerebellar afferents in V.o.p, the homunculus of endurance run, of slow and enduring muscle action induced by pallidar locomotor afferents in V.o.a, and in the rostral pole (*L.po*) the homunculus of turning to the contralateral side with loud vocalizing. The half asleep and half awake face symbolizes the sleep-waking regulation by the intralaminar (*i.La*) nuclei. The homunculus in the inner segment of the pallidum (*Pall.i*) is in fiber connection and functional relation with the homunculus of enduring slow muscle action in V.o.a. The homunculus in the internal capsule (*Ca.i*) stands for the voluntary muscle action of the fast pyramidal fibers. (Modified after Hassler, 1972)

influence them therapeutically. The areas concerned are those that have a direct functional relationship with the Parkinson syndrome and its signs.

The most important target system for stereotaxic therapy is the thalamic relay nuclei of different sensory systems and of afferent motor systems, which project to the sensorimotor cortex. They are represented in a horizontal section through the human thalamus (Fig. 28). Each of these relay nuclei is somatotopically subdivided for the different regions of the body. The subdivisions are indicated and the functional role is symbolized in a schematic drawing in which distinct homunculi are located in the precise topographic sequence, somatotopic arrangement and functional differentiation (Fig. 29). A description of the pertinent afferent and efferent connections of each of these nuclei and their topographic relation is unavoidable, because this information is of practical importance for stereotaxic inactivation. The individual ventral nuclei will be dealt with from rostral to caudal.

2.5.1. Nucleus Lateropolaris (L.po or VA). This nucleus forms the rostral pole of the thalamus and is architectonically different from the caudally following nucleus ventro-oralis anterior (V.o.a). The L.po receives fibers directly from the ext. pallidum; these fibers run through the internal capsule in the frontal plane of the beginning of the lamella pallidi (Fig. 5b) without making a detour through the subthalamus (HASSLER, 1949, 1972b). This group of fibers splits up in the frontal plane 3–7 mm behind the interventricular (Monro) foramen in the L.po. Some of the individual fibers from the pallidum converge again after they have passed through the internal capsule in the lamella lateralis and the nucleus reticulatus on the lateral and rostral sides of the L.po (Fig. 5b). The part of the reticulate nucleus lying close to the L.po on its rostral and lateral sides therefore receives impulses from the pallidum in the same way as the L.po itself. The L.po receives from the basal side additional fibers from the pallidum out of the fasciculus thalamicus (H_1). A number of the efferent fibers from the L.po project to the cerebral cortex. There are several indications that they project into the supplementary motor area (see Fig. 6) on the medial aspect of the brain (BERTRAND, 1958). Its stimulation effects turning the head to the other side, raising the arm, and loud vocalization (PENFIELD and WELCH, 1951) as does stimulation of L.po.mc (Fig. 61).

Following complete decortication, some of the nerve cells of the L.po disappear. This means that, while some of them depend on the cortex, they also have a subcortical projection area, which is not yet known. Hemiatrophy of the brain leads to retrograde degeneration of all thalamic nuclei that project onto the cortex by the systematic destruction of the third and fourth cortical layers in one hemisphere. These experiences clearly show that the L.po contains many remaining nerve cells (HASSLER, 1950b, 1959a). It therefore seems correct to regard the L.po as a "semispecific" nucleus in contrast to the other ventral nuclei, which project exclusively and specifically to the cortex.

Many animal experiments have been performed by stimulating the nucleus VA, but because most of these experiments were performed on anesthetized animals for electrophysiologic purposes, they are only briefly mentioned here. Stimulating the nucleus VA in anesthetized cats or rhesus monkeys quickly induces unspecific cortical responses in the same hemisphere with frontal parietal predilection (Fig. 6). However, they spread to most of the other cortical areas. Stimulating at low frequencies results in typical recruiting phenomena in the cortex after a latency of only 10 msec (SLOAN and JASPER, 1950). Destruction at the base of the VA, which probably also destroys the bundles of fibers from the pallidum in H_1, blocks the way for recruiting phenomena from other unspecific nuclei in the thalamus, e.g., from the *centre médian* to the cortex. There is no other circumscribed area in the diencephalon where the pathways for the recruiting responses can be interrupted (SLOAN and JASPER, 1950; STARZL *et al.*, 1951). Bilateral coagulation in the rostral pole of the thalamus in monkeys, destroying in addition to the L.po parts of the lamella medialis of the medial nucleus of the V.o.a and of the anterior main nucleus, leads to a severe akinetic state in which the animals lose even hunger and thirst drives (CHOW *et al.*, 1959).

The L.po nucleus comprises parts with pallidum fiber supply that are relevant for stereotaxic therapy of slow, cramp-like hyperkinetic movements; other parts form the efferent truncothalamic activity- and consciousness-regulating system. Bilateral coagulation must be avoided in stereotaxic therapy.

2.5.2. Nucleus Ventro-oralis Anterior (V.o.a) or Anterior Part of the Ventral Half of the Nucleus Ventralis Lateralis (VL). The relays of neurons going from the pallidum internum through H_1 (fasciculus thalamicus) to the motor cortex lie in the V.o.a (Fig. 5a). Although no other afferent fibers of this nucleus are known, their existence cannot be excluded. In man this nucleus does not receive dentato-thalamic fibers (HASSLER, 1950). The V.o.a is a nucleus projecting only to the cerebral cortex. The area into which it projects is the motor cortex and probably corresponds to area 6aα in man (Figs. 1 and 2); it is unlikely that it projects to the primary motor area 4γ.[3] Very few physiologic animal experiments have been performed stimulating this nucleus because even in the cat it is very slender. Marchi's method enables fiber connections from the nucleus entopeduncularis to the V.o.a to be demonstrated in the cat (MONTANELLI and HASSLER, 1964). It is therefore possible to relate the turning movements to the opposite side, which can be obtained from this thalamic nucleus by Hess's method, to the excitation of this pallido-thalamic system (Fig. 3a–f), since stimulation of both segments of the pallidum results in contraversive turning (HASSLER, 1957b; MONTANELLI and HASSLER, 1964). Stimulation by an 18-pole probe electrode at 1 cps produces, in a unipolar recording from different points on the scalp of man, only longlatency cortical responses that appear like a phase reversal of post- and precentral recordings if points 12 and 13 in V.o.a or in zona incerta, respectively, have been stimulated (Fig. 30a, b). This indicates that the responses are from nonspecific systems. If, however, a 1-cps stimulus of 1 ms duration is applied via a concentric bipolar electrode with a pole distance of 3 to 5 mm, a phase reversal of the early components of the cortical response between pre- and postcentral recording can be detected (Fig. 30c) with a 5-ms latency of the first positive wave (GANGLBERGER, 1962). This could be confirmed by recording with Ag/AgCl electrodes on the cortex in triangular-switching (GANGLBERGER, 1970), whereas the latency of the negative and second positive response wave is difficult to determine with averaging. It is remarkable that, in the cat, recruiting responses can be evoked electrophysiologically and in field potentials at first exclusively in area 6aβ (lower lip of the sulcus cruciatus) by low-frequency stimulation, with a secondary non-specific spread of cortical responses to many other areas (according to MORISON and DEMPSEY, 1942 and SASKJ *et al.*, 1970). The effect of stimulation in the human patient before therapeutic inactivation of V.o.a will be extensively described below, with post-mortem localization of the point of stimulation. V.o.a belongs to the system of tonic and enduring slow special movements, such as grasping and turning the head and body to the contralateral side, with a somatotopic arrangement. V.o.a is therefore represented diagrammatically (Fig. 29) by a homunculus of enduring-run in bent posture with hip, knee, and elbow flexed and tongue hanging out. The functional role of the pallidum internum, the next neuronal station before V.o.a, is almost the same. It is not adequately expressed in Figure 29, as it is not the psychomotor function of V.o.a and pallidum.

[3] Preliminary results of retrograde labeling of the afferent projections of distinct thalamic nuclei of the cat to the subareas of the somatomotor cortex by the horseradish peroxidase method indicate that areas 6aβ, 6aδ, and 6aα receive thalamic projections from V.o.a, L.po, and Z.o. The subfields of primary motor area 4 receive specific afferents from V.o.p and V.o.i nuclei (NAKANO and HASSLER, in prep., 1979).

2.5.3. Nucleus Ventro-oralis Posterior (V.o.p) or Caudal Part of the Ventral Half of VL. This nucleus receives afferent input via the dentato-thalamic fibers (Figs. 1 and 2). Before reaching the V.o.p they pass through the red nucleus and then run in a rostral direction along the base of the caudal (somatosensory) ventral nuclei and finally split up in the V.o.p. Many animal experiments have been performed on this nucleus, which is the termination point of the dentato-thalamic fibers in the thalamus, but its exact architectonic definition only became possible when, in human cases with localized lesions, differences were found in the endings originating from the pallidum and the cerebellum (HASSLER, 1949a). The V.o.p is one of the *oral* ventral nuclei because its nerve cells are unequivocally of middle size, while the ventral nuclei lying more caudally (V.im and V.c) all contain in addition very large nerve cells. The V.o.p projects exclusively to area 4γ of the motor cortex. This connection between cerebellum and motor cortex has also been demonstrated several times by electrophysiologic methods since WALKER and GREEN (1938) did it by recording evoked potentials. Stimulation of this nucleus with the modified Hess method in unrestrained cats results in twitching of the contralateral limbs or the muscles of the contralateral side of the face in rhythm with the stimuli (BÜRGI, 1943; HASSLER, 1949b, 1956a). If the stimulation is protracted, the rhythmic twitching at the distal ends of the limbs can be combined with maintaining flexion of the same limbs. As a rule, the twitching is restricted to cer-

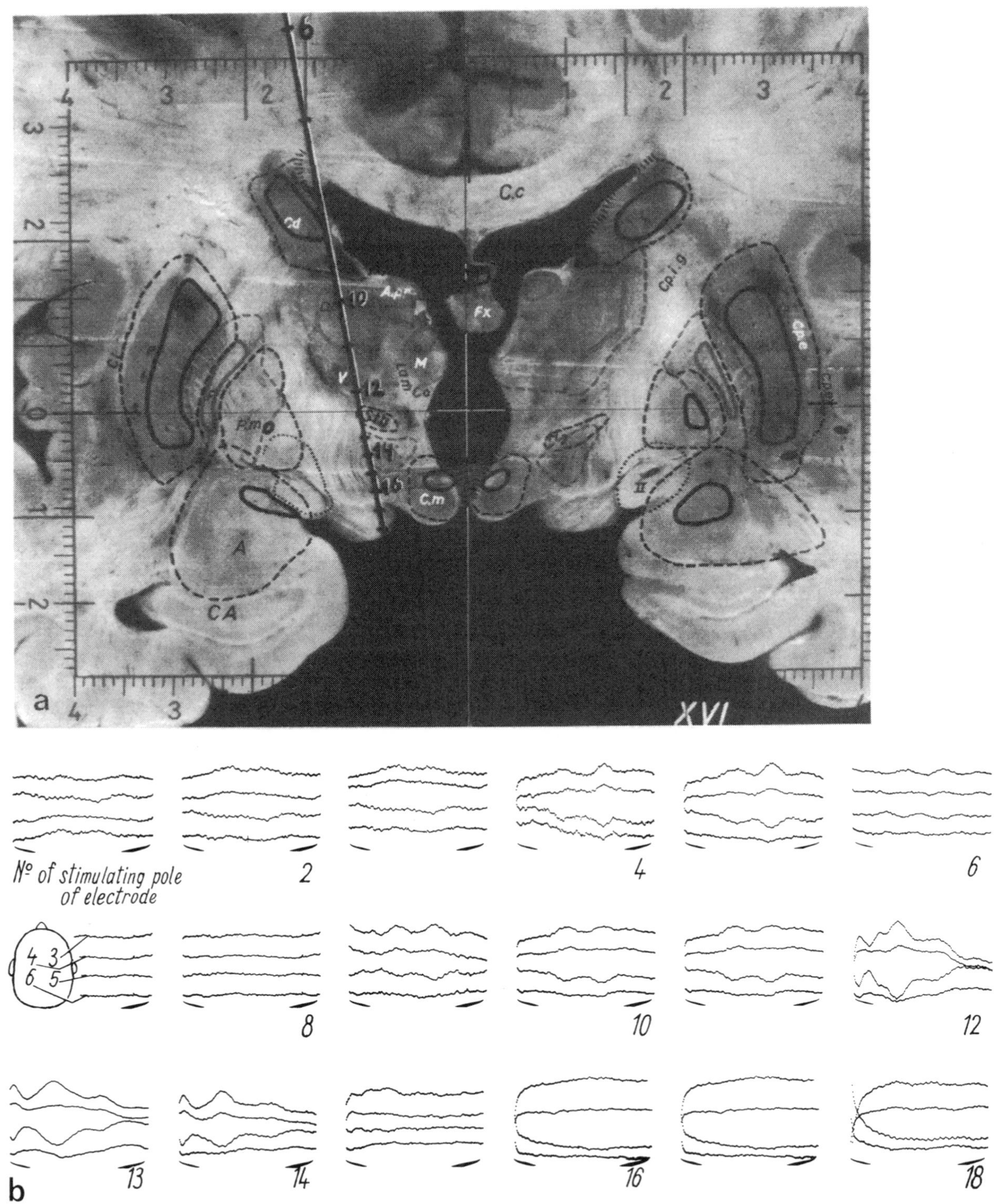

Fig. 30. (a) In frontal section of human diencephalon (In: SCHALTENBRAND and BAILEY, Vol. 2, Fig. 16) the location of multiple probe electrode with 18 poles constructed by RAY and VOGEL is demonstrated. Lowest recording site 18 is located in cerebral peduncle. Angle of approach of this multiple-lead electrode to midline is 10°. The various recording poles with their numbers are marked in picture. Notice that pole number 12 is located in base of V.o.a (*V*) and recording site number 13 is located in zona incerta and subthalamic region.

Fig. 30. (b) Unipolar records from scalp places 3, 4, 5, 6 (see insert) against ipsilateral ear. Cortical responses during stimulation are averaged from 10 records. During 1 cps stimulation of poles No. 12 and 13, average cortical response shows clear reaction. On ipsilateral side, note phase reversal of precentral and postcentral scalp recording sites (3.5). Circumscribed cortical responses demonstrate nonspecific projections of stimulated thalamic or subthalamic point to motor cortex. (After MUNDINGER and VOGEL, 1973)

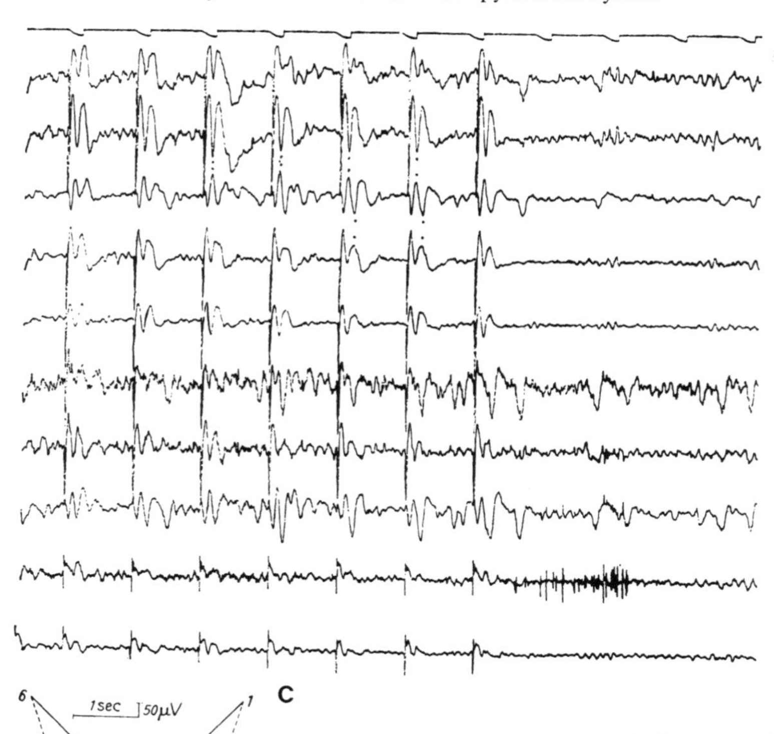

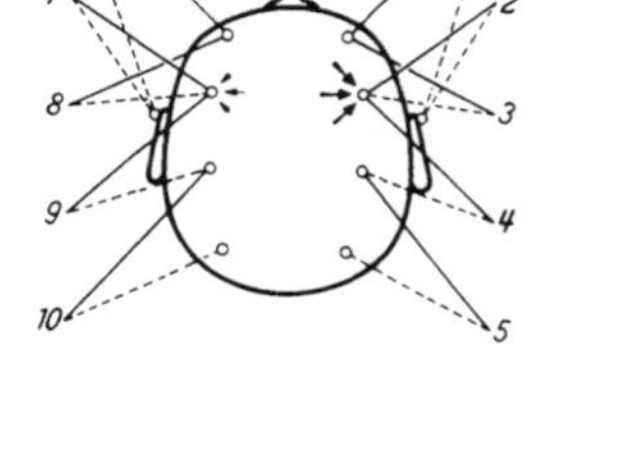

Fig. 30. (c) Scalp recordings during 1 cps stimulation (1 msec; 40 V) of the right V.o.a nucleus in a Parkinson patient. The localisation of the maximum of cortical response is demonstrated by the phase reversal of the early waves between the 2. and 3. lead and much less pronounced between the 3. and 4. lead. The waves are marked by points. The maximal projection corresponds to the area 6aα in accordance with the anatomical projection of V.o.a. Symmetrical but much weaker phase reversal in cortical response in the non-stimulated hemisphere (GANGLBERGER, 1962)

tain parts of the body and rarely affects the whole contralateral half of the body.

In the experimental and clinical literature, this nucleus is usually called VL, although this term, first used by von MONAKOW (1885), then by VOGT (1909) and later by CROUCH (1934) and WALKER (1936, 1938), is used either as a general name for the whole rostral part of the lateral thalamus or for individual parts of it. According to Walker's definition, both ventral oral nuclei—V.o.p (with cerebellar afferents) and V.o.a (with fiber supply from pallidum)—as well as all parts of the lateral thalamus dorsal to them, belong to VL, with no differentiation and with no definite separation from VA, which is primarily confused with the V.o.a because it also receives pallidar fibers. The numerous discrepancies and confusion concerning the results of electrophysiologic studies of VL are due to the confusing of at least two different functional systems: V.o.a and V.o.p. At the cortical level, no one hesitates to differentiate area 4γ and area 6aα or 6aβ. In this monograph, the V.o.p belongs to the system originating from the nucleus dentatus parvocellularis (Fig. 32a, b) and projecting by upper cerebellar peduncle fibers that cross the midline and terminate in V.o.p. After degeneration of the dentate nucleus (Fig. 31a), the dentatothalamic fibers no longer enter the V.o.p, whereas the afferent fibers from other sources remain (Fig. 31b). The projection of V.o.p goes to area 4γ and is somatotopically arranged (NAKANO and HASSLER, 1979). This system controls quick, jerky, starting, and acceleration movements and is therefore symbolized in Figure 29 by a sprinter or a high jumper.

2.5.4. Nucleus Ventrointermedius (V.im). Structurely the V.im resembles the caudal more than the oral ventral nuclei since it contains, among other neurons, very large nerve cells. By using methods of myelin sheath degeneration in man and the Marchi method in the cat, a tractus vestibulo-reticulothalamicus has been identified as an afferent supply of fibers that do not cross (HASSLER, 1948/49, 1956b). They are intermingled with ipsilateral trigemino-thalamic fibers (LANDGREN *et al.*, 1967). In contradiction to previous statements, new experiments on cats with the Nauta-Gygax method have shown (HASSLER, 1972b) that the termination of the vestibulothalamic pathways is limited to the medial half=V.m.i of the V.im.

However, ANDERSSON *et al.* (1966), using electrophysiologic methods, have shown an afferent

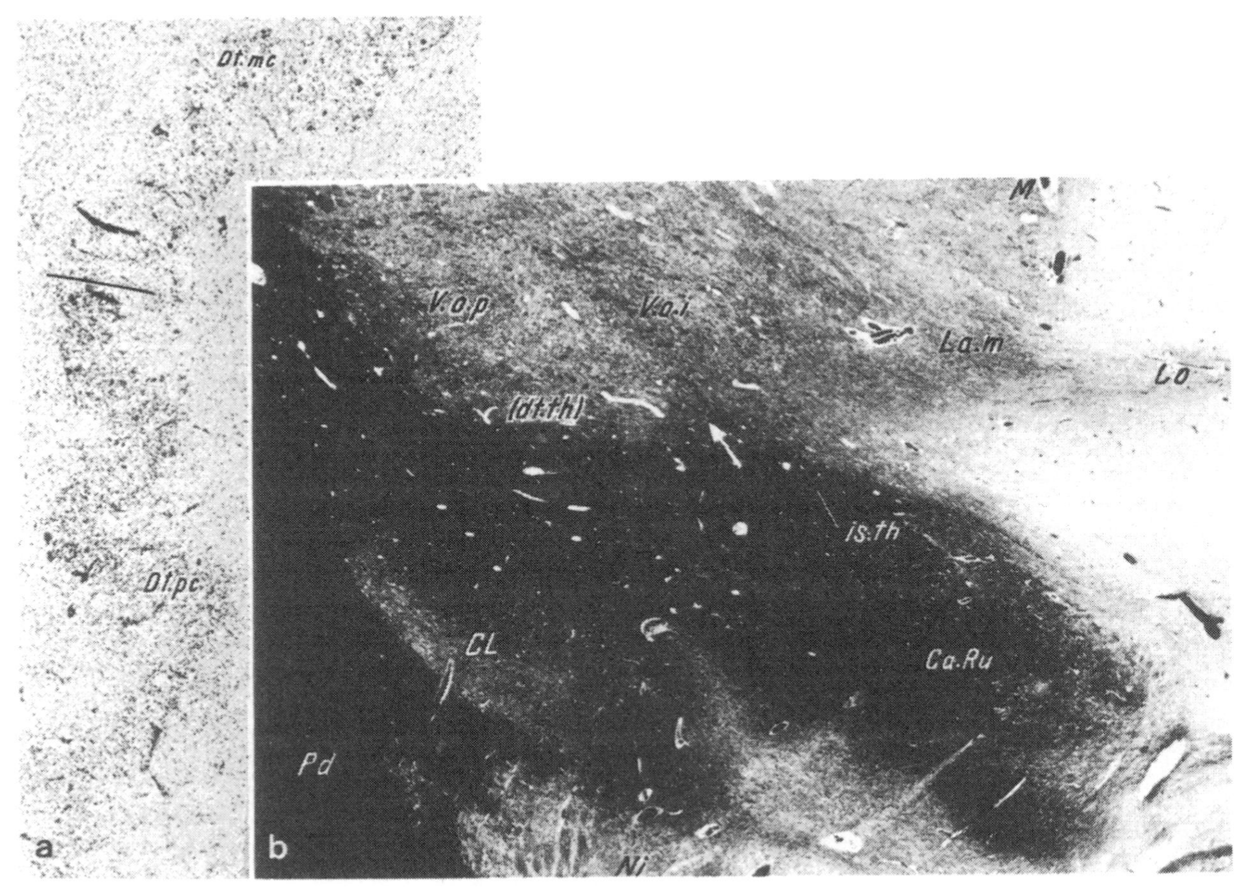

Fig. 31. (a) In case of juvenile amaurotic idiocy, main part of dentate nucleus (parvocellular: *Dt.pc*) is almost completely deprived of intact nerve cells with glia proliferation in contrast to better preserved magnocellular part (*Dt.mc*). Nissl stain. ×30.

(b) Consequently all dentato-thalamic fibers (*dt.th*) are degenerated and lost in base of V.o.p nucleus, whereas interstitiothalamic fibers (*is.th*) are completely preserved on medial border of V.o.p where they enter V.o.i nucleus (*white arrow*). (After HASSLER, 1950a)

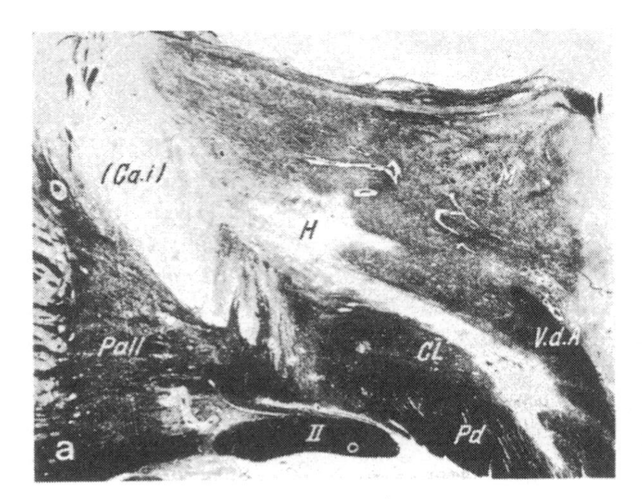

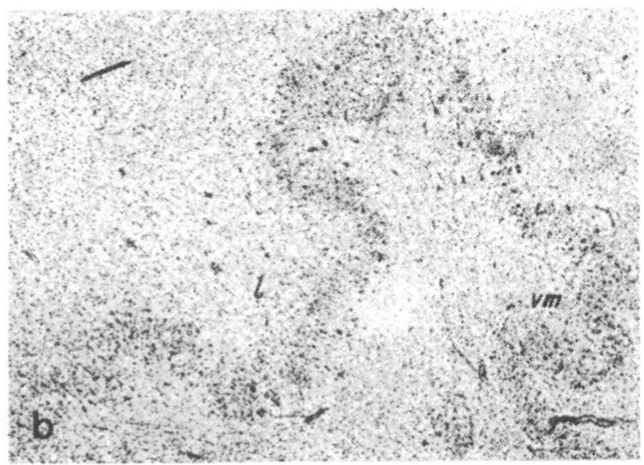

Fig. 32. (a) Transverse section of thalamus in human case of vascular foci, here in diencephalon with destruction of internal capsule (*Ca.i*), of the V.o.p nucleus (*H*), and of the zona incerta above the subthalamic nucleus (*CL*). Fiber stain. ×4.

(b) Nerve cells of parvocellular part of dentate nucleus [mainly in its lateral part (l)] are retrogradely degenerated, due to destruction of their terminals in V.o.p nucleus. Demonstration of direct cross connection from dentate nucleus to V.o.p. Nissl stain. ×20. (After HASSLER, 1950a)

representation of the Ia muscle spindles in the lateral half of the V.im. In the cat, this is analagous to the cortical representation of the afferent pathways of the Ia muscle spindles in the gyrus sigmoideus posterior in the region of the so-called postcruciate dimple. This field corresponds in its architectonic characteristics to area 3a (HASSLER and MUHS-CLEMENT, 1964). PHILLIPS *et al.* (1971) recently showed the same phenomenon in the baboon. In cat, monkey, and man the V.im projects into the area of the sensorimotor cortex in the order of projection to area 3a (Fig. 2). It follows that in the V.im there is a lateral part (V.im.e), which serves the central representation of the muscle spindles and conducts to area 3a, in additional to the vestibular representation in V.im.i.

Using stimulation in man, we observed that the head and anterior part of the body turned to the side of stimulation from the V.im, but mainly its medial half. Stimulating experiments have, however, not confirmed the earlier assumption based on these findings that the V.im, which has a vestibular supply, projects to area 3a. Ipsiversive turning in response to cortical stimulation in the cat occurs only when part 6aδ in area 6 in the lower lip of the sulcus cruciatus is stimulated by the Hess method (HASSLER, 1960, 1972). Inside the V.im the inner and outer parts have been found to have two different functions, represented in the cortex by two separate areas. It must therefore be regarded as proven that the V.im has a double functional significance with separate cortical projections. The pathway from the vestibular nerve goes through the V.im to area 6aδ, and that from the Ia muscle spindle goes through the V.im.e to area 3a. Consequently they are symbolized in Figure 29.

In the last 15 years, the term V.im has often appeared in stereotaxic literature in relation to a 5–7 cps rhythm, which can be recorded from the base of V.im. Thus the pacemaker of tremor has been localized in V.im.e without post-mortem check, and the same area has been claimed to be the best target for tremor therapy. Post-mortems of 23 Parkinson patients with tremor relief of different degrees up to 100% showed (HASSLER *et al.*, 1970) unequivocally that the therapeutic lesion involved a very tiny border of V.im in only a few cases and in most cases left it completely undamaged, even with 90–100% tremor relief. The explanation is that the afferent bundles to V.o.p run, still bundled, through the base of V.im.e, where they can best be recorded.

2.5.5. Nucleus Ventrocaudalis (V.c.e = VPL and V.c.i = VPM). This nuclear complex receives the terminations of the medial lemniscus and tractus spinothalamicus as well as the corresponding pathways of the trigeminal nerve (Figs. 1 and 7). The fibers of the medial lemniscus split into two bundles before entering the thalamus. The thicker fibers of the caudal bundle originating in the posterior column nuclei terminate in the V.c.p (= VPLc). Perception of joint movements is the main quality conducted by lemniscal fibers (MOUNTCASTLE and POWELL, 1959; MOUNTCASTLE, 1966). They are symbolized in Figure 29 by a joint homunculus in V.c.p, in contrast to a tactile homunculus (sensory hairs) in V.c.a. The anterior bundle containing fibers of the cervicothalamic tract terminates more anteriorly in front of V.c.p in V.c.a (HASSLER, 1972). Tactile sensations excited by hair bending are the main modality conducted in these fibers (HORROBIN, 1966). The spinothalamic tract conducting pain fibers among others has its main terminal in the most basal part of the V.c, which is characterized by its smaller nerve cells (V.c.pc) (HASSLER, 1960a; 1964, 1970).

All three nuclei project to the cortical fields of the posterior central gyrus. The special projection of the V.c.p is area 2, that of the V.c.a seems to be area 1, and the fibers from the V.c.pc terminate (Fig. 2) predominantly in area 3b (HASSLER, 1960a), which is hidden in man in the central sulcus.

Stimulation of the V.c.p and V.c.a in unanesthetized cats and monkeys leads to twitching in rhythm with the stimuli in the contralateral muscles of limbs or of the face. They have to be interpreted as reflexes in response to sensory stimulation. Experimental coagulation of these somatosensory nuclei produces a sensory defect on the contralateral side while in all other ventral nuclei (V.im, V.o.a, V.o.p) coagulation does not produce any somesthetic defects.

Stimulation of the V.c.a in unanesthetized patients during stereotaxic operation evokes a prickling sensation, paraesthesia, and twitching of muscles of the contralateral side. If the parvocellular nucleus of the caudal-ventral complex (V.c.pc) is stimulated, the unanesthetized patient feels a severe circumscribed pain, often cramplike, on the contralateral side if the frequency exceeds 20 cps. Circumscribed coagulation of the parvocellular ventral–caudal nucleus results in contralateral analgesia and thermanesthesia, as in a dissociated sensory defect (HASSLER and RIECHERT, 1959).

III. Clinical and Pathophysiologic Findings Related to Autopsy Data in Cases of Parkinsonism Operated on by Stereotaxis

1. Methods

Using our own stereotaxic apparatus[1] and our own method of localization[2], we have performed 5664 stereotaxic operations since 1950. Of these, 3486 operations were for the Parkinson syndrome, which is the most frequent indication for stereotaxic operations (Table 1).

Table 1. Number of stereotaxis operations (1950–31. 3. 1977)

Parkinsonism	3 486
Hyperkinesia	844
Curie therapy	717
Other untractable painful states	175
Psychiatric surgery	155
Epilepsy	76
Angiomas	17
Combined stereotaxic–open operations on hypophysis	105
Others	89
Total	5 664
Clinic mortality	0.82%

Clinical mortality rate in operations for Parkinson was 0,82%. As is customary, this figure includes all patients who died in hospital following the operation, including those who died of noncerebral complications (e.g., pulmonary embolism). Table 2 shows in greater detail the causes of death for those cases whose post-mortem findings are described here. We distinguish here between deaths due to cerebral and to extracerebral complications. The cerebral complications include two cases of allergic hemorrhagic reactions of the brain to antibiotics.

Tissue destruction was performed by high-frequency current. In the first 475 operations we used a commercial spark-gap diathermy apparatus (Erbotom, Tübingen, and Hellige, Freiburg i.Br.). Then, in 202 cases, we used the Wyss coagulator (J. Monti, Genève) (Wyss, 1945), which enabled us to obtain better limitation of the coagulated area by using a high-frequency current of pure frequency. Since 1959 we have employed a coagulating instrument that we developed specially for stereotaxic operations (Mundinger *et al.*, 1960). It has been used in 2 700 Parkinson cases. The instrument always produces a pure sinus high-frequency current. In addition, it measures tissue temperature during coagulation and records most important factors in the process of coagulation (current, voltage, resistance, and power). The three groups in Table 2 were formed because the different methods of coagulation affected the number of complications. The method of coagulation at controlled temperature (temperature was regulated to reach 70° C at the point of the electrode) only denatures protein. If there are no rapid increases of current, this method prevents hemorrhage at the point of coagulation.

Below we compare the autopsy findings in the brain with the clinical findings according to the apparatus used in seventeen cases involving 23 operations on patients who had Parkinsonian disorders of movement.[3] We have already published post-mortem findings following thalamic operations for pain (Hassler and Riechert, 1959) and for a case of temporal epilepsy after bilateral stereotaxic fornicotomy (Hassler and Riechert, 1957). A few cases of Parkinsonism were also briefly described elsewhere (Hassler, 1961; Hassler *et al.*, 1965, 1967, 1969a, b, 1970). We have also included three patients who died at varying intervals, some after several months, following discharge from treatment.

[1] The stereotaxic apparatus and its use have been described in detail by Riechert and Mundinger (1956) and in Schaltenbrand and Bailey (eds.), (1959, pp. 437–470). See also discussion on pp. 180–182.

[2] See discussion pp. 146–159; pp. 252–253.

[3] Two cases of functional inactivation by implanting ^{192}Ir are published separately.

Table 2

Case No.	Name	Age in years	Diagnosis and Cause of Death
1	S.G.	55	**Paralysis agitans** Death occurred 23 days after operation due to lung embolism
2	F.J.	59	**Paralysis agitans** Death occurred 21 days after 3rd operation due to acute circulatory failure with extensive thrombosis of the thigh veins with recurrent lung embolism and lung infarction. Final lung embolism fatal
3	R.L.	59	**Paralysis agitans** Death occurred 6 days after operation due to lung embolism and right cardiac insufficiency and chronic arteriosclerotic circulatory failure and chronic recurrent pyelonephritis
4	H.J.	58	**Postencephalitic Parkinsonism** Death occurred 7 days after operation due to large lung embolism with bilateral venous thrombosis of the thigh
5	T.W.	63	**Paralysis agitans** Death occurred 21 days after operation due to a massive lung embolism with thrombosis of the thigh veins
6	P.T.	50	**Parkinson syndrome due to hamartoma of the hypothalamus** Death occurred 82 days after operation due to purpura cerebri (penicillin allergy)
7	C.Z.	57	**Paralysis agitans** Death occurred 45 days after 3rd operation due to putrid bronchopneumonia after a fracture of the neck of the femur
8	S.H.	65	**Postencephalitic Parkinsonism** Death occurred 28 days after operation due to putrid bronchitis with aspiration pneumonia
9	A.I.	61	**Paralysis agitans** Death occurred 24 days after 2nd operation due to lung embolism (post-mortem of the body not allowed)
10	B.W.	66	**Paralysis agitans** Death occurred 5 days after operation due to pulmonary embolisms with pneumonia and cardiac insufficiency with genuine hypertension
11	B.H.	64	**Paralysis agitans with severe arteriosclerosis and atrophy of the brain** Death occurred 7 days after operation due to lung embolism (Infarction of left lobe, extensive bronchopneumonia in right lobe, acute lung edema)
12	R.B.J.	63	**Paralysis agitans combined with cerebral arteriosclerosis** Death occurred 15 days after 2nd operation due to confluent bronchopneumonia and kidney failure
13	F.L.	58	**Paralysis agitans** Death occurred 29 days after 2nd operation due to circulatory failure following a massive lung embolism
14	H.P.	57	**Postencephalitic Parkinsonism** Death occurred 32 days after operation due to severe leukocytic, partially absceding bronchopneumonia
15	B.L.	62	**Postencephalitic Parkinsonism** Death occurred 52 hours after operation due to acute cardiac failure with bilateral bronchopneumonia (from gunshot wound with scarred empyema)
16	D.M.	59	**Postencephalitic Parkinsonism** Death occurred 5 years after operation due to cardiac and circulatory failure with lung emphysema
17	O.A.	65	**Postencephalitic Parkinsonism** Death occurred 7 days after operation due to hypostatic pneumonia (circulatory insufficiency, post-mortem of the body not allowed).

Transverse sections were made through the brain after fixation with formalin. In addition, in twelve cases we made complete serial sections of the area of the basal ganglia. We paid particular attention to the exact measurement of the size and boundaries of the coagulated areas and their location.

Five patients had bilateral operations, so that we were also able to examine areas of coagulation that did not cause complications. We could therefore make statements regarding the course of the illness in connection with the formation of scars in the coagulation areas and the compensatory processes.

2. Case Histories

Case 1. S.G., aged 55; paralysis agitans treated by ventro-oral thalamotomy with marginal involvement of the internal capsule; stereotaxic operation in the left V.o.a and V.o.p (no. 1733); patient died 23 days after operation from pulmonary embolism.

History. Parkinson disease started at age 48 without any indication of earlier encephalitic illness. There had been nocturnal states of clouding of consciousness during which the patient looked at his wife without speaking. He first developed rigidity in the right arm more than in the leg. In 1957 a moderate tremor appeared on the same side. In 1960 the left side became involved.

Clinical findings. Moderate rigidity in arm and leg, right a little more than left. There was only occasional tremor, but this disturbed the patient considerably. Gait with small steps, delay in starting, and marked propulsion were observed, as were akinesia and abnormal postures of trunk with backache. His face was expressionless. He needed assistance for dressing and many ordinary activities. He was less alert and often was dissatisfied. His memory for recent events was impaired. However, his reflexes were normal, as was the cerebrospinal fluid (C.S.F.). The electroencephalogram (EEG) was normal, with α rhythm at 8–9 cps (Fig. 33a, b).

The physicians (Medizinische Poliklinik of the University Freiburg, Director Prof. Dr. H. SARRE) recommended a two-week preparatory treatment of the circulation because of essential hypertension without decompensation and because of pulmonary emphysema and obesity. This was carried out before the operation.

Lumbar encephalography with helium demonstrated a moderate enlargement of the interior C.S.F. spaces with an enlargement of the whole

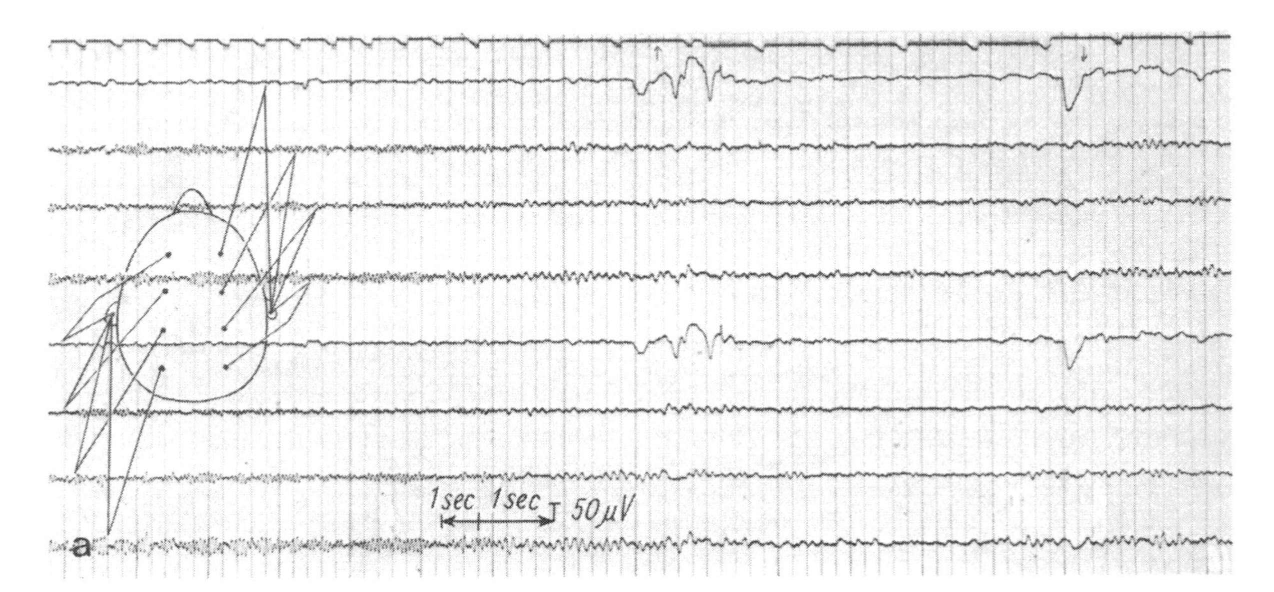

Fig. 33. (a) Electroencephalogram in paralysis agitans, case No. 1. Bipolar recording from scalp before operation shows slightly irregular α-type EEG (8–9 waves/sec) and slight, through not clearly abnormal, side difference, amplitudes being somewhat higher on left; reduction of α rhythm after eye opening (*upward arrow*) with eye movement artifact in frontal leads; increase of α activity after eye closure (*downward arrow*).

left exterior angle of the ventricle and dilatation of the third ventricle to a diameter of 10 mm.

A *stereotaxic intervention* was performed 3 days after the encephalography. Stimulation in the V.o.a on the left side with rectangular waves of 1 msec duration with a distance of 5 mm between the two poles of the electrodes at a rate of 8 cps led to jerks of the right arm synchronous with the stimuli. Stimulation at the rate of 50 cps evoked very rapid movements and a spasmodic turning of the right hand. Active alternating movements were slightly retarded. Counting backward was inhibited during the stimulation. Increasing the strength of the stimulation led to additional apnea in inspiration, widening of the palpebral fissure and tetanic contracture of the face on the right side.

Coagulation was performed with the thermoelectrode regulated for temperature with a noninsulated point of 4×2 ϕ mm. Coagulation of 30 sec duration at a regulated temperature of 70° C was performed in three places. One was the target point (0), one 2 mm (position +2) in the ventrocaudal extension of the axis of the electrode, and one 5 mm (+5) deeper. After the first coagulation in position +2, the rigidity in the leg as well as in the arm was slightly improved. After the second coagulation in position +5, the rigidity in the arm was improved by 80–90% and abolished in the leg. After the last coagulation at the target point (position 0), no rigidity could be found on the right side. After the first coagulation the patient became slightly irritable, but at the end of the operation he seemed slightly sluggish; consciousness was unimpaired.

Postoperative course. When discharged from the hospital 14 days after the operation, the patient's tremor and rigidity on the left side had disappeared. The delays in starting and propulsion had disappeared. Gait and abnormal postures had improved. There was no more backache. The patient showed no psychosyndrome and no depressive change.

The EEG (recorded by Dr. GANGLBERGER) 7 days after the operation showed repeated short runs of slow δ waves of medium voltage on the left side in the frontotemporal to parietotemporal areas and an α rhythm of 8–10 cps (Fig. 33a, b).

During the period of treatment when he received physiotherapy as an outpatient, he was found twice by his wife lying unconscious next to the bed. This raised the suspicion of major epileptic attacks. On the 23rd day after the operation the wife found the patient motionless, having fallen out of bed. Admitted immediately to the hospital, he was pronounced dead.

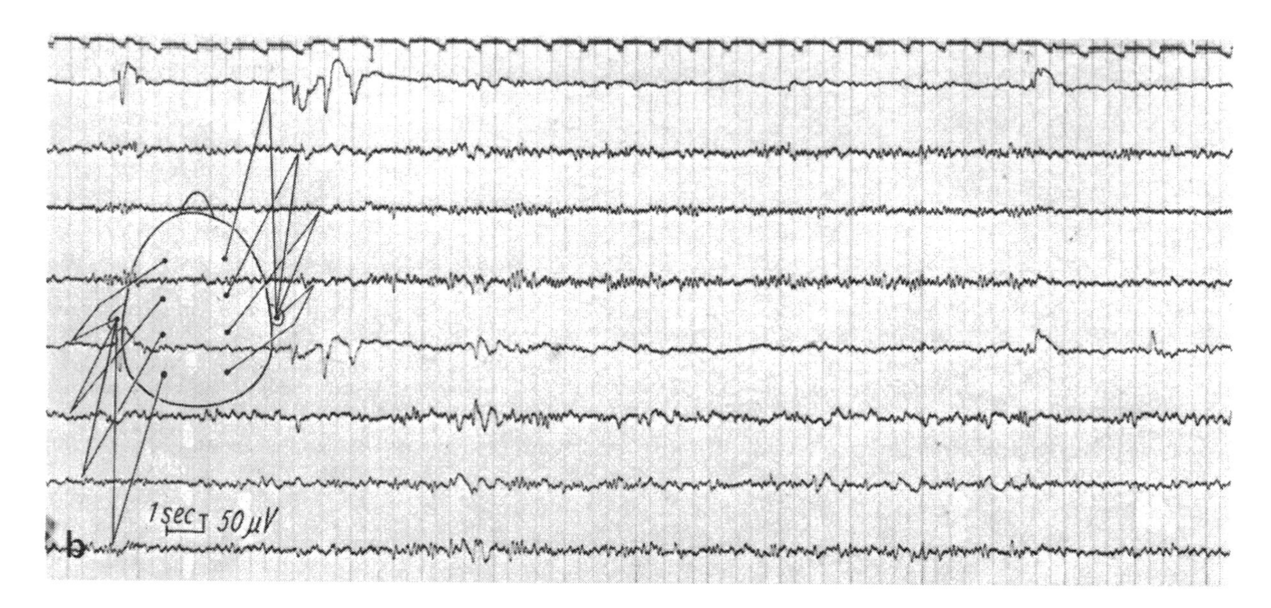

(b) Electroencephalogram of same patient on 7th day after left-sided stereotaxic destruction in V.o.a. Fairly irregular α-type EEG with slowing and reduction of α rhythm in occipital and parietal areas on left side and intermediate waves appearing intermittently. A δ focus on left side with occasional steep parietal waves with accent in anterior areas. Similar changes, but of much lesser degree, are found over unoperated right hemisphere. Dominant rhythm is slower than before operation

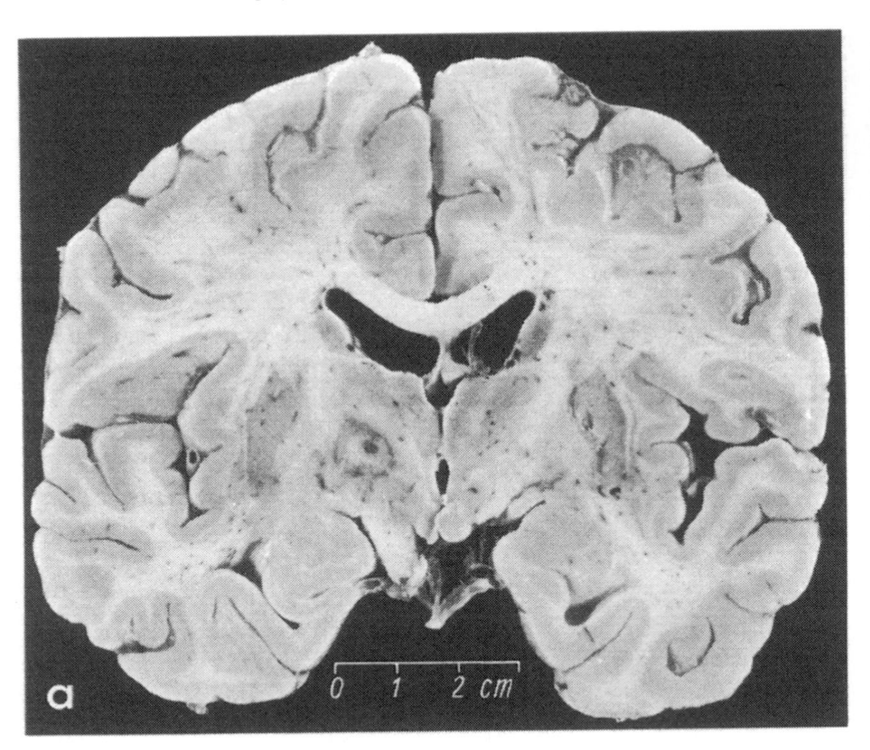

Fig. 34. (a) Unstained transverse section of brain of case No. 1. In plane of massa intermedia, left thalamus is somewhat enlarged by recent coagulation in left V.o.a. Track of electrode (*black*) lies eccentrically in coagulation, more medially than in middle. Area of coagulation is enlarged toward internal capsule, which is partially involved. Rigid walls of the medial cerebral artery.

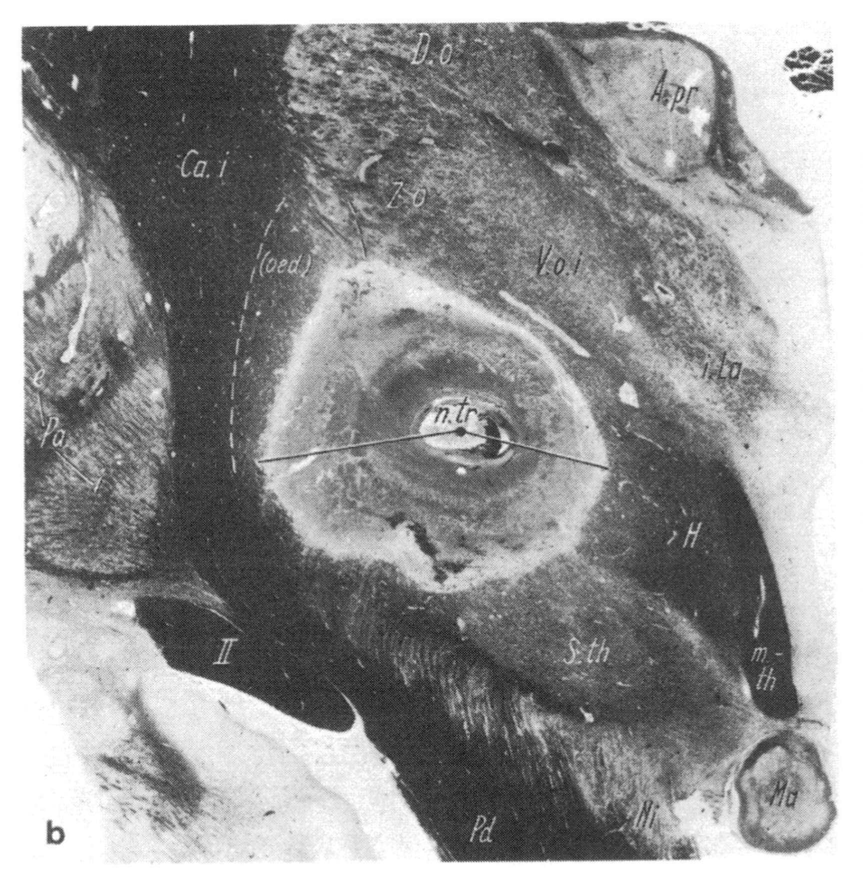

(b) Transverse section of thalamus, case No. 1, approximately same plane as Fig. 34a. Fresh lesion of coagulation in basis of V.o.a pushes internal capsule (*Ca.i.*; *oed*, oedematous zone in *Ca.i*) laterally. Electrode track (*n.tr*) lies eccentrically in middle of lesion with small hemorrhage surrounded by concentric layer of darker coagulation necrosis, nearly 2 mm diameter, then by hemorrhagic colliquation necrosis and lighter zone of demarcation. Because coagulation causes greater heat in fibrous structures, all layers are wider in direction of internal capsule. In horizontal plane of base of V.o.a, lesion extends into medial parts of internal capsule. Ventrally, lesion has attacked dorsal parts of nucleus subthalamicus (*S.th*) and has completely interrupted bundle H_2. Tendency of recent coagulation lesion to displace structures can also be seen dorsomedially in area of V.o.i. Fiber stain. ×6

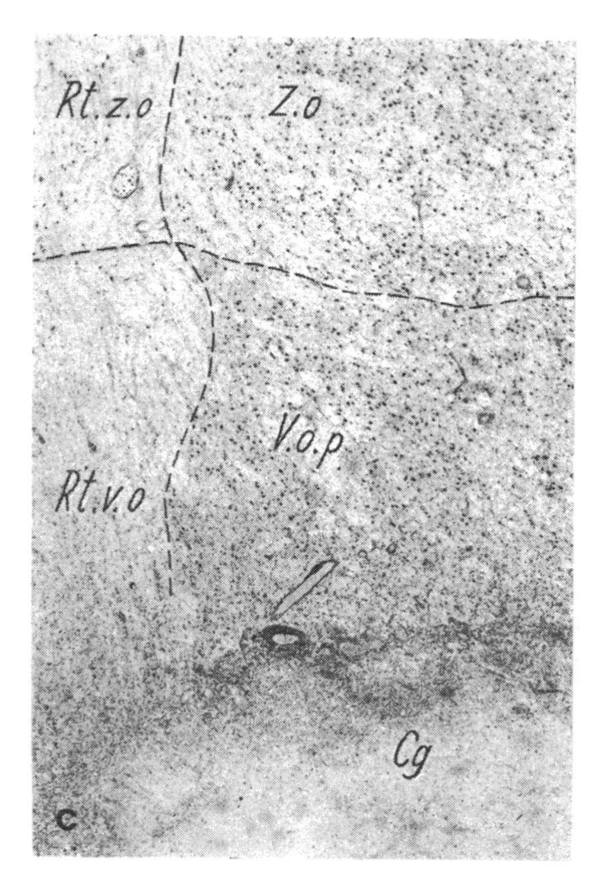

Fig. 34. (c) Above fresh coagulation focus (*Cg*) in case No. 1, nucleus V.o.p is characterized by increased numbers of glia cells due to degeneration of dentato-thalamic fibers, which had passed through coagulation focus below. In contrast to V.o.p, increase of glia cells above in Z.o nucleus is much less, demonstrating that only a few collaterals of dentato-thalamic fibers terminate in this nucleus (*Z.o*). In contrast to preserved nucleus reticulatus zentralis oralis (*Rt.z.o*), nerve cells in reticulate nucleus (*Rt.v.o*) around V.o.p are lost or atrophied, also in regions that are not directly involved by coagulation focus. Nissl stain. ×20

Post-mortem on the body (Prof. Dr. H. NOETZEL, Chairman, Neuropathological Department, Pathologisches Institut, University of Freiburg) showed considerable arteriosclerosis reaching to the periphery, marked hypertrophy of the heart with arteriosclerosis of the coronary vessels, which caused miliary areas of damage to the cardiac muscle, endocarditis of the mitral and aortic valves (hypertension had existed since 1952), multiple older small pulmonary embolisms, and one large acute pulmonary embolism on the right. Cause of death was pulmonary embolism and cardiac failure.

Anatomical findings in the brain. Laterally in the first frontal gyrus, a defect in pia and cortex the size of a pea (5×5 mm), slight atrophy of the frontal cortex. Severe arteriosclerotic changes of the carotid and of the circle of Willis (Fig. 34a), with large plaques and rigid walls. In the area of the posterior cerebral artery there were hardly any arteriosclerotic changes. Dilatation of the right and considerable dilatation of the left lateral ventricles were evident. The pathway of the electrode had left a discoloration of 2×1 mm at the upper edge of the internal capsule.

The area of coagulation was situated in the anterior ventral nucleus of the thalamus (V.o). The track of the electrode was surrounded by hemorrhagic necrosis with an inner zone of undissolved tissue and an external zone of demarcation. The area of coagulation (Fig. 34a) was 8.5×8 mm, with a sagittal diameter of 11 mm. Its medial edge lay 7 mm and its lateral edge up to 15.5 mm away from the wall of the third ventricle. Ventrally, the coagulation area reached as much as 3 mm below the baseline and dorsally up to 6 mm above it. The path of the electrode was 10.5 mm from the wall of the third ventricle and 1 mm dorsal to the baseline. The frontal plane was 8.5 mm behind the interventricular foramen. The lesion extended from the needle track more in direction of the internal capsule, than medially into the thalamic gray matter. The lesion destroyed the nucleus ventro-oralis anterior (V.o.a) and the neighboring nucleus reticulatus (*Rt.v.o*) completely (Fig. 34b), as well as the nucleus ventro-oralis posterior (*V.o.p*), including large parts of the zona incerta (Figs. 34b and 35) and encroaching on the subthalamic nucleus below and the V.im caudally. 10 mm behind the interventricular foramen, the lesion did not extend into the internal capsule.

Nerve cell loss occurred in the most dense nerve cell groups of the posterior main section of the substantia nigra (Fig. 36a). The most medial cell group of the anterior main section of the substantia nigra was severely involved by nerve cell loss and intensive gliosis (Fig. 36b). The locus caeruleus shows a lightening of the black pigmented nerve cells (Fig. 37a). Some of the nerve cells contain intraplasmic inclusion bodies. The dorsal nucleus of the vagus (*Nc. X. d*) is reduced in its nerve cells mainly on the left side (Fig. 37b).

Clinical-anatomical correlation. As intended, the lesion had destroyed almost all of the left V.o.a and V.o.p; narrow dorsal segments of the V.o.p were preserved. The bases of both nuclei, where the afferent fibers enter, were also completely destroyed. On the caudal end the coagulation had

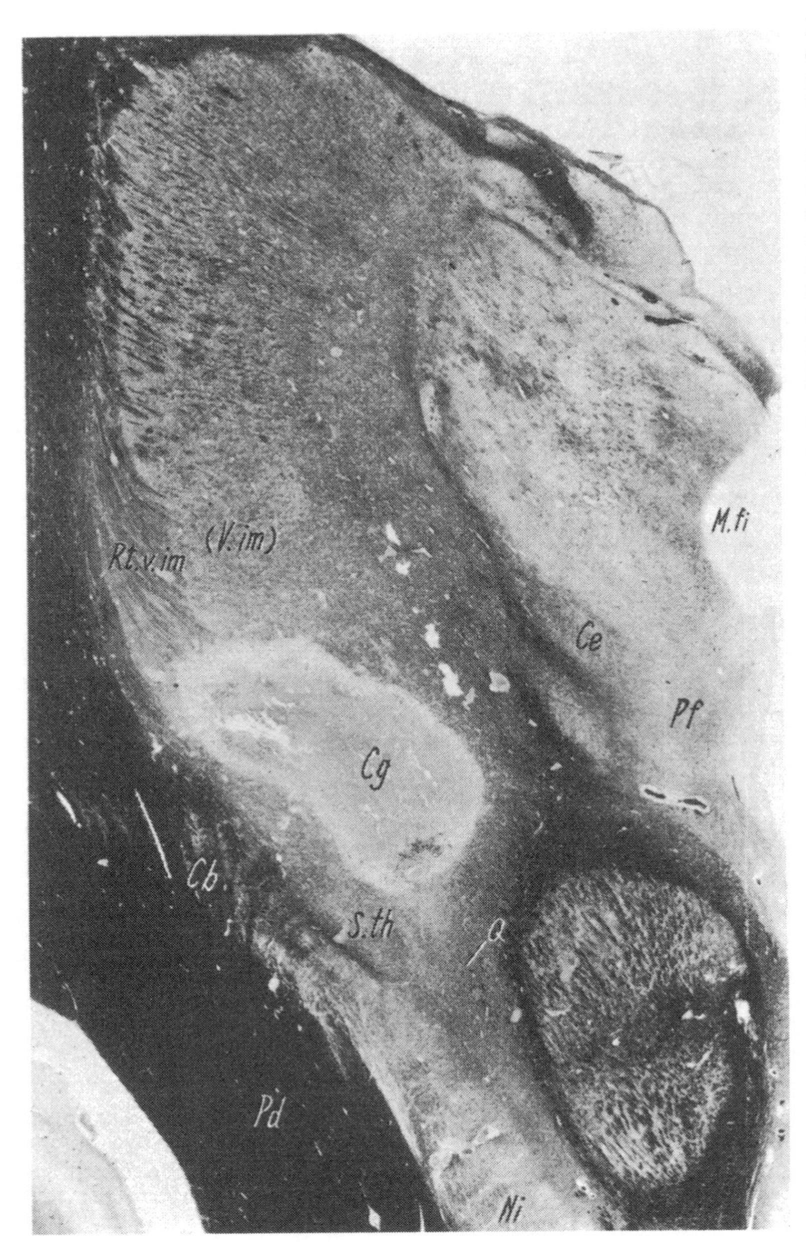

Fig. 35. Cross section through thalamus of case No. 1 (paralysis agitans). Occipital end of coagulation focus in V.o.a and V.o.p appears in base of ventral nuclei. Caudal enlargement of coagulation focus, produced by moving string electrode forward medioventrally and caudally, with intention of reaching dentatothalamic fibers, is located in base of ventro-intermedius nucleus (*V.im*). Coagulation focus (*Cg*), with small hemorrhage around arteriole in its most ventral expansion (black points), involves dorsal margin of subthalamic nucleus (*S.th*). V.im nucleus above coagulation focus shows demyelination due to interruption of afferent fibers of V.im and due to swelling around coagulation focus. Q bundles between subthalamic nucleus (*S.th*) and red nucleus are partially preserved. *Cb*=comb system of the foot passed by dark pallidosubthalamic fibers. Ventrally they are accompanied by lighter strionigral fibers. Fiber stain. ×4

affected marginal by the S.th and the zona incerta (Fig. 35). Only at the lateral side of the V.o.a did the coagulated area include about 1 mm of the internal capsule. At the level of the V.o.a, the internal capsule and even part of the neighboring nucleus reticulatus had been spared (Fig. 34c). The puncture track of the coagulation corresponded to the point aimed at, having a distance of 10.5 mm from the wall of the ventricle. The distance shown in the x-ray was 11.1 mm. The track was in a plane 10 mm behind the interventricular foramen, only a fraction of one millimeter dorsal to the baseline—that is, the radiologically determined target on the baseline. The coagulation had, therefore, succeeded in inactivating the target thalamic nuclei V.o.a and V.o.p of left side (Figs. 35, 38a, b).

The autopsy showed that the therapeutic effect was the result of inactivating the aimed-at structures. The operation completely abolished the tremor, which had been present occasionally, and the contralateral rigidity. At the same time, the starting delay, propulsion, and disorders of gait and posture disappeared or were much improved. This effect can also be ascribed to functional loss of nuclei V.o.a (Fig. 34b) and V.o.p (Fig. 35). The disappearance of backache may have been due, as a secondary effect, to the disappearance of rigidity and to the improved posture. The abolishment

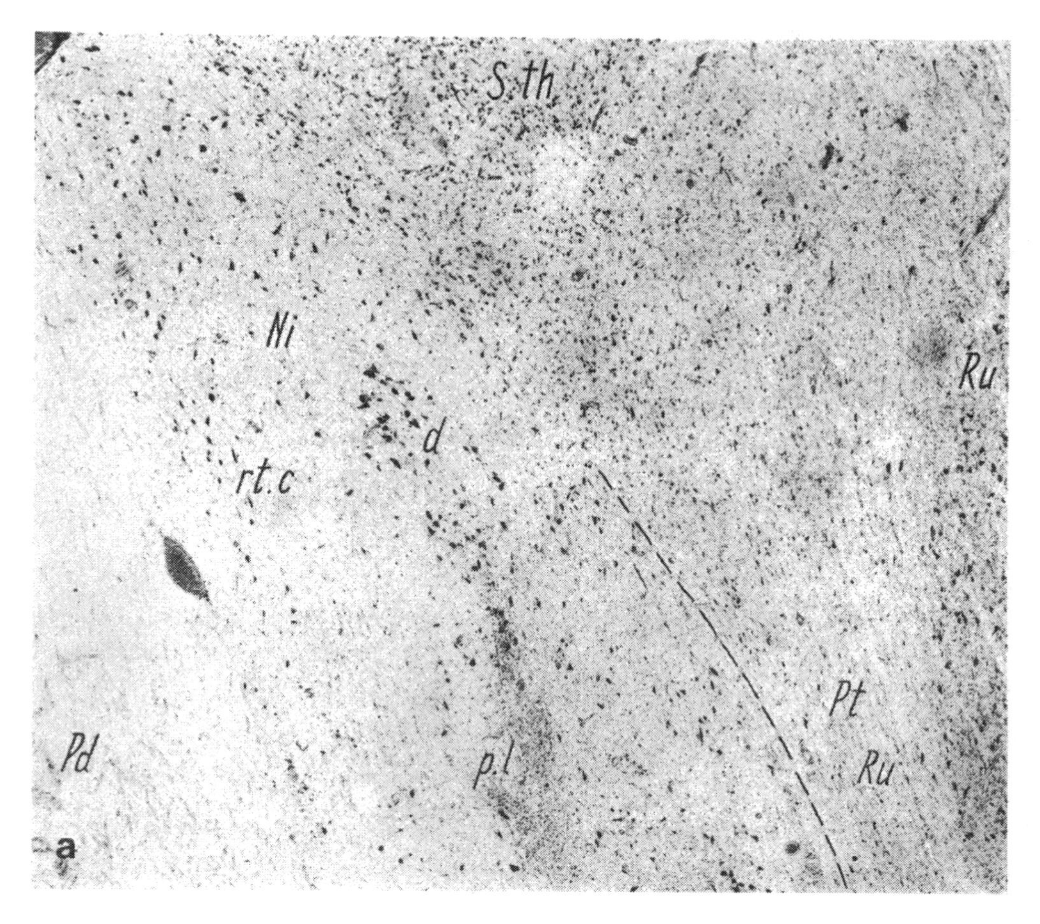

Fig. 36. (a) Frontal section through substantia nigra of case No. 1. Lateral groups of cells of posterior main part of substantia nigra have disappeared and have been replaced by dense glia scar (*p.l*). In dorsal groups (*d*), reduction of cells is considerably less. Nerve cells of zona reticulata (*rt.c*) without melanin are well preserved. *Pt.Ru*= putamen (or shell) of Ru (red nucleus); *S.th*=subthalamic nucleus. Nissl stain. ×18.

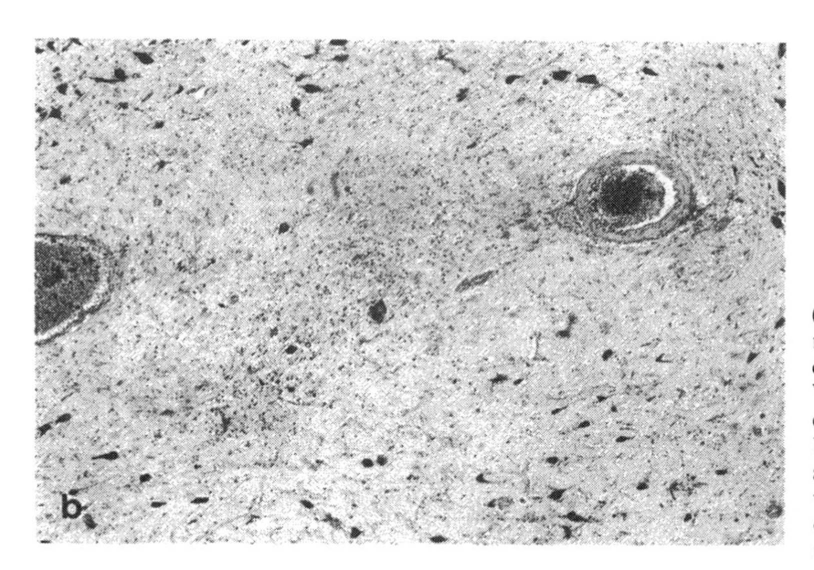

(b) Glia scar in medial cell group of anterior main section of substantia nigra of case No. 1 adjacent to a small vein (left). Wall of small arteriole (right) is thickened, showing that arteriole is affected by arteriosclerosis. There is mild proliferation of adventitia of artery, so that a vascular factor in degeneration of medial cell group cannot be excluded. Nissl stain. ×16

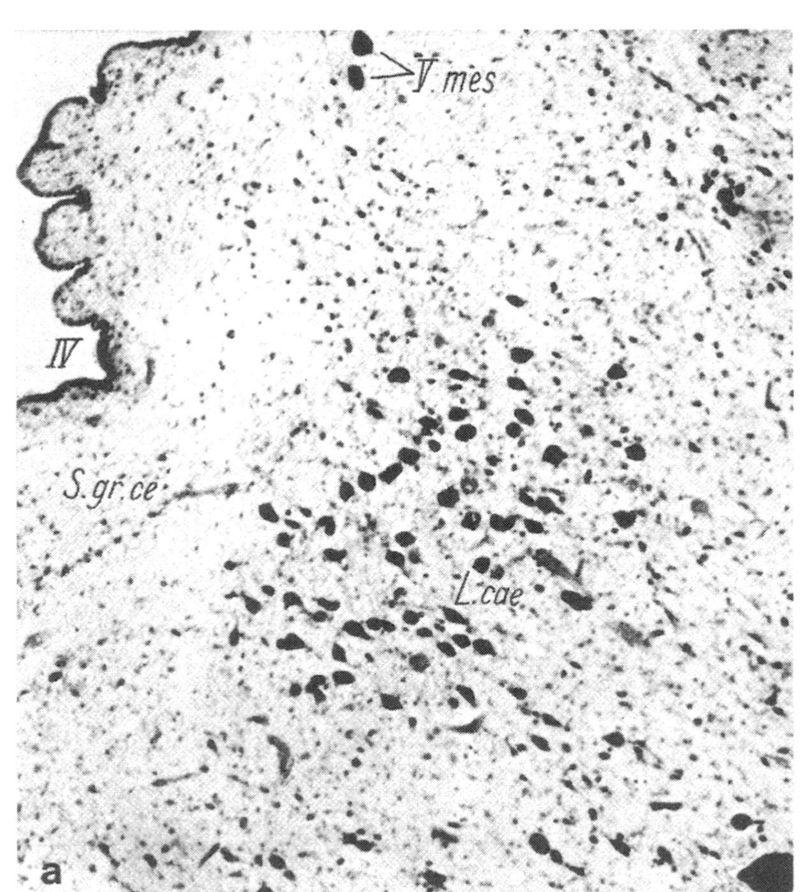

Fig. 37. (a) Section through locus caeruleus of case No. 1. Slight reduction of number of cells with slight replacement growth of glia. *V.mes* = mesencephalic root of Vth nerve; *IV* = 4th ventricle; *L.cae* = locus caeruleus; *S.gr.ce* = substantia grisea centralis. Nissl stain. ×64.

(b) Frontal section through medulla oblongata of case No. 1. Over left nucleus of hypoglossus (*XII*) there occurred losses of nerve cells and growth of glia in left nucleus intercalatus (Staderini) (*Ic*). Nucleus dorsalis vagi (*X.d*), lateral to this, shows some reduction of cells and a few inclusion bodies on left side; on right side it is less damaged. Severe cell loss only in left prepositus hypoglossi (medial to *Ic*). Typical of finding in genuine paralysis agitans. *Nc.Ro* = nucleus Roller. Nissl stain. ×24

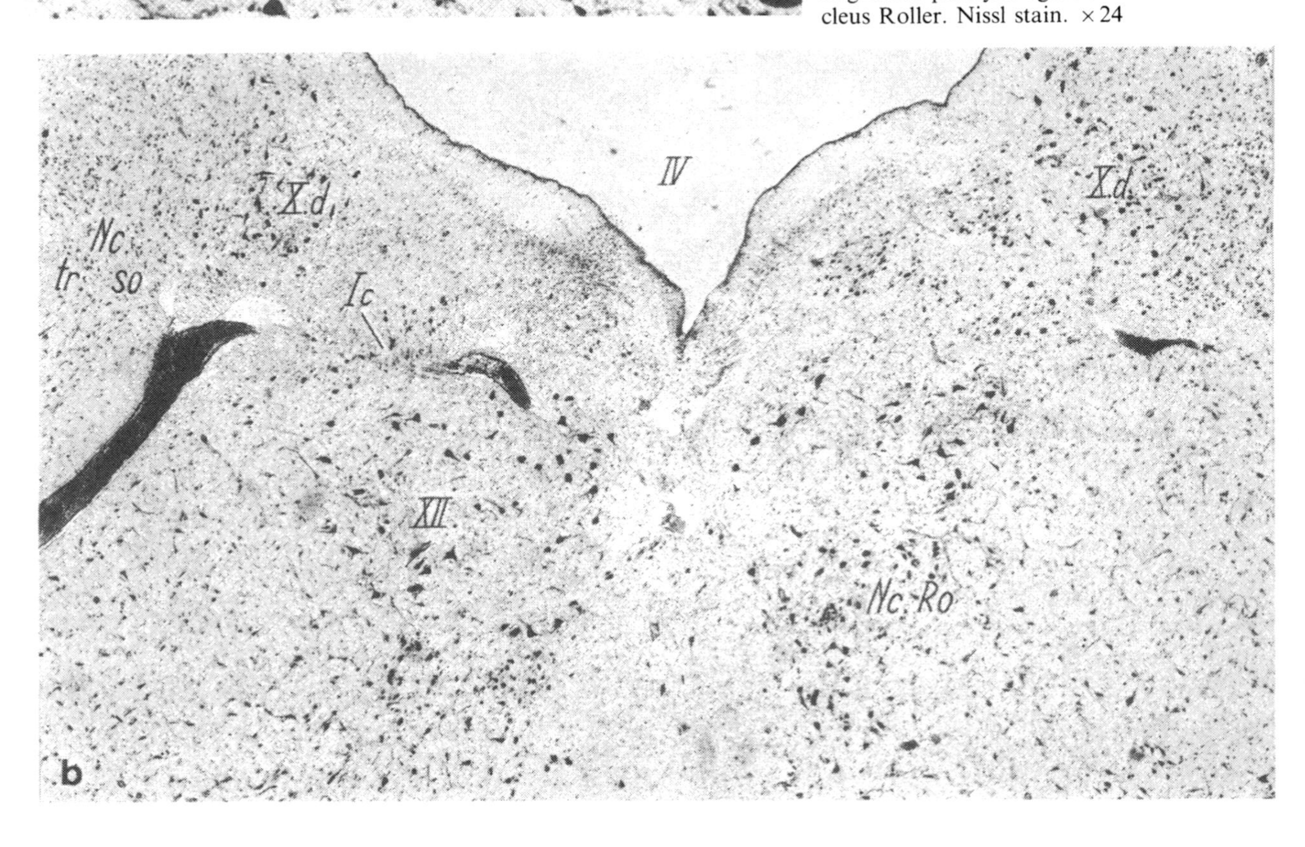

of rigidity and tremor lasted until the death 23 days after the operation and was, therefore, not due to a transient edema in the neighborhood of the lesion. The patient's depressive mood, present before the operation, was absent afterward. This was probably not an immediate effect of the operation but rather the result of the abolishment of the Parkinson symptoms, which had provoked a reactive depression. The left-sided frontotemporal δ focus in the EEG (Fig. 33b) observed 7 days after the left-sided operation was a direct expression of the destruction of the V.o.a and V.o.p.

The results of the stimulation can be regarded as characteristic for the left V.o.a since it has been shown that the stimuli reached the target. This applies particularly to the jerks of the contralateral arm, which occurred synchronously with the stimuli at a rate of 4 and 8 cps. The stimuli at higher frequency (50 cps) caused frequent hand movements, mydriasis, and a slight slowing of active alternating movements by excitation of the thalamic substrate. Only an increase of the stimulating current evoked the spasmodic turning of the right hand and tetanic contracture of the muscles on the right side of the face. These two latter effects were probably due to the stimulating current spreading further and reaching the internal capsule, which was 3 mm away. Neither a paralysis of the face nor of the limbs occurred although a narrow stretch of the internal capsule on the lateral side of V.o.a (Fig. 34b) was affected by the coagulation. From this, one can conclude that the higher-frequency stimulating current affects an area larger than that of the lesion set by coagula-

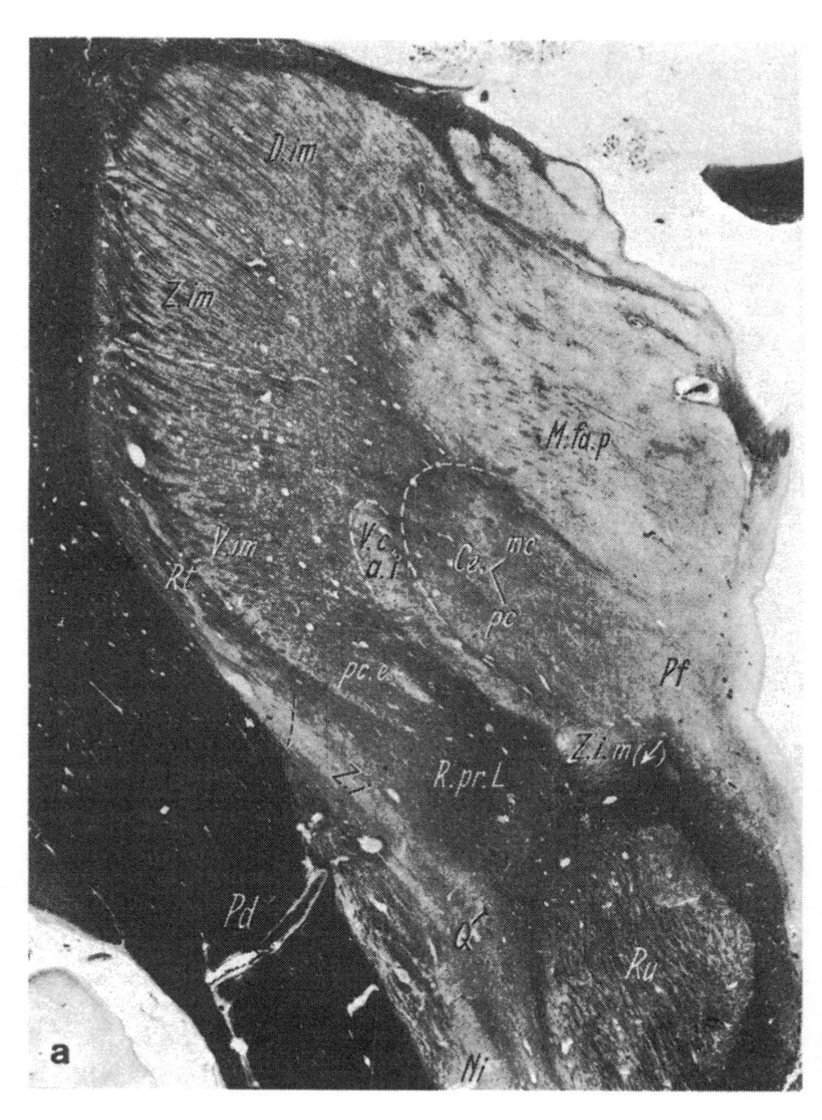

Fig. 38. (a) Transverse section through thalamus of case No. 1, 3.5 mm further posteriorly. At this level coagulation focus has disappeared. Zona incerta (*Z.i*) dorsomedial to pedunculus cerebri (*Pd*) and its continuation by the (ventral) reticulate nucleus (*Rt*) are remarkably demyelinated. In medioventral part of ventro-intermedius nucleus (*V.im*), there is a small island of gray substance that is part of ventrocaudal parvocellular nucleus (*pc.e*). Between this nucleus, which is continuous with prelemniscal radiation (*R.pr.L*), and *centre médian* nucleus (*Ce.pc*), is internal part (*V.c.a.i*) of V.c.a nucleus, much reduced in size by nerve cell loss. Heidenhain-Wölcke stain. ×4

tion. It is probable that the vegetative signs – dilatation of the palpebral fissure and inspiratory apnea – were also due to the thalamic substrate.

Death was due to repeated pulmonary embolisms and not to a cerebral complication. The pulmonary embolisms caused incidents of acute loss of consciousness. This, together with the story of nocturnal impairment of consciousness with inability to speak, led to the suspicion of epileptic fits. In view of the normal EEG and the post-mortem findings, the impairment of consciousness can be interpreted as having been due to reduced cerebral activity of circulatory origin. This is supported by the post-mortem findings of coronary sclerosis with scarred myocardium, endocarditis of the mitral and aortic valves and numerous small and larger pulmonary embolisms.

Case 2. F.J., aged 59; paralysis agitans with scarring in the basal ganglia after pallidotomy and bilateral ventro-oral thalamotomy. The 1st stereotaxic operation was performed in left pallidum internum (no. 238); the 2nd, in left V.o.p and left pallidum internum (no. 338); the 3rd, in right V.o.a (no. 864). Patient died 21 days after 3rd operation from massive pulmonary embolism.

History. Parkinson's disease started at age 51 with trembling of the right hand when writing. There was no history of epidemic encephalitis or another infectious disease. At age 55 the tremor affected the right leg and at times the left limbs. A weakness in the right leg was noticed a year before the operation.

Clinical findings. Marked tremor of the right arm, especially in the distal parts. When the patient

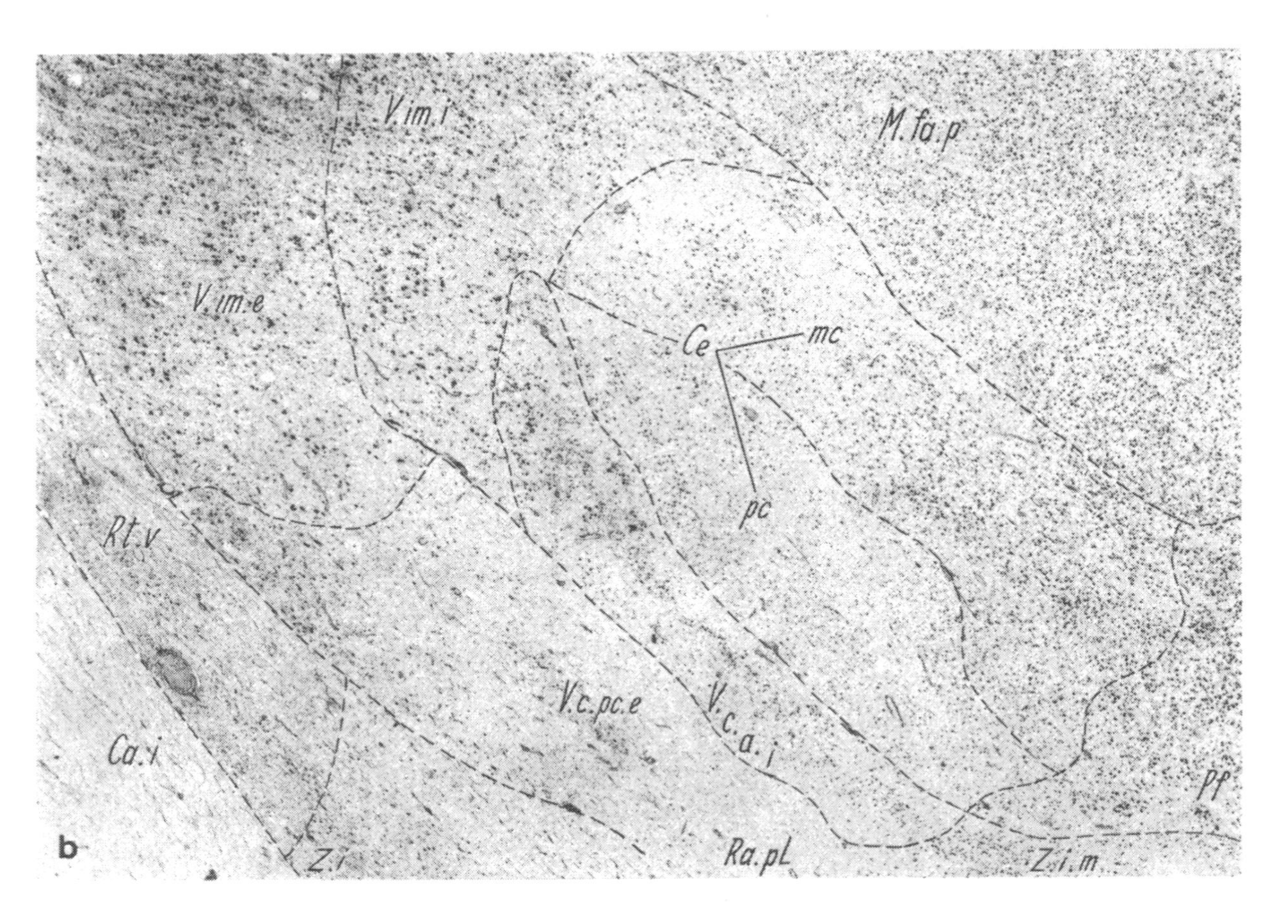

Fig. 38. (b) Adjacent section to Figure 38a. Nerve cell loss and compensatory gliosis in internal part of nucleus ventrocaudalis internus (*V.c.a.i*). Shrinkage and some patchy nerve cell losses mainly in medial part of parvocellular part of *centre médian* nucleus (*Ce.pc*) due to interruption of centro-striatal fibers. Also, ventral part of thalamic reticulate nucleus (*Rt.v*) shows glia growth substituting nerve cell loss and nerve cell atrophy. This is less in adjacent zona incerta (*Z.i*). These alterations are marginal effects of thermal coagulation in V.o.a and V.o.p nuclei, which also interrupted parts of efferent fibers of V.c.a.i and Ce.pc, whereas efferent fibers of Ce.mc are also spared. Nissl stain. ×9

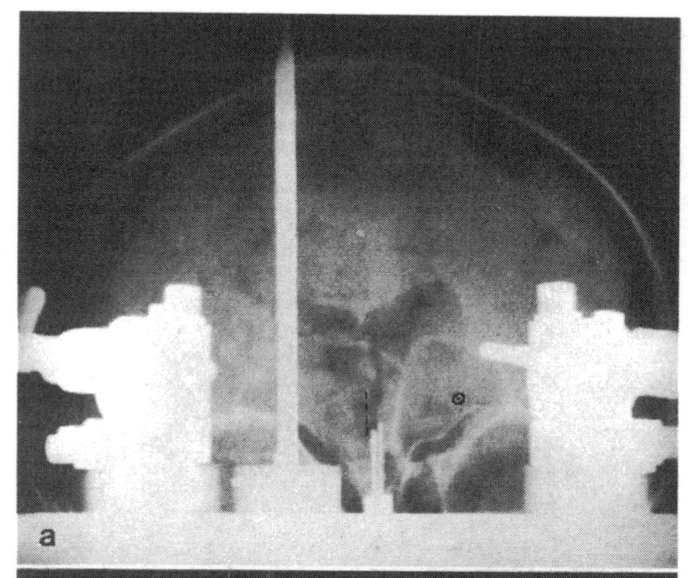

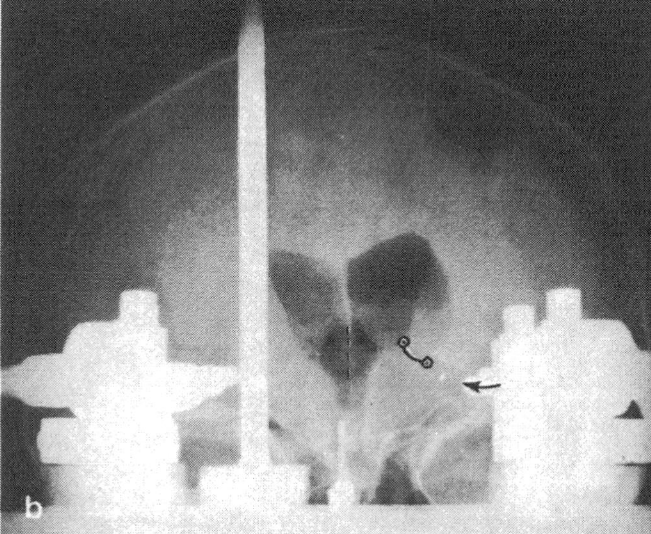

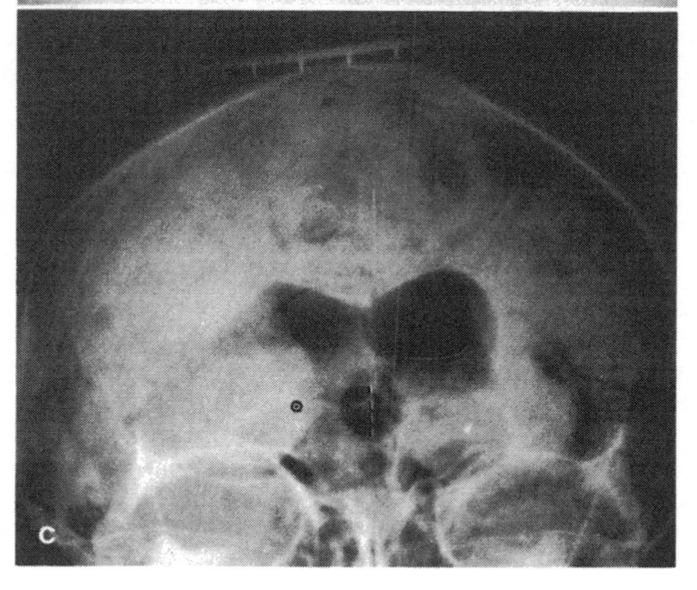

Fig. 39a–c. Pneumoencephalogram of case No. 2 in anterior-posterior view.

(a) Before 1st operation, ventricular system lying approx. in middle, not enlarged, with target point in left pallidum internum.

(b) 11 months later, before 2nd operation. In correct size, it shows definite enlargement of both lateral ventricles and considerable dilatation of 3rd ventricle. On left side, outline of thalamus has dropped laterally due to 1st operation. Left pallidum internum is marked by silver clip (*arrow*). Target point of 2nd operation is in left V.o.a, from which string electrode goes ventrolaterally to pallidum internum.

(c) Due to 2nd operation, in which V.o.a was coagulated and in which there was additional coagulation by string electrode through internal capsule into pallidum internum anterior, left lateral ventricle is much more dilated 3 years after 1st operation, and outline of thalamus has sunk basally and is concave. Further increased dilatation of thalamic part of 3rd ventricle. The external air shows widening over left hemisphere in area of insula and in parietal region. Target point of V.o.a is entered on right side. Notice clip in left pallidum

moved, the tremor affected the entire right leg. Also present were a slight tremor of the chin and at times a "no-tremor" of the head; a cogwheel phenomenon in the right arm, rigidity in the right arm and leg; expressive weakness of face muscles on the right side; weakness of the right leg; exaggerated tendon jerks in the right leg; and dysdiadochokinesia on the right; tremor was provoked right more than left during the finger-nose test.

The *first stereotaxic operation* was performed on the left pallidum internum. Stimulation was carried out with thyratrone impulses at 20 cps stimuli interrupted the rhythm of the tremor. Coagulation was performed with the spark-gap diathermy apparatus and an uninsulated electrode of $1 \times 1.1\ \phi$ mm at the target and with an electrode of $4 \times 1.1\ \phi$ mm in position +2 and an additional string electrode[4] up to position +2. After this the rigidity in the right arm and leg almost disap-

[4] A string electrode of piano wire is used for stimulation as well as for coagulation. It can be pushed forward out of its cannula and is constructed with a predetermined radius, so that it can be moved a distance of 15 mm away from the axis of the cannula.

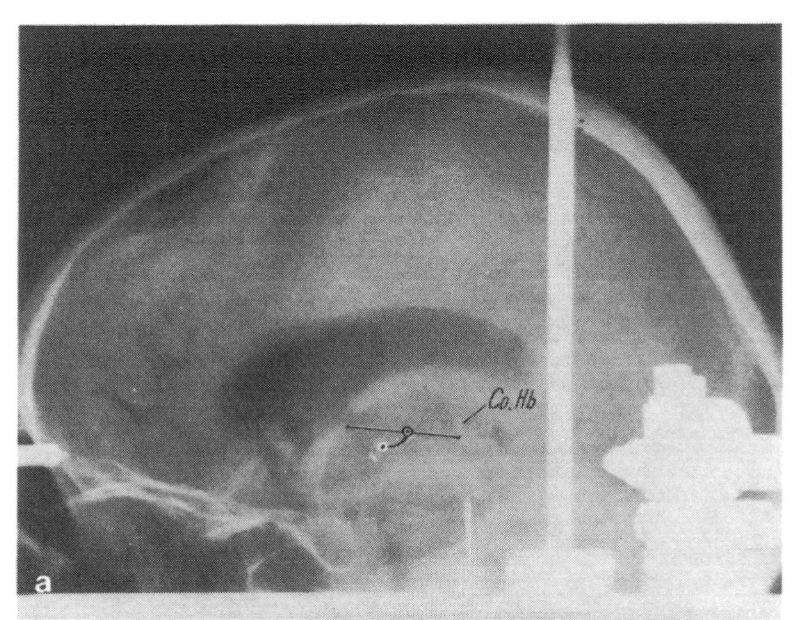

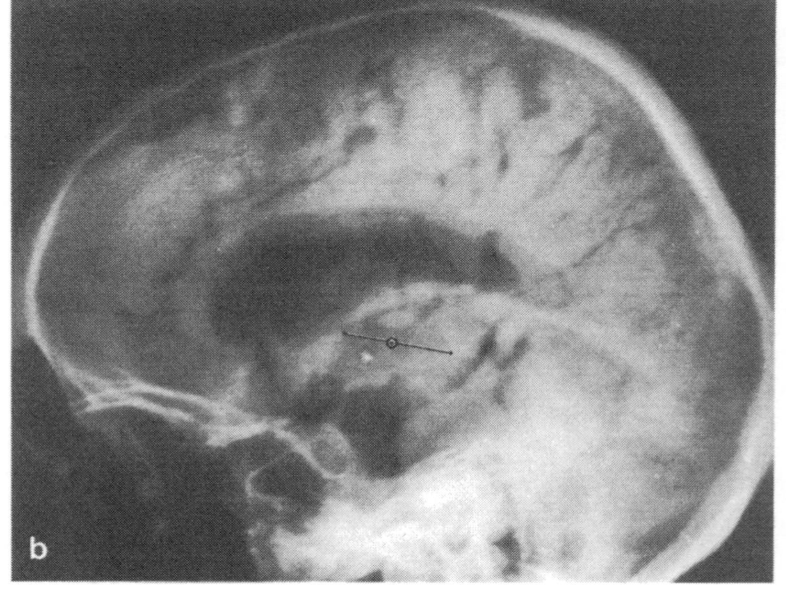

Fig. 40a and b. Lateral pneumoencephalogram of case No. 2.
(a) Before 2nd operation, with dilatation of both anterior horns (left more than right) and marked dilatation of 3rd ventricle. Silver marking clip implanted into pallidum internum anterius projects into lower part of 3rd ventricle. Baseline for 2nd operation is drawn from lower edge of interventricular foramen to posterior commissure and to target point in V.o.a from which caudal pallidum internum was reached with bent string electrode (2 mm caudal to silver clip). Denser shadow of commissura habenularum (*Co.Hb*) lies directly above posterior commissure.
(b) Coarsening, particularly of left lateral ventricle, has progressed 3 years after 2nd operation. Basal cisterna is much enlarged, external air is particularly widened in parietal subarachnoidal spaces. Baseline of thalamus has been entered with target point V.o.p on right side. There is no displacement of silver clip, compared to (a)

peared, but the tremor was only slightly improved. During further coagulation with the string electrode up to +6 caudal to the target, a short fit occurred in which the eyes turned upward. Subsequently the rigidity on the contralateral side completely disappeared and the tremor was abolished to a considerable degree. The right arm and leg were largely neglected in their proximal parts; the patient was more somnolent. Stimulation at the target point following the coagulations again evoked a tremor that was synchronous with the stimuli. The target point was marked with a little silver clip (Figs. 39, 40a, b).

After the operation no tremor was observed on the right side. Rigidity had also disappeared. Complete abolishment of the tremor lasted for only 2 days. The patient was discharged from the hospital 7 days after the operation. Although the tremor increased progressively during the next months, the patient could use his right hand well, apart from being unable to write. He was able to drive the car every day. The rigidity remained completely absent from the right arm and almost completely from the leg so that the gait was not impaired.

Because of the recurrence of the tremor and the inability to write, the patient asked for a second operation for the right side of the body. The EEG taken 11 months after the first operation on the left pallidum showed no abnormalities apart from the α frequency varying between 8 and 10 cps. A drop of the left thalamis outline after 11 months (Fig. 39b) and dilatation of the left more than of the right anterior horn (Fig. 40a) were observed.

The *second stereotaxic operation* was performed December 11, 1957, in the left V.o.p and medial pallidum. Stimulation at rates of 4–8 cps led to an increase of the tremor synchronous with the stimulation. Stimulation at 25 cps led to a high-frequency tremor after a short period of blocking. Coagulation was carried out with the spark-gap diathermy apparatus (electrode 1×1.1 mm and $4 \times 1.1\ \phi$ mm and string electrode). Contractions on the opposite side could be seen during coagulation in the V.o.p. Afterward the resting tremor disappeared but could still be provoked. The coagulation in medial pallidum (Figs. 39b, 40a) resulted in a transitory increase of the tremor, which then disappeared without power being diminished. The next day the patient's consciousness was clear, and he was able to eat his breakfast unaided. Neither differences of the reflexes nor diminution of power could be demonstrated. Tremor could not be provoked even by holding the arms forward.

The effect on rigidity and tremor on the right side persisted. The EEG had been normal 5 days before the operation, the α frequency being not constant, without difference as to side. On the second postoperative day the α rhythm was slower frontally and frontotemporally. Over the hemisphere on the side of the operation, localized θ and δ waves occurred. On the tenth day after the second operation the patient left the hospital at his own request. A telegram was received that he had arrived home in good condition. During the next days the tremor in the left arm and leg, for which the operation had not been performed, gradually increased but moderately. Definite mild rigidity developed in these left limbs.

Even after 2 years, the patient was satisfied with the effect on the right side although on readmission in January, 1960, a slight weakness of the oral branch of the facial nerve on the right side was found. He asked for an operation for the tremor of the left arm.

The *third stereotaxic operation* was performed on January 11, 1960, on the right V.o.a. Single stimuli clearly accentuated single tremor phases. Stimulation at a rate of 4 cps slowed the tremor and made it more rhythmic; at a stimulation rate of 8 cps, the left hand spontaneously grasped the base ring without the patient being able to explain this movement. Prolonged stimulation at 8 cps led to dilatation of the pupils and looking to the left together with laughter. The latter two responses and reddening of the face occurred at higher frequency (50 cps) of stimulation. At the same time he made regular grasping movements upward with his left arm, like the supplementary area 6 effect.

Coagulation with the Wyss coagulator ($4 \times 2\ \phi$ mm electrodes) was performed in positions +3 and +5, each lasting more than 20 sec. After the first coagulation the patient shut his eyes spontaneously and said he was rather tired. The left side was hypotonic after the second coagulation. The tremor in the left arm decreased but did not disappear. Facial paresis appeared on the left side especially during expressive movements. He could no longer count backward.

Postoperative course. Following the operation he developed an organic psychosyndrome with transient urinary incontinence. He had considerable difficulty with concentration and remembering. The rigidity of the left side changed into hypotonus; the tremor diminished but did not disappear. The patient became increasingly somnolent and on the fourth postoperative day developed pulmonary congestion with a rise of temperature, which may have corresponded to the first embolism. The patient became increasingly apathetic and comatose and did not respond to antibiotics or to drugs intended to improve circulation. On the 21st day after operation, death occurred due to acute circulatory collapse.

Post-mortem examination of the body (Prof. Dr. H. NOETZEL) showed considerable arteriosclerosis of the aorta with ulceration, stenosing sclerosis of the coronaries, extensive femoral venous thrombosis reaching on both sides into the iliac vein, and recurrent pulmonary embolisms with infarcts and pulmonary edema. Death was caused by a massive pulmonary embolism.

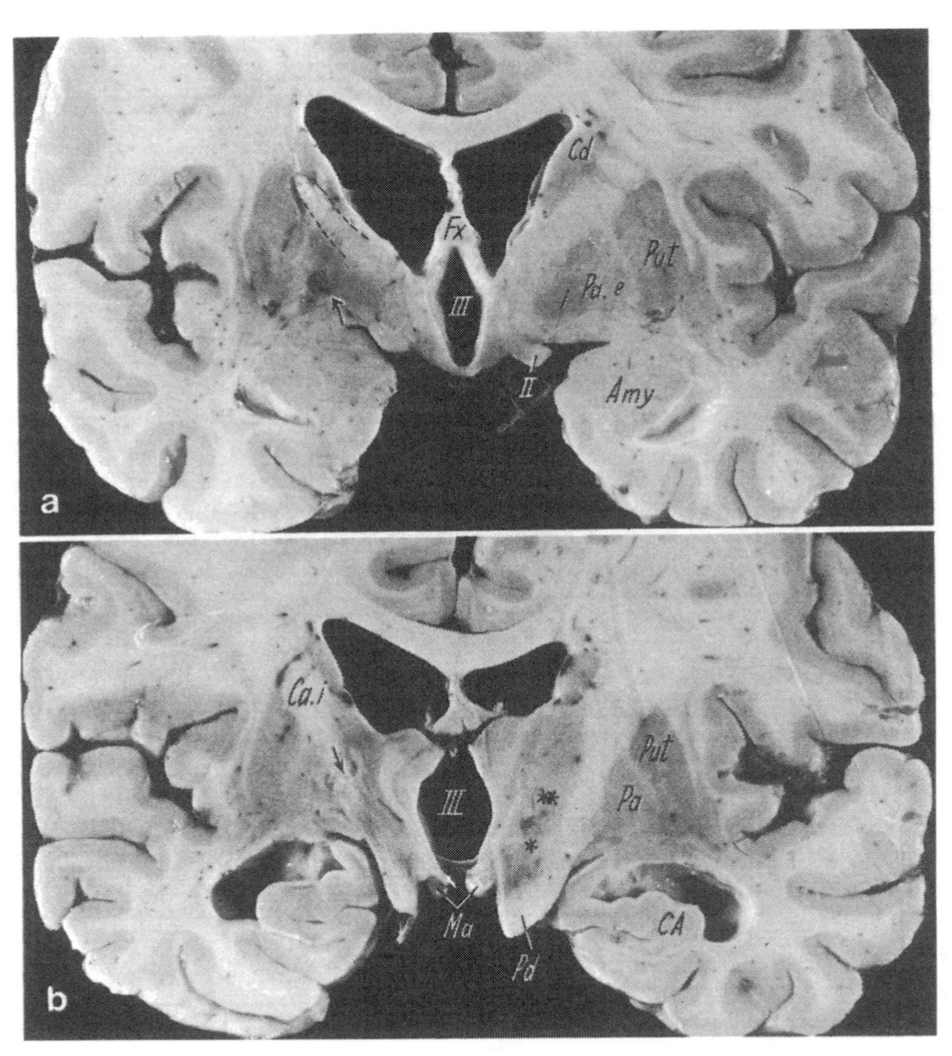

Fig. 41. (a) Obvious shrinking of left basal ganglia on transverse section of brain of case No. 2 in plane of interventricular foramen. Three necrotic lesions in left pallidum, one above the other; the middle one is largest and appears particularly dark. Oxidized silver clip can be recognized as black oblique structure (*arrow*). Compared to uncoagulated right pallidum, left one is reduced by operation to about 1/3 of its size. Above it, internal capsule (between *rows of dots*) is narrowed to less than half the size and because of fiber degeneration, is in parts gray.

(b) Transverse section of brain of case No. 2, 11 mm behind that in Figure 41 a. Left basal ganglia are shrunken and contour of thalamus is deformed. Scar in left internal capsule (*arrow*) is located between coagulation focus in V.o.a of thalamus and internal segment of pallidum. Third ventricle (*III*) is much dilated mainly toward left side. In right thalamus there is swelling of ventral nuclei at level of V.o.a and V.o.p (*double star*). Subthalamic nucleus is marked by one star. Right basal ganglia are completely preserved, separated from thalamus by internal capsule. ×1

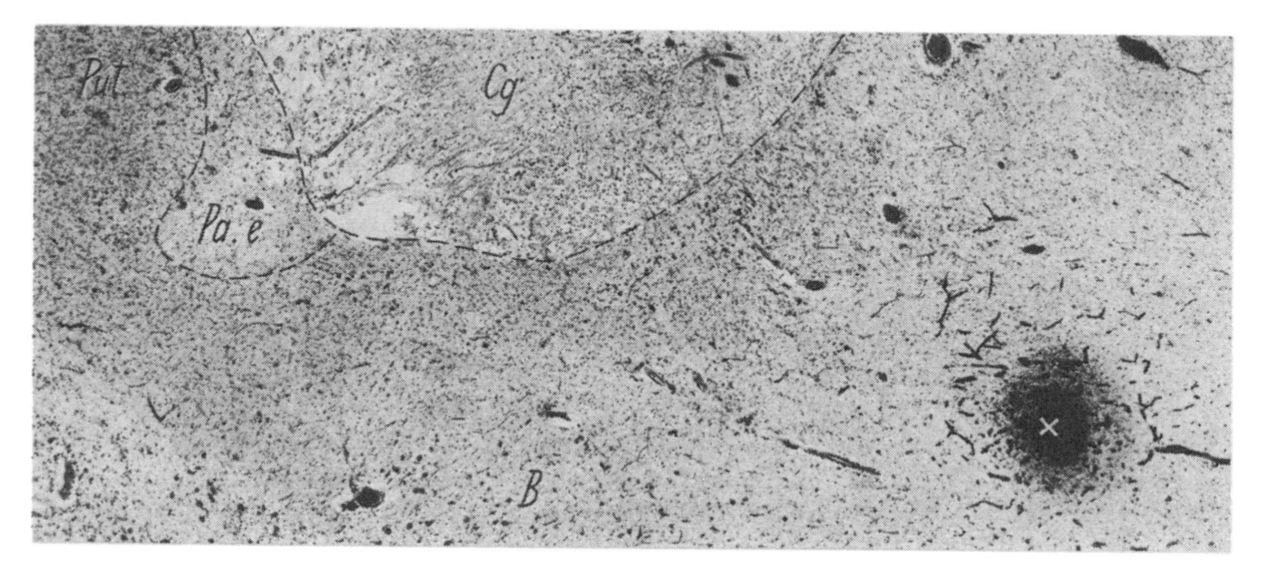

Fig. 42. Histologic section caudal to silver clip in case No. 2 shows around point of clip black deposit of metal in tissue of pallidum (×). Oxidized metal was stored in glia cells but also was deposited in capillary walls; consequently, capillaries in surroundings are prominently black. Left coagulated lesion (*Cg*) with connective tissue organization lies to left above black spot. A small triangle of outer part of pallidum (*Pa.e*), still intact, can be found only at boundary to putamen. On ventral side of Pa.e lies basal nucleus (*B*) with little scarring. Nissl stain. ×23

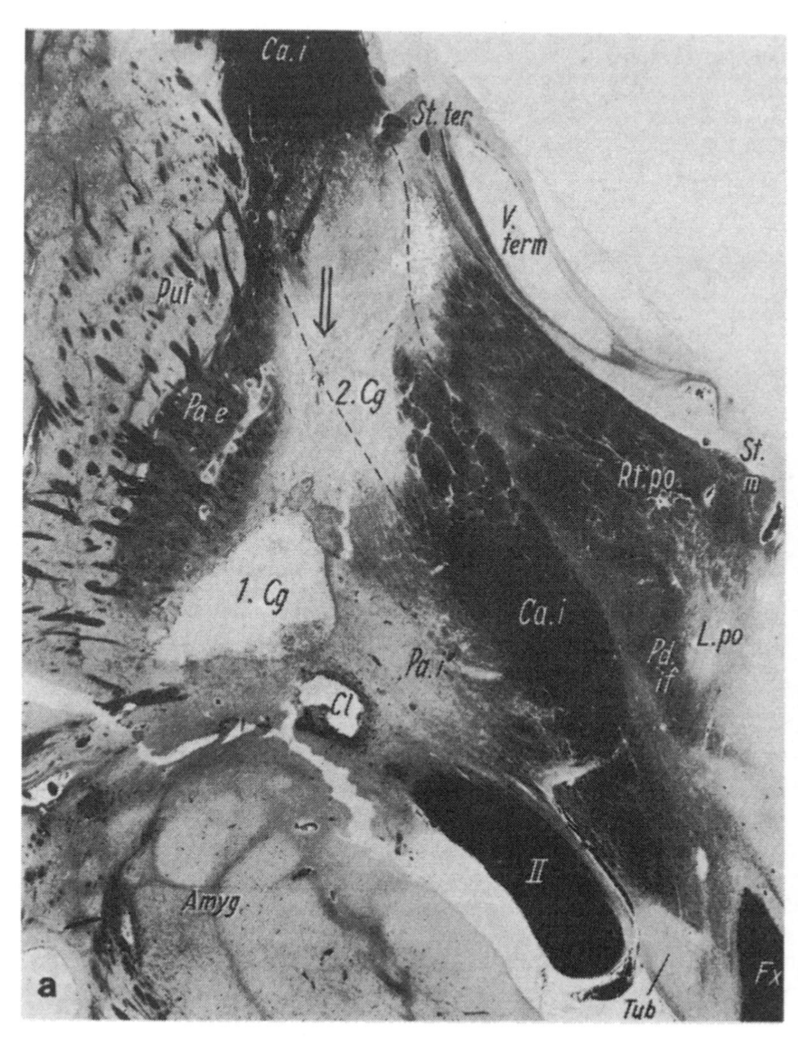

Fig. 43. (a) Transverse section of left thalamus of case No. 2. Electrode track (⇓) is located in internal capsule, already interrupted by 1st operation and shows degeneration. In internal part of pallidum is cyst of coagulation from 1st operation performed 11 months previously (*1.Cg*). Next to it is a cystic defect in tissue from silver clip (*Cl*), surrounded by black tissue due to oxidation of silver and deposit of silver oxide in surrounding tissue and in capillary walls. Despite severe degeneration, one can see small bundles of fine myelinated fibers at medial edge of internal part of pallidum. 1st coagulation during 2nd intervention also destroyed internal capsule but at a more dorsal level and somewhat more medially. Most lateral part of reticular nucleus of pole of thalamus (*Rt.po*) is involved in coagulation. *V.term* vena terminalis; *Tub* nucleus tuberis lateralis; *Pd.if* inferior peduncle of thalamus. Fiber stain. ×4

Anatomical findings in the brain. Severe atrophy of both frontal poles, left more than right; two older puncture marks on the second frontal gyrus on the left; the more lateral mark left a cortical defect of 5 × 5 mm, a smaller puncture mark on the right first frontal gyrus.

The ventricular system was considerably more dilated left than right (Fig. 41 a, b). The two tracks joined in the left white matter where a softening of 4 mm had developed. Because of various lesions the left basal ganglia shrank so much that they were 4 mm lower than the right ones (Fig. 41 b). A disintegrated area of 3.5 mm, which extended into the internal capsule, was found in the left caudatum (Fig. 41b). The rostral end of the internal capsule had shrunk considerably and at the level of its genu had lost a large proportion of its fibers (Fig. 43a). The area of coagulation extended over the lamella pallidi interna and into the left pallidum internum. The silver clip (Fig. 43a) was located in the base of the left internal pallidum (Fig. 42). The clip caused an oval defect of tissue 1.6 mm broad and edged by pigment due to oxidation of metal. The left thalamus shrank most in a plane 8.5 mm behind the interventricular foramen and was displaced toward the scarred internal capsule (Fig. 41b). The dorsal nuclei were so atrophied that the thalamus had lost 6 mm of its height. The center of the scar of the first lesion lay 12 mm from the wall of the third ventricle, 3 mm below the baseline. The second series of coagulations led to a vertical lesion across the capsule; its center was 10 mm lateral to the wall of the third ventricle (Fig. 43b; 41 b).

In the right thalamus a lesion due to coagulation lay with its center about 6.5 mm from the wall of the third ventricle and 3 mm below the baseline (Fig. 44). The lesion destroyed part of the V.o.a and extended medially to the mamillothalamic bundle including the posterior part of Latero-

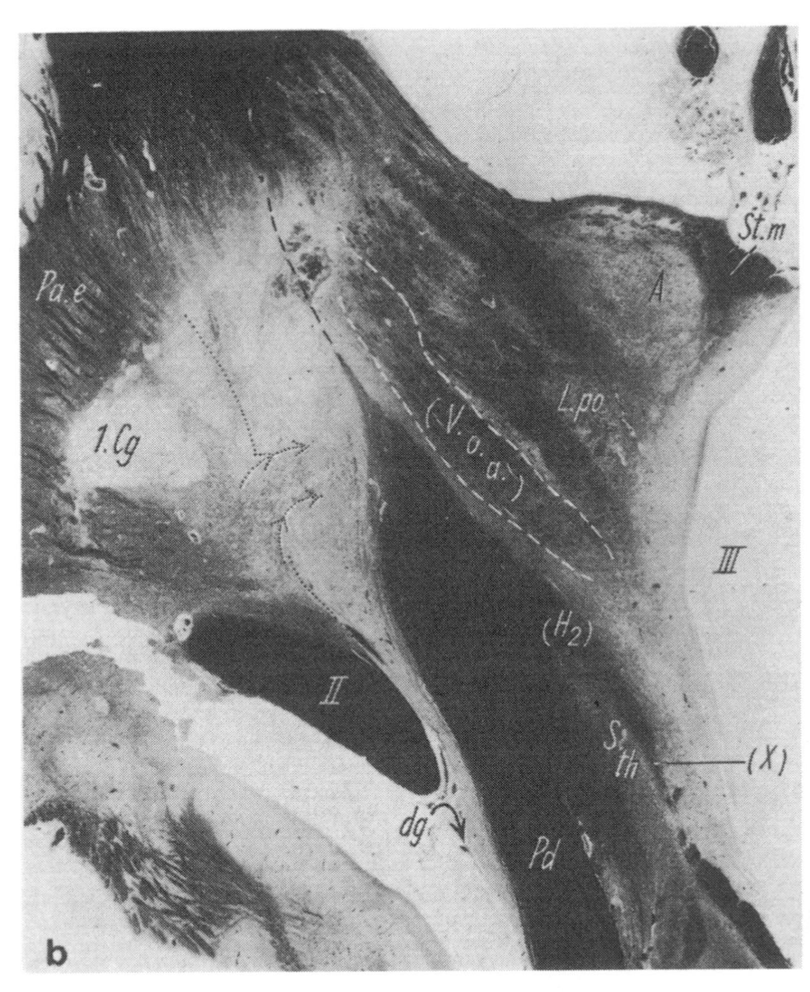

Fig. 43. (b) Transverse section of left thalamus of case No. 2 in plane of V.o.a, 2.7 mm further caudally. In lateral part of V.o.a is a cystic lesion with centrally increased growth of connective tissue that extends slightly into internal capsule. Most cells of V.o.a are degenerated; V.o.a and subthalamic nucleus (*S.th*) are shrunken. This lesion merges lateroventrally with lesion in pallidum produced during 1st operation (*1.Cg*). The strionigral fibers are interrupted in the medial and basal pallidum. Degeneration of internal capsule continues into peduncle as thin white strip. At upper edge of nondegenerated peduncle is a curving upward of thinnest fiber bundles crossing degenerated part of internal capsule to enter H_2 bundle. These are remaining efferent pallidosubthalamic and pallidothalamic fibers medial to peduncle. Fiber stain. ×6

polar nucleus above (Fig. 45c). Because of the V.o.a lesion a retrograde degeneration of the internal segment of the pallidum with intensive gliosis (Fig. 45b) occurred. The internal capsule and lateral and dorsal parts of the V.o.a had been spared and with them the strionigral fibers in the comb system of the foot (Fig. 46). Consequently, the reticular zone of the right nigra was intact. In contrast, the left strionigral fibers were destroyed mainly in the pallidum internum and medial to it (Fig. 43a, b). As a result they are completely lacking in the comb system and in the left reticular zone of the substantia nigra except in the most medial part (Fig. 47a), so that the left nigra is sharply narrowed. The degeneration of left strionigral fibers is continued in the degenerated segment of the cerebral peduncle (*Pd*) (Fig. 47a, *arrow*). The nerve cell loss in the substantia nigra shows the typical distribution pattern of idiopathic paralysis agitans with severe involvement of the lateral and ventral cell groups of the posterior main section and the most medial cell group of the anterior main section (Fig. 47a, b). Rarely, intraplasmic inclusion bodies are found. The left substantia nigra is much more shrunken than the right and has lost most of the fiber bundles of the pars reticulata (*rt*) (Fig. 47a). This fiber bundle defect is in direct relation with the tapelike fiber defect in the cerebral peduncle (*arrow*). On the other hand, the lesion extended just into the zona incerta.

Clinical-anatomical correlations. The underlying illness was a typical paralysis agitans. Its endogenous etiology could be ascertained from the presence of a few inclusion bodies and from the typical cell loss in the medial cell group of the anterior main section of the substantia and in the lateral and ventral groups of the posterior main section, left considerably more than right, corresponding to the more severe tremor and rigidity on the right side. The nerve cell loss is superimposed on the

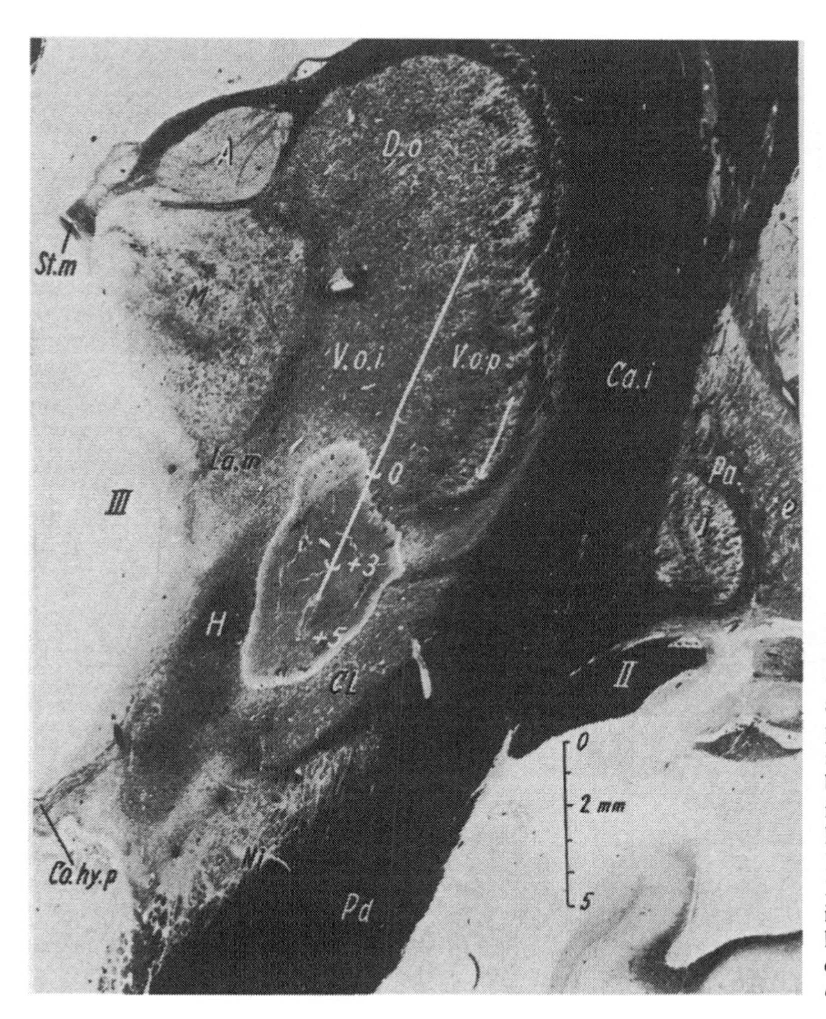

Fig. 44. Lesion of coagulation in case No. 2 at base of right V.o.a. Position of electrode has been drawn in at +3 and +5, because no coagulation was performed at position 0. Coagulated lesion lies in base of V.o.p, in zona incerta below it, and in bundle H_2 and extends into dorsomedial part of nucleus subthalamicus (*CL*). Medially it extends to Forel's field H. Electrode was positioned 3.6 mm medially to target point (drawn in white); according to pneumoencephalography, this was probably due to later drying out of tissues. (After HASSLER *et al.*, 1969)

left side (Fig. 47a, b) in contrast to the right by secondary degeneration due to interruption of strionigral fibers.

During the first operation, the 20-cps thyratrone stimulation of the left pallidum internum only resulted in an interruption of tremor rhythm. Since restimulation of the target point after complete destruction of the pallidum internum resulted in the shaking of the right hand in synchrony with the stimulus, this must be ascribed to a spreading of the current to resistant parts of the internal capsule. The extensive inactivation and scarring of the left pallidum with shrinking of the pallidum internum was the cause of the therapeutic success of the first operation, which permanently eliminated the rigidity. In contrast, the tremor of the right side had been abolished for only two days. It worsened in a month's time, although the use of the right hand improved from severe neglect (due to functional loss of pallidum) to almost perfect functional capacity, including driving a car. However, he was unable to write. The use of the right leg was unimpaired when walking and rigidity did not reappear because the anterior two-thirds of both pallidar segments were completely inactivated.

During the second operation, performed only for relief of the recurrence of the right tremor, the directly stimulated V.o.p tissue produced a synchronous tremor upon 4, 8, or 25 cps stimulation, although most of the pallidum internum was destroyed. The contractions of contralateral muscles during spark-gap discharges must be ascribed to the stimulation effect on capsular fibers by electric and thermal stimuli. Complete relief of the recurrence of tremor on the right side resulted from the almost complete destruction of only the left V.o.p in the second operation. The subsequent coagulation in remaining parts of the internal pallidum immediately lateral to the capsule again triggered a transitory tremor. Postoperatively the additional destruction of the parts of the pallidum internum that survived the first operation and of almost all of the V.o.p did not interfere with consciousness, with the use of the right hand or leg, with the display of strength of the right limbs,

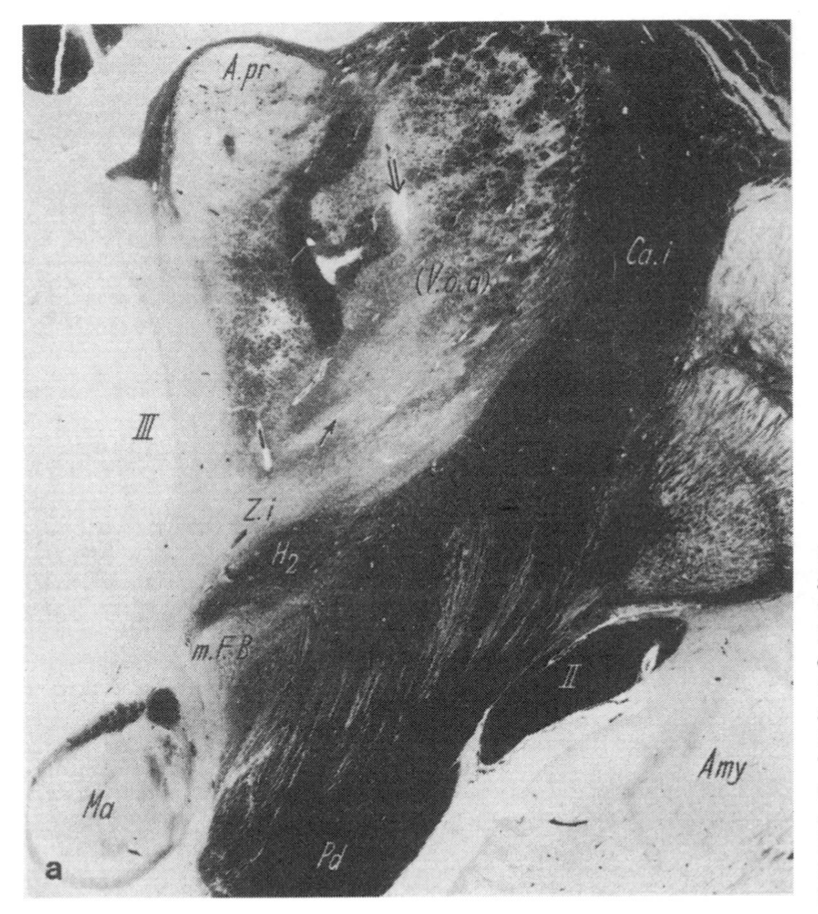

Fig. 45. (a) The transverse section cut about 6 mm more rostral to Figure 46 transects the rostral border of the coagulation focus, made three weeks before death. The coagulation led to a loss of the pallidothalamic fibers (*solid arrow*) and demyelination of the ventral fiber bundles of their terminal nucleus V.o.a. The open arrow (⇓) above marks the track of the coagulation electrode. The postoperative relief of rigidity results from this destruction of pallidothalamic fibers and the base of V.o.a (corresponding to the endurance run homunculus)

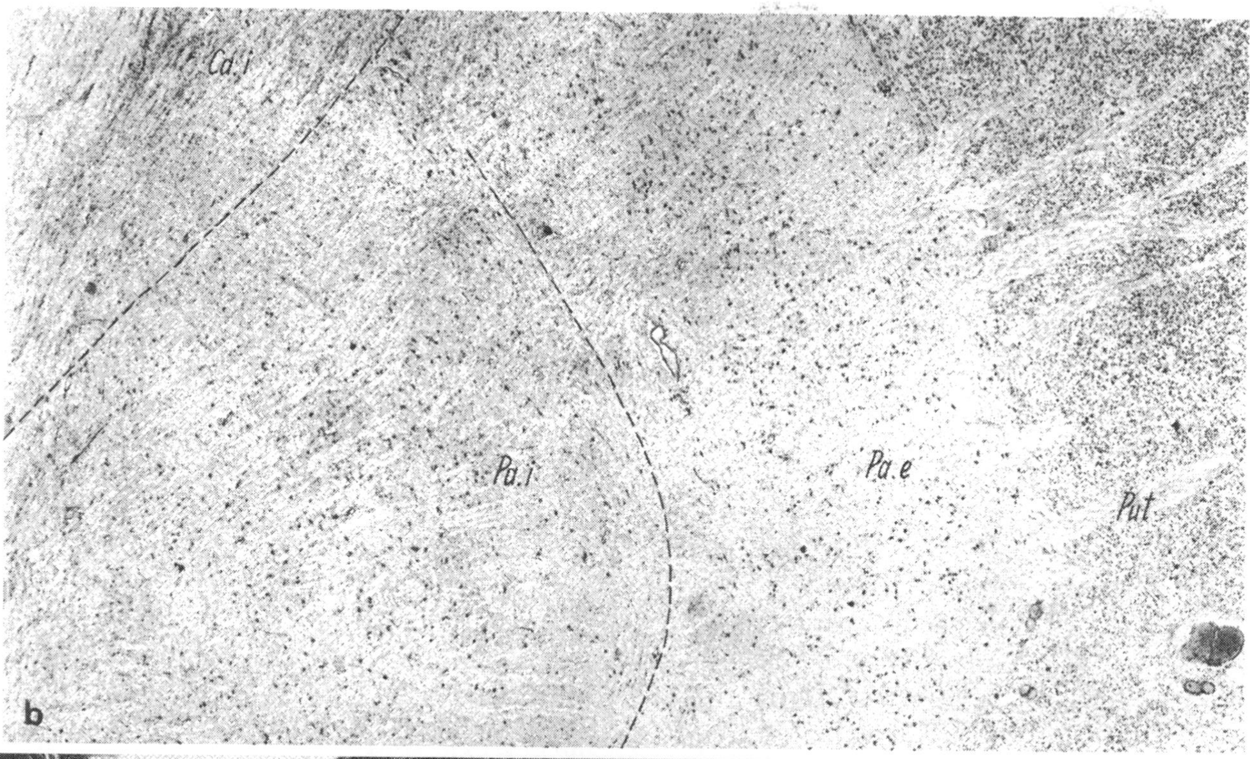

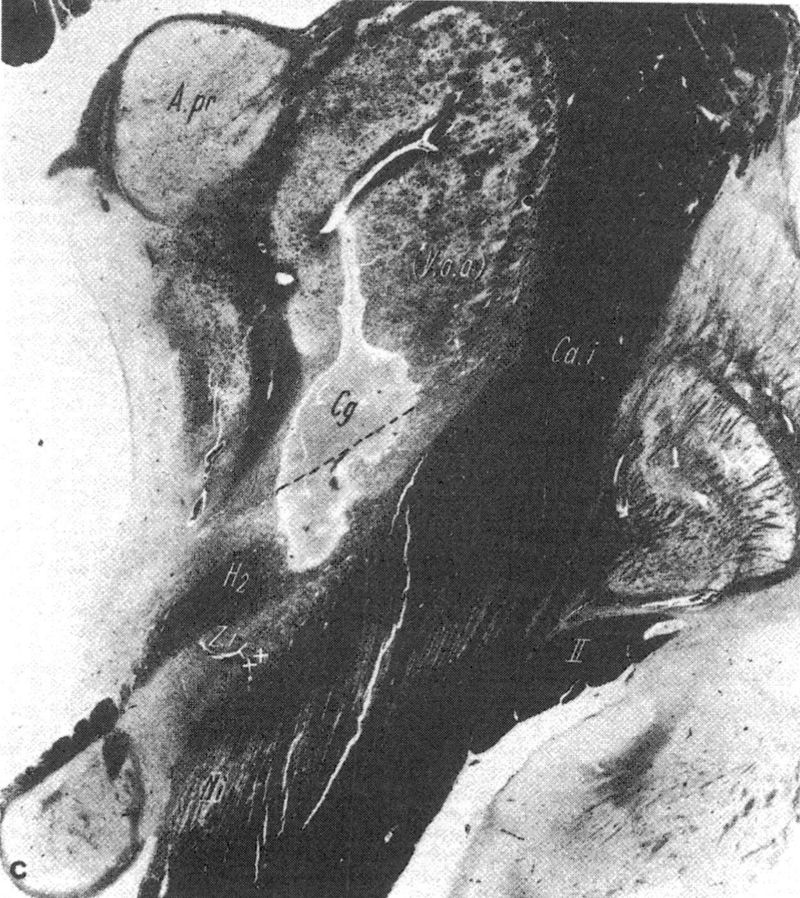

Fig. 45. (b) Transverse section of right pallidum of case No. 2. External part of pallidum shows normal number of cells without increased glia, but number of cells in internal pallidum has diminished, they have become smaller, and dense increase of glia has developed. Degeneration of pallidum internum is due to complete interruption of pallidothalamic fibers in H_1 and consequently to retrograde nerve cell degeneration in internal pallidum (*Pa.i*), in contrast to preserved Pa.e, 3 weeks after coagulation of H_1. Nissl stain. ×25
(c) Cross section through thalamus, 1 mm behind Figure 45 a. Needle track of coagulation electrode, widens into a coagulation focus (*Cg*) that passes over lower border of V.o.a nucleus into zona incerta and H_2 bundle. Further extension of coagulation focus in basomedial direction has been performed with string electrode. A slender subthalamic nucleus, marked by two crosses (+ +), lies below zona incerta (*Z.i*). Lateral part of pallidum internum and medial part of pallidum externum are lightened by demyelination and disappearance of basal fiber work. In contrast, most internal part of pallidum is well preserved. Fiber stain. ×5

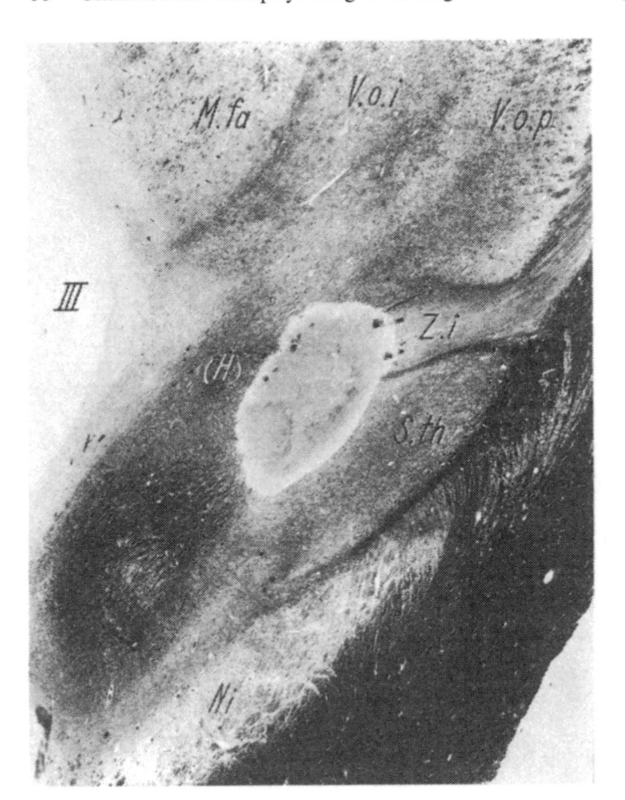

Fig. 46. In a Parkinson patient the tremor at rest has been relieved by a coagulation focus in the base of the right V.o.p nucleus (corresponding to the sprinter homunculus). This section shows the most caudobasal extension of this focus in the zona incerta, interrupting the dentato-thalamic fibers, which terminate in the V.o.p nucleus. The subthalamic nucleus (*S.th*) is minimally involved. Fiber stain. ×6

Fig. 47. (a) Frontal section through midbrain of case No. 2. Right substantia nigra (*Ni*) is approx. normal in width and structure with a compact reticular (*rt*) zone. Coagulation area lying in base of right V.o.a and V.o.p ends about 0.5 mm rostral to this section. It destroyed base of thalamus and zona incerta (*Z.i*) below it, which now contains very few fibers. Dorsal edge of nucleus subthalamicus (*S.th*) has also lost many myelinated fibers. Despite these lesions, this magnification shows no gross change in right substantia nigra. Left substantia nigra (*Ni*) is considerably narrowed, lateral parts contain no fiber bundles or isolated fibers, and there is no reticular zone, due to interruption of strionigral fibers. This fiber-free zone merges into degenerated segment of cerebral peduncle (*arrow*). Because it is at the end of a block, the section is incomplete, e.g., in red nucleus (*Ru*). Fiber stain. ×7.5

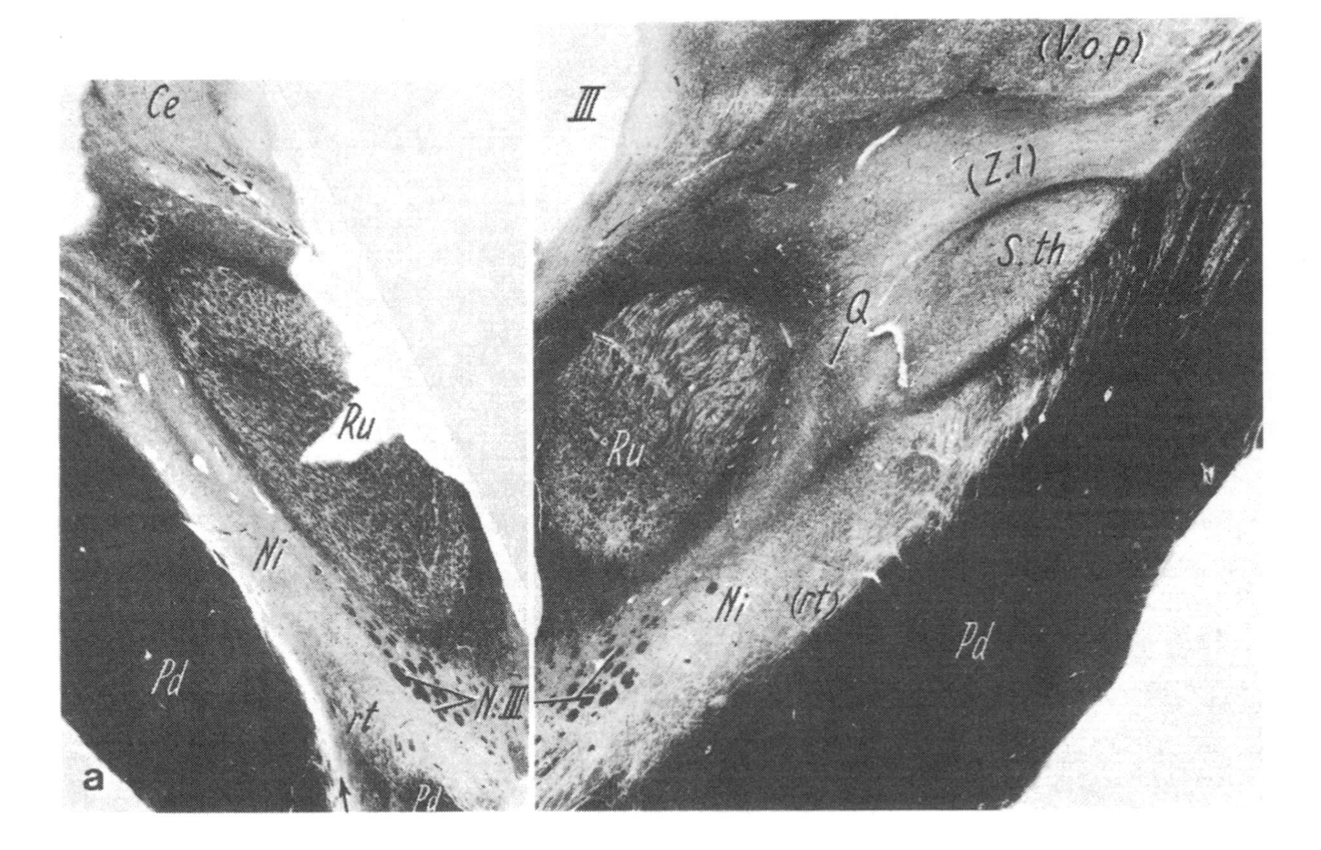

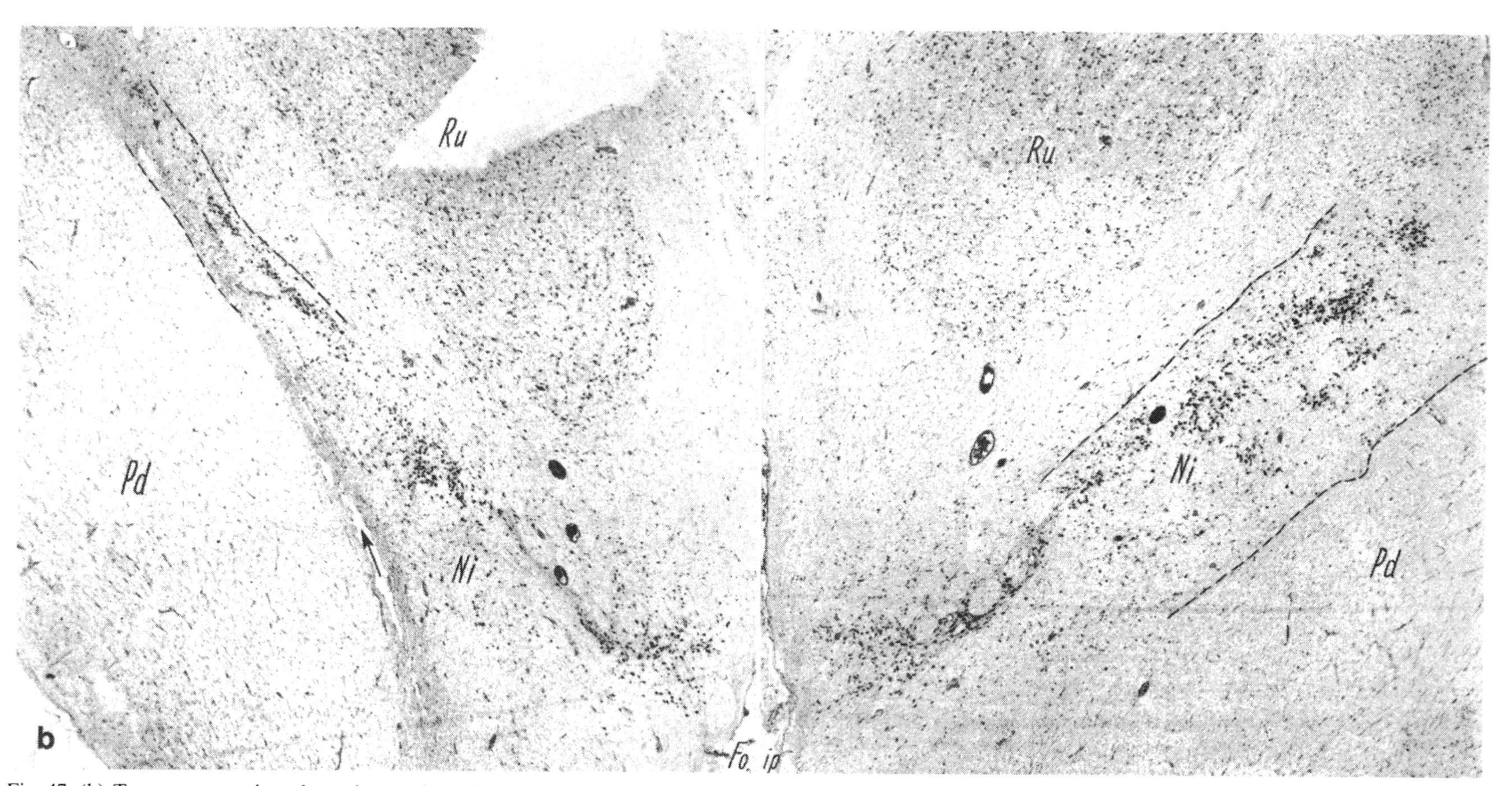

Fig. 47. (b) Transverse section through anterior midbrain of case No. 2. *Left* substantia nigra (*Ni*) is considerably narrowed particularly in its lateral reticular parts, due to degeneration of strio-nigral fibers (s. Fig. 43b). Increased glia at medial part of peduncle replaces reticular zone. Glial scar has shrunk. Ventrally a striped increase of glia becomes a sector of peduncle with degenerated fibers (*arrow*). Considerable shrinking of left substantia nigra is due to destruction of pallidum internum, externum, strionigral connections, and parts of internal capsule in first 2 operations. There is greater loss of nerve cells with increased growth of glia in medial cell groups than in right hemisphere. Main anterior part of *right* substantia nigra has not changed in size. One sees only considerable loss of cells with compensatory growth of glia in medial cell group, which is crossed by root bundles of oculomotor nerve. Slight loss of nerve cells can also be seen in group lying on lateral side, while smaller medial cells have not been affected. No additional change in right nigra caused by coagulation at base of V.o.a and V.o.p can be seen. Nissl stain. ×12

or with walking. The narrow band of fiber loss in the cerebral peduncle caused by passage of the string electrode through the internal capsule (seen post-mortem; Fig. 47b) was responsible for a slight weakness in the oral branch of the facial nerve.

The slowing of the α rhythm from 8–10 to 7–9 cps 2 days after this inactivation and, in particular, the focus of the Θ and δ activity in the frontal and frontotemporal areas of the hemisphere operated on, were due to the functional loss of V.o.p. These conclusions are only partly invalidated in view of the fact that the neighboring internal capsule had been destroyed to a considerable extent.

During the first operation, soon after onset of the most ventroposterior spark-gap coagulation in the lamella pallidi interna, an attack occurred lasting about 2–3 sec in which the eyes turned upward and the patient did not respond. This attack can be explained by irradiation to the lamella pallidi interna at this level with a spark-gap discharge that did not resemble a pure sine wave, because it was produced by the spark-gap diathermy apparatus. Following this attack a slight latent paresis was observed on the right side. It is remarkable that the extensive destruction of the left internal capsule reaching 9 mm back from the genu left no motor disorder during the next 2 years, apart from a weakness of the oral branch of the facial nerve (Fig. 47a).

In the third intervention performed to relieve the left arm tremor, according to the anatomical findings, the stimulation in the right V.o.a was located close to the mamillo-thalamic bundle (Fig. 44). Eight- and 50-cps stimulation had a peculiar effect consisting of an upward jerking of the contralateral arm with grasping toward the face, eyes moving to the other side, pupils widening, face flushing, and laughing. Except for the laughter, this effect of stimulation corresponded to the description that PENFIELD and WELCH (1951) first made of the supplementary motor area. In this patient it appeared to have been caused by stimulating the thalamic nucleus projecting into the supplementary area 6. In view of the specificity of this effect of stimulation, it seems probable that we reached the caudomedial portion of the lateropolar nuclei, which is the thalamic nucleus projecting into the motor supplementary area. In this case the effect of the supplementary area was combined with laughter that could not be suppressed. Its substrate must lie close to the nucleus projecting to the supplementary area (lateropolaris magnocellularis). It was only after the rostral part of the right thalamus (Fig. 45a) had also been inactivated by the third operation that the patient from the fourth day on showed an impairment of consciousness with incontinence, disorientation, and increasing somnolence. It is probable that the extensive, bilateral destruction of all pallido-thalamic connections in addition to the damage of the left pallidum externum and internum (Fig. 42, 43, 45) was the substrate of the postoperative clouding of consciousness. This view is supported by the influence of the pallidum systems on the α rhythm, as demonstrated by electrophysiologic means by GANGLBERGER (1959b, 1960, 1961a, b) during stereotaxic operations. In addition, one can point to previously published experiences (HASSLER, 1957a, b; HASSLER and RIECHERT, 1961).

The anatomical findings showed that death was caused by repeated pulmonary embolisms that obviously started on the fourth day after the last operation. The increasing clouding of consciousness of the patient may have been partly responsible for the embolisms and for the development of the pulmonary congestion.

A silver clip 2×1 mm in diameter was left at the target point in the left pallidum during the first operation following the coagulation. Eleven months later, encephalography showed the clip to be 1.4 mm further in a caudal and 0.9 mm further in a ventral direction without having moved in the frontal plane. This small deviation was due probably to the fact that the clip was implanted into coagulated tissue of reduced consistency. In this tissue it shifted somewhat. In an encephalogram taken 2 years later, the clip was found only 0.2 mm further caudally but 2.1 mm dorsal to the place of implantation (Figs. 39b, 40b). The dilatation of the basal cisterns by the shrinking of the diencephalon could already be seen on the encephalogram (Figs. 39c, 40b). The anatomical findings explained this dorsal movement of the metallic marker by the extensive shrinking after 2 years in the rostral part of the basal ganglia.

In the first two operations, which were performed 11 months apart, the electrodes were introduced through the same opening in the bone, one 6.5 mm behind the other in a parallel sagittal plane, directed toward the basal ganglia. The post-mortem showed that the closeness of the tracks led to a softening of the tissue between them. As a result, a 4-mm wide, 4-mm high defect was left in the white matter of the frontal brain. It was only near the basal ganglia that the distance between the electrodes increased as they approached their targets. This finding demonstrates the danger which occurs if parallel electrodes situated only a few millimeters apart are introduced deep in the brain. It is improbable that the defect in the white matter of the frontal lobe had any significance, but the destruction and scar formation in the genu and anterior part of the internal capsule were due to this damage (Fig. 43b). The retrograde degeneration and atrophy of the medial nucleus of the left thalamus (Fig. 131) were caused by this severe damage to the internal capsule.

Case 3. R.L., aged 59; paralysis agitans; stereotaxic operation performed in right V.o.a (no. 1599); capsular lesion without hemiparesis; death occurred 6 days after operation, due to pulmonary embolism.

History. Parkinson's disease started at age 47 with a left-sided tremor. There was no indication of previous encephalitis. Later, marked rigidity of the left limbs developed that extended onto the right side. Before the operation the patient could no longer work, she could no longer write, and she needed help to get dressed, to use the lavatory, and to turn in bed. She suffered from cramplike pains in both legs, especially in the left.

Clinical findings. Marked rigidity and tremor in all limbs, left more than right. Gait disturbed, typical Parkinson posture; because of her impaired movements she was completely helpless. There was dysphagia. Mentally she was markedly slowed down and lacked initiative. According to the physicians (Medizinische Poliklinik, University of Freiburg), her general condition was poor, her heart was enlarged to the left, and she had massive pretibial edema, which was evidence of right cardiac failure. Palpation indicated that her liver was considerably enlarged and hardened. Her sedimentation rate was increased to 38/66 mm Westergren. Radiologic investigation showed no evidence of a neoplasm. The aorta was markedly dilated and sclerosed. A retrosternal goiter was seen, with the trachea being displaced to the right; emphysema was also present. The EEG was slightly abnormal (Fig. 49), the occipital α rhythm being somewhat slow (7–9 cps).

Despite the medical findings and the evidence of central arteriosclerosis, the patient and her relative insisted on an operation. Following helium encephalography (Fig. 48: dilation of parietal sulci) she went into a nocturnal confusional state that apparently had already occurred several times before. Because of the psycho-organic syndrome and the circulatory disturbances, she was prepared during the next 10 weeks for the operation.

A *stereotaxic operation* was performed on January 31, 1962; the target was the right V.o.a (Fig. 48). Stimulation at the target point caused twitchings in the left lower lip and the left forearm, replacing the tremor. Because of the poor general condition, stimulations with higher frequencies were not performed. Four coagulations were

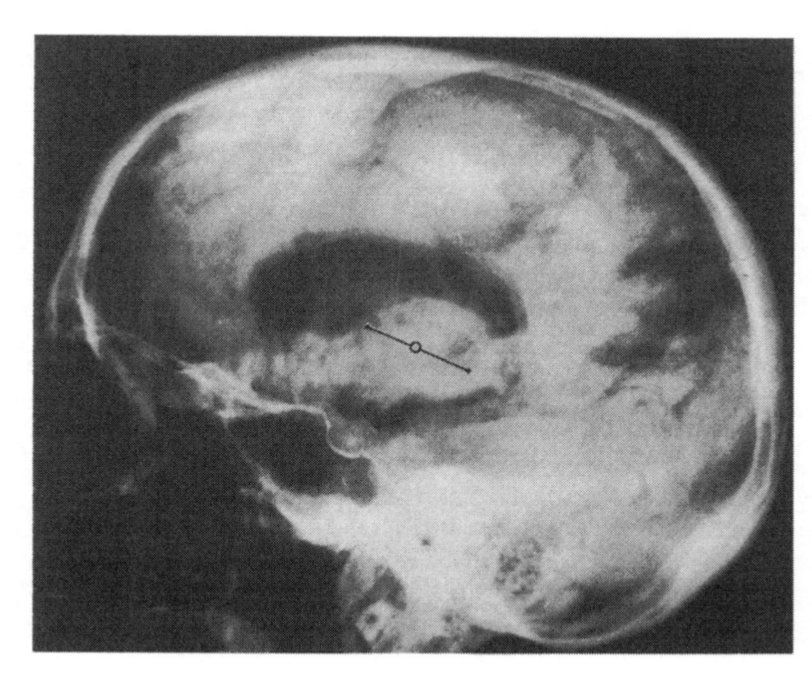

Fig. 48. Lateral encephalogram of case No. 3. It shows considerable dilatation of lateral ventricles, frontal subdural air, and broad subarachnoidal markings in parieto-occipital region as evidence of cerebral atrophy. Baseline with target point V.o.a/p has been marked

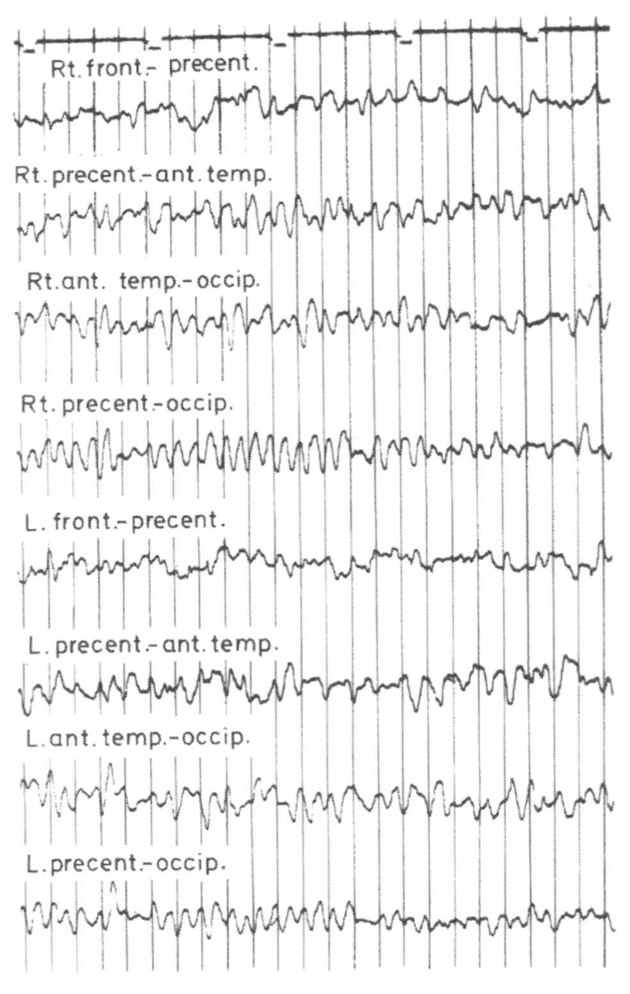

Fig. 49. Preoperative EEG of case No. 3 in bipolar recording in a row. Slowing of α-rhythm on right more than on left side, mainly in precentral, occipital, and temporo-occipital regions. The α-rhythm of left frontal precentral lead is replaced in right frontal-precentral lead by a δ-rhythm

performed with the thermoprobe from positions 0 to +5. Following this, the tremor was completely abolished in the left leg and almost completely relieved in the left arm. The rigidity of the left side was also much reduced.

Postoperative course. During the evening after the operation the temperature rose with tachycardia, rapid breathing, and clouding of consciousness. During the next days the patient became negativistic and eventually could no longer respond. Despite intensive treatment with drugs and tracheostomy, death occurred from cardiac failure on day 7 after the operation.

Post-mortem of the body (Prof. Dr. H. NOETZEL) showed one old recurrent and one fresh pulmonary embolism starting from a left femoral venous thrombosis. Additional findings were: bronchopneumonia with mixed flora; right half of the heart dilated with arteriosclerosis but without stenosis of the coronaries; moderate arteriosclerosis of the aorta and the other vessels; chronic recurrent pyelonephritis; and nodular enlarged thyroid. The multiple pulmonary embolisms together with the bronchopneumonia were the cause of death by cardiac failure.

Anatomical findings in the brain. A subarachnoid hemorrhage (11 × 9 mm) due to the trephining was found over the first frontal sulcus. The ventricular system was symmetrically dilated. The cerebral arteries showed a slight thickening of the walls and arteriosclerosis, most marked in the area of the vertebral and the basilar arteries. The right basal ganglia were enlarged through swelling, particularly around the track made by the electrode. There also was swelling in the head of the right caudatum. The area of coagulation in the right V.o.a and V.o.p involved just a fringe of the neighboring capsula interna (Fig. 50a, b). The lesion was 8.5 mm wide, 11.5 mm high, and had a sagittal length of 12 mm. In the plane of the V.o.a its lower edge reached 3.5 mm below the baseline. Due to the severe swelling of the thalamus on the 7th postoperative day, the middle of the coagulated area was 13.5 mm away from the wall of the third ventricle. The coagulation had destroyed the V.o.a and V.o.p as well as the medioventral part of the V. im (Fig. 50b) and the capsule close to the V.o.a. Also involved in the coagulated area were the dorsal half of the subthalamic nucleus (Fig. 50b) and parts of the zona incerta below the V.o.p (Fig. 50c). Most of the nerve cells of the lateral cell groups of the substantia nigra were destroyed and replaced by dense gliosis (Fig. 50c), indicating an idiopathic form of Parkinsonism.

Clinical-anatomical correlations. The length of the lesion in the V.o.a and V.o.p was 12 mm. This corresponded to the electrode, which was bare for 4 mm, having been moved a distance of 5 mm so that in front of the point of the electrode an additional area of 3 mm had been coagulated. The fresh coagulation cylinder had a width of 8.5 mm, which correlated with the other parameters. The fact that the coagulation cylinder was 11.5 mm high – that is, 3 mm higher than it was wide – was due to the cut going at an angle to the cylinder. The area of coagulation had the typical formation for thermocoagulation, with tissue being preserved around the track and having a clearly demarcated periphery. The size of the coagulated lesion was probably about the same as it was on the day of the coagulation. Considerable swelling of the right thalamus occurred around the coagulated area, as is typical for the 7th postoperative day. As a result, the anatomically determined coordinates of the coagulated area were not identical to those taken from the x-ray picture made the day of the operation. The coordinates were determined from points in the brain whose positions changed because of the postoperative swelling of the thalamus. However, the anatomical findings showed that the aimed-at structures, particularly the V.o.p, had been reached and been put out of action completely (Fig. 50a). On the medial side of the V.o.a, 2 mm had been included in the coagulation.

Inactivation of the V.o.p corresponded clinically to abolishment of the tremor of the contralateral limbs while the inactivation of about three-fourths of the V.o.a was in keeping with the big reduction in rigidity. The patient said spontaneously after the first coagulation that the bothersome pain in the left limbs had already become more tolerable. An extension of the coagulated area into the left internal capsule next to the V.o.a could be found only rostrally from the target point. This was caused by the oblique direction of the axis, from rostrolateral to caudomedial. In spite of this lesion and of the marginal involvement of subthalamic nucleus, no paresis or ballismus of the limbs on the other side could be found either on the operating table or a week after the operation.

Death was caused by multiple pulmonary embolisms. The pulmonary embolisms were surrounded by areas of bronchopneumonia. The thromboses and the rapid death were due to the simultaneous right-sided cardiac failure with arteriosclerosis of the systemic vessels and a chronic recurrent pyelonephritis.

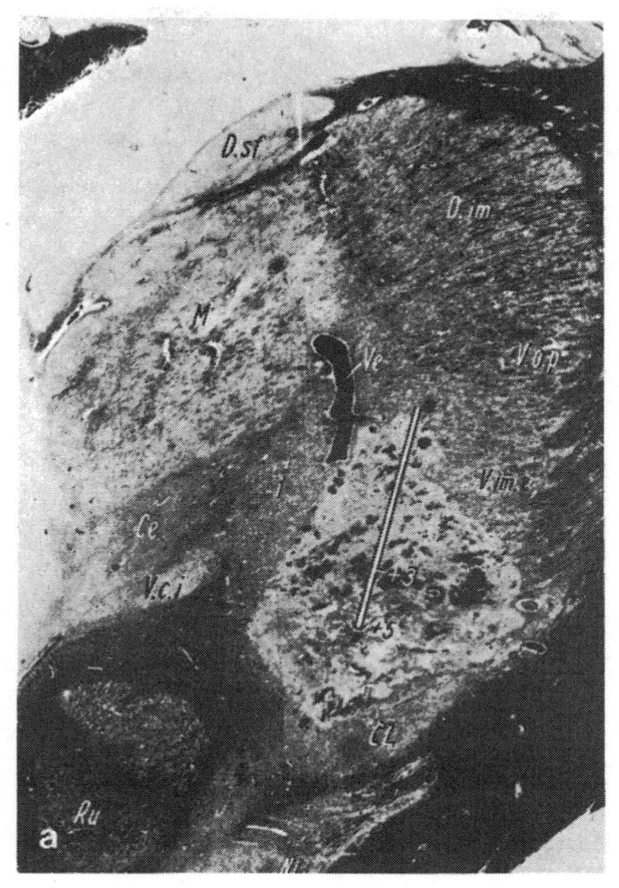

Fig. 50. (a) Coagulation lesion of case No. 3 in base of right V.o.p and V.im, with tracing of electrode tract. Because of narrow rostrocaudal angle of entry of electrode, latter has reached base of V.im in positions +3 and +5, without involving internal capsule. Base of V.im and V.o.p has been completely destroyed as well as zona incerta below and caudal end of bundle H_2. Lesion includes small part of dorsal edge of subthalamic nucleus (*CL*). Above lesion is necrotic area with confluent petechial hemorrhages. Hemorrhages are linked to very full vein (*Ve*) and to many other vessels filled with blood. Thalamic radiation is interrupted where it leaves red nucleus (*Ru*). (After HASSLER *et al.*, 1969)

Case 4. H.J., aged 58; postencephalitic Parkinsonism; stereotaxic operation in left V.o.a (no. 1199) and extension of the lesion into the cerebral peduncle; death occurred 7 days after operation, due to a large pulmonary embolism originating in a bilateral femoral venous thrombosis.

History. At age 43, the patient, who at age 19 had a severe influenza and doubtful encephalitis, developed rigidity and tremor in the right arm, later also in the right leg. A year later the tremor extended to the left arm and the chin. From age 52 on, considerable salivation, sweating, and greasiness of the face supervened. For 3 years he had been unable to work and required some assistance.

Clinical findings. The physicians (Medizinische Poliklinik, University of Freiburg) found a retrosternal goiter, emphysema, and bronchitis. An electrocardiogram (EKG) showed right ventricular failure and tendency of the blood pressure to be reactively lowered. Treatment of the cardiac and circulatory condition was suggested and carried out for 10 days prior to the operation.

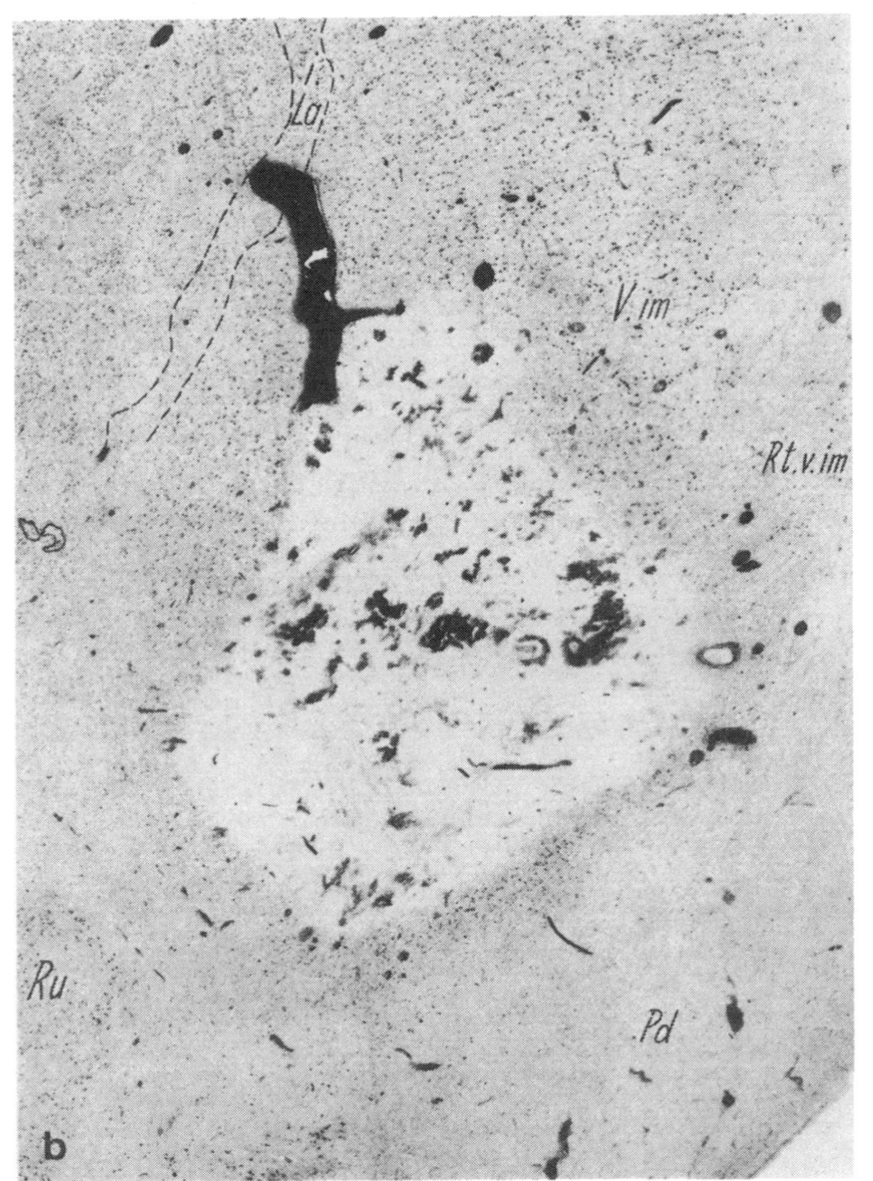

Fig. 50. (b) Adjacent section to Figure 50a. Veins and arteries filled with blood in area surrounding fresh coagulation focus are prominent. Notice glia wall around fresh coagulation focus, which includes the dorsal half of subthalamic nucleus. Nissl stain. ×9

There was a moderate disturbance of gait on both sides with a lack of associated movements and the typical posture. Rigidity and tremor of medium strength were found on the right side; tremor of the chin was also discerned. There was moderate rigidity and tremor equally in the left arm and leg. He had pain in the right limbs. There were contractures in the basal joints of the fingers of the right hand. The voice was aphonic.

Following encephalography with helium the patient became confused and restless for 2 days, so the operation was postponed for an additional 10 days.

A *stereotaxic operation* was performed on November 15, 1960. The target was the left V.o.a. Stimulations at 4–8 cps made the tremor on the right side rhythmic. Stimulations at a rate of 50 cps caused a more pronounced mydriasis on the contralateral side than that of the stimulation, and mistakes in counting, while the tremor continued. While the temperature was being controlled by a second probe with a thermocouple in the neighborhood of the thermoprobe (Fig. 51), coagulations were performed in 3 steps. These lasted 20 sec each at a temperature of 70° C from positions 0 to +4. Following the coagulation in position +4,

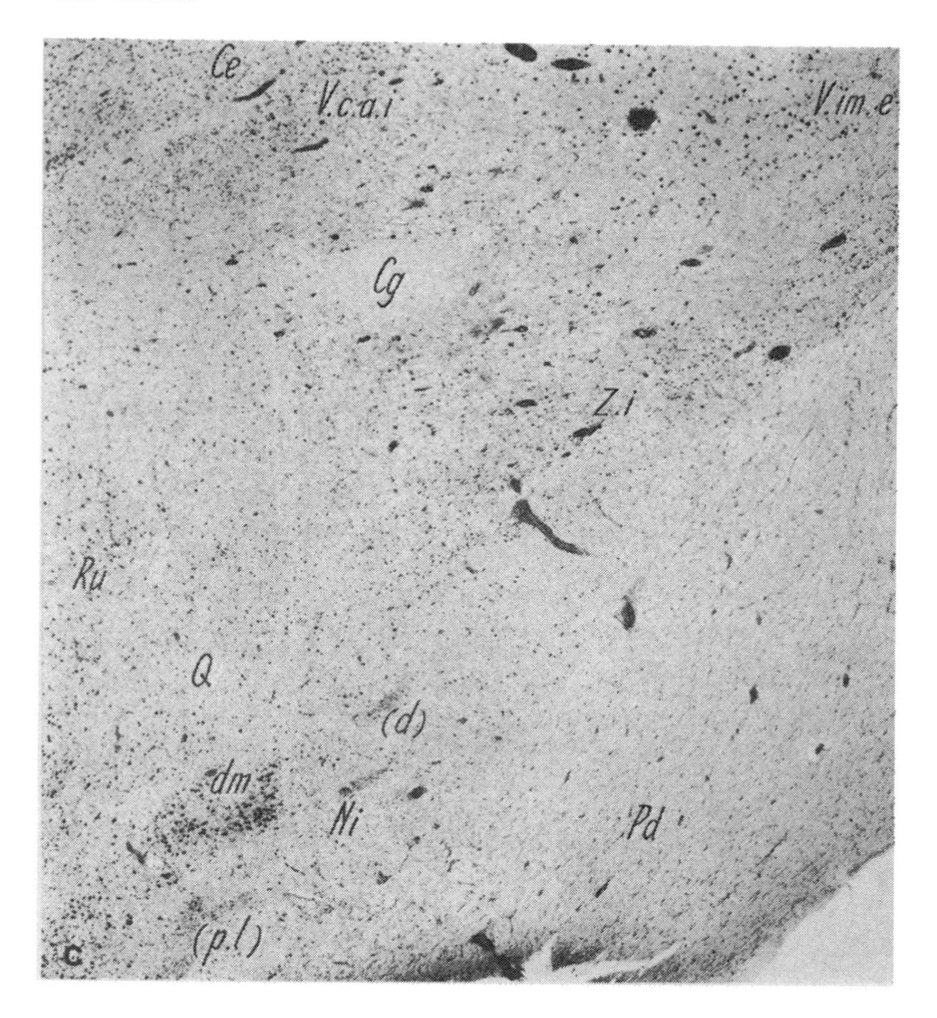

Fig. 50. (c) Transverse section of thalamus of case No. 3, 1.9 mm further caudally than Figure 50b, shows end of coagulation focus (*Cg*) just in basis of V.im nucleus. Zona incerta below V.im.e nucleus shows filled blood vessels and increase in glia cells. In lower left corner is posterior main section of substantia nigra with complete cell loss with intensive gliosis in p.l group and in d group, whereas dorsomedial cell group (*dm*) has only slight cell loss and light glia proliferation. Nissl stain. ×8

---

the tremor in the right hand was very diminished and in the right leg, noticeably diminished. The contractures in the joints of the fingers persisted. After the operation, the speech disorder was more marked and was more aphonic than before.

Postoperative course. The day after the operation, the patient had moderate rises of temperature, remained lucid, and had no psychosyndrome. Five days after the operation he got up. On the 7th postoperative day, in the early hours of the morning, sudden disappearance of pulse and arrest of respiration occurred. Despite artificial respiration and intracardial injections, the patient died.

Post-mortem of the body (Prof. Dr. H. NOETZEL) revealed large acute pulmonary embolisms on both sides, deep femoral thrombosis on the right side, thrombosis of the plexus prostaticus, and moderate sclerosis of the aorta.

Anatomical findings in the brain. Atrophy at the frontal pole and on the convexity, small cortical necrosis on the left frontal gyrus below the trephine hole. In the white matter of the frontal lobe two hemorrhagic puncture tracks were found that converged in the direction of the basal ganglia. The left lateral ventricle was markedly more dilated than the right. The area of coagulation was

7 mm wide, 6.5 mm high; its lower edge was 6 mm below the baseline (Fig. 52). It could be seen in the frontal plane of the V.o.a and ended about 11 mm behind the interventricular foramen. The lesion destroyed the basis of the V.o.a and extended into the internal capsule at the point where it becomes the rostral part of the cerebral peduncle. The lesion also destroyed the fasciculus lenticularis (H_2) and medially extended into the zona incerta and H-field. The fasciculus mamillo-thalamicus was interrupted.

Clinical-anatomical correlations. The coagulated area at the base of the V.o.a also involved the subthalamus and the internal capsule where it becomes the peduncle (Fig. 52). This deviation of the center of the lesion by 4.5 mm in the basal direction was caused by a 5-mm deviation of the point of the electrode toward the base, as was seen on the x-ray picture (Fig. 51). The deviation of the point of the electrode from the target point had been found radiologically before the coagulation. It was, however, not changed for the coagulation because the 4- to 8-cps stimulation synchronized the tremor, as occurs with V.o.a stimulations, and when the electrode was pushed further in a caudal direction, no additional extension of the lesion in a ventral direction was expected because of the very flat way the electrode was introduced. However, since the direction of the electrode formed a much larger angle with the median plane than is usual, the area of coagulation did not reach the base of the V.o.p. Abolishment of rigidity in the opposite arm and the marked reduction in the opposite leg could be explained by the coagulation sited in the base of the V.o.a and and in the bundles H_1–H_2 of Forel lying below it. The worsening of speech together with the increased aphonia corresponded anatomically to the unintentional extension of the coagulation into the basal part of the internal capsule where it merges into the cerebral peduncle.

Although the stimulation was performed in the subthalamus 5 mm below the target point and probably also affected rostral fibers of the internal capsule, the tremor in the right arm was made more rhythmic. The fact that stimulation at higher frequency had no influence on the tremor could be related to the absence of stimulation of structures in the V.o.p and its cerebellar afferent fibers. In this case, the tremor was more synchronized when stimulated at 4–8 cps because of a stimulation of the efferent fibers of V.o.a and V.o.p to the capsule, which are not the upper motor neurons for the limbs. Making the tremor more rhythmic need not be regarded as a certain sign of correct localization of the electrode in the V.o.p. In this case the radiologic findings would have permitted a more correct localization of the coagulation, despite good tremor relief for 6 days.

The extension of the lesion by coagulation into the rostral fibers of the peduncle does not appear to be the cause of death, since the patient devel-

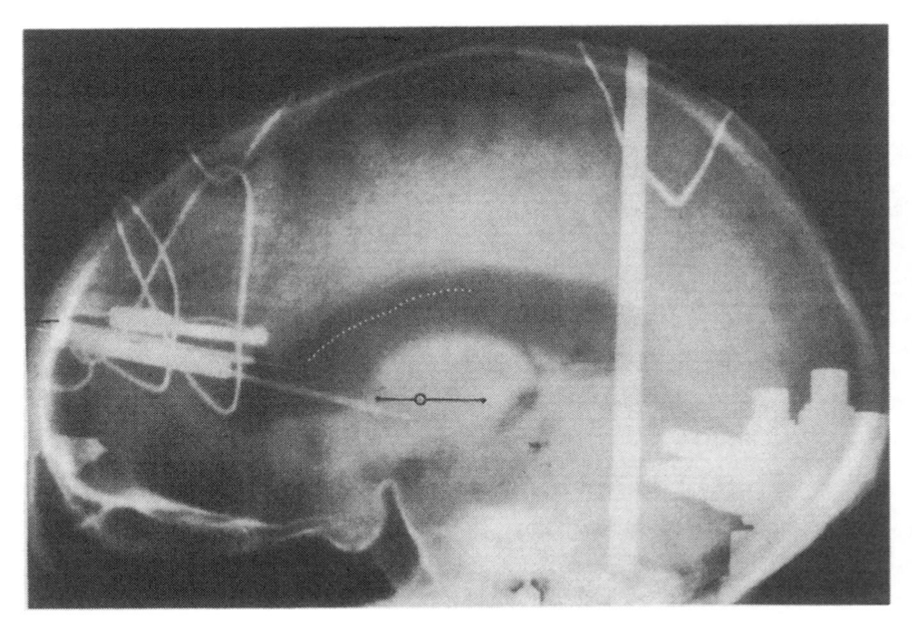

Fig. 51. In case No. 4, lateral picture of pneumoencephalogram with its calculated target point and x-ray taken during operation to check position of electrode were superimposed; distortion is same in both pictures (3 m distance to plates). Precalculated target point V.o.a in baseline is projected into massa intermedia. Electrode point deviates 5 mm basally. This electrode goes obliquely down, while multipolar recording electrode has correct position axially. Point of recording electrode lies within pale outline of caudatum (outline marked by dots)

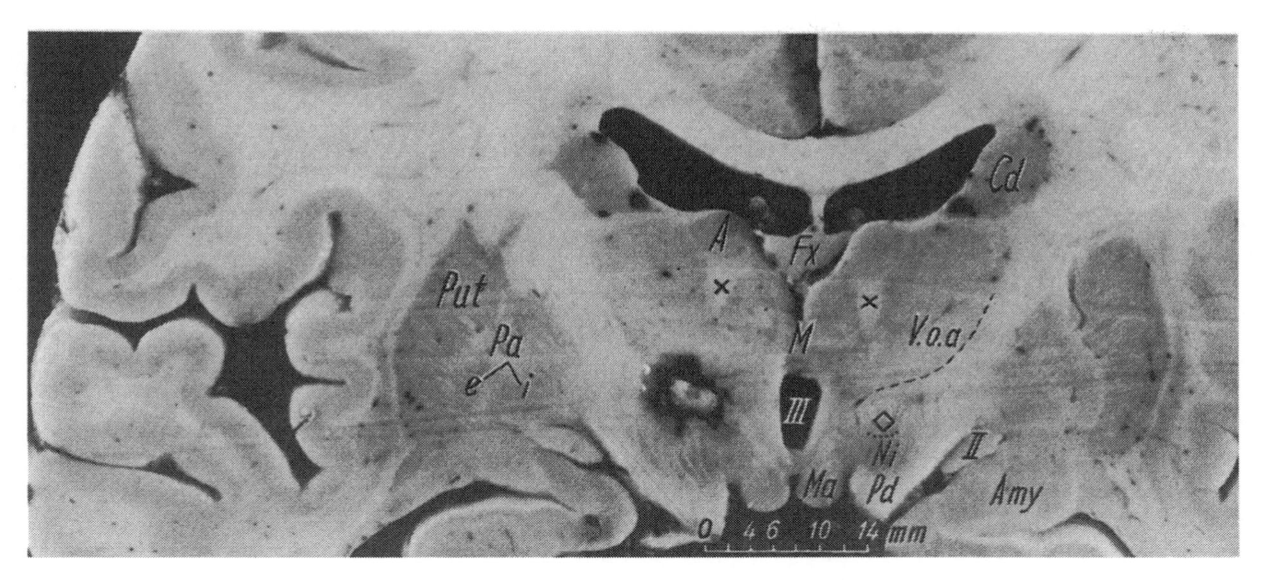

Fig. 52. Transverse section through unstained brain of case No. 4. Center of coagulation lesion lies in base of V.o.a and below it. Lesion consists of white-yellowish coagulate, pea-sized, with dark electrode track in center. It is surrounded by a colliquation necrosis appearing black-gray (with hemorrhages). It had destroyed subthalamic nucleus and its ventral end extends into fibers of cerebral peduncle. Note slight swelling of left coagulated thalamus. Diamond marks 1st cut of right subthalamic nucleus. × marks mamillo-thalamic tract

oped no paresis apart from the increase of aphonia and dysarthria postoperatively (Fig. 52). Nor was his consciousness clouded, and he got up on day 5 following the operation. The cause of death was a massive pulmonary embolism originating from a femoral thrombosis.

Peculiarities of this case include the 24-year latency period after severe influenza with encephalitis in 1921, before the appearance of postencephalitic Parkinsonism with tremor of the chin, contractures of basal joints of the right fingers, and nine years later, strong sialorrhea, hyperhidrosis, and seborrhea faciei. The confusion with restlessness following helium encephalography is ascribed to the external brain atrophy.

Case 5. T.W., aged 63; paralysis agitans; first stereotaxic operation in right pallidum internum (no. 408); second, in left V.o.a (no. 876); fluctuating state of consciousness after bilateral damage of pallidum; death occurred 27 days after second stereotaxic operation, from pulmonary embolism.

History. Parkinson symptoms started at age 52 on the left side. There was no history of previous illness. A year later they extended to the right side. Simultaneously existing hyperpiesis could be satisfactorily treated with drugs.

Clinical findings. At the first admission in 1958, there was marked rigidity in both legs and tremor in both arms and legs, left more than right. The handwriting was distorted by tremor, propulsion was very pronounced. He had dragging pains in the muscles of the back. He reacted to the illness with depressive symptoms (see Fig. 53). No other neurologic abnormalities were found. The C.S.F. was normal. During the night following encephalography (Fig. 54a, c) the patient was confused; the confusion recurred during the next 2 nights.

A *stereotaxic operation* was performed in the pallidum internum on the right side on June 6, 1958. During stimulation the very pronounced tremor of the left hand and the left leg became more rhythmic and more severe. Stimulation at a rate of 50 cps blocked the tremor. Alternate movements of the left arm were disturbed; when counting backward, some numbers were left out.

Coagulations with the spark-gap diathermy instrument were performed in positions 0 to +8. At first the tremor in the left arm was reduced, then mobility of the trunk and rigidity of the left limbs improved at the same time. At the end of the operation the patient became increasingly tired, yawned, was disoriented, and repeated words said to him. With the last coagulation the left arm went down a little. Rigidity and tremor on the left side

with the exception of a trace in the left ankle joint were abolished, while they continued on the right side, i.e., on the unoperated side.

Postoperative course. During the night after the operation the patient was again confused and experienced his double (*Doppelgänger*). After 6 days

Fig. 53. Portrait of case No. 5 with left-sided facial paresis contralateral to the operated side. The scar is seen in the right frontal region. Typical rigid Parkinson posture and hypomimia

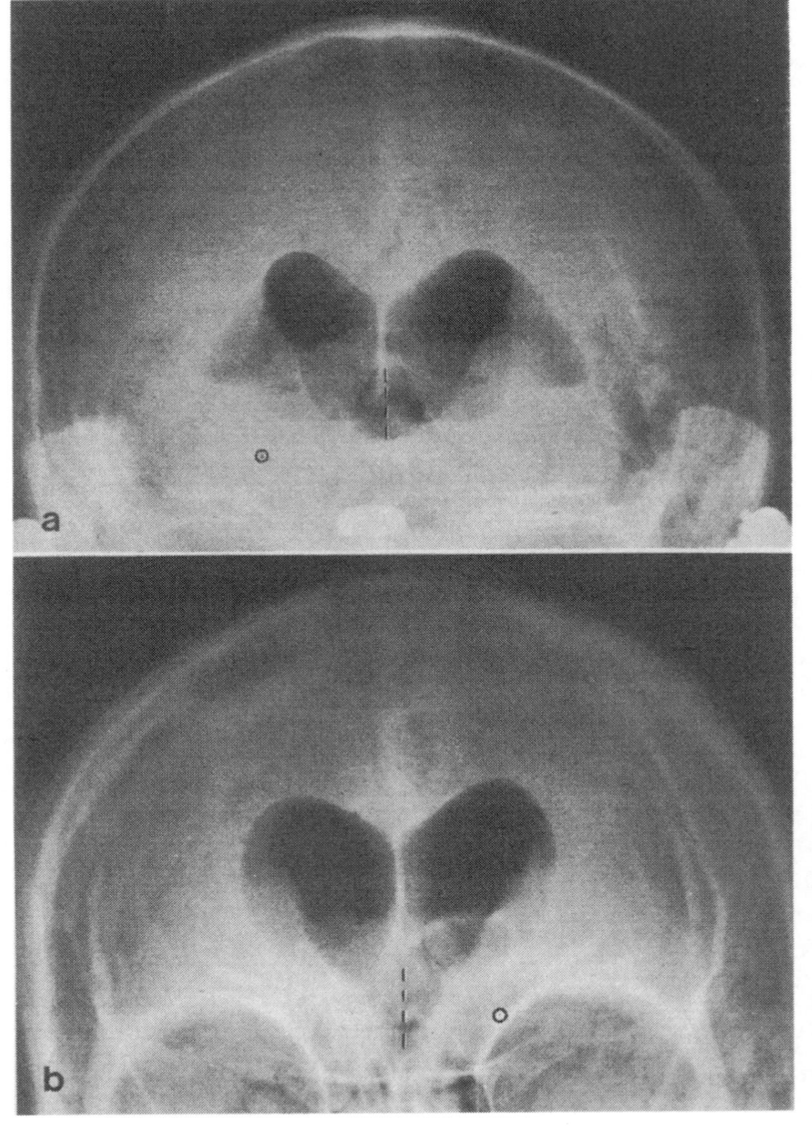

Fig. 54a and b. Anterior-posterior pneumoencephalogram of case No. 5 before 1st operation and 1.7 years later. Lateral ventricles are very dilated, particularly right one, which was already somewhat wider than left before operation. Since 1st operation was performed on right hemisphere, dilatation on right side has increased more after 1.7 years. Considerable dilatation of 3rd ventricle before operation did not increase after 1st operation (14 mm)

the postoperative confusion disappeared; the tremor was barely present, and rigidity was considerably diminished on the left side; there was, however, severe mimical facial paresis (Fig. 53). This effect on tremor and rigidity continued beyond the period of observation. The patient's mood improved considerably, and he could therefore again work part time.

Ten months later, rigidity was absent and a slight tremor was present on the left, while the signs had become much worse in the right limbs. Nineteen months later the right side had deteriorated further, but on the left there still was no rigidity. The patient had no psychosyndrome.

Repetition of encephalography before the second operation showed an increase of hydroceph-

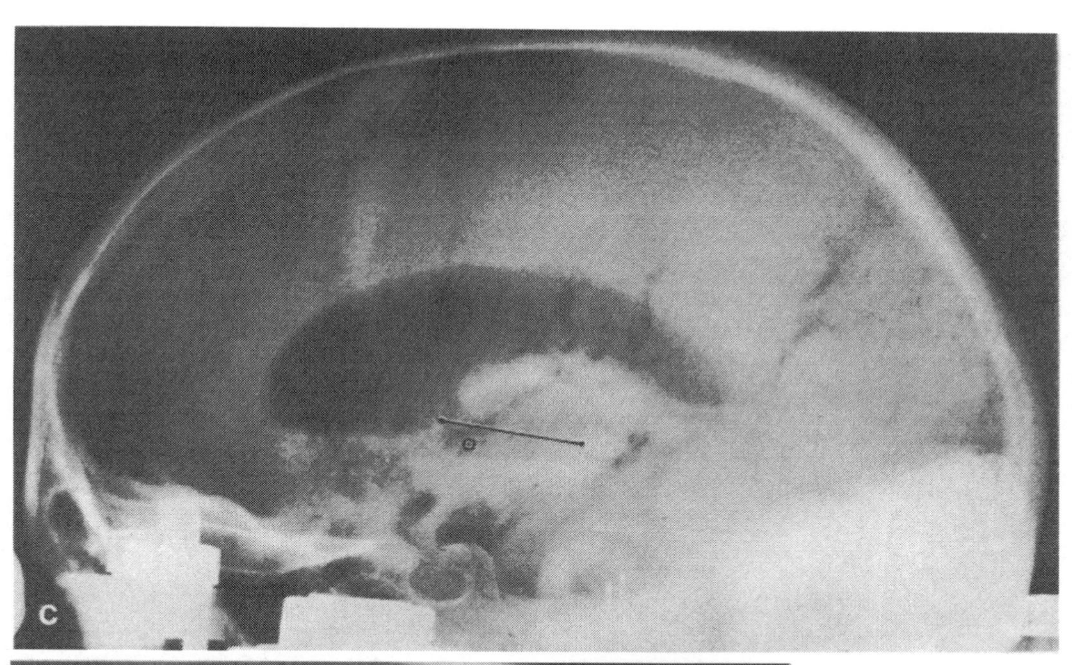

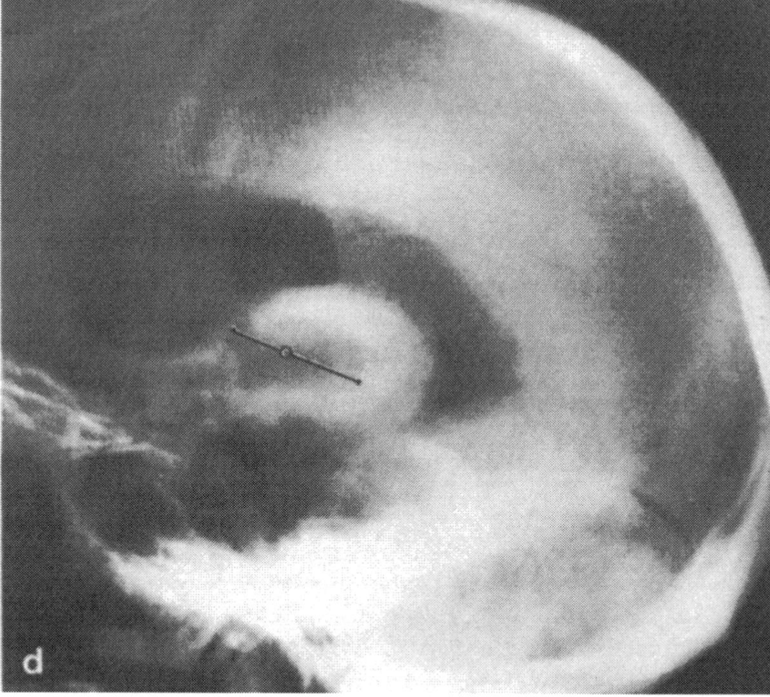

Fig. 54c and d. Lateral encephalogram before 1st operation shows considerable dilatation of lateral ventricles. Target point in pallidum has been marked. The right-sided hydrocephalic dilatation 1.7 years after 1st operation has increased considerably, so that both anterior horns appear ballooned. More marked contrast in anterior part of cella media is caused by superimposed projection of two radiographs

alus internus (Fig. 54a–d) and a marked coarsening of both anterior horns. The patient was again confused for 2 days after encephalography.

A *second stereotaxic operation* was performed on January 1, 1960, in the left V.o.a. Stimulation made the tremor rhythmic. Single stimuli increased individual phases of the tremor. During stimulation at a rate of 50 cps counting was inhibited, which was explained by a feeling of tightness in the chest. Occasionally there was pressure in the upper part of the abdomen; even stimuli of low intensity led to inspiratory arrest of respiration and widening of the pupils.

Coagulations with the Wyss coagulator were performed in positions 0, +2, and +5 with an electrode with a point bare for 4×2 mm. After the second coagulation (+5), there was transitory confusion. Although subjectively the patient could

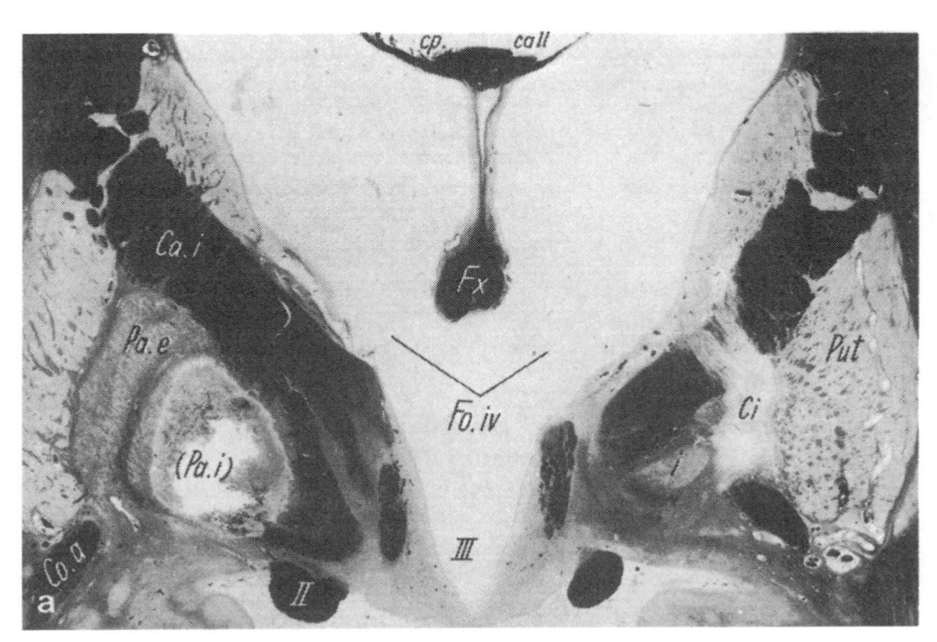

Fig. 55. (a) Transverse section through diencephalon at height of interventricular foramen (*Fo.iv*) of case No. 5. Both lateral ventricles appear blown up, particularly dorsally (see Fig. 54a–d). Scarred coagulated lesion (*Ci*) in right basal ganglia is in external part of pallidum with involvement of lamella pallidi interna. Internal capsule lying immediately over external pallidum has lost its myelin sheaths and is somewhat displaced ventrally toward lesion. In left basal ganglia coagulation extending from ansa lenticularis to lower surface of internal capsule has increased size of internal pallidum (*Pa.i*). This area is conspicuous because of its lighter color compared to surviving fiber structures. Fiber stain. ×2.3

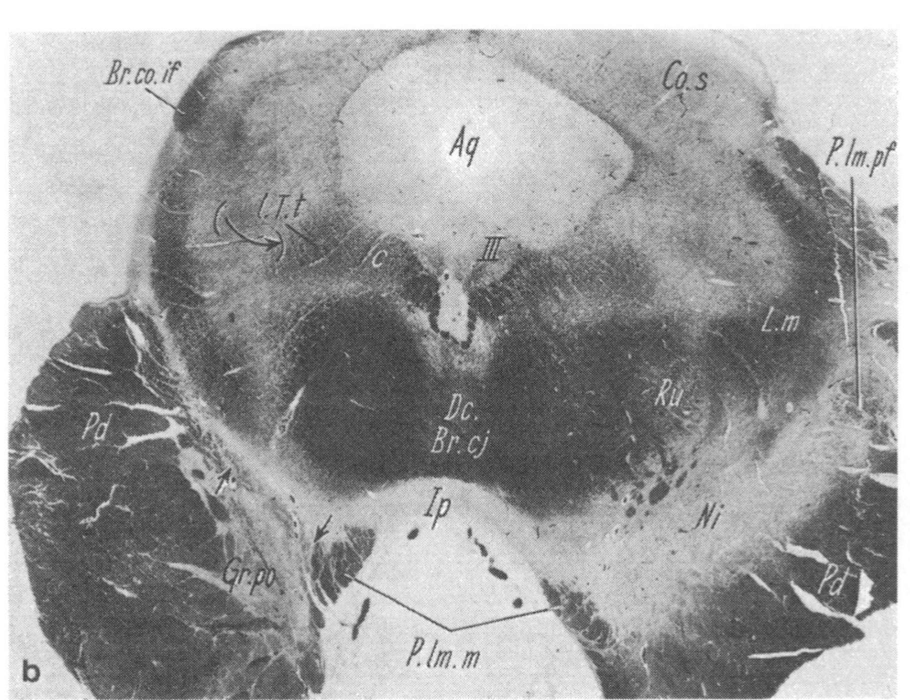

(b) Cross section through midbrain of case No. 5, reversed (!). Section is asymmetrical so that left ruber (*Ru*) is just hit and left substantia nigra is considerably larger than right. Right contains shrinkage because of lesion of internal capsule which is here continued in medial part of pontine gray (*Gr.po*); only most medial part of cerebral peduncle is spared as fully preserved fiber structure and is not demyelinated (*P.lm.m*). *P.lm-pf* Pes lemnisci profundus. Fiber stain. ×4

move better, the result of the operation was only partly satisfactory. There was slight improvement of the rigidity and slight diminution of the tremor in the right arm. At the end of the operation the patient was no longer disoriented.

Postoperative course. The day after the operation the patient frequently choked when swallowing. Because of the danger, a permanent gastric tube was inserted. On day 3 after the operation the patient could get up. A bronchopneumonia developed, which could initially be controlled, so that the patient walked about without right-side tremor twice a day. Following the operation, the patient frequently "struck out" with the right hand. On day 10 after the operation he vomited and simultaneously coughed severely. Breathing became difficult (38 per minute), blood pressure rose from 140/80 to 235/80 mm Hg. Vomit was sucked out. Two days later the condition necessitated a tracheostomy. Clarity of consciousness varied and the general condition gradually deteriorated; death by circulatory failure occurred 27 days after the operation.

Post-mortem of the body (Prof. Dr. H. NOETZEL) showed thrombosis of the femoral vein with a massive pulmonary embolism on the right side; considerable arteriosclerosis of the aorta; coronary sclerosis without stenosis; and slight pyelitis.

Anatomical findings in the brain. Moderate cortical atrophy near the frontal pole; shallow cortical defect measuring 9 × 8 mm in the first frontal gyrus on the left side due to the second operation; on the right frontal gyrus a cortical defect the size of a pea from the first operation; severe hydrocephalic dilatation of the ventricular system; and a coagulated lesion in the right pallidum with shrinking of the surrounding tissue (Fig. 55a). The center of this area lay 3.5 mm below the baseline, i.e., only 0.5 mm ventral to the radiologic target point. The scarred coagulated lesion started in the plane of the interventricular foramen and stopped about 12 mm further caudally. This distance corresponded to the shortening of the advancing range of the electrode in that the angle of entrance measured about 30° to the baseline. The exterior part of the pallidum had been destroyed, in some parts reaching the medial edge of the putamen. The internal pallidum was considerably shrunken but less so than the external pallidum (Fig. 55a).

Because of fiber degeneration in the right cerebral peduncle, the right substantia nigra was much

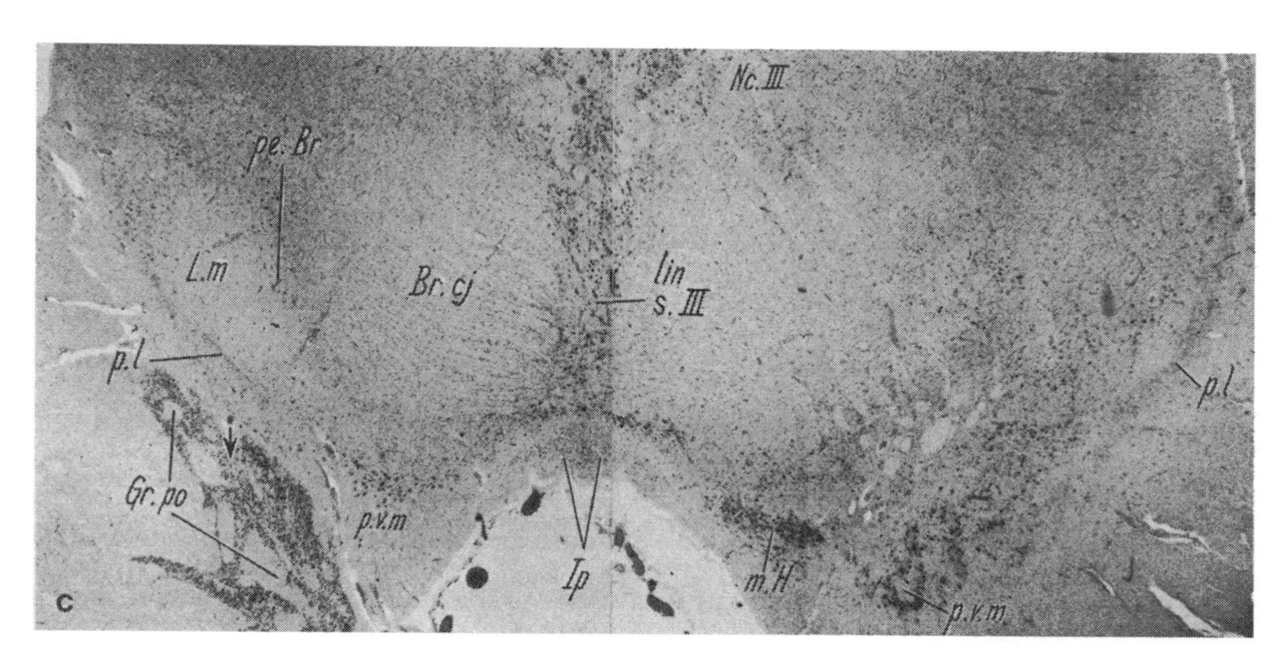

Fig. 55. (c) Adjacent transverse section of midbrain of case No. 5, reversed (!). Reduced size of right substantia nigra is even more impressive here. Compensatory glia growth is so thin that it is difficult to see, but it is present in *pl* group of substantia nigra. *p.v.m* group of substantia nigra is better preserved on both sides, mainly right. The *pl* group is more sharply reduced and more completely compensated by glia growth on right side. Notice that pontine gray matter to right of arrow appears darker than lateral nondegenerated part, because of degeneration of cerebral peduncle fibers medial to arrow. Peribrachial (*pe.Br*) and linear suboculomotorius nucleus (*lin s.III*) are preserved, as is the most medial pervocellular group m.H. Nissl stain. ×6.5

more shrunken than the left (Fig. 55b). The nerve cell groups of the posterior main section of the right substantia nigra lost nerve cells almost completely in the neighborhood of the peduncular degeneration, whereas the nerve cell groups of the left substantia nigra are less involved (Fig. 55c). The pattern of nerve cell loss indicates an idiopathic Parkinson syndrome.

The center of the coagulation in the left diencephalon was 11.5 mm from the wall of the third ventricle. The target point adapted to the individual's measurements was 0.9 mm in a lateral direction. Three weeks after the coagulation there was a slight shrinking of the tissue. 10 mm behind the interventricular foramen the lesion had a diameter of 6.5 mm in a horizontal and 10.5 mm in a vertical direction, but its center lay 6 mm below the baseline (Fig. 56a). This deviation in the basal direction was caused by an error of 5.0 mm in reading the vertical coordinate in the x-ray control. The sagittal extension of the coagulated area was 11 mm, which corresponds to the three coagulation points (0, +2, +5).

The lesion destroyed the peduncle at its transition into the internal capsule in a plane 10 mm behind the interventricular foramen (Fig. 56b). Also destroyed was the fasciculus H_1, rostrodorsal parts of the nucleus subthalamicus, the zona incerta, and the V.o.a was strongly shrunken not the V.o.p (Fig. 123a). The lesion invaded the internal capsule in a rostral direction and destroyed medial parts of the pallidum (Fig. 56c) and on its ventral side, the basal nucleus. Due to interruption of dentato-thalamic fibers, the V.o.p has a dense gliosis (Fig. 123b).

Clinical-anatomical correlations. The right pallidum lesion affected the external more than the internal part. The autopsy showed that the results of stimulation in the pallidum at the first operation – the slight increase of the tremor in the left arm and leg and the blocking of the very strong tremor of the left hand by stimulating at a rate of 50 cps – were due to the predominant excitation of the pallidum externum. After the pallidum had been inactivated, there was increasing tiredness, with frequent yawning, slight disorientation, and echolalia. These signs were related to the extensive coagulation of the pallidum, particularly at its rostral part. The extensive one-sided coagulation of the pallidum together with the proved insufficiency of the cerebral blood vessels caused acute narrowing of consciousness. This showed itself during the

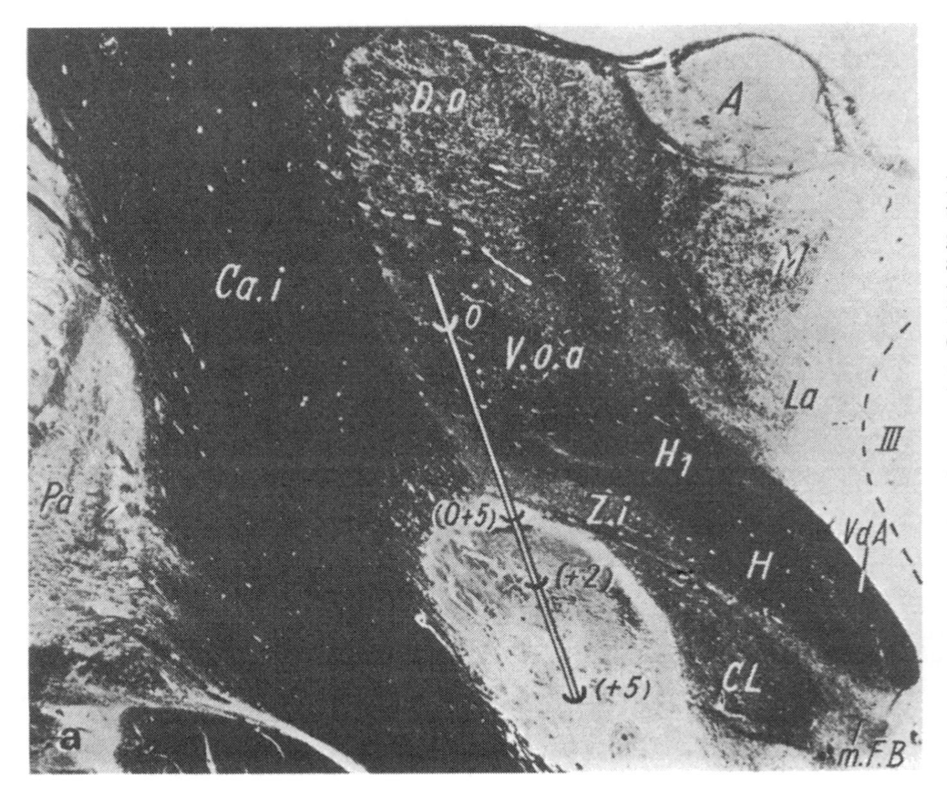

Fig. 56. (a) Transverse section through left basal ganglia of case No. 5, with coagulated lesion in cerebral peduncle and in subthalamic nucleus, lateral part (*CL*). Displacement of lesion is due to reading error that caused intended target point V.o.a (marked *0*) to be set 5 mm too low: (0+5). Two coagulations were placed ventral to this in positions +2 and +5 corresponding to cerebral peduncle and CL.
Z.i zona incerta; *CL* subthalamic nucleus; *H* Forel's field H; *La* lamella medialis; *VdA* mamillothalamic tract; *m.F.B* medial forebrain bundle.
(After HASSLER *et al.*, 1969)

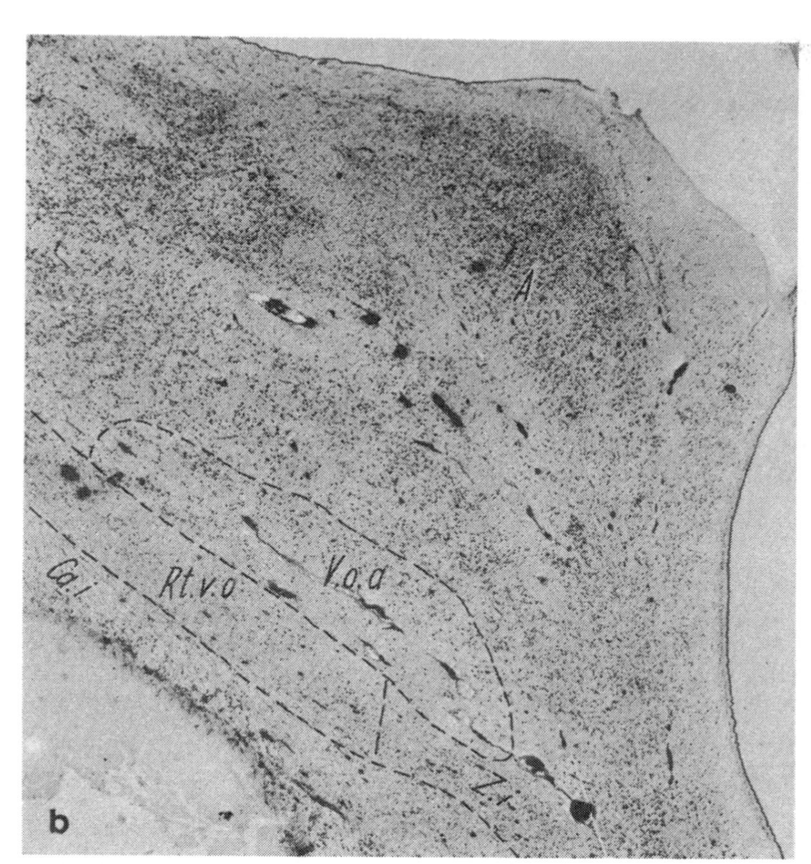

Fig. 56. (b) Adjacent section more rostral to Figure 56a shows cell reduction in left V.o.a nucleus, which is greatly reduced. Slight reduction of nerve cells with glia proliferation in Rt.v.o and more severe cell loss in zona incerta (*Z.i*). Medio-ventral to the anterior principal nucleus (*A*) the nerve cells of the medial part of intralaminar nucleus are shrunken or lost and replaced by glia cells. This is probably due to destruction of left pallidum. Nissl stain. ×9

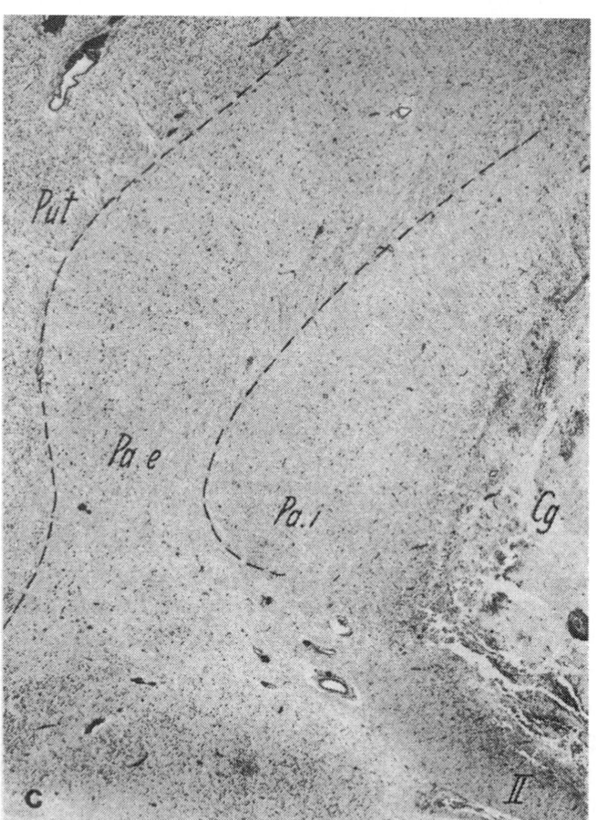

Fig. 56. (c) Both segments of left pallidum and surrounding putamen in case No. 5 lateral to coagulation (*Cg*). Nerve cell density in external segment of pallidum (*Pa.e*) is almost normal. Internal segment (*Pa.i*) has lost most of its nerve cells. Remaining tissue is shrunken and replaced by glia. This picture gives evidence that efferent connections of pallidum internum are mainly directed medially, while efferent connections of pallidum externum are either directed to pallidum internum or leave in base of pallidum in caudal direction. Nissl stain. ×9.5

first 6 days after the operation with the experience of his double, although there had been no pulmonary complications. In the further course following the first operation the improvement of the patient's mood was quite remarkable and allowed him to start work again, at least partially, after a long interval. The pains that he had been complaining about before the operation were absent after it. The rigidity of the left arm was abolished and much reduced in the left leg, due to the right-sided inactivation of the pallidum; this continued until death. The tremor in the left limbs was reduced but not abolished.

The much less effective therapeutic result of the second operation was due to the incorrect localization of the vertical coordinate. A check of the records made during the operation and the x-ray pictures showed that a mistake of 5 mm was made in reading of the vertical coordinate in the x-ray picture: 61 instead of 66 mm was found in the protocol. It was not possible to uncover this error during the operation because the stimulation synchronized the tremor, dilated the pupils, and arrested respiration as expected for V.o.a.

Because of this deviation the coagulation in +5 did not destroy but deafferented the V.o.p, and destroyed only the ventral edge of the V.o.a. The coagulation affected the rostral peduncle at its transition to the internal capsule, the zona incerta, the fasciculus H_1, parts of H_2, and rostrodorsal parts of the Subthalamicus (Fig. 123a, b). The length of the coagulation cylinder – 11 mm – corresponds to the distance covered by moving the electrode for the three coagulations (positions +5, +2, 0) including the 4 mm of the electrode that were not insulated, and the 2 mm in front of the point of the electrode.

The stimuli applied 5 mm ventrally to the target point affected the internal capsule, the zona incerta, and the fasciculus H_1. It was difficult to decide which individual responses could be related to which structures. It is possible that the increase of individual phases of tremor during stimulation at low frequencies was related, to excitation of dentato-thalamic fibers in the base of V.im. It is improbable that the individual jerks of the fingers were due to stimulation of the pyramidal tract, since on the one hand the fibers of the pyramidal tract go through the internal capsule about 10–14 mm further caudally, and on the other hand inactivating the rostral parts of the internal capsule did not cause a paresis or reduction of movement in the contralateral hand. Stimuli at the higher rate of 50 cps were followed by vegetative signs, like dilatation of the pupils when the stimuli were fairly weak, a tightening feeling in the chest, respiratory arrest and a feeling of pressure in the abdomen. Based on only this one case, it was not possible to determine which of the stimulated structures was responsible for the vegetative signs. However, since feeling of tightening and widening of the pupils are observed frequently or even regularly when the pallidum internum or its efferent fibers in H_1 are stimulated, the signs in this case were probably also related to this structure.

Coagulation in position +5 was accompanied by a transient confusion that made it impossible for the patient to count and calculate correctly. While it is remarkable that this nonperformance was produced only by coagulation in this position, the confusion can probably be explained by the particular location of coagulations in both pallida and cerebral peduncle which results in interference with articulation.

The incomplete effect on the tremor corresponds to the fact that V.o.p was only deafferented, while the positive effect on rigidity corresponded to inactivation at the basis of the V.o.a and in the fasciculi H_1 and H_2. There were no severe special signs due to the lesion of the internal capsule located about 10 mm behind the interventricular foramen. This lesion involved fibers that originate in the more rostral parts of the cerebrum and run in the horizontal plane of the internal capsule near its genu. This position of the lesion explains why the lesion in the internal capsule did not clinically show an involvement of the pyramidal tract; except for the left facial paresis (see Fig. 53), which is due to the involvement of the upper motor neurons of the facial nerve in the right internal capsule (see Fig. 55a) and its secondary degeneration in the cerebral peduncle (Fig. 123a) up to its entrance into the pontine gray matter (see Fig. 55b, c).

The right substantia nigra is much more shrunken and its nerve cell loss more complete than on the left side, because of the additional interruption of corticonigral fibers in the internal capsule above pallidum externum and of strionigral fibers during their passage through the pallidum externum. The interruption of parts of the left cerebral peduncle (2nd operation) occurred less than 4 weeks before death, so that the secondary degeneration in the left substantia nigra (see Fig. 55b, c) had less time to operate on the nigra nerve cells and fiber structure. During the second operation the H_2 bundle in and above the cerebral peduncle was interrupted, as was the transitional zone of the internal capsule up to the optic tract (II) and up to the inner segment of the pallidum (Pa.i. in Fig. 56c). The consequence of the H_2 destruction is the severe cell loss in the pallidum internum (Fig. 56c), whereas the nerve cell stock of pallidum externum (Pa.e) is almost maintained. This finding supports the idea of the projection of pallidum externum through the lower part of "comb system of the foot" to midbrain structures and the projection of the pallidum internum through the bundle H_2 and H_1 to the thalamic nucleus V.o.a. This pallidothalamic fiber system has been interrupted and thereby the pallidum internum nerve cells retrogradely degenerated, whereas the V.o.a nucleus is shrunken (see Fig. 56b) because of the complete loss of its main afferents from pallidum internum.

The observation during the first postoperative days of frequent striking with the right hand corresponded to the damage of the rostrolateral part of the left nucleus subthalamicus, which had been shown anatomically (see Fig. 56a).

The postoperative confusion disappeared relatively quickly, and the patient was able to walk about with little tremor from the third day on, despite a small aspiration. On day 7 after the operation there was an acute onset of excitement, disorientation, coughing and vomiting, and severe rise of blood pressure and rate of respiration. In retrospect one can say that this was certainly related to the massive-left-sided basal ganglia edema (Fig. 56c). Recurrent pulmonary embolisms together with the extension of the pneumonia did probably lead to the patient's death.

Case 6. P.T., aged 50; Parkinson syndrome, with hamartoma of the hypothalamus; 1st to 3rd operations, chemopallidectomy left (another hospital); 4th stereotaxic operation, right V.o.a (no. 758); death occurred 82 days after operation from purpura cerebri (allergy to penicillin).

History. At age 44, tremor and rigidity in the right arm and leg appeared without preceding illness. At age 49 the left half of the body became involved. Since the onset of illness, the patient was only partly fit to work. Nine months before the second stereotaxic operation he needed assistance.

In another hospital, left-sided "pallidectomies" were performed by injection of alcohol and by an electrolytic lesion on January 24, January 30, and February 5, 1959. These were followed by a reduction of rigidity and tremor on the right side. A fourth operation was planned in order to increase the lesions since the improvement lasted for only one month. Three months after these operations the patient developed nocturnal episodes of trembling and impairment of the general condition lasting 10 min. The doctor reported that these were regarded as epileptic equivalents and treated them as such.

Clinical findings. Severe rigidity of both arms and legs; tremor on the left side was of medium severity, on the right it was severe; writing, speaking, and walking were severely disturbed. The patient perspired profusely. Neurologic findings included paresis of the oral branch of the facial nerve, weakness of the right 6th cranial nerve, and exaggerated knee jerk on the right side. The total protein in the C.S.F. was increased to 3.9 Kafka units without increase of the albumin/globulin ratio. Of the proteins, the albumins were particularly increased. The hydrochloric acid–collargol curve was of a transudative type. In the EEG the α rhythm was slow (see Fig. 112b).

The examining physicians (Medizinische Poliklinik, University of Freiburg) found moderate pulmonary emphysema and general osteoporosis of the vertebral column, otherwise no abnormalities. A helium encephalogram showed a hydrocephalus permagnus of the inner and outer spaces with particularly large basal cisterns (Fig. 57a, b) without dislocation or intrusion. Filling with air did not lead to any complications.

A *stereotaxic operation* was performed September 10, 1959, in the right V.o.a. Stimulation 1 mm ventrally to the target: single stimuli produced synchronous jerks of the left hand. Stimuli at rates of 4–8 cps caused a slow shaking of the left arm. Stimuli at higher frequencies (25–50 cps) led to tetanic contraction of the left arm that caused the active alternate movements to stop. When stimuli at a rate of 50 cps were applied, the left arm was rapidly raised to the base ring of the stereotaxic instrument. There was also facial contracture on the left side and slow opening of the eyes.

A total of four coagulations were carried out with the Wyss coagulator from -2 to $+4$. After the second coagulation the rigidity was almost completely abolished in the left limbs, and the tremor was much reduced. After the last (-2) coagulation he suddenly stopped speaking, he could not be made to open his eyes, and he actively resisted movement. He did not respond when

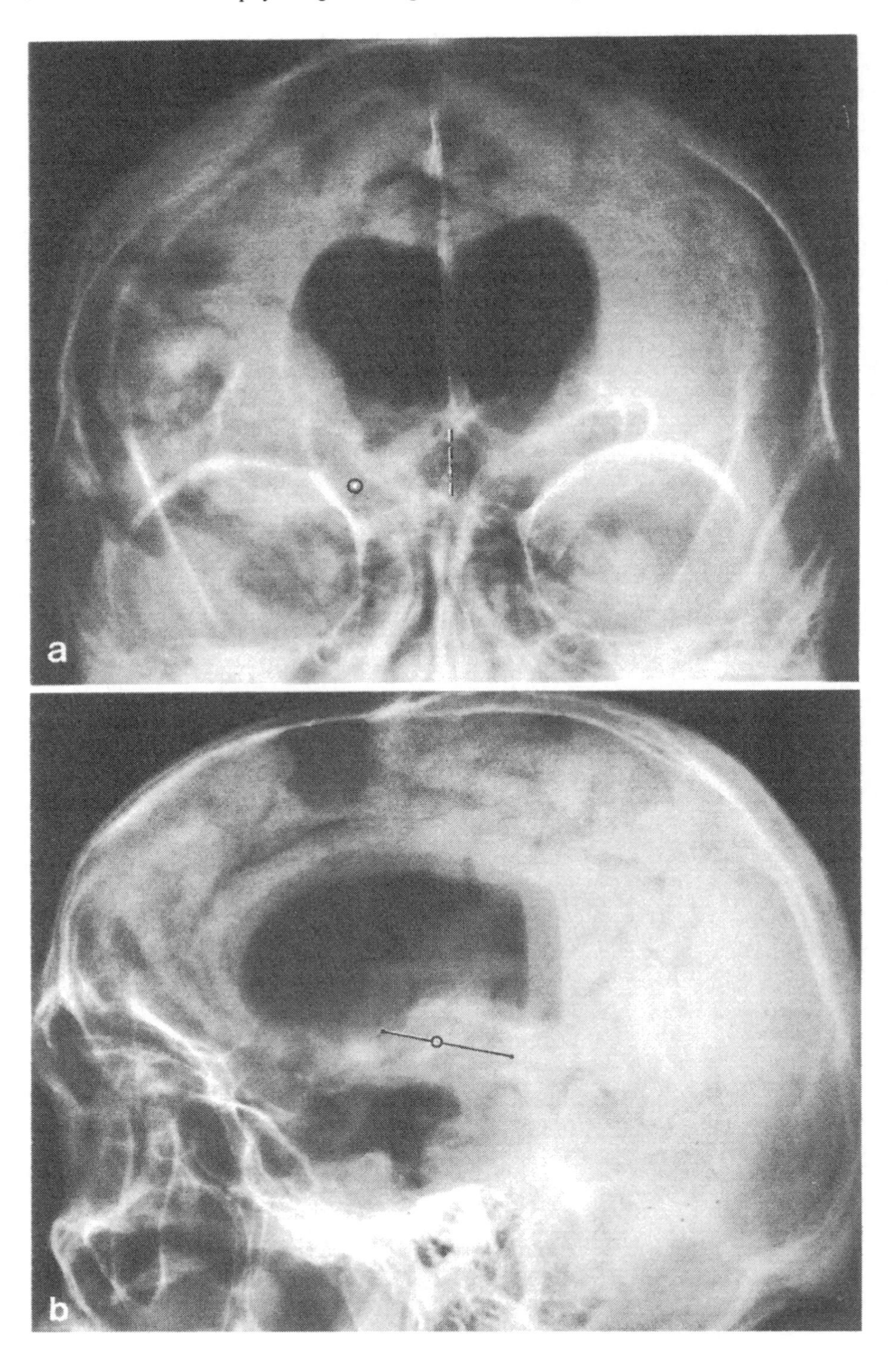

Fig. 57a and b. Pneumoencephalogram of case No. 6.
(a) Anterior-posterior picture shows severe hydrocephalic dilatation of both lateral ventricles, particularly the thalamus contour is displaced basally. 3rd ventricle, lying in midline, is also dilated but shows less contrast. Increase of subarachnoidal air, particularly of right hemisphere. Marking of target point in right V.o.a. Right cisterna is not shown.
(b) Lateral pneumoencephalogram shows ballooning of both anterior horns and considerable increase of height of both cellae mediae, so that thalamus seems basally displaced. Basal cisterna is considerably dilated at the same time, because 3rd ventricle is slightly higher. No tumor outline can be seen in posterior hypothalamus or in basal area of cisterna. Doubtful tumor outline can be seen in shrunken anterior part of 3rd ventricle

spoken to, nor did he carry out instructions. After the stereotaxic apparatus was removed, the patient spontaneously thanked those present for the successful operation. His consciousness appeared to be clear and he had no paralysis.

Postoperative course. After the patient had been returned to the ward, he no longer answered when spoken to, he reacted negativistically to all instructions, and he kept his eyes tightly shut. There was no difference between the pupils. On the day after the operation he reacted for a short time when spoken to and seemed a little livelier, but again became stuporous, similar to a vigil coma, with open eyes (see Fig. 112a). For 30 min he experienced tonic-clonic jerks of the right arm. Because of the start of bronchopneumonia in both lower lobes, he was given antibiotics. A month after the operation, an attempt was made to stop his negativistic behavior with an electric shock, but to no avail. Six weeks after the operation, his consciousness became clearer, he was friendlier, he understood what was said to him, he reacted to it, and he spoke spontaneously. Later there was again an impairment of consciousness (see Fig. 112a, b) and of circulation. Despite continuous treatment with antibiotics, the temperature could not be held in check. Eighty-two days after the operation, the patient died. Clinically, death seemed to be due to pneumonia and circulatory failure. A post-mortem of the body was not permitted.

Anatomical findings in the brain. Severe atrophy of the pole of the frontal lobe and of the right orbital cortex; cortical necrosis extending over 20×20 mm and having a depth of 5 mm on the convexity of the left frontal lobe (chemopallidectomy). From the last stereotaxic operation there remained a cortical necrosis of 6×3 mm on the first frontal gyrus on the right side. There was severe hydrocephalic dilatation of the entire ventricular system (Fig. 58a, b), right more than left; the septum pellucidum was lengthened and thickened. In the left basal ganglia the internal capsule was narrowed and the caudatum very flattened; at the lower edge of the left putamen was a necrosis of 8×2 mm in the area of the limen insulae (Fig. 58a).

In the anterior hypothalamus there was an infiltrating hamartoma that had shifted both basal ganglia laterally (Fig. 58b, c). In the right anterior thalamus extending into the medial part of the internal capsule was a hemorrhagic lesion 11 mm in diameter (Fig. 58a).

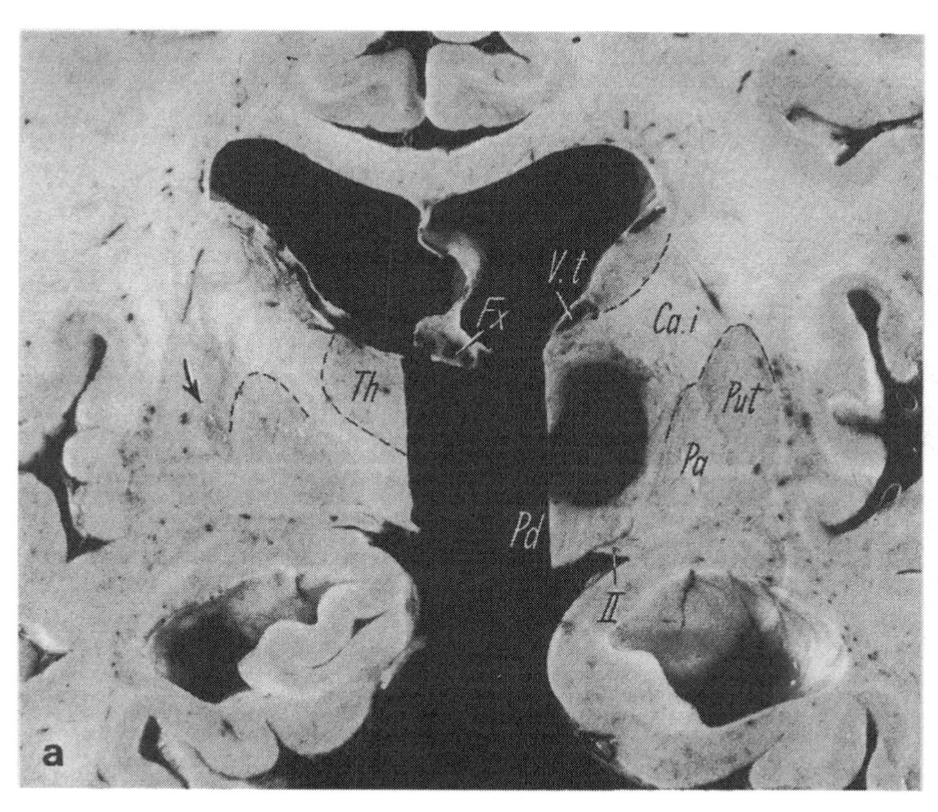

Fig. 58. (a) Unstained frontal section through anterior diencephalon of tumor case No. 6. Severe hydrocephalic dilatation of both lateral ventricles and lower horns is remarkable. Right inferior horn appears to be particularly affected. Block of diencephalon around 3rd ventricle has been removed. To right of this is 14-mm high hemorrhage that is barely organized. It extends from thalamus into internal capsule. Note confluent purpuric hemorrhages in transversely cut fornices (*Fx*) due to hypersensitivity to penicillin. Remainder of a former alcohol injection (*arrow*)

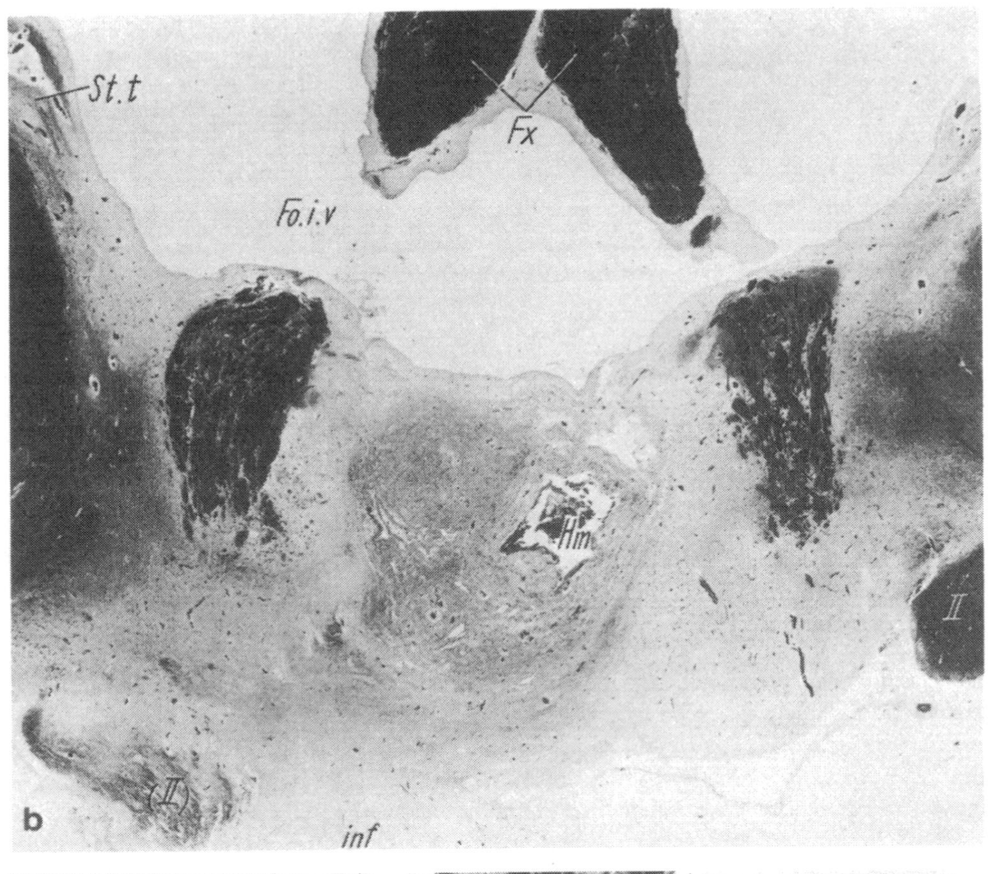

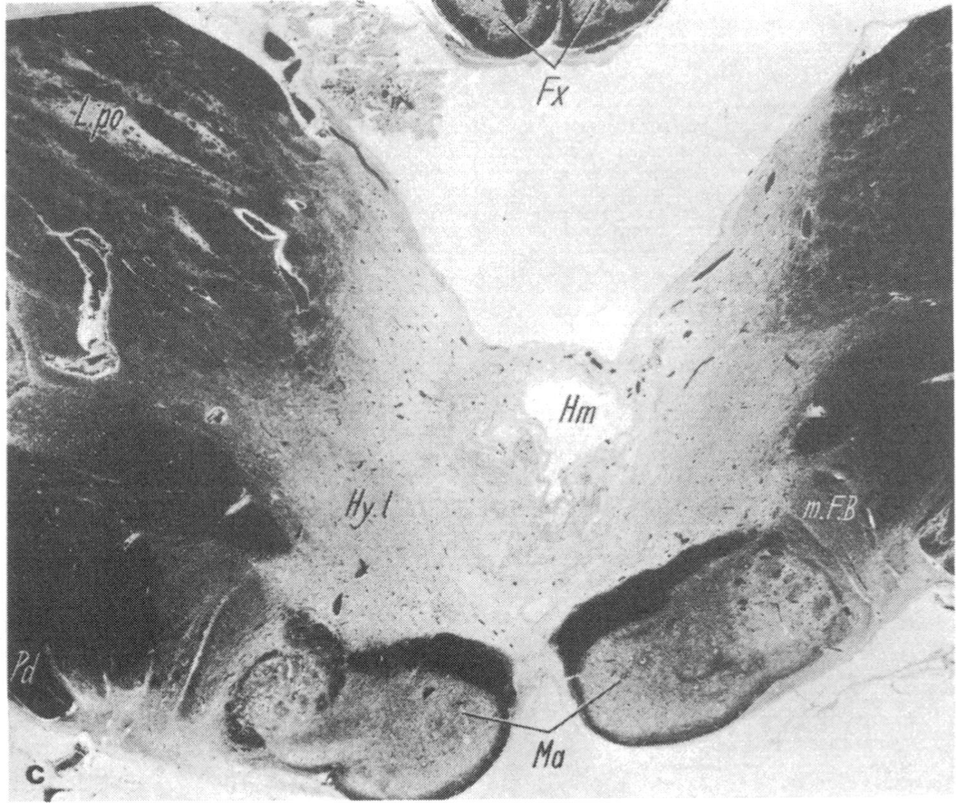

Fig. 58b and c.
Figure caption see
opposite page

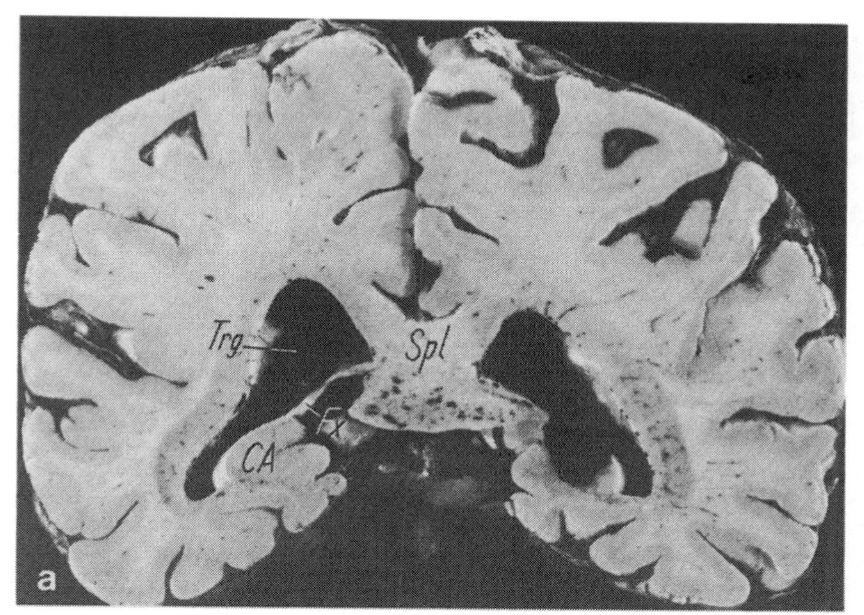

Fig. 59. (a) Frontal section of case No. 6 through unstained brain in plane of splenium of corpus callosum. Purpura cerebri. Ventral part of corpus callosum (*Spl*) and neighboring fornix commissure show numerous small fleabitelike hemorrhages. In part of corpus callosum lying above it there are only a few small purpura lesions but marked vascular dilatation.
(b) Longitudinal section of fornix of same case shows purpura lesions with larger central areas of necrosis, which always have a stalk of a capillary. They are surrounded only by small amount of blood. Areas of stasis and purpuric lesions are found in remarkable predilection in other parts of cornu Ammonis and fornix system, i.e., in molecular stratum and alveus. Goldner stain. ×35

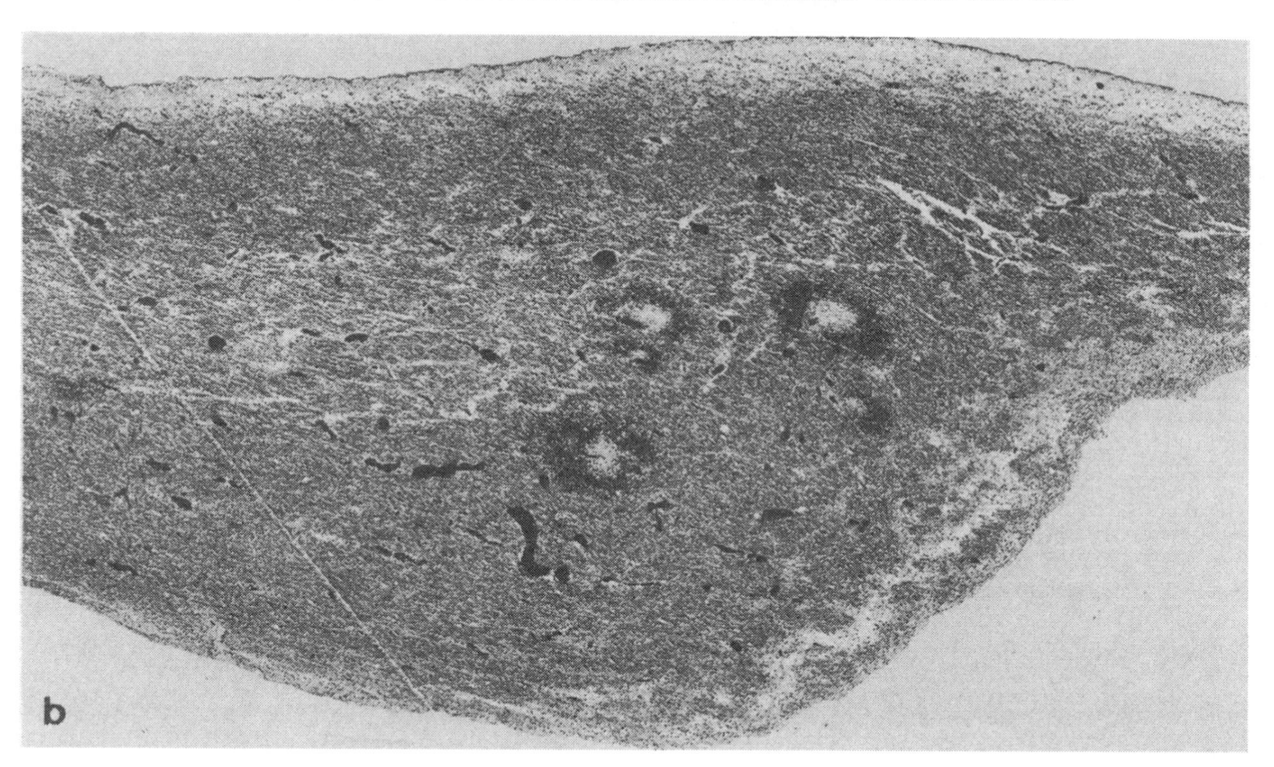

◀ Fig. 58. (b) Transverse section through 3rd ventricle at level of interventricular foramen in case No. 6. 3rd ventricle is filled by a tumor in midline that extends a little further toward left side. Tumor tissue is very vascular and contains blood-filled cyst. (*Hm*). Tumor has infiltrated hypothalamus basally on left side. Left optic tract (*II*) is badly damaged by degeneration of myelin sheaths, infiltrated by tumor and frayed, while right optic tract [*II*] is unchanged.
Fx fornix; pars tecta as well as pars libera. *Fo.i.v*=interventricular foramen. Fiber stain. ×4.

(c) Section through diencephalon of case No. 6 in plane of mamillary bodies (*Ma*). Lower part of 3rd ventricle is filled by cystic part of tumor, which is smaller here than further rostrally. There is no sharp demarcation against hypothalamus, which contains little myelin. Infiltration by spongioblastoma tissue caused an unusual deformity of lateral parts of both mamillary bodies. Lateral structures of diencephalon as well as anterior pole of thalamus and cerebral peduncle (*Pd*) seem unchanged. Corpora fornicis (*Fx*) over 3rd ventricle are filled by purpura lesions (black) and to some extent show partial demyelination. Fiber stain. ×4

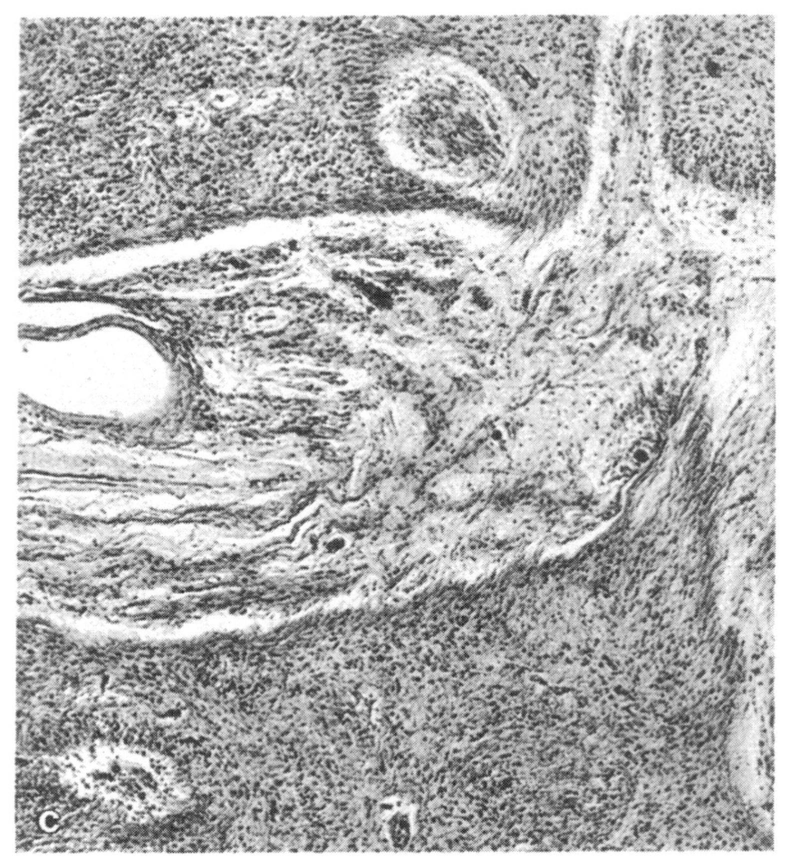

Fig. 59. (c) Histologic section through hamartoma of 3rd ventricle in case No. 6. Light-colored vessels are surrounded by numerous spaces in which tumor tissue is ordered in green tissue strands. Typical bipolar spongioblasts with glia fiber extensions intrude into perivascular spaces; strands of tumor tissue are also found in perivascular walls. No necrosis. In view of typical distribution as in subependymal glia, this is spongioblastoma. Goldner stain. ×50

Fig. 60a. Figure caption see opposite page

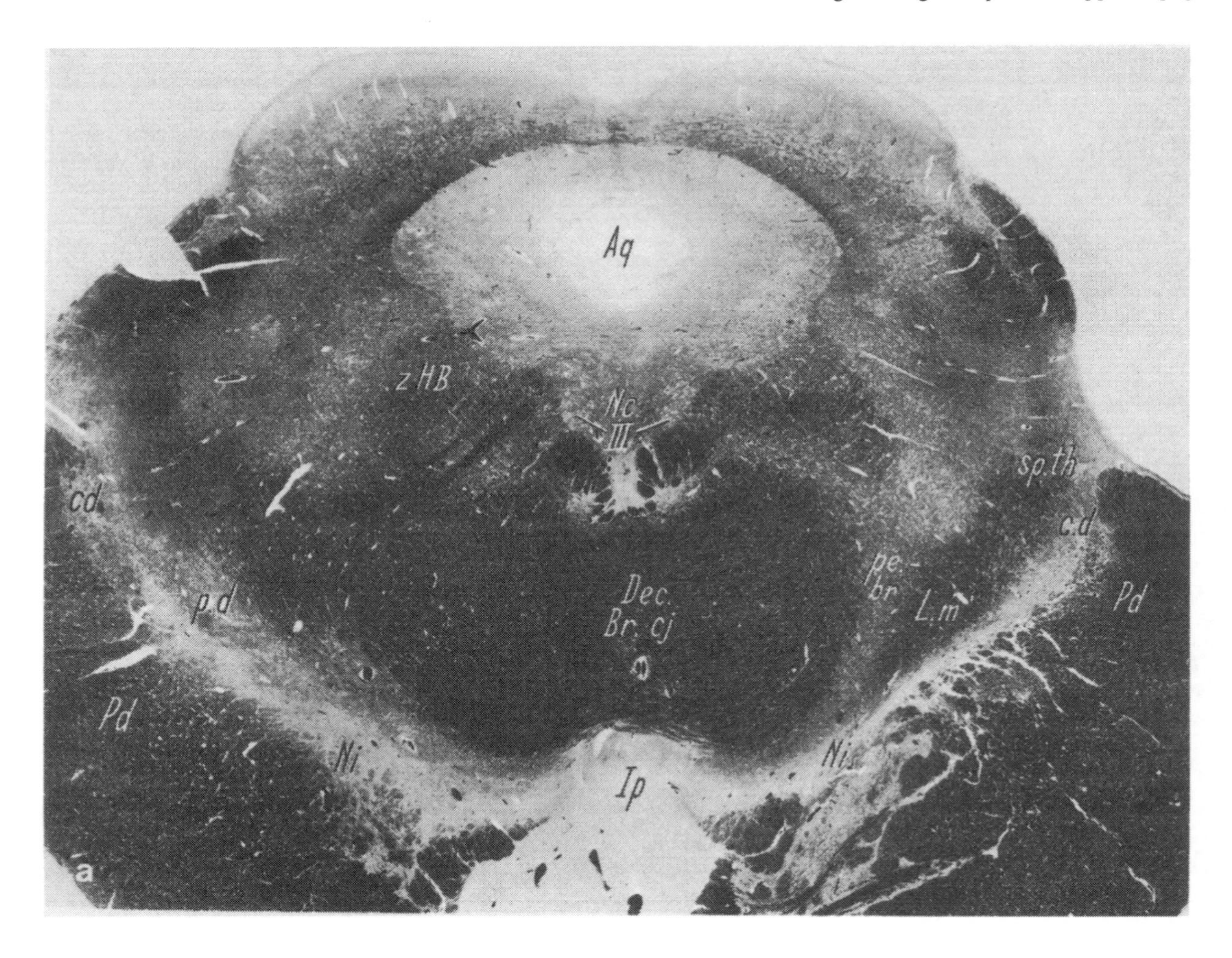

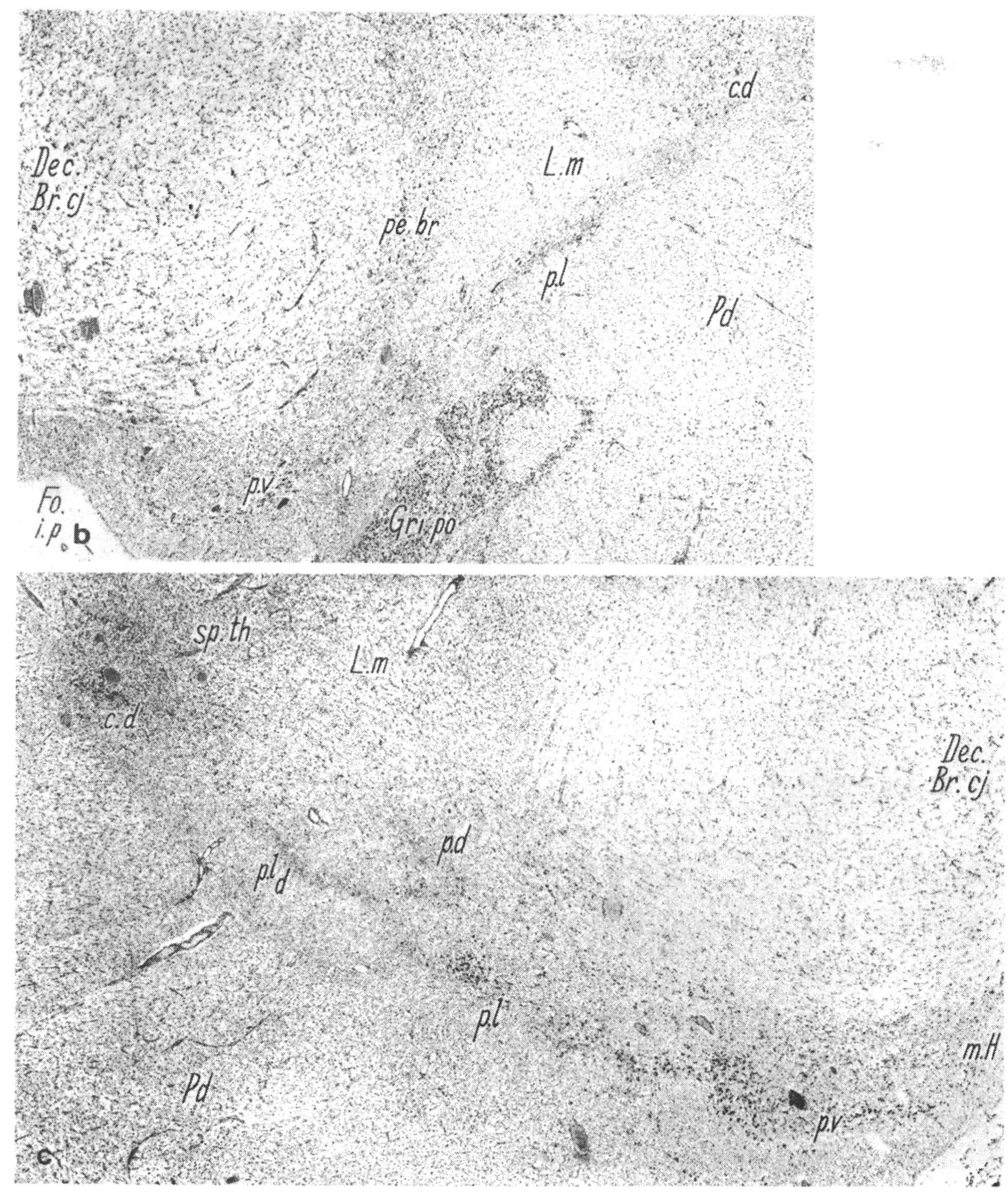

Fig. 60. (a) Cross section through anterior midbrain of case No. 6, who had spongioblastoma in 3rd ventricle. Bilaterally, area of substantia nigra is greatly shrunken and some fibers and fiber bundles are degenerated, right more than left. Pressure of tumor in 3rd ventricle and thalamic hemorrhage is recognizable by strip of demyelination in right cerebral peduncle produced counter pressure of free edge of tentorium. In right cerebral peduncle (*Pd*) small purpuralike hemorrhages can be noticed with minimal necrotic center. Fiber strain after Woelcke.
(b) Right substantia nigra is strongly shrunken in all posterior nerve cell groups from medioventral (*p.v*) to lateral (*p.l*), with severe nerve cell losses replaced by glia proliferation. Caudodorsal group (*cd*) is less shrunken. Preserved stock of nerve cells in peribrachial nucleus (*pe.br*) despite gliosis is present. Nissl stain. ×11. *Dec.Br.cj*=crossing of brachia conjunctiva
(c) In left substantia nigra medial cell groups (*p.v.; m.H*) are partially preserved; involvement increases from middle to lateral and dorsal groups (*p.ld; p.d*) and to caudodorsal group (*cd*) with inflammatory glia proliferation and severe shrinkage. Pattern of nerve cell loss does not coincide with the idiopathic type because it is due to increased intraventricular pressure, particularly in tentorial slit. ×11

The track of the last operation, 11 mm from the wall of the third ventricle had relieved rigidity and tremor; it corresponded to the aimed-at structure. Remaining parts of right diencephalon had been pushed dorsally or ventrally by the hemorrhage. The bleeding destroyed particularly V.o.a, V.o.p, Z.o, H_1, H_2, subthalamic Nc, and Zona incerta. Next to it was an 8×9 mm area of softening (Fig. 58a).

Punctiform hemorrhages were found in the walls of both lateral ventricles, and in those parts of the thalamus that were near the ventricles, even at a considerable distance from the coagulated area. They were particularly numerous below corpus callosum in the fornix commissure (Fig. 59a). Histologically, there was severe stasis with spheroid hemorrhages around the small vessels (Fig. 59b). The walls of the arterioles were swollen. The hamartoma had the structure of a spongioblastoma (Fig. 59c). Because of pressure in the tentorium slit, the right substantia nigra was more severely damaged than the left and was reduced in size (Fig. 60a). There was very extensive, and in some groups almost complete, nerve cell loss in the right substantia nigra as well as in the left, whereas the medial cell groups of the left substantia nigra were partially preserved; left dorsolateral cell groups were destroyed with fresh gliosis. The pattern of nerve cell loss in the substantia nigra does not correspond to idiopathic Parkinsonism (Fig. 60b, c).

Clinical-anatomical correlations. The severe hydrocephalus internus was due to temporary occlusion of the 3rd ventricle by a hamartoma of the hypothalamus with transudative C.S.F. alteration. The dilatation of the basal cisterns and the displacement of the hypothalamus in a dorsal direction, which was discovered only at the post-mortem, had been caused by the hamartoma at the base of the hypothalamus (Fig. 58b). The cisterns were so dilated around the tumor that they hid its outline in the encephalogram (Fig. 57a, b). The necrosis caused by the three former chemopallidectomies was situated medially and in the base of the left putamen and extended slightly into the pallidum externum (Fig. 58a). Because these lesions had inactivated only small parts of putamen and pallidum, their effect on rigidity could not be permanent. In fact, the symptoms were reduced only during the first 4 weeks.

During the fourth operation the point of electrode reached the target in the right thalamus. Although the neutral pole of the electrode reached the area of the internal capsule about 4 mm behind the interventricular foramen, the motor effects of bipolar stimuli of low frequency were obviously due to the stimulation of the V.o.a structures. It is probable that the unusual result of 25–50 cps the stimulation – upward jerking of the contralateral arm, grasping at the stereotaxic instrument, and twitchings in the facial region – was a complete syndrome of the supplementary area. A better defined localization in the oral-ventral nuclei for these results of stimulation could not be made in view of the subsequent hemorrhages in the coagulated area (Fig. 58a) which were due to the abnormal pressure around the 3rd ventricle.

The use of the Wyss coagulator led to the almost complete abolishment of rigidity of the left limbs, but the tremor did not completely disappear. The inability to abolish the tremor completely was in this case correlated with the incomplete destruction of the V.o.p. An exceptional state lasting a few minutes, i.e., negativistic behavior, arrest of speech, and compulsive shutting of the eyes, occurred after the last coagulation by hemorrhagic pressure in the thalamus (position -2). After removal of the stereotaxic instrument, the patient again behaved in a manner appropriate to the situation and expressed thanks for the successful operation. The next day, normal behavior again alternated with negativism and shutting of the eyes, which increased when an attempt was made to open the eyes passively. The negativistic episodes alternated with vigil coma (Fig. 112a, b), also in an EEG, and were unaffected by an electroshock applied after 4 weeks. The unusual negativism and vigil coma must be regarded as being due to the hemorrhage in the rostral thalamus in combination with the severe damage to both hypothalami caused by a large hamartoma and by the hydrocephalic dilatation of the third ventricle with intermittent increase of pressure. As had already happened after the chemopallidectomy, the transient increase of pressure in the third ventricle caused one focal fit with tonic-clonic convulsions of the right arm on the day after the operation. The central focus must, therefore, have been localized in the left hemisphere and not in the right, where the coagulations of the last operation and the hemorrhage had occurred. The focal attacks are interpreted as being due to herniation at the tentorium affecting the left peduncle. The pressures at the edge of the tentorium also led to the abasia, agraphia and *anarthria* and a weakness of the right 6th cranial nerve.

Among all of the cases undergoing post-mortem examinations, a similar hemorrhage into the thalamus occurred only in case 12. The following fact is remarkable from the point of view of the technique of coagulation: During the first coagulation with the Wyss coagulator in position +2, the current dropped after 7 sec from 120 to 60 mA. This drop was indicative of a sharp rise of resistance in the tissues, which caused an undesirable rise in temperature, followed by involuntary resistance negativistic compulsive closure of the eyes. Because the current did not drop again, it can be assumed that the blood vessel damage was caused by the first coagulation, the result being a hemorrhage. This hemorrhage could have been avoided by checking the temperature by a thermocouple during the operation, which we have consistently done since 1961.

The death 82 days after the operation was caused by a combination of noxae. The function of both hypothalami had been severely damaged by the hamartoma as well as by the hydrocephalus. The diencephalic damage was increased by the hemorrhage in the right thalamus. The day after the operation a bronchopneumonia developed that worsened because of the persistent impaired consciousness; treatment with repeated large doses of antibiotics was necessary. At the post-mortem, purpura cerebri (Fig. 59a, b) was found in the corpus callosum and in the fornix as well as around ventricles. Such hemorrhagic reactions of the cerebral tissue occurred as a result of damage by toxins, by avitaminosis, and by allergy to antibiotics. Though all the factors mentioned were contributory, special importance must be attached to the multiple cerebral lesions caused by the fundamental cerebral abnormalities—hamartoma of the hypothalamus, Parkinson's syndrome, and thalamic hemorrhage. In view of these lesions, the intensive treatment by drugs and physiotherapy remained unsuccessful, and the patient died with a clinical picture of circulatory failure after recurrent bronchopneumonia.

The severe Parkinson syndrome was caused by the chronic intermittent pressure of the hamartoma in the third ventricle exerted on the cerebral peduncles against the edges of tentorium. The loss of nigra cells occurred without formation of tangles or intracytoplasmic inclusions, more complete on the left side where the hamartoma originated from (Fig. 60b, c). This corresponded to the more severe tremor of the right arm, with complete nerve cell destruction in the dorsolateral cell groups responsible for the extreme rigidity of both legs, so that the patient was unable to walk. Only in the medial cell groups of the left nigra were some nerve cells spared (Fig. 60c); their functional equivalent was the patient's ability to speak and use his mouth before the last operation. The intensive glia proliferation in the left caudodorsal cell group (cd) is interpreted as an inflammatory reaction to the left-sided chemopallidectomies, involving the medial putamen, performed elsewhere.

Case 7. C.Z., aged 57; paralysis agitans; first stereotaxic operation in left V.o.a/p; 2nd, in right pallidum internum (no. 188); 3rd, left pallidum internum (no. 292); amentia occurred after hemorrhage into the basal ganglia; death came 45 days after last stereotaxic operation, with purulent bronchopneumonia following a fracture of the femoral neck.

History. The first signs of Parkinsonism occurred at age 48 in the left shoulder and elbow. Moving the arms became more difficult. A year later the pains were mistakenly regarded as being rheumatic and the patient was treated with massage and histamine iontophoresis. Six months later the impairment of movement involved the left leg. After four years the right half of the body was badly affected; the patient trembled and could no longer write legibly. Despite treatment with anti-Parkinson drugs, his condition deteriorated so much that he strongly wished to have an operation in order to remain able to work as an academic teacher.

Clinical findings. Three and one-half years before his death, he was admitted to another hospital. There he exhibited a masklike face with rare blinking; quiet monotonous speech that was difficult to understand; rigidity of the trunk, neck, and proximal joints; typical posture; and coarse tremor at rest that was more marked right than left and that increased to shaking when he was excited. Synergic movements were lacking when he walked. He experienced difficulty in moving, propulsion, retropulsion, and in similar sideward movements. His reflexes and the C.S.F. were not abnormal.

The *first stereotaxic operation* was performed on February 24, 1954, in the left V.o.a/p. Single stimuli with thyratron discharges resulted in brief interruptions of the tremor of the right hand. Stimuli at the rate of 8 cps reduced the tremor, sometimes stopping it altogether. Stimuli at higher frequencies caused an upward hitting of the arm, a looking to the right, and the uttering of incom-

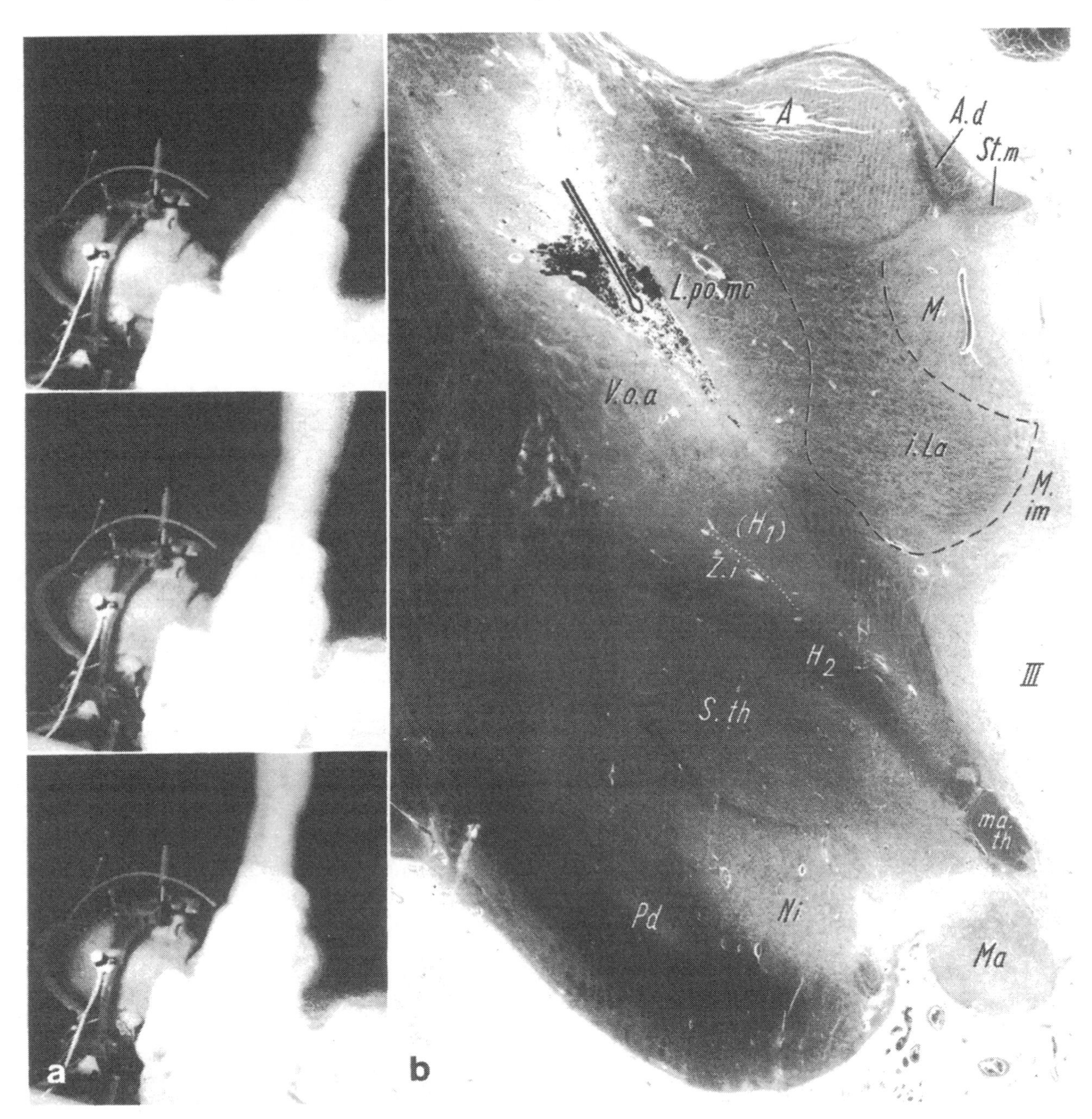

Fig. 61. (a) Individual frames from film strip illustrating stimulation effects in Parkinson case No. 7. During stimulation of magnocellular part of L.po nucleus, strong deviation of both eyes to right with fast uttering of unintelligible words and sudden raising of right arm results. The turning of head is prohibited because it is fixed in base ring of stereotaxic apparatus and cannot be moved to any side. This corresponds to stimulation effect of supplementary motor area first described by PENFIELD and WELSH (1951). (b) Position of stimulating electrode in case No. 7 between V.o.a and magnocellular part of lateropolar nucleus (*L.po.mc*), near medial lamella (*i.La*). Stimulation spot is included in coagulation focus, performed later. Coagulation focus in L.po.mc and in inner part of oral ventral nuclei has shrunk and become scarred by time of death 3 years after this first operation

prehensible sounds (Fig. 61). The patient himself noted this and could not suppress it when the stimulation was repeated. When asked afterwards why he had spoken, he could not give an explanation.

Coagulation was performed with the spark-gap diathermy apparatus. The coagulations had a favorable effect on the tremor. During a repeat of the stimulation in the coagulated area, a focal attack occurred on the right side with loss of consciousness. At the end of the operation the rigidity on the right half of the body was considerably reduced, the active movements had improved, and the tremor was much ameliorated without paresis or abnormal reflexes having appeared.

At a follow-up examination one year and 9 months later, the patient stated that due to the therapeutic effect he had been able to give a few lectures, had been successful in his scientific work, and had published a monograph. He could frequently walk without help; but the steps on the right side were much longer than on the left. The pulsions still interfered with his standing and walking; the rigidity was now much less on the right than on the left side, the tremor rather more marked right than left.

The *second stereotaxic operation* was performed January 25, 1956, in the right pallidum internum. Stimuli at a rate of 8 cps increased the left-sided tremor. Stimuli at the rate of 50 cps decreased the slow tremor when the arms were stretched forward.

Coagulations were performed with the spark-gap diathermy instrument in positions 0 and +2. During the last coagulation the insulation of the electrode developed a fault and the current decreased. Immediately, the patient became more somnolent and reacted less when spoken to. Paresis of the facial nerve appeared on the left side, and the left arm gradually sank lower when it was stretched forward. When asked to make a fist, he did not do so. The muscle tone was diminished in the left arm and leg without there being paresis, but the plantar reflex became extensor and the abdominal reflexes on the left side became less marked.

Postoperative course. The next day he was less spontaneous and in a bad mood. Eighteen months later, the paresis of the left arm had improved. The tremor on the right side had recurred. The tremor at rest on the left side could be suppressed when walking.

The *third stereotaxic operation* was performed August 2, 1957, in the left pallidum because of the recurrence of tremor on the right side. Stimulation had no obvious effect. Coagulations with the spark-gap diathermy apparatus were performed from positions −4 to +8. Because of the rigidity at rest, an additional mechanical severing was performed at the base of the pallidum with the string electrode at an angle of 45° toward the medial and lateral sides. Following this, consciousness was so impaired that the patient no longer opened his eyes spontaneously and hardly answered. The right arm sank downward; he yawned repeatedly.

Postoperative course. The afternoon of the day of the operation, he reacted only to painful stimuli. He was in sopor, he exhibited involuntary jerks of the right forearm, and his eyes deviated toward the right side. The next day consciousness was temporarily improved; he got up. The tremor was less but had not disappeared; it had become much slower. He had no distal paresis. Rigidity had disappeared completely. Three weeks after the last operation he sustained a fracture of the femoral neck when walking along the corridor. This fracture was surgically treated by insertion of a pin. While still in bed with the leg fixed in extension, he suggested to a visitor that they go for a walk. In addition, he did not even recognize a book he had written. Forty-five days after the last operation he died of circulatory failure.

Post-mortem examination of the body showed a fracture of the neck of the femur, bilateral bronchopneumonia, pulmonary edema, and slight arteriosclerosis of the aorta. The surgical wound at the neck of the femur was clean.

Anatomical findings in the brain included cortical atrophy in the area of the frontal and occipital poles; a small cortical defect in the left second frontal gyrus due to the puncture at the first operation; dilatation of the lateral ventricles and their anterior horns. Due to the 3rd operation a hemorrhagic lesion in the head of the left striatum, 16 × 18.5 mm, destroyed the rostral internal capsule, most of the caudatum, half of putamen and half of pallidum. In their base there are hemorrhagic fissures, remnants of mechanical lesions with the string electrode (Fig. 118). They were the starting point of the large, 12 × 11 mm, hemorrhage in both parts of pallidum (Fig. 62) which extended by a narrow stalk through the capsule to the caudatum. The other parts of the left basal ganglia, particularly the internal capsule on the caudal side of the hemorrhage, were swollen. The left thalamus was narrower due to a thin scar (6 × 0.5 mm), a remnant of the first operation, in the V.o.a running parallel to the internal capsule (Fig. 63). Because of this

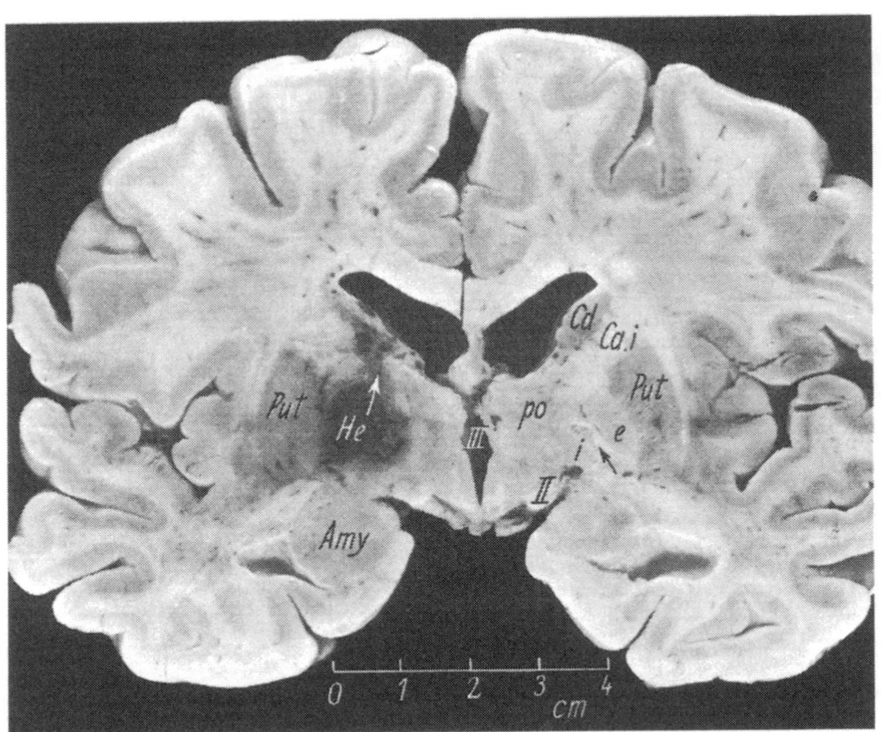

Fig. 62. Unstained transverse section through brain of case No. 7 in plane of anterior thalamus. Only slight dilatation of lateral ventricles with marked lowered outline of thalamus on right side due to coagulation of pallidum internum (*i*) about 20 months previously. Right pallidum has changed into a cyst and shows some scar formation (*black arrow*). In this plane external part of pallidum (*e*) has been accurately spared. Left basal ganglia are filled by hemorrhagic lesion (*He*). Its largest extent is found in both parts of pallidum, but it extends narrowly through internal capsule and becomes wider in caudate nucleus. Hemorrhage has sharp boundary against putamen but several extensions medially. Left basal ganglia are somewhat swollen including internal capsule

scar, the left thalamus in the area of the V.o.a was 1.5 mm narrower than the right. Near the posterior commissure the narrowing of the thalamus measured about 4 mm. The main narrowing was in the plane of the intermediate ventral nucleus (V.im). The scar had destroyed large parts of the V.o.a and parts of the neighboring nucleus reticulatus thalami. Atrophy of the dorsal nuclei caused a lowering of the upper thalamic surface (Fig. 63 right).

A cystic defect 5.2 mm high and 2 mm wide was found in the right pallidum, a consequence of the second operation (Figs. 63, 64). This lesion extended only slightly over the lateral edge of the lamella pallidi interna. The lower edge of the lesion in the pallidum was 1.4 mm further ventrally than the radiologic target point and, therefore, still in the aimed-for area. The same pallidum lesion was 3.5 mm wide and 13 mm high in the plane 9 mm behind the interventricular foramen; it extended from the lateral surface of the optical tract through both parts of the pallidum (Fig. 64), and spared only the lateral 3 mm of the pallidum internum. In the plane of the V.o.a the lesion extended into the internal capsule, which was completely softened above the cyst (Fig. 64).

Clinical-anatomical correlations. The aim of the first operation was to inactivate the left V.o.a. The autopsy showed that this aim had been completely fulfilled: most of the tissue of the V.o.a (Fig. 63) had been destroyed. The entrance of the fasciculus H_1 was totally included in the necrosis. The point of the stimulating electrode was sited in a coagulated area that became a scar in the medial part of the oral ventral nucleus, in the V.o.i, and more rostrally (Fig. 61 b) in the medial part of the lateropolar nucleus (L.po.mc), the VA. It was excitation of these structures by the stimulating current that explained the remarkable supplementary area effect – the excited uttering of unintelligible words, the upward jerking of the contralateral arm, the turning of the eyes to the other side, and the contraction of the crossed facial muscles (Fig. 61 a). Corresponding to the scar formation in the V.o.a, the rigidity of the right limbs remained reduced.

The initial good improvement in the right-sided tremor, which 22 months later was hardly less than the left-sided tremor, corresponded to the less complete destruction of the V.o.p. In particular, it seemed as if the dentato-thalamic fibers radiating into the V.o.p had not been completely interrupted by the operation. Because the nucleus was located in the thalamus, it had not led to a cystic degeneration of the tissues, although in these coagulations temperatures exceeding 100° C were still employed.

The second operation was performed to inactivate the right pallidum internum. According to the autopsy, this objective was achieved. Almost all of the right pallidum internum, the entire capsule above, and about two-thirds of the outer pallidum (Fig. 64) had been converted into a cystic cavity. The 8-cps stimulations of the pallidum reinforced the left tremor. The inactivation led to a paresis of the right upper and to a lesser degree of the lower limb so that the patient at first could not move his fingers; the left abdominal reflexes were diminished. The right limbs were markedly hypotonic and the tremor had at first disappeared completely from these limbs. However, because of the considerable reduction of active mobility, the patient at first was dissatisfied with the result.

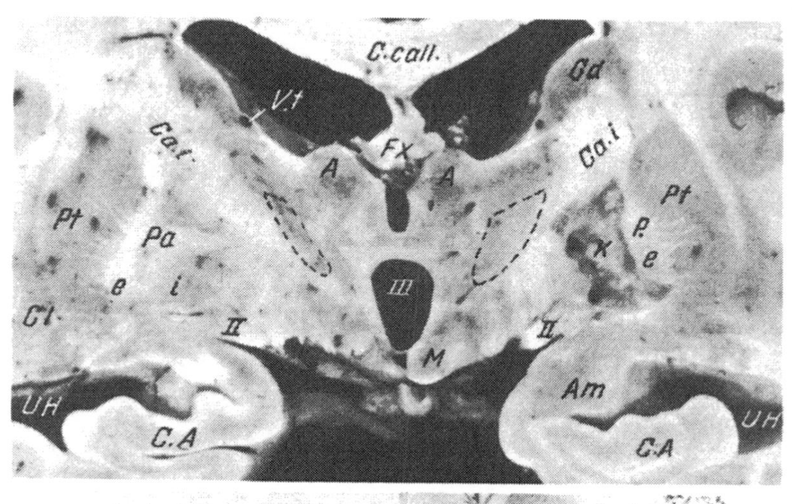

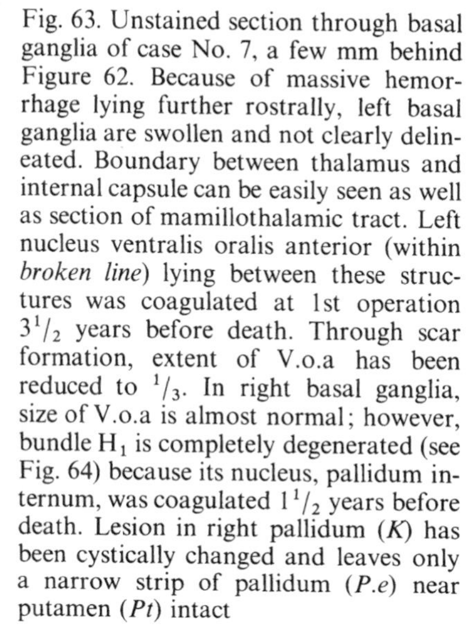

Fig. 63. Unstained section through basal ganglia of case No. 7, a few mm behind Figure 62. Because of massive hemorrhage lying further rostrally, left basal ganglia are swollen and not clearly delineated. Boundary between thalamus and internal capsule can be easily seen as well as section of mamillothalamic tract. Left nucleus ventralis oralis anterior (within *broken line*) lying between these structures was coagulated at 1st operation $3^1/_2$ years before death. Through scar formation, extent of V.o.a has been reduced to $^1/_3$. In right basal ganglia, size of V.o.a is almost normal; however, bundle H_1 is completely degenerated (see Fig. 64) because its nucleus, pallidum internum, was coagulated $1^1/_2$ years before death. Lesion in right pallidum (*K*) has been cystically changed and leaves only a narrow strip of pallidum (*P.e*) near putamen (*Pt*) intact

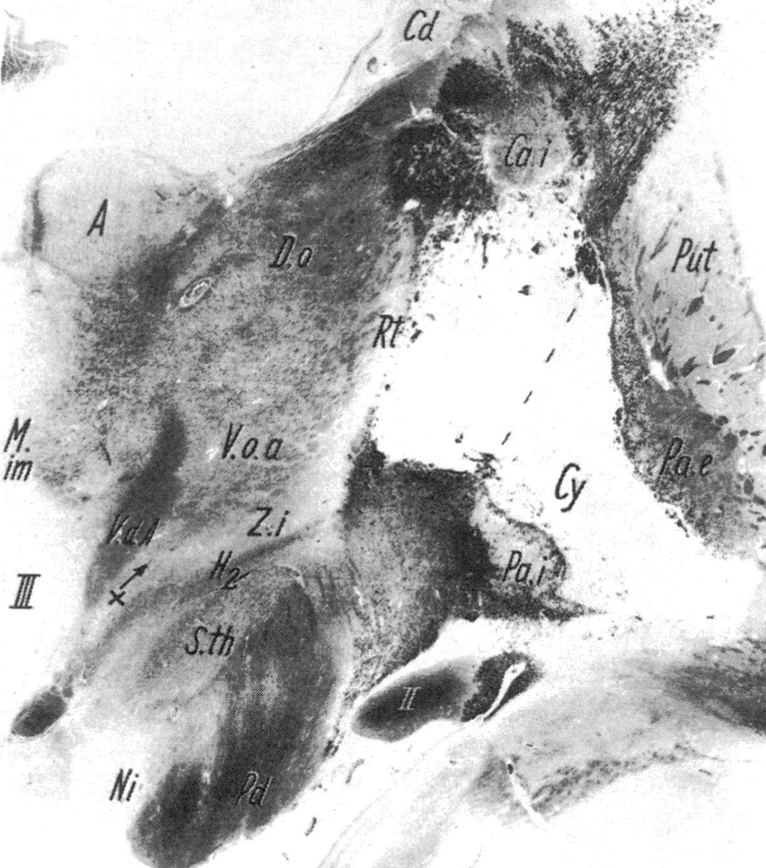

Fig. 64. Transverse section through right basal ganglia of case No. 7. Pallidum is replaced by large cyst (*Cy*) that has only spared traces of internal and external parts of pallidum (*Pa.e; Pa.i*). Dorsally and medially, cyst extends into internal capsule and destroys it up to nucleus reticulatus thalami (*Rt*) in dorsal-ventral middle plane. Because of previous destruction of pallidum 3 years before death. fiber bundle H_2 has become thinner and Forel fields H and H_1 have disappeared. In V.o.a all H_1 bundles are missing. However, V.o.a has not shrunk much. Note myelin degeneration in H_2 and surrounding cystic lesion. ×4

In 18 months the active mobility improved and only traces of arm and facial paresis remained despite the capsule lesion (Fig. 64). The tremor in the left leg could be suppressed during walking and returned only to a mild degree in the left arm. The good therapeutic effect on rigidity and active mobility could be explained by the inactivation of the pallidum internum and of the medial parts of the pallidum externum. When a hemiparesis due to the additional capsular lesion improved, there was partial recurrence of the tremor; the facial difference remained.

That the coagulated area was transformed into a cyst while a scar was formed in the left V.o.a is remarkable (Fig. 63). It is possible that this different organization of the coagulated tissue was due to varying high coagulation temperatures. However, the decisive point is the following: In the pallidum, destruction of tissue by high-frequency coagulation without controlling the temperature led to a complete interruption of the arterial supply by blocking larger vessels; this did not happen in the smaller vessels of the thalamus because of a better collateral vascular supply.

In the right internal capsule an area of softening formed above and was slightly more extensive than the coagulated area in the pallidum. This lesion destroyed the fibers of the internal capsule in a circumscribed area lateral of V.o.a and V.o.p. This area of softening also developed into a cyst.

The third operation was directed mainly against the recurrence of tremor in the right limbs. In this operation, several coagulations were set in the pallidum internum. At the autopsy, they could be differentiated by their exact localization. Because the tremor was not completely abolished after these coagulations, a mechanical severing was performed with the string electrode at the base of the rostral end of the pallidum internum on the left side, with the object of interrupting the efferent fibers of the pallidum in the ansa lenticularis. At autopsy, this mechanical severance could be seen in the area of the ansa and in the passing anterior commissure (Fig. 118); however, it led to damage to a larger arterial branch from the substantia perforata anterior. As a consequence, a profuse hemorrhage occurred that went through the coagulated pallidum into the caudatum, passing through the pallidum medial to the putamen and through the internal capsule, causing a spherical hemorrhage in the head of the caudatum with the immediate consequence of sopor and conjugated deviation of the eyes to the right. The impairment of consciousness with closing of eyes and reduced reactions characterized by confusion since the operation was producing to the bilateral destruction of both pallidal segments, parts of the capsule, and two-thirds of the left caudatum. After this the patient had proximal paresis of the contralateral arm. The most prominent signs were amential disorder with impaired consciousness and disturbed perception up to occasional complete disorientation. This condition persisted for 45 days, until the patient's death. The transient postoperative paresis of the right arm may have been caused by pressure of the hemorrhage on those pyramidal fibers in the internal capsule that lie close to the caudal side of the hemorrhage. Although there was good motor improvement, impairment of consciousness persisted, and 3 weeks after the operation the patient sustained a fracture of the neck of the femur while walking. Following the surgical insertion of a pin, he developed bilateral bronchopneumonia that resulted in death 45 days after the last operation.

Case 8. S.H., aged 65; postencephalitic Parkinsonism; stereotaxic operation performed in left pallidum internum (no. 690); death occurred 28 days after the operation, from purulent bronchitis with aspiration pneumonia and with apoplexy in the basal ganglia due to cerebral arteriosclerosis.

History and clinical findings. The illness started at age 56 with rigidity and tremor in the right arm and disorders of gait and posture. Soon the left limbs were also affected; speech became indistinct. The EEG showed a slowing of the α rhythm to 7–8 cps without localizing signs. An encephalogram showed a bilateral internal hydrocephalus of considerable severity, and there occurred simultaneously a confusional state of several days' duration. As a result, the operation was postponed until 5 months later, when the patient had fully recovered.

A *stereotaxic operation* was performed June 24, 1959, in the left pallidum internum. According to the radiologic check, the pallidum internum was reached by an electrode of 4×1.1 mm diameter. The point of the needle lay 2 mm below the target point.

When the tremor was slight, single stimuli produced only single jerks at a scale reading of 100. Stimuli at a rate of 8 cps blocked the tremor. Stimuli at higher frequencies accelerated it while reducing its amplitude. An inhibition of counting occurred together with the patient looking to the right. During the stimulation, alternating movements of the arms became slower. During coagulation with the spark-gap diathermy instrument in position +4, the rigidity of the right arm clearly diminished. After further coagulations lasting 10 sec each, in positions +6 and +8, the rigidity in the right leg disappeared. Tremor in the right hand stopped after coagulations in positions 0 and −2. After coagulation in position −4 the patient no longer promptly obeyed instructions. Because some rigidity remained in the right arm, a further coagulation was performed with the string electrode 6 mm ventral and lateral to position 0. The point of the electrode showed some blood. After the last coagulation the patient went to sleep and could be roused only with difficulty.

Postoperative course. A few minutes after removal of the stereotaxic instrument, the clouding of consciousness became more severe but the patient remained rousable and gave relevant answers (but with an aphonic voice). Paresis of the right arm and leg with increased muscle tone developed quickly and became considerably worse in the arm within a short time. The impairment of consciousness increased until the patient was in coma, to the point that, 3 days after the operation, a tracheostomy became necessary. Six days after the operation, when the patient tried to open his eyes upon request, his eyes were more firmly closed (negativism). Movements of the lips were recognizable but no speech was produced. The clinical course was suggestive of a hemorrhage into the basal ganglia. This was confirmed 10 days after the operation by the brown-red C.S.F. On day 23 after the operation the EEG showed low θ and δ activity, particularly marked on both sides in the frontotemporal regions while unequivocal focal lesions were lacking. His state of consciousness improved to a certain degree for a time but he never reacted to speech. Atonia of the stomach led to vomiting and aspiration on day 25. An aspiration pneumonia developed rapidly and led to death 28 days after the operation.

Post-mortem of the body (Prof. Dr. H. NOETZEL) showed confluent bronchopneumonia with purulent bronchitis and pulmonary emphysema as cause of death. Moderate arteriosclerosis was found.

Anatomical findings in the brain. The pial veins over the right frontal pole were congested and contained thromboses. On the medial side of the left frontal pole a hemorrhagic area 10 × 4 mm was found. There was a slight subdural hemorrhage on its surface; 11 mm lateral to this was the site of the puncture without cortical necrosis. The brain was generally swollen, and the gyri were flattened. At the sagittal sulcus the right hemisphere was higher than the left because of considerable increase of volume. Both cerebellar tonsils showed marked signs of pressure.

The needle track in front of the massive hemorrhage in the frontal pole was surrounded by an area of softening that measured 8 × 5 mm. The filling of the left anterior horn by blood shifted the midline 7 mm to the right. The contralateral ventricle was enlarged. Laterally and ventrally, the hemorrhage extended over the edge of the anterior horn. The head of the striatum was enlarged to 43 × 38 mm because of a large hemorrhage leading to a narrowing of the left cella media, and the insula was displaced to the ventrolateral side. The hemorrhage partly softened the caudatum, and it destroyed the putamen and pallidum completely. Because of the large hemorrhage with necrosis of the basal ganglia, only the hypothalamus and the tractus opticus remained recognizable. The parts over the hematoma were softened and contained many small hemorrhages. Toward the ventral side the hematoma showed a distinct boundary in the area of the substantia perforata anterior. Only a few millimeters of the wall of the third ventricle of the thalamus remained. The hemorrhage also led to considerable vacuolization of the corpus callosum; in the parietal white matter it ended in the plane of the splenium of the corpus callosum. It extended sagittally for 90 mm. On both sides there was a distinct diminution of the black pigmentation in the substantia nigra, including the medial cell groups, but without a particular pattern.

Clinical-anatomical correlations. The coagulation with the spark-gap diathermy apparatus in the pallidum, 2 mm ventral to the target point and therefore at the boundary of the substantia perforata anterior, had caused a massive hemorrhage which, at the last two coagulations, showed itself via the patient's increasing somnolence. Within a few minutes hemiparesis developed and led eventually to complete paralysis of the right arm. The next day the patient became comatose. The following factors were contributory causes of death: (1) the arteriosclerosis of the cerebral vessels, shown clearly in the basal vessels (clinically demonstrated by the slowing of the α rhythm and the long confusional state following air encephalography; the confusional state was so severe that the operation was at first refused and was not performed until the patient had recovered, 5 months later); and (2) the damage to the vessel wall at the base of the pallidum caused by the high-frequency current when the temperature was not checked.

Blocking of the tremor by stimulating at a rate of 8 cps, and looking to the opposite side and slowing of active alternating movements when stimulating at a rate of 50 cps before the coagulations, were related to the pallidum. The first few coagulations up to position +4 led to considerable diminution of the rigidity in the arm. The coagulations up to +8 reduced the rigidity in the contralateral leg. Though the coagulated area and parts of the puncture track disappeared in the massive hemorrhage, the middle of the hemorrhagic area with its well-defined lower boundary in the substantia perforata anterior corresponded to the pallidum. The reduction of rigidity during the operation must therefore be ascribed to the pallidum. The ability to respond and the speed of reaction of the patient were reduced only after the anterior parts of the pallidum externum had been inactivated up to position −4. Although damage to the basal ganglia was only one-sided, speech became aphonic. The hemiplegia resulted from the destruction of the internal capsule, which had been completely interrupted even in the plane of the posterior commissure, i.e., in the caudal part of its posterior limb. The large left-sided hemorrhage in the basal ganglia showed itself in the EEG as the uniform slowing of activity to low θ and δ waves with frontotemporal accent on both sides without a focus.

The typical apoplexy in the basal ganglia with the subsequent hemorrhagic softening was the consequence of damage to the vessel walls at the base of the pallidum caused by the lack of temperature check during the coagulation. The hemorrhage into the left lateral ventricle acted like a space-occupying process leading to a displacement of the midline into the right half of the skull.

The cause of death 28 days after the operation was aspiration pneumonia caused by continuing coma, with dilatation of the stomach. The pattern of cell loss in the substantia nigra indicates a postencephalitic etiology.

Case 9. A.I., aged 61; paralysis agitans; first stereotaxic operation in right V.o.a (no. 775); 2nd, in right pallidum internum (no. 806); death came 24 days after the 2nd operation, from pulmonary embolism with subdural hematoma.

History and clinical findings. The symptoms of Parkinsonism began at age 57. There was no history of a previous illness. The patient developed a tremor in the left arm, slight rigidity on the left side, and disturbance of gait, with festination. A year before the operation the right side was affected and tremor of the chin developed. The

Fig. 65. Electroencephalogram recorded during operation while stimulating in right V.o.a in case No. 9. Each 1-cps stimulus is followed by several multiphasic cortical responses with maximum amplitude in precentral frontal region ipsilaterally. Slow negative components toward end of response to stimulus are somewhat larger occipitally than parietally. Cortical response is only moderate on contralateral side with maximum in frontoprecentral area (increasing stimulus leads to increase of amplitude; there is no phase reversal between different channels)

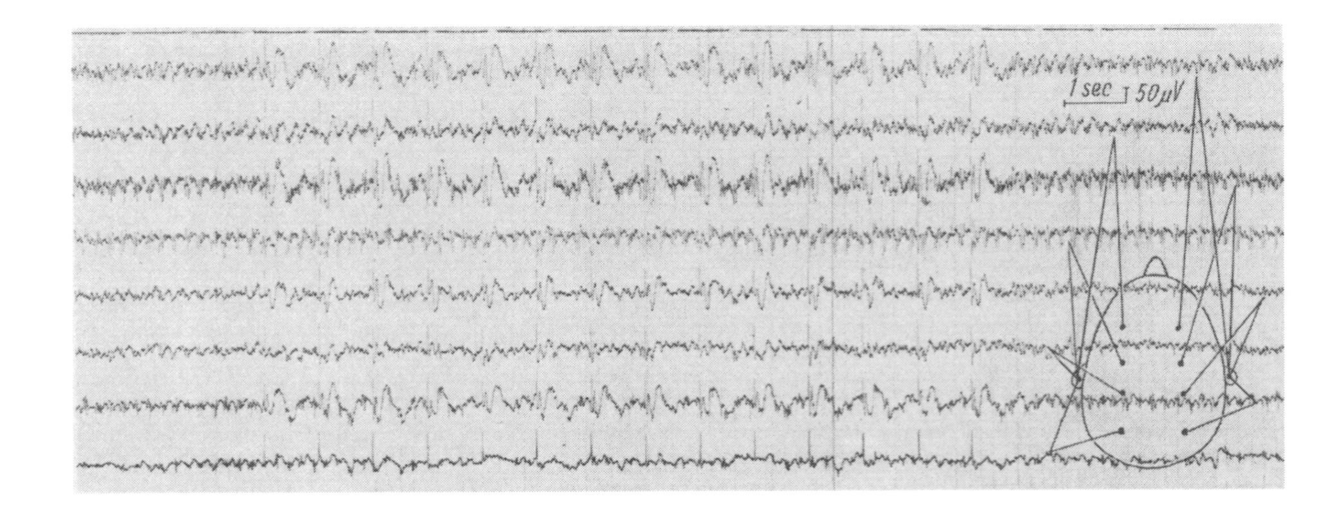

physicians found a moderate degree of arteriosclerosis, emphysema with bronchitis, and early disorders of the coronary circulation. In the C.S.F. there was a moderate increase of protein and clear widening to the left of the mastix curve.

The *first stereotaxic operation* was performed on October 5, 1959. According to the radiologic check, the point of the electrode made precise contact with the right V.o.a. Single stimuli at a rate of 4 and 8 cps increased the tremor considerably and made it more rhythmic. Stimuli at 25 cps blocked the tremor but did not influence active alternating movements. Stimulation at the rate of 50 cps led to fluttering of the eyelids, opening of the eyes, and inhibition of counting. EEG was recorded during the operation. In response to stimuli it showed precentrally a biphasic response with a slow wave (the latency of the spike potential being 200 msec). In response to stimuli in the V.o.a of 4 and 8 cps the potentials varied but showed recruitment. After stimuli at 25 cps fewer slow waves were seen after the end of the stimulation (Figs. 65, 66).

The coagulations were made with the Wyss coagulator, in the right V.o.a with a coaxial electrode from positions +5 to −3; with the string electrode two more coagulations were performed 6 mm toward the dorsal, caudal and medial sides. During the coagulations the resistance rose so much after 5 sec that the current dropped and the coagulations were, therefore, stopped. At the end of the operation, rigidity and tremor on the left side had disappeared and could not be provoked even by painful stimuli. The patient yawned spontaneously. A few days later the left-sided tremor recurred.

Therefore, a *second stereotaxic operation*, in the right pallidum with pure sinus waves was performed 22 days after the first operation. Individual stimuli increased individual phases of the tremor on the left side. Stimuli at 4 and 8 cps made the tremor more rhythmic. Stimuli at 25 cps increased the tremor considerably. Inhibition of counting occurred. Active alternating movements on the left side were suddenly arrested in the middle position. During the first coagulation with the Wyss coagulator in position +2, the current dropped after 8 sec and the coagulation was interrupted. After the second coagulation the tremor decreased. After the coagulations from +4 to −4, the left arm and leg became hypotonic; the tremor, including that of the chin, stopped. When making aimed movements, the left hand trembled only minimally in contrast to the severe tremor of the right side, which had not been treated.

Postoperative course. Ten minutes after the patient had been put to bed, a weakness of the left arm started with a tendency for the arm to sink down. When he attempted to sit up in bed, he hit his head against the bedside table. The doctor on duty found that the previously existing slight paresis had become a more severe paresis of the whole side. In the following days the patient became increasingly drowsy. Thirteen days after the operation a right angiogram showed the typical picture of a subdural hematoma of medium size; the cortical vessels had been displaced by about 2 cm (Fig. 67) and the anterior cerebral artery was displaced in the shape of an arc by about 9 mm without the corpus callosum being affected.

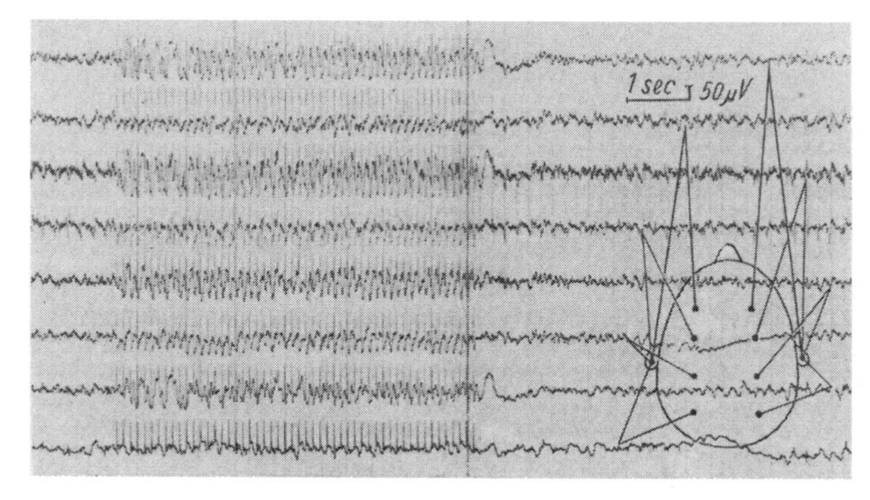

Fig. 66. Electroencephalographic recording during operation in case No. 9, during 8 cps stimulation in right V.o.a. In all channels, cortical responses show waxing and waning of amplitudes, particularly marked in right frontal and precentral responses. In left hemisphere, picture resembles α rhythm corresponding to frequency of stimulation, particularly frontally where one sees considerable delay and low amplitudes. A rapid rudimentary cortical recruiting phenomenon can be seen

On the following day, 14 days after the second stereotaxic operation, a temporal longitudinal opening was made and the hematoma evacuated. The biopsy (Prof. NOETZEL) of the dura showed histologically a more recent subdural hemorrhage that originated in a thickened inner layer of the dura (slight pachymeningopathy). Six days after removal of the hematoma the patient was more awake; he opened his eyes spontaneously, reacted to instructions, and took an interest in his surroundings. The improvement continued during the following days. When sitting up on day 24 after the second stereotaxic operation, the color of his skin changed; he lost consciousness immediately and developed severe circulatory shock, consequent upon a pulmonary embolism. He died within 3 min from respiratory arrest, before therapeutic measures could be taken. The clinical picture indicated as cause of death an acute severe pulmonary embolism (post-mortem of the body was not permitted).

Anatomical findings in the brain. The area surrounding the puncture mark, which was 4×4 mm in the right frontal pole, showed a marked depression from the subdural hematoma (Fig. 68a, b). A smaller defect from the first operation was seen further dorsally and occipitally. In the right basal ganglia a large ellipsoid hemorrhage, 44 mm wide and 19 mm high, reached from the caudatum to insular cortex (Fig. 68b). The hemorrhage caused a 6 mm displacement of the midline into the other half of the skull and the narrowing of the anterior part of the right lateral ventricle. It also destroyed the basal ganglia above the anterior commissure. Only small medial parts of the caudatum remained. In the region of the anterior horn in front of the massive hemorrhage, a softening took place in the white matter of the frontal lobe. The third ventricle was narrowed to a small oblique slit. All that remained of the thalamus were the anterior nucleus and a 6 mm wide band of tissue next to the third ventricle. The preceding coagulation could be recognized by a double protrusion of the medial edge of the hemorrhage in the thalamus (Fig. 68b). The anterior part of the internal capsule was displaced by the hemorrhage toward the medial side. In the plane in front of the posterior commissure areas of softening were found only at the boundary of the hemorrhage—for instance, in the intermediate ventral nucleus (V.im). In this plane the hemorrhage was confined to the pallidum. The posterior part of the internal capsule was swollen, which caused a part of the thalamus to be pushed across the midline to the left (Fig. 68b). Even in the plane of the pineal gland and the splenium, the brain was swollen.

Clinical-anatomical correlations. The coagulation of the right oral ventral nucleus (V.o.a) abol-

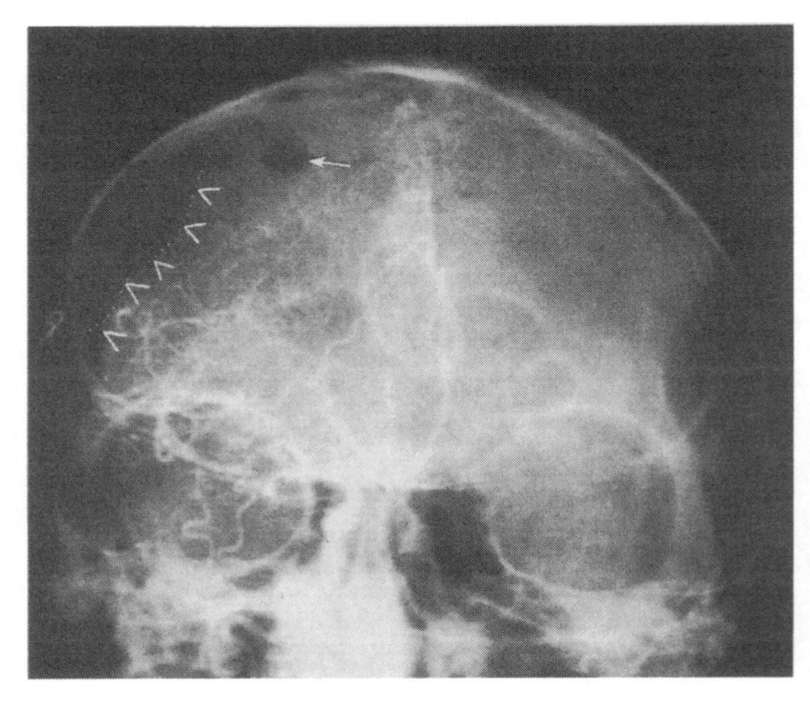

Fig. 67. Right-sided anterior-posterior carotid angiogram of case No. 9, 14 days after 2nd operation. Anterior cerebral artery is arched toward other side. The right cortex is displaced toward left (<) as if due to a space-occupying process. In area of skull convexity, particularly in temporoparietal region, there is an area without blood vessels corresponding to extent of hematoma. Trephine hole made during stereotaxic operation projects parasagitally (*arrow*)

ished for a few days the primary symptoms of the illness—the severe tremor of the left arm. It recurred, however, though the abolishment of rigidity of the left limbs remained. Because of the destruction of the thalamus by the massive hemorrhage, it was not possible to correlate the clinical effects with the inactivation of individual thalamic nuclei. Stimuli of 4 and 8 cps (Fig. 65) in the V.o.a evoked cortical responses that were accentuated frontoprecentrally and ipsilaterally with recruitment (gradual increase and decrease of the responses). These were due to excitation of the V.o.a and V.o.p complexes. The same applied (Fig. 66) to biphasic wave after individual stimuli. The second operation in the pallidum, performed because of the recurrence of tremor, was again immediately successful and even achieved a stopping of the tremor of the chin. Half an hour after this operation, however, a paresis of the left arm began that became complete after a short time. The cause of the paralysis and of the severe impairment of consciousness was shown 14 days later by the angiogram: a displacement of the right anterior cerebral artery considerably across the midline because of a hemorrhage in the basal ganglia. At the same time there was a subdural hematoma over the right frontotemporal region with histologic evidence of pachymeningiosis. The condition, including the left-sided paresis, did not improve after surgical evacuation of the hematoma, which substantiated the clinical conclusion that the hemiplegia was due not to the subdural hematoma but to the massive hemorrhage in the depth of the brain.

The massive hemorrhage in the right basal ganglia (Fig. 68b) was linked with the pallidotomy. At the first coagulation with the Wyss coagulator, a drop in current appeared after 8 sec, which probably caused damage to an arterial wall in the thalamus. This in turn led after an hour to a large hemorrhage in rostal thalamus (Fig. 68b), which showed itself by the impairment of consciousness and hemiplegia. Anatomically this was a typical apoplexy in the striatum, with bleeding into the lateral ventricle and hemorrhagic softening of the surrounding tissues. Judging by the extent of the damage, the responsible vessel was the arteria lenticulostriata. The anterior part of the internal capsule was medially displaced, its middle part was destroyed, and its posterior part was affected by swelling. The damage to the internal capsule and the fact that the brain tissue had not assumed its normal configeration after removal of the subdural blood clot explained the hemiplegia. Death was caused by an acute massive pulmonary embolism after a period of prolonged impairment of consciousness due to hemorrhage in the basal ganglia and pressure on the right basal ganglia due to flattering of the central and parietal cortex (Fig. 68a, b).

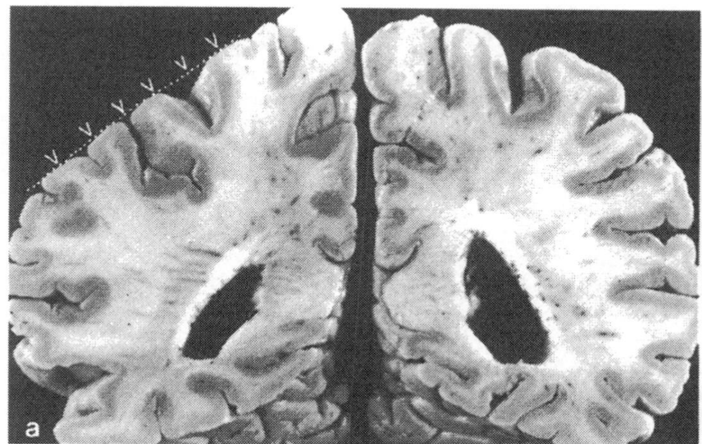

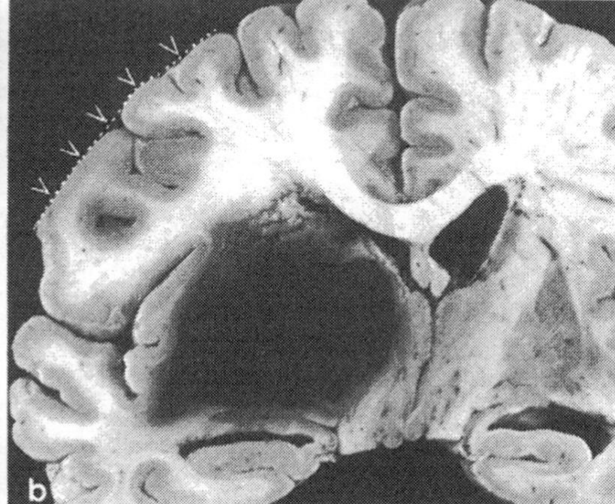

Fig. 68. (a) Reversed section through unstained brain of No. 9 at level of parietal lobe. Because of swelling of white matter of right hemisphere, due to hemorrhage in right basal ganglia, posterior horn is narrowed. Parietal lobe of right hemisphere has a flat surface because of subdural hematoma evacuated 10 days before death. The flat surface, however, remained.
(b) Reversed section through unstained brain of case No. 9. Large subcortical hemorrhagic area has caused right basal ganglia to be blown up, and medial thalamus has been displaced across midline toward left. Hemorrhagic lesion has destroyed lateral thalamus, internal capsule, pallidum, and putamen and extends to external capsule. Softening of hemorrhagic area high up, probably with necrotic material breaking into right lateral ventricle. Due to the evacuated subdural hematoma in spite of its evacuation the surface of the right central gyri is flat

Case 10. B.W., aged 66 years; paralysis agitans; stereotaxic operation performed in left V.o.a (no. 1660); hemorrhage in frontal lobe; death 5 days later due to repeated embolisms, pneumonia, and cardiac failure in a case of genuine hypertension.

History. The patient had genuine hypertension (270/180 mm Hg) and severe general arteriosclerosis. This led to cardiac failure with congestion of the liver and elephantiasis like edemas of both legs. In her 60th year she developed clumsiness of both legs and difficulties in walking. Since her 63rd year, she had tremor of the left limbs.

Clinical findings. Of the Parkinson signs, rigidity exceeded tremor considerably. All signs were more marked right than left. The C.S.F. showed slightly increased total protein and a trace of a meningitis curve in the hydrochloric acid–collargol reaction. The tendon reflexes were more marked right than left. The EEG exhibited an α rhythm with a tendency to slowing at about 7–10 cps; its frequency varied considerably. Single frontotemporal δ waves were present. The internist's opinion was that the very obese patient was not suitable for operation because of cardiac failure, hypertension with general arteriosclerosis, pulmonary emphysema, infection of the urinary tract, and goiter with narrowing of the trachea. The heart was enlarged on the left side and showed abnormal pulsations. The pulmonary vessels were engorged; the liver was enlarged and hard due to congestion. Marked spongy, almost elephantiasislike edema of both lower legs was evident. It was suggested that she be treated with digitalis and, for her urinary infections, with antibiotics.

After medical treatment of 9 months' duration the patient and her husband insisted on an operation in view of her repeated depressive mood swings and suicidal ideas. Encephalography revealed moderate left-sided hydrocephalic dilatation of the lateral ventricle (Figs. 69, 70), with much air showing outside the ventricles frontally in the region of the insula and of the sulcus cinguli. The patient reacted badly to encephalography, and the operation was not performed until 10 days later.

The stereotaxic operation was performed April 6, 1962, in the left V.o.a. The point of the electrode passed through the target point by 4.5 mm in the direction of the axis basally, 2.5 mm medially. Stimuli at a rate of 4 cps in this position evoked jerks of the arm and leg that were synchronous with the stimuli but that went in varying directions. Acceleration of the tremor was also seen. Stimuli at a rate of 50 cps caused blocking of the tremor with tetanic jerks of the right arm

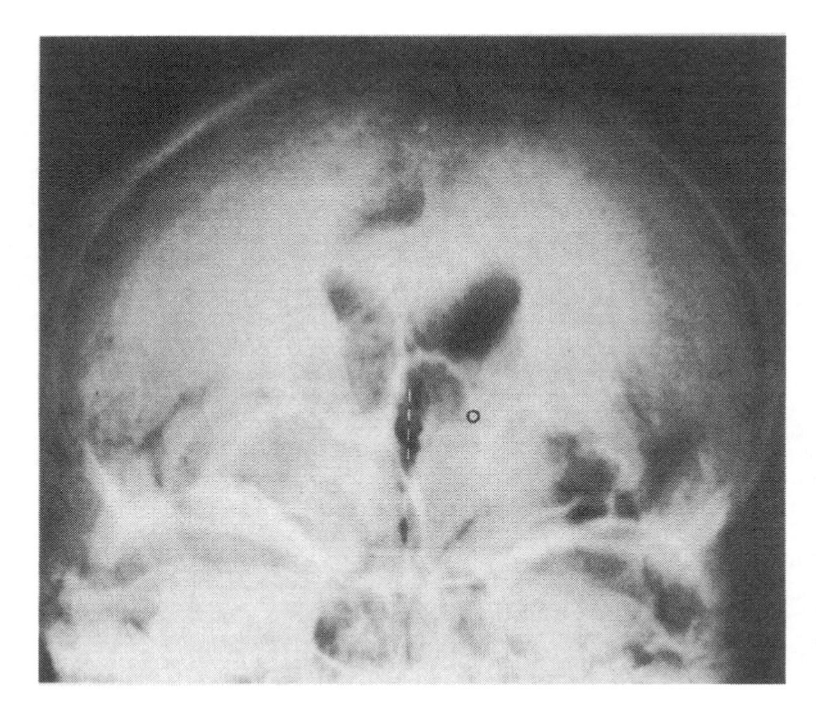

Fig. 69. Preoperative pneumoencephalogram of case No. 10. Anterior-posterior picture shows marked dilatation of left lateral ventricle, and anterior horn, to twice size of right lateral ventricle. 3rd ventricle is slightly widened toward left side. Subarachnoid collection of air at left insula shows severe atrophy compared to almost normal markings of insula on right side. Sulcus cinguli on right side is also widened. Left V.o.a target is marked

and inhibition of counting. Five coagulations with the thermoprobe followed in positions +2.5 and +6.5. The patient said after the first coagulation that movements in the right arm had become easier. The fourth coagulation was followed by an improvement of the rigidity in the right leg. After this the patient became tired. At the end of the operation some rigidity remained in the right leg, while rigidity and tremor had disappeared from the right arm.

Postoperative course. In 24 h the patient's consciousness became increasingly impaired. The C.S.F. pressure was above 300 mm H_2O and the C.S.F. was bloody. The next day, mydriasis appeared first on the left side, later on the right. Cerebral arteriography showed a clear displacement of the anterior and the middle cerebral arteries as produced by a left-sided frontal mediobasal space-occupying lesion. A hemorrhage was evacuated surgically from the medial frontal white matter 2 days after the stereotaxic operation through the enlarged trephine hole. This hemorrhage came from a branch of the anterior cerebral artery. However, consciousness was not improved (although the mydriasis disappeared). Two days later the patient had two pulmonary embolisms followed by pneumonia. She died 5 days after the stereotaxic operation from respiratory and circulatory failure.

Post-mortem of the body (Prof. Dr. H. NOETZEL) confirmed the hypertrophy of the left side of the heart with a coronary sclerosis going partly into the perphery, miliary scars in the myocardium, and sclerosis of the aorta and the renal vessels. Bronchopneumonia with an infarct the size of a walnut was seen in the left lower lobe. An embolism the thickness of a little finger was found in the main branches of the pulmonary artery; the remains of the thrombus were found in the right iliac vein. The heart was dilated on the right side; tracheitis, pulmonary edema, and nutmeg liver were also found.

Anatomical findings in the brain. A large surgical defect was seen on the left convexity of the frontal brain. The white matter of the frontal lobe was destroyed and full of blood. At the pole there was a hemorrhage below the pia. The basal cerebral vessels showed slight arteriosclerotic changes. In the white matter of the gyrus rectus a hemorrhage the size of a bean was found that was not connected with the operation field. Through swelling the whole left hemisphere was enlarged by 8 mm, with pressing of the corpus callosum and displacement of the massa intermedia into the right half of the skull (Fig. 71). Through the cerebral swelling the right lateral ventricle was narrowed. The left basal ganglia were also enlarged through swelling. The left gyrus cinguli showed a hemorrhagic infarct,

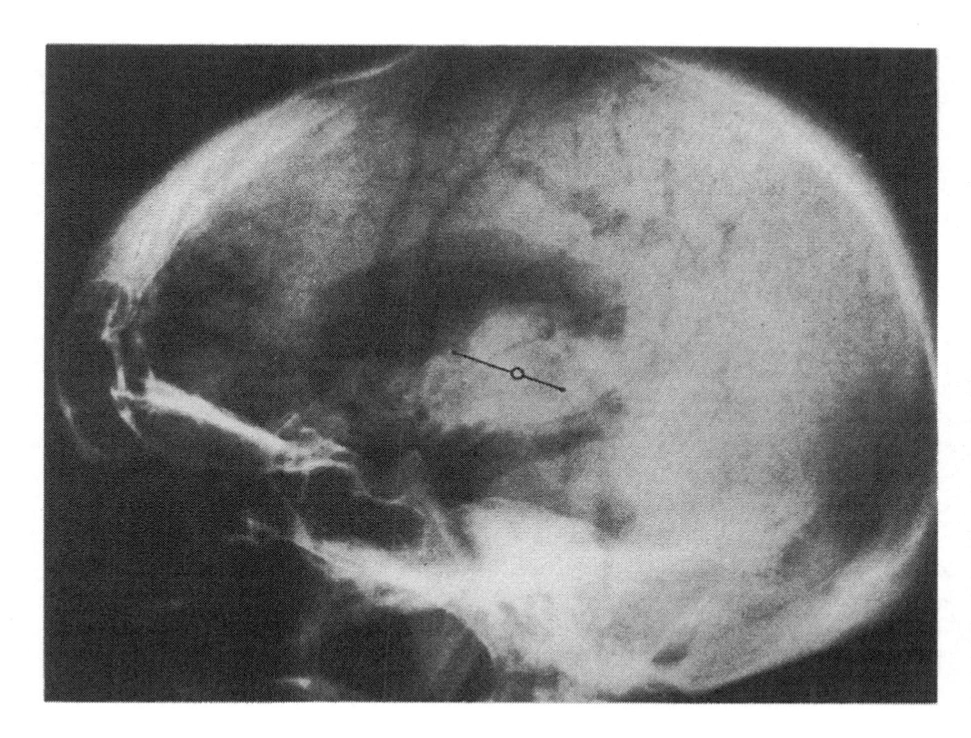

Fig. 70. Lateral pneumoencephalogram of same case shows difference in height of anterior horn and lateral ventricles; frontal bone shows marked hyperostosis frontalis interna. Tentorium is partly calcified, and there is basilar impression. Slightly dilated central parietal and frontal subarachnoid spaces. Baseline of thalamus with V.o.a target is marked

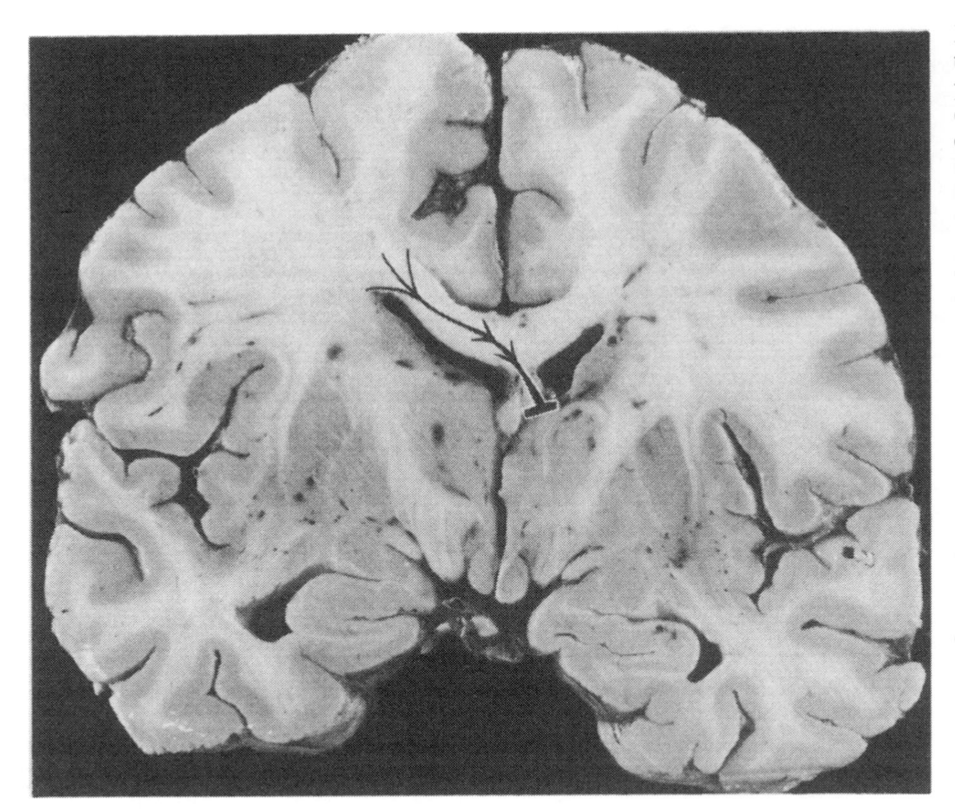

Fig. 71. Transverse section through unstained brain in plane of mamillary bodies of case No. 10. There is considerable swelling of left basal ganglia, due to a hemorrhage into the frontal lobe. This leads to narrowing of left lateral ventricle and displacement of columns of fornices into right lateral ventricle. Direction of pressure is indicated by multiple arrow line. Pressure is exerted on right anterior thalamic nucleus. Thalamus as well as white matter of left hemisphere is swollen. Hemorrhagic electrode tract in left thalamus is seen. Pressing of corpus callosum and thalamus in right half of cranium

cortical necrosis, and hemorrhages into the white matter in the adjoining sulcus. The cerebral swelling affected the left thalamus remarkably, so that its height and width had increased by 3 mm (Fig. 71). In the needle track in the left thalamus there was an area of softening in the V.o.a 9 mm from the ventricular wall and 3 mm above the base.

The area of coagulation was in the V.o.p; its center was 10.5 mm away from the wall of the ventricle. Although the internal capsule had not been damaged by the coagulation, it appeared swollen and enlarged.

A large blood clot filled the anterior part of the fourth ventricle. The cerebellar tonsils showed a 7-mm high impression of the occipital foramen. Caudal parts of the substantia nigra showed a loss of pigmented nerve cells.

Clinical-anatomical correlations. The stimulating electrode lay in and below the left V.o.p (Fig. 71) 8.5 mm lateral to the wall of the third ventricle. The stimuli caused jerks of the arms and legs synchronous with the stimuli but in different directions. The variability of the direction was typical for the afferent system of the motor cortex. The tremor was accelerated by stimuli at 4 cps, which is unusual and perhaps explained by the addition to phases of tremor of jerks evoked by stimulating the cerebral peduncle. Stimuli at 50 cps blocked the tremor and led to tetanic convulsions of the right arm that were typical for the extension of the stimulating current to peduncular fibers. Because the stimulating electrode was 4.5 mm basal to the target point, the current could involve the fibers in the peduncle. The coagulations were performed on the oral ventral nuclei 1.5 mm more caudal than usual. These coagulations, which were performed up to position +6.5 at an angle of 45°, abolished rigidity and tremor in the right arm. Even multiple coagulations improved rigidity in the right leg but did not abolish it, which is in accordance with the medial position (2.5 mm) of the coagulated area. The region representing the leg, the inactivation of which was responsible for the reduction of rigidity, had obviously been less completely destroyed.

Examination of the brain showed the center of the coagulated area to be in a plane about 13 mm behind the interventricular foramen and 10.5 mm from the wall of the third ventricle. These measurements would fit the usual target point. In this case, however, because death occurred 6 days after the stereotaxic operation and 3 days after an additional open brain operation, the thalamus was swollen to an unusual degree. On the operated side, its width and height were each larger by 3 mm than on the other side. Therefore, at post-mortem, the distance of the center of the lesion from the wall of the third ventricle was greater than it would have been in the brain at the time of the operation. The increase in distance, as a result of the swelling of left thalamus, was probably about 2 mm, since the whole thalamus was about 15 mm wide and was swollen by 3 mm. Thus, before the swelling occurred, the distance from the wall of the third ventricle would have been 8.5 mm. This measurement agreed with that obtained from the radiologic check of the position of the electrode.

The increasing impairment of consciousness with mydriasis but without hemiparesis occurring 24 h after the operation was due to a hemorrhage in the area of the anterior cerebral artery. A part of the hemorrhage was evacuated 2 days after the stereotaxic operation, but the hemorrhagic lesion in the region of the anterior cerebral artery could be found by autopsy in the white matter of the gyrus rectus that was not reached by the operation and in the white matter of the gyrus cinguli. The autopsy excluded a link with the deep-seated coagulated area. The hemorrhagic lesion in the area of the anterior cerebral artery was independent of the coagulation in the basal ganglia. These anatomical findings showed with certainty a tendency toward hemorrhages in the area of the anterior cerebral artery. The autopsy showed no evidence that the main trunk or even a branch of this artery had been damaged by the coagulation in the depth and only superficial damage where the electrode had entered the frontal lobe convexity in the area of the anterior cerebral artery; this had probably caused a functional impairment of the blood supply in the area of the artery. This supposition was supported by the subarachnoid hemorrhage at the left frontal pole. In the more occipital part of this supply area hemorrhagic infarcts were found in the gyrus cinguli and in the corpus callosum. These were also the consequences of disorders of the blood supply with small petechial hemorrhages from end arteries of the anterior cerebral artery.

Another important factor in regard to the hemorrhage was the fact that the patient suffered from severe hypertension (up to 270 mm Hg systolic) with cardiac failure and very severe signs of congestion, including the liver. (It is known that massive hemorrhages occur almost exclusively in cases of hypertension.) Even the preoperative lowering of the blood pressure was not sufficient to prevent the massive hemorrhage after the stereotaxic brain operation.

Death was due to a combination of cerebral events—(1) brain damage and swelling with considerable rise in cerebral pressure, the severity of which was shown by the pressure cone of the cerebellar tonsils and (2) two pulmonary embolisms followed by pneumonia and circulatory collapse in the presence of severe damage to the myocardium.

The physicians thought that this patient could not be operated on because of her bad general condition together with cardiac failure due to genuine hypertension. These considerations were clearly explained to the relatives and the patient after the 9 months' medical treatment. This case showed that the high operative risk was due mainly to the diseases of the internal organs despite specialist preoperative treatment; the stereotaxic operation is not particularly taxing for a patient who has good circulation.

Case 11. B.H., aged 64; paralysis agitans, with severe cerebral arteriosclerosis and cerebral atrophy; massive hemorrhage in the parieto-occipital lobe with displacement of the basal ganglia following air encephalography; stereotaxic operation performed in right V.o.a (no. 452); death from pulmonary embolism 7 days after the operation.

History and clinical findings. Without etiologically significant previous illness, Parkinson signs started at age 62 with tremor in the left arm followed by tremor in the right leg, the left leg, and head. Rigidity remained less marked than the tremor. The legs were more affected than the arms. Because of a sleep disorder, the patient had been treated regularly with dilaudid-atropine. His EEG was normal, having an α rhythm of 8–10 cps. A slight pulmonary emphysema and blood pressure values in keeping with the patient's age were not considered by the physicians to constitute a contraindication to the operation.

Encephalography showed an internal and external hydrocephalus of moderate degree with marked dilatation of the basal cisterna (Fig. 72). The C.S.F. was normal. After encephalography, the patient was at first confused at night, later also during the daytime. He was disoriented in regard to time and only poorly oriented in regard to space. Six hours after encephalography, he became very restless and complained about pains in the left upper arm.

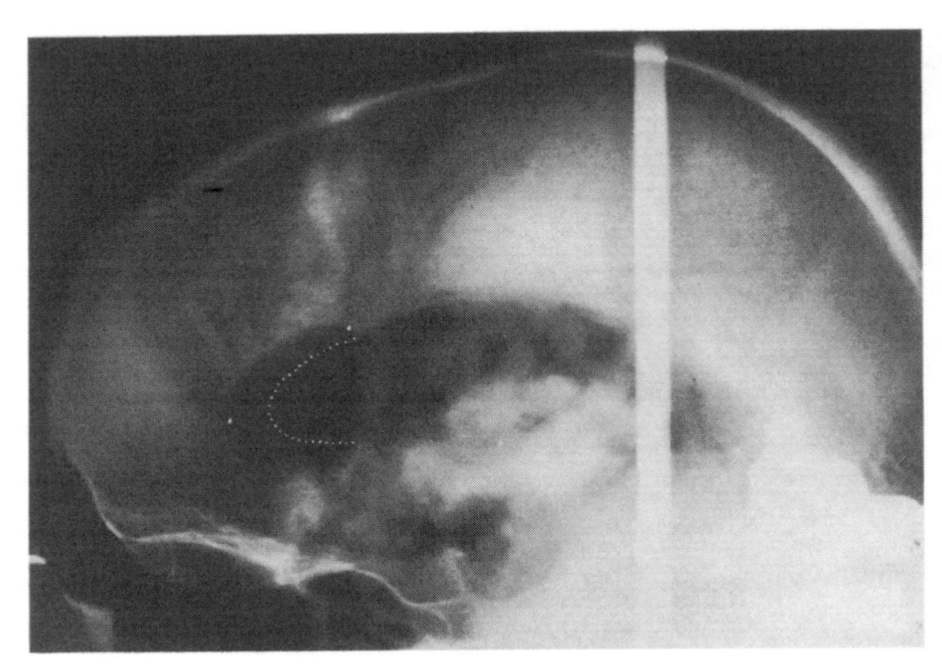

Fig. 72. Correctly measured superimposed lateral radiographs of skull of case No. 11 during pneumoencephalography and 5 days later. Pole of anterior horn is displaced 10 mm rostrally, and anterior horn is so dilated and displaced toward base that lower outline is 4.5 mm lower. Extent of anterior horn during encephalography 5 days before operation is marked by row of dots

On the day of the operation the anterior horn was displaced 10 mm rostrally and 4.5 mm ventrally (Fig. 72). Stimulation of the right V.o.a at rates of 4, 8, and 50 cps 1.5 mm rostrodorsal to the target point increased the tremor of the left arm considerably and also caused inhibition of counting and of iterative active movements. Using the spark-gap diathermy apparatus, the electrode was moved in a line from +1.5 to −6.5. During the coagulation in position +1.5, twitchings in the area of the left facial nerve and in the left arm appeared immediately along with spontaneous unintelligible utterings. During coagulation in position −2.5, the left arm went down. This also happened during the penultimate coagulation at −6.5. The last coagulation was performed with the string electrode from position −2.5 in a ventral, rostral, and medial direction of 4 mm. At this point there was a recurrence of spontaneous sounds and severe twitchings of the face that caused interruption of the coagulation after only 12 sec. At the end of the operation the rigidity had disappeared from the left leg and was improved in the left elbow. The tremor of the left hand was now less than that of the right; power in the left leg and arm was reduced and a slight extensor response appeared. However, when firm instructions were given he could, at the end of the operation, raise his left limbs and keep them in position.

On the afternoon of the day of operation he developed a fever with abnormal central regulation and moderate dullness on percussion over both lungs. During the night the patient showed psychomotor restlessness. He had already experienced on the operating table definite motor weakness in the left limbs, and the next day, he was unable to move them. The patient was confused. In view of the deterioration of his condition, the poor ventilation of the lungs, and the impairment of consciousness, a tracheostomy was performed. Motor restlessness showed itself mainly in the nonparalyzed right limbs and continued until death.

On day 7 after the stereotaxic operation, the patient died with the signs of a massive pulmonary embolism with pneumonia and cardiac failure.

Post-mortem of the body (Prof. Dr. H. Noetzel) showed infarct of the left lower pulmonary lobe after a massive embolism. The thrombus from which it originated could no longer be demonstrated. There was extensive bronchopneumonia in the right lower lobe, acute pulmonary edema, and moderate sclerosis of the coronary vessels and the aorta.

Anatomical findings in the brain. Included marked arteriosclerosis of the basal vessels, considerable cortical atrophy (especially on the frontal convexity), and a large cerebellar pressure conus. On the right frontal pole a subarachnoid hemorrhage (14 × 22 mm) was found around the puncture. A small hemorrhagic lesion (20 × 18 mm) with some stratification was found medially in the white matter of the first frontal gyrus. At the lower end of this hemorrhage the needle track could be seen. Below the ependyma a hemorrhagic lesion resembling a fleabite was found in the right anterior horn. The track in the head of the right striatum was filled with blood. The terminal vein was dilated and thrombosed. The right hemisphere was swollen, leading to diminution of both lateral ventricles. Because of the swelling, the right cingulate gyrus was pushed to the left under the falx. Near the third ventricle the midline was displaced into the left side of the skull by about 6 mm (Fig. 73b).

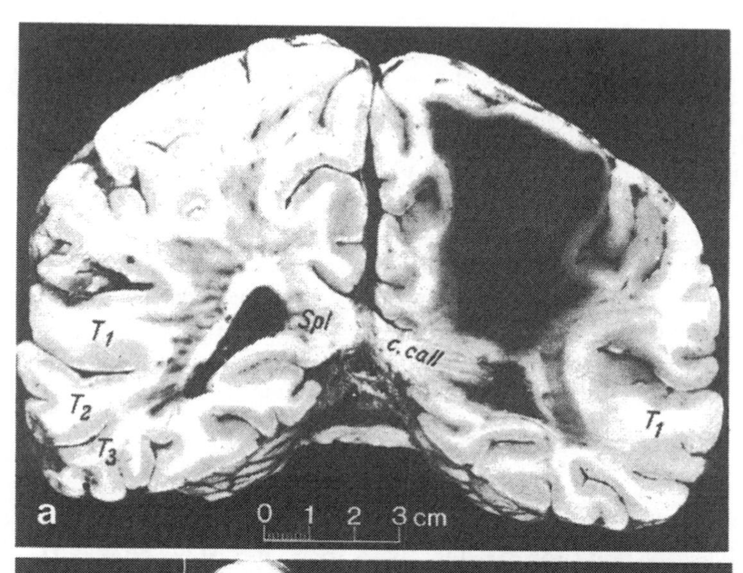

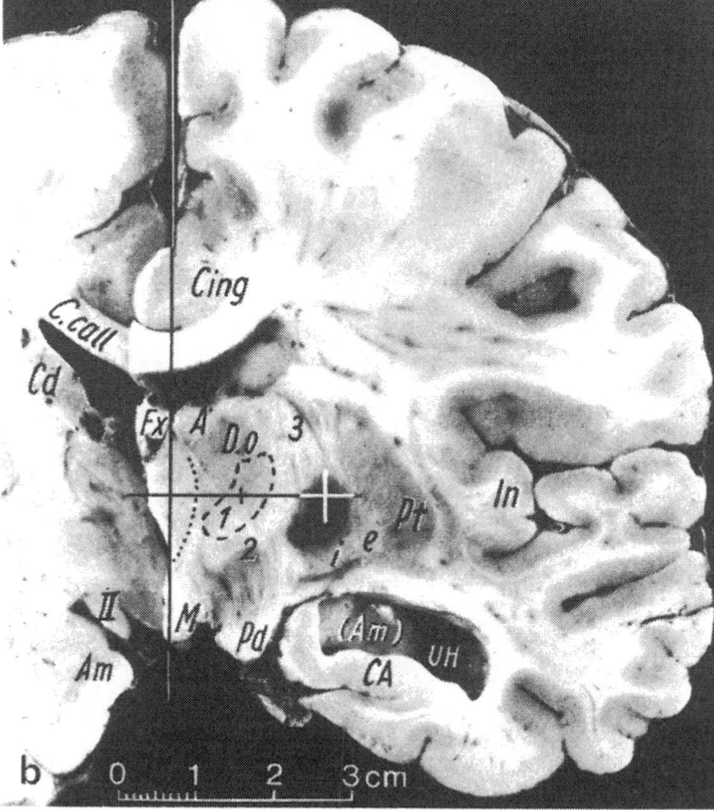

Fig. 73. (a) Unstained transverse section through brain of case No. 11. White matter of right occipitoparietal lobe caudal to corpus callosum contains large hemorrhage that has pushed right posterior horn basally. Surrounding hemorrhage are smaller hemorrhagic lesions in white matter and in cortex. This hematoma developed in the time between pneumoencephalography and operation (b) Right basal ganglia of case No. 11 are shifted across midline by swelling of white matter; at the same time, right gyrus cinguli is shifted below falx into left half of cranium. Position of wall of 3rd ventricle before hemorrhage at time of pneumoencephalography is marked by row of dots, 3.5 mm right of midline. Right wall of ventricle is now situated 4.5 mm left of cranium midline. Intended target in V.o.a is marked but not reached by electrode because of swelling of white matter and shifting of basal ganglia. Instead, coagulation focus is located in inner segment of right pallidum, 2.3 mm below target, due to basal shifting of corpus callosum by pressure

The coagulated area was found at the lower edge of the right internal capsule and affected the external and internal parts of the pallidum equally (Fig. 73b). The lesion involved the internal capsule in the plane 10 mm behind the interventricular foramen about 0.5 mm only from the ventromedial side. The lesion had destroyed the dorsal half of the internal pallidum and the dorsal third of the external pallidum.

In the plane of the posterior commissure, the right hemisphere was – because of swelling – 12 mm wider than the left. The corpus callosum was tilted so that the left lateral ventricle was 9 mm higher than the right. A small hemorrhagic area was found in the right fornix. A large hemorrhage (21 × 40 mm) filled the right superior parietal lobe and the entire white matter of the hemisphere above the trigonum (Fig. 73a). This lesion went as far as the interparietal fissure. Above the hemorrhage the cortex had been preserved. The dorsomedial part of the stratum sagittale externum and internum was destroyed. In the occipital direction the height of the hemorrhagic lesion increased up to 53 mm; sagittally it extended for about 34 mm.

Clinical-anatomical correlations. The autopsy showed that the area of coagulation was located in the pallidum and not, as intended, in the oral ventral nucleus of the thalamus. The cause of the incorrect localization and of the coagulation in the wrong place was not a fault in technique but was due to the large massive hemorrhage in the white matter of the right superior parietal lobe. The massive hemorrhage had obviously occurred after the air encephalography, for it did not show a displacement of the ventricular system across the midline (Fig. 72). However, the patient had withstood the encephalography very badly; 6 hours later, confusion and motor restlessness occurred that lasted for several days. He also complained about pains in the left upper arm. On the other hand, the x-ray taken on the day of the operation (6 days after encephalography) showed a lowering of the upper edge of the right anterior horn by 4 mm and a displacement of the septum to the left by 3.5 mm. On the lateral x-ray the outline of right anterior horn was pushed down (Fig. 72). In retrospect one could conclude with certainty that the hemorrhage occurred between encephalography and operation, obviously immediately after the encephalography.

The coagulated area was in the pallidum (Fig. 119), although the x-ray check during the operation showed that the point of the electrode had, so far as measurement goes, reached the thalamic target point calculated on the basis of the encephalogram. That the electrode had entered the pallidum was due to the considerable swelling of the right hemisphere, which had pushed the right wall of third ventricle 4.5 mm across midline into the left side (Fig. 73b). The distance of 16.5 mm from the midline of the skull is explained as follows: the target was 13 mm from the wall of the third ventricle, but also 3.5 mm away from the midline of brain and skull, i.e., 16.5 mm. The post-mortem finding that 21.0 mm separated the coagulated area and the needle track on the one hand and the midline of the ventricle on the other, was due to swelling of the brain, which had caused the distance between the basal ganglia to be increased by 4.5 mm. The displacement of the right basal ganglia was not detected during stimulation and coagulation because the pallidum reached by the electrode (Fig. 119) is the next neuronal station before the intendet target V.o.a. It has almost the same stimulation and coagulation effects as the V.o.a. If no displacement had occurred between encephalography and the operation, we would have reached the target using our aiming methods. However, this was realized only by the post-mortem with demyelination of field H(H) and bundles H_1 (Fig. 119).

Case 12. R.B.J., aged 63; paralysis agitans combined with additional cerebral arteriosclerosis; interruption of pallidothalamic fibers in internal capsule. First stereotaxic operation performed in right V.o.a (no. 1453); second, in left V.o.p (no. 2383). Death occurred 15 days after operation, due to confluent bronchopneumonia and kidney failure.

History. The patient developed Parkinson syndrome in his 54th year. In 1958 he was refused a stereotaxic operation in France because of a doubtful cardiac infarct sustained when he was 57. The Parkinson signs were restricted during the first few years to the left limbs. He started having a tremor in the right limbs only at the age of 60. Mentally he had slowed down generally; his handwriting was very small and tremulous. However, he did not yet need help with everyday activities.

Clinical findings. Upon the first hospital admission on August 15, 1961, the predominant sign was tremor of the left arm and of the head. The left leg and, to a lesser extent the right limbs, also shook. There was only slight rigidity in the left limbs and none in the right arm. He showed a

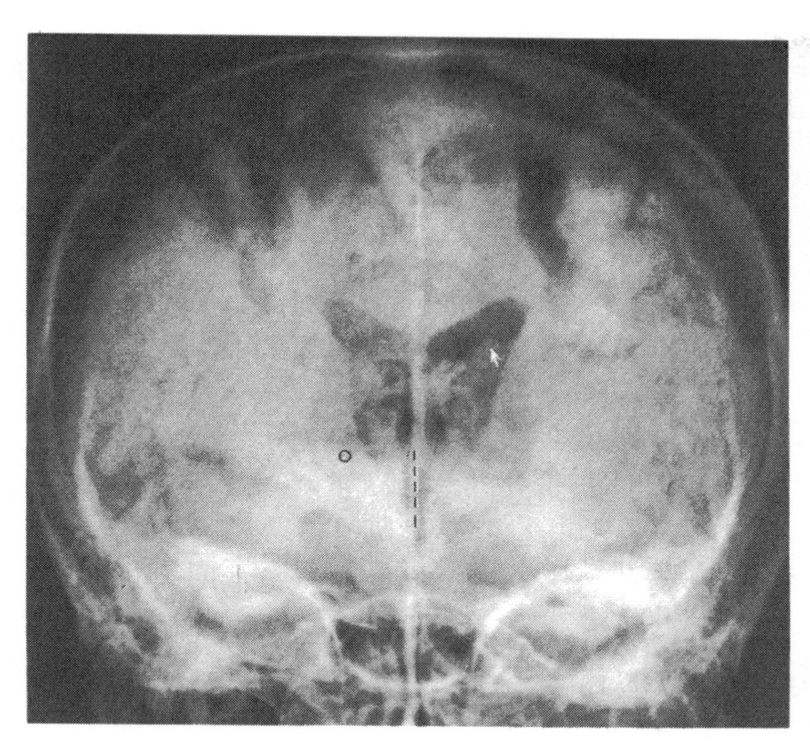

Fig. 74. Anterior-posterior pneumoencephalogram of case No. 12. Encephalography before 1st operation shows slight dilatation of left-sided ventricle without lowering of thalamus outline; tuberculum anterius is clearly prominent in lateral ventricle (*arrow*). Only slight dilatation of 3rd ventricle more toward left than right. Extensive collection of air in subarachnoid spaces in parietal regions as evidence of severe cortical atrophy

marked arcus senilis. The palpable vessels were hardened. The neurologic examination showed no changes in the reflexes and none in the C.S.F. The findings in the internal organs included pulmonary emphysema, generalized arteriosclerosis, blood pressure of 155/90 mm Hg, and no signs of cardiac failure, although the EKG showed a right bundle-branch block. The prostate was hypertrophic. Air encephalography showed considerable parietal cortical atrophy with big air bubbles (Fig. 74) and dilation of the anterior horns (Fig. 79).

The *first stereotaxic operation* was performed August 21, 1961, in the right V.o.a. The point of the electrode reached the target accurately, according to the x-ray. Individual stimuli led to doubtful twitches in the left arm. Stimuli at a rate of 4 cps enhanced and synchronized the tremor. Stimuli at 50 cps caused a mydriasis. Counting backward was first accelerated together with forced breathing; then counting was interrupted while respiration was arrested in inspiration for a few seconds and the eyes were turned to the left. Stimulation at this frequency was experienced as so unpleasant that he moaned, "It cuts my breathing off very unpleasantly."

The coagulation was performed with a 4 × 2 mm thermoprobe in positions 0 to +5. In addition, three coagulations were set with the string electrode which was moved 6 mm occipitally and 30° laterally, to reach the cerebellar fibers. Following the coagulation in position +5, the tremor was markedly reduced. The rigidity in the left arm and leg disappeared after coagulation in position 0. The remaining tremor was almost completely abolished by the seventh coagulation by the string electrode. The patient was satisfied with the result.

When the patient was discharged 10 days after the operation, all symptoms, including backache and rigidity, had disappeared. However, a trace of tremor remained in the left arm.

At the first follow-up examination on December 7, 1961, all symptoms on the left side, including the remaining tremor, had disappeared. However, a considerable rigidity and tremor had appeared in the right arm and leg.

Further follow-up examinations, on November 14, 1962 and April 24, 1963, showed continued good results on the operated left side. The patient had improved mentally and had no memory disorder. Although the effects of the operation on tremor and rigidity on the left side persisted, the tremor and rigidity on the right side had increased.

On August 18, 1964, the patient was readmitted for surgical treatment of the right-sided signs. The EEG (August 5, 1964) showed an α rhythm of 8–9 cps that sometimes slowed down to 7–8 cps.

The *second stereotaxic operation* was performed on August 19, 1964, in the left V.o.p. Individual stimuli 1 mm above the target point caused irregular jerks of the right hand in different directions. Stimuli at a rate of 4 cps made the tremor synchronous and increased it after a short latency. Stimuli at a rate of 8 cps also caused an increase of the tremor with acceleration. Stimulating at a rate of 50 cps blocked the tremor. At the same time there occurred a slight flexion of the right arm, mydriasis, and arrest of counting.

The first coagulation with the thermoprobe (bare point of 4 × 2 mm at position +2) had to be interrupted because of a rapid drop of current. It was repeated at the same spot with a controlled temperature of 73° C and a duration of 25 sec. A third coagulation was performed in the same place but the temperature rose very slowly to 70° C despite high voltage and current. Following these coagulations, the patient yawned spontaneously, became very tired, and shut his eyes. The right arm no longer shook, but on holding it forward it gradually sank. Some carbonized tissue was found on the electrode. On the operating table the patient became increasingly drowsy; he developed a hemiparesis on the right side with pyramidal signs in the leg. When he no longer reacted to being spoken to, he was given coagulants and dehydrating drugs. After this he temporarily opened his eyes.

During the following day, the patient reacted when spoken to but showed a tendency towards preservation. At rest the right arm showed a tremor although the right limbs were paretic. In view

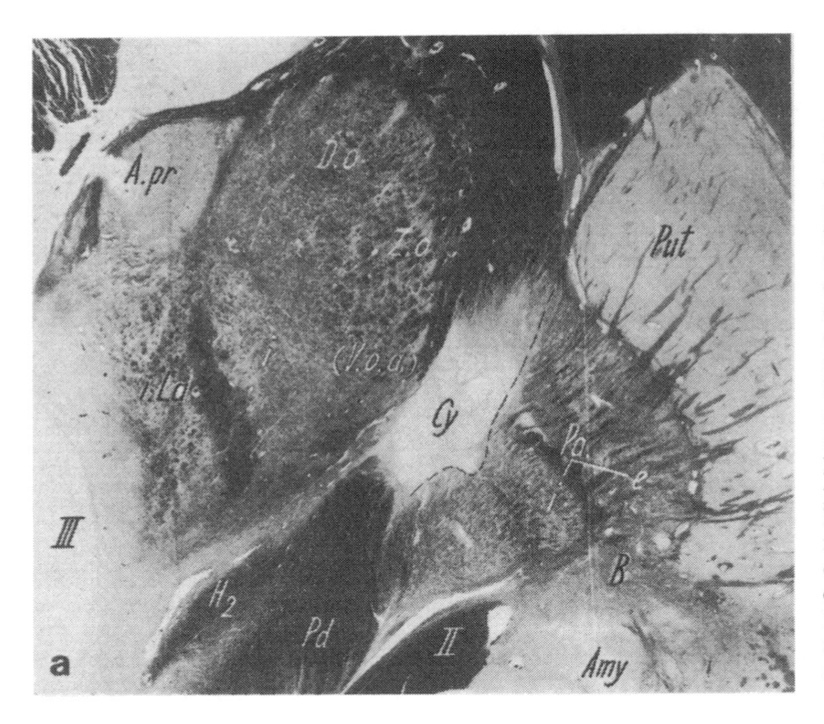

Fig. 75. (a) Cross section of right diencephalon in case No. 12. Internal capsule between dorsomedial border of pallidum and lateroventral border of thalamus is occupied by coagulation focus. Middle of focus is completely destroyed and shows cystic degeneration (*Cy*). Dorsal border of focus shows only demyelination of fiber bundles of internal capsule. Only small parts of dorsal border of internal segment of pallidum are involved. Coagulation focus stops in lamella lateralis thalami; that at this level borders V.o.a nucleus, which is deprived of H_1 bundles. Parts of lamella lateralis are demyelinated but tissue is not destroyed. This focus is cause of secondary degeneration of fiber bundles in cerebral peduncle (*Pd*) (see Fig. 75d). In the grey matter of the 3rd ventricle and above H_2 bundle are remnants of a subependymal hemorrhage. Fiber stain. ×4

Fig. 75. (c) Cross section through thalamus of case with so-called striatum apoplexy, in which all fibers of internal capsule are interrupted and degenerated so that internal capsule (*Ca.i*) and cerebral peduncle (*Pd*) appear white in fiber preparation. Consequently, pallidosubthalamic and pallidothalamic fibers are no longer hidden by fibers of internal capsule. H_2 bundle is marked by two stars and H_1 bundle by one star. Fiber structure of subthalamic nucleus (*CL*) is preserved and clearly shows connection to posterior hypothalamic decussation.

(d) Cross section of thalamus of case No. 12, 3.3 mm behind Figure 75a. Coagulation focus is just in transition zone of internal capsule and cerebral peduncle. From this focus, coagulation was performed by string electrode laterally, reaching internal part of pallidum, causing comma-shaped coagulation through internal capsule. Fine myelinated bundles (*arrow*) below coagulation focus in degenerated part of cerebral peduncle are remnants of pallido-subthalamic fibers. Double arrow indicates degenerated fibers of caudal H_2. Most H_1 bundles are demyelinated. Fiber stain. ×4 ▶

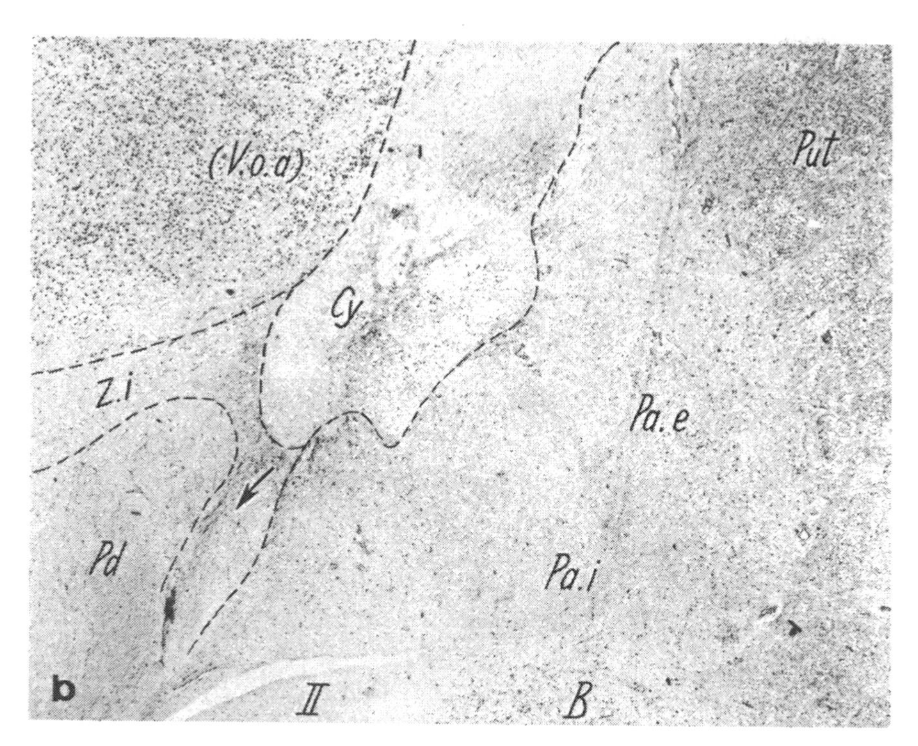

Fig. 75. (b) Adjacent section through diencephalon of case No. 12. Lowest part of internal capsule is destroyed by coagulation and shows signs of cystic degeneration (*Cy*) with mesenchymal tissue proliferation. Fiber bundles of most dorsal part of cerebral peduncle (*Pd*) are degenerated and replaced by dense gliosis (*arrow*). Due to interruption of all dorsal pallido-thalamic and pallido-subthalamic fibers in this part of the internal capsule (Fig. 75a, d) the dorsal parts of internal and external segments of pallidum (*Pa.i; Pa.e*) exhibit cell degeneration and cell loss. Nissl stain. ×8

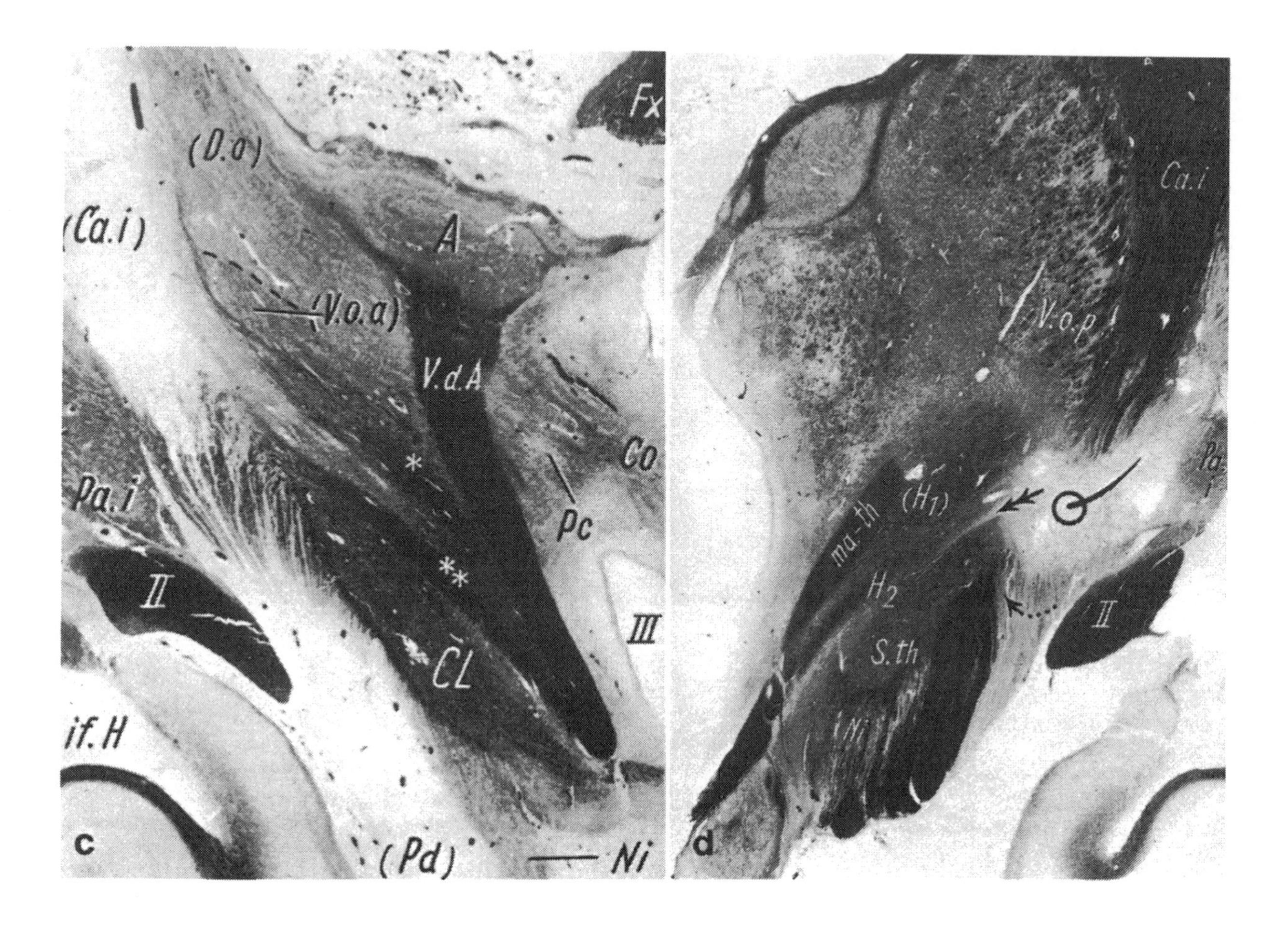

of the impaired consciousness, a tracheostomy was performed 2 days after the operation in order to improve the pulmonary ventilation. Despite antibiotic treatment, very high temperatures occurred on day 4 after the operation and bronchopneumonia developed. Eleven days after the operation, in addition to the already diagnosed pyelitis, there was a nonprotein nitrogen of 228 mg/100 ml. The day before death, generalized clonic twitchings appeared; all further therapeutic measures failed. The patient died 14 days after the operation from circulatory failure.

Post-mortem of the body (Prof. Dr. H. NOETZEL) showed severe fibrinous-purulent bronchopneumonia in all parts of the lungs, chronic pulmonary emphysema, cardiac dilatation (particularly on the right with endocardial hemorrhages), patchy coronary sclerosis, minute diffuse fibrotic areas in the myocardium, acute congestion of the internal organs, pulmonary edema, erosions of the mucous membranes in stomach and duodenum, and hemorrhagic pyelitis with a minimal pyelonephritic scar.

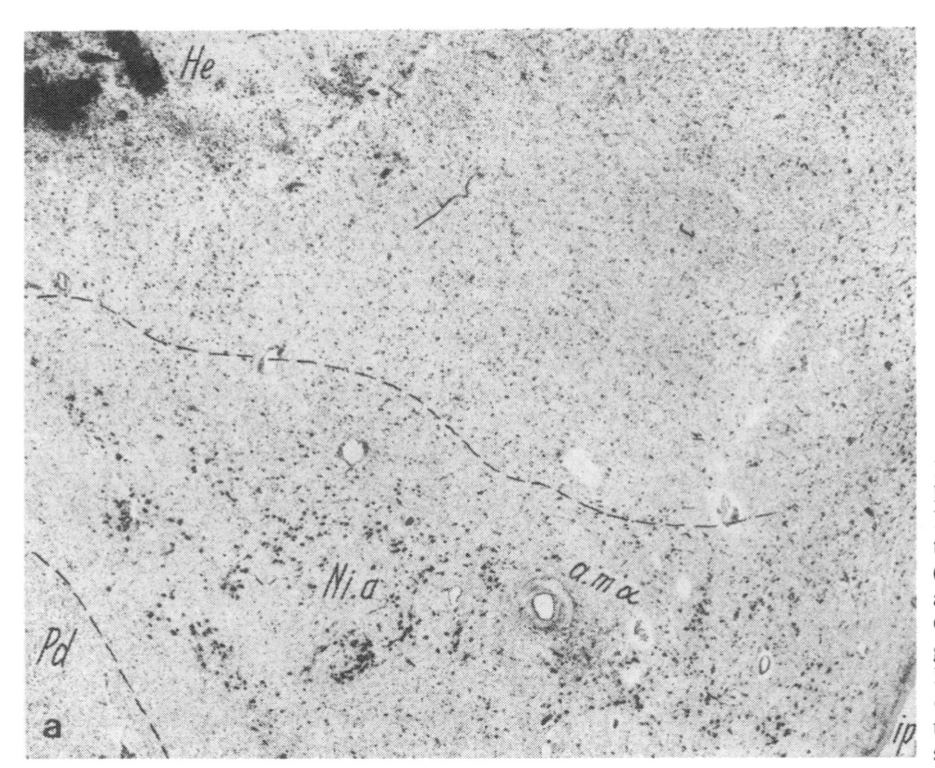

Fig. 76. (a) Cross section of left substantia nigra of case No. 12, showing deterioration of medial cell group (am. α) around arteriole, with arteriosclerotic changes. Other black pigmented cell groups are almost free of cell loss. Lower border of hemorrhage (*He*) appears in upper left corner. Nissl stain × 15

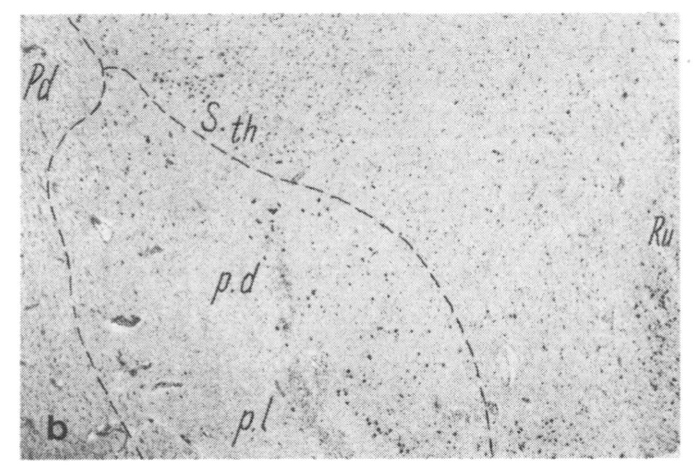

(b) Dorsal part of posterior main section of substantia nigra between subthalamic nucleus (*S.th*) and cerebral peduncle (*Pa*) of case No. 12, adjacent to Figure 78c. Severe cell loss in lateral (*p.l*) and dorsal (*p.d*) cell groups, with compensatory glia proliferation. Nissl stain × 5

Anatomical findings in the brain. The brain was of medium size. It showed considerable arteriosclerotic changes of the basal arteries (Fig. 99) without areas of arteriosclerotic softening being visible on the surface. Due to cerebral swelling, the left hemisphere was a little broader (Fig. 78a); slight atrophy was found in the frontal and middle parietal lobes. A lesion due to puncture with a superficial defect of the cortex was seen in the first right frontal gyrus; on the left a small navel-like depression caused by the second stereotaxic operation was found in second frontal gyrus (Fig. 78a). The needle track of the first operation could easily be seen in the white matter of the right frontal lobe and caudatum. The coagulated area was 7.5 mm wide and destroyed the area between the V.o.a and the pallidum internum including the internal capsule (Figs. 75a; 78a).

The middle of the coagulated area was 2 mm lateral (Fig. 75a) to the point intended on model brain. The coagulated area showed neither marked scar formation nor cavity changes, but it was surrounded by a narrow demyelinated zone. The lesion destroyed the lateral 2 mm at the base of the V.o.a, the neighboring nucleus reticulatus, and a band of the internal capsule at the level of the internal part of the pallidum and medial apex of the internal pallidum (Fig. 75a, c, d). On the ventral side of the area of destruction in the internal capsule, thin bundles of myelinated fibers that crossed the internal capsule from the lateral to the medial side were recognizable (Fig. 75d).

Loss of nerve cells had occurred in two of the cell groups in left substantia nigra (Figs. 76a, b): (1) in the anterior main part, in the large medial group surrounding the root bundles of the oculomotor nerve, with slight substituting growth of glia (Fig. 76a); and (2) in the posterior main part, in the lateral groups (Fig. 76b). The loss of nerve cells, therefore, showed the typical distribution pattern of paralysis agitans (HASSLER, 1938); only rarely does one find a spheroid cytoplasmic inclusion with a hint of layer formation for typical genuine Parkinson. The increased growth of the adventitious connective tissue around the arterioles with their thickened walls was remarkable (Fig. 76a). Increased capillaries surrounded the slight substituting growth of glia in the areas where cell groups had disappeared. Despite investigating serial sections, areas of complete softening or parenchymatous softening were absent in the substantia nigra. The gliosis due to complete nerve cell loss in the right shrunken posterolateral cell groups (Fig. 77c) overlapped with the glia growth due to the degeneration of the pes lemnisci profundus (P.Lm.pf) and fused with the continuation of the fiber bundle degeneration in the cerebral peduncle (Fig. 77b, c). This was secondary, descending from the coagulation in the transition zone of the internal capsule and cerebral peduncle (Fig. 75d), which had its largest extent (Pd) in Figure 77a. Despite the complicating cerebral arteriosclerotic focus in the medial substantia nigra (Fig. 76a), this case was not one of arteriosclerotic muscular rigidity but a constitutionally determined paralysis agitans occurring at the typical age.

The left thalamus had been largely destroyed by a massive hemorrhage, 21 × 22 mm (Fig. 78a). Its dorsal extension caused the lateral ventricle to be narrowed while its medial extension caused the left thalamus to extend over the midline. The hemorrhage communicated through the medial aspect of the thalamus with the third ventricle (Fig. 78b). The massa intermedia, present in the air encephalogram (Fig. 79), had been destroyed by the hemorrhage. A small subependymal hemorrhage was found (Fig. 78b) below the interventricular foramen, rostral to the large hemorrhage. A symmetrical subependymal hemorrhage was also found under the right interventricular foramen where the fornix enters the hypothalamus (Fig. 75a). The hemorrhage had destroyed the ventral thalamus and subthalamus (see Fig. 127a, c) up to the region of the posterior commissure and the medial thalamus except for a mediodorsal band. The dorsal thalamic nuclei had not been displaced laterally and ventrally. Where the hemorrhage bordered the thalamus, it destroyed only the nucleus reticulatus thalami (Fig. 127c).

Clinical-anatomical correlations. According to the x-ray, the V.o.a, which was the target, had been reached by the point of the electrode during the first stereotaxic operation. However, the anatomical examination showed that the middle of the coagulated area was 9 mm behind the interventricular foramen and 2 mm lateral to the calculated target point (Fig. 75a). It therefore was located in the nucleus reticulatus on the medial edge of the internal capsule. This deviation was explained by the fact that the target point was reached somewhat below the anatomical baseline, where the internal capsule was nearer to the wall of the third ventricle (Fig. 75d).

The effect of stimulation during the first stereotaxic operation must be attributed mainly to the internal capsule or to the fibers crossing it next

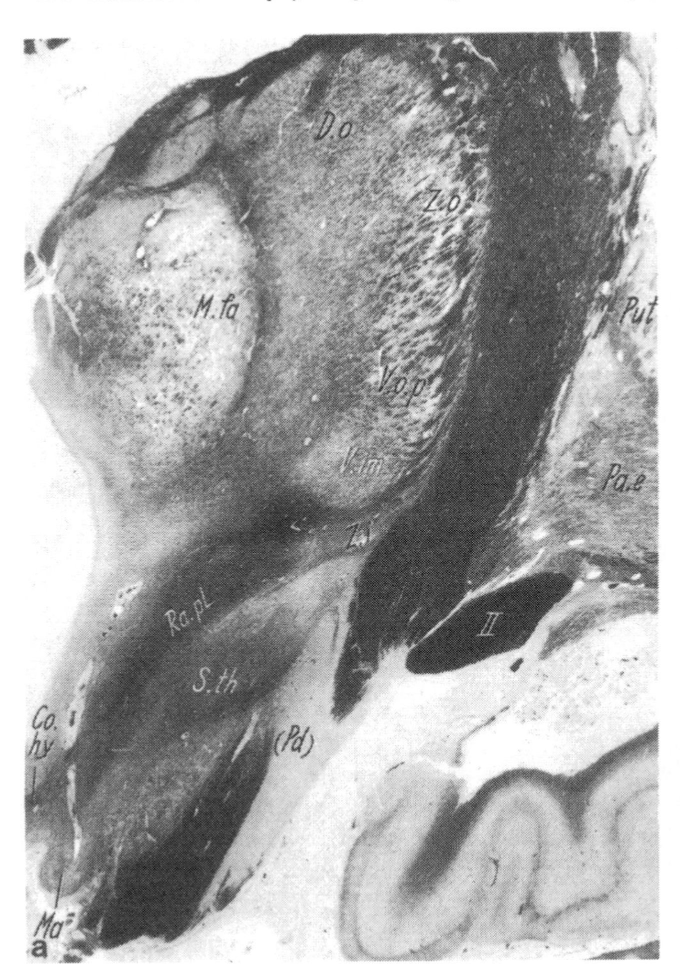

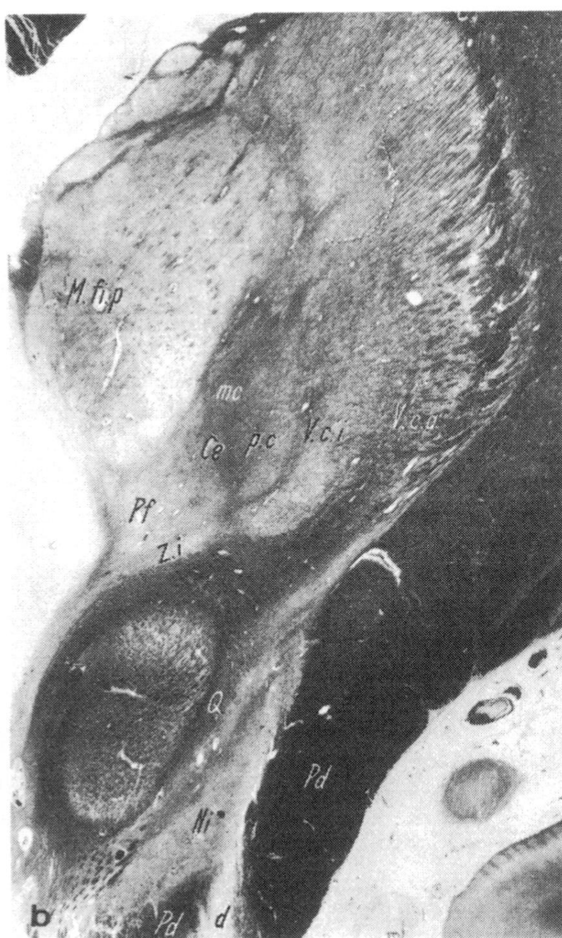

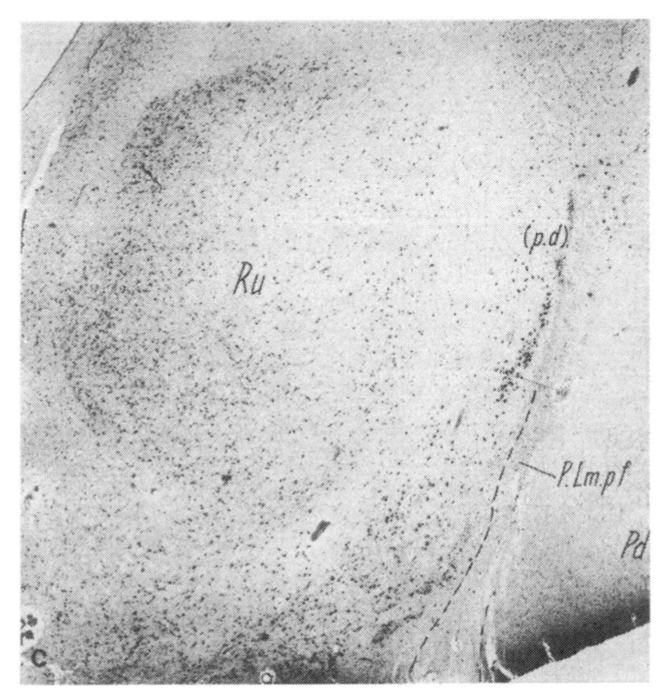

Fig. 77. (a) Cross section through right thalamus 1.8 mm behind Figure 75d shows degenerated portion of corticospinal fibers in cerebral peduncle (*Pd*). Part of subthalamic nucleus (*S.th*) above degenerated part of cerebral peduncle is atrophied and demyelinated, involving zona incerta above it. External pallidum (*Pa.e*) is only demyelinated in its dorsal sixth. Ra.pL prelemniscal radiation. Fiber stain. ×4
(b) Cross section of right thalamus of case No. 12, 5 mm behind Figure 77a. Degenerated part (*d*) of cerebral peduncle is descended further and is now in connection with lateral part of substantia nigra, which normally contains pes lemnisci profundus, whose fiber structure is lost (above *d*). Myelin stain. ×4
(c) Cross section through midbrain of case No. 12, 1.7 mm behind Figure 77b, shows dense glia proliferation in zona reticulata that replaces pes lemnisci profundus (*P.Lm.pf*). It continues downward to degenerated part of cerebral peduncle. Dorsolateral cell groups of substantia nigra (*p.d*) have deteriorated and lateral groups are replaced by glia growth that is fused with glia proliferation of pes lemnisci profundus. Nissl stain. ×10.3

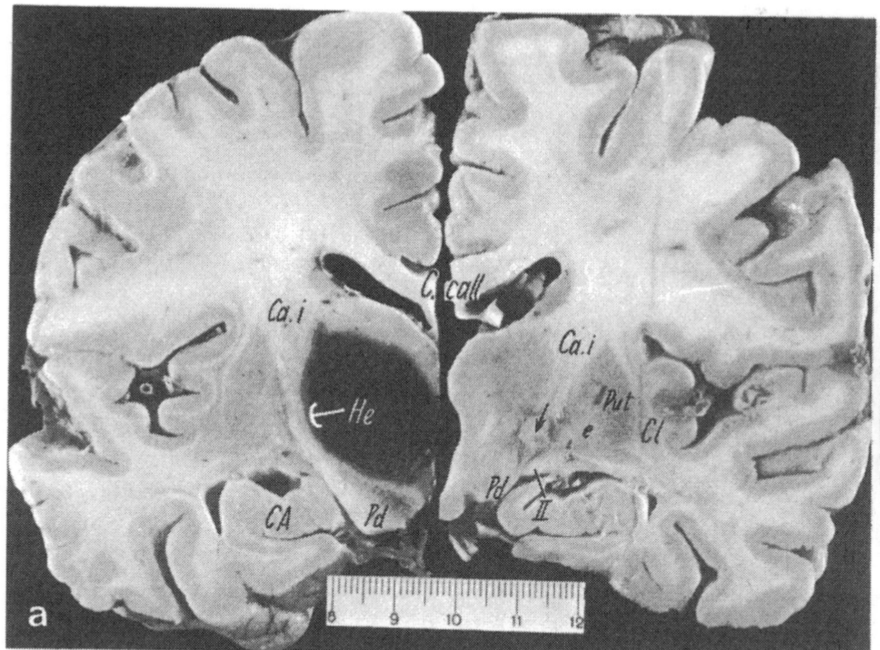

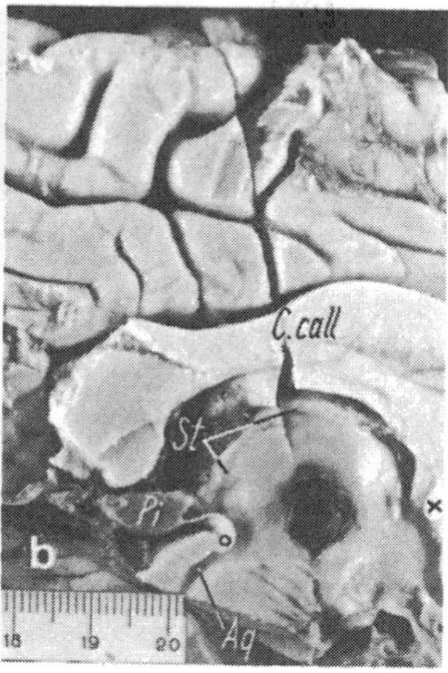

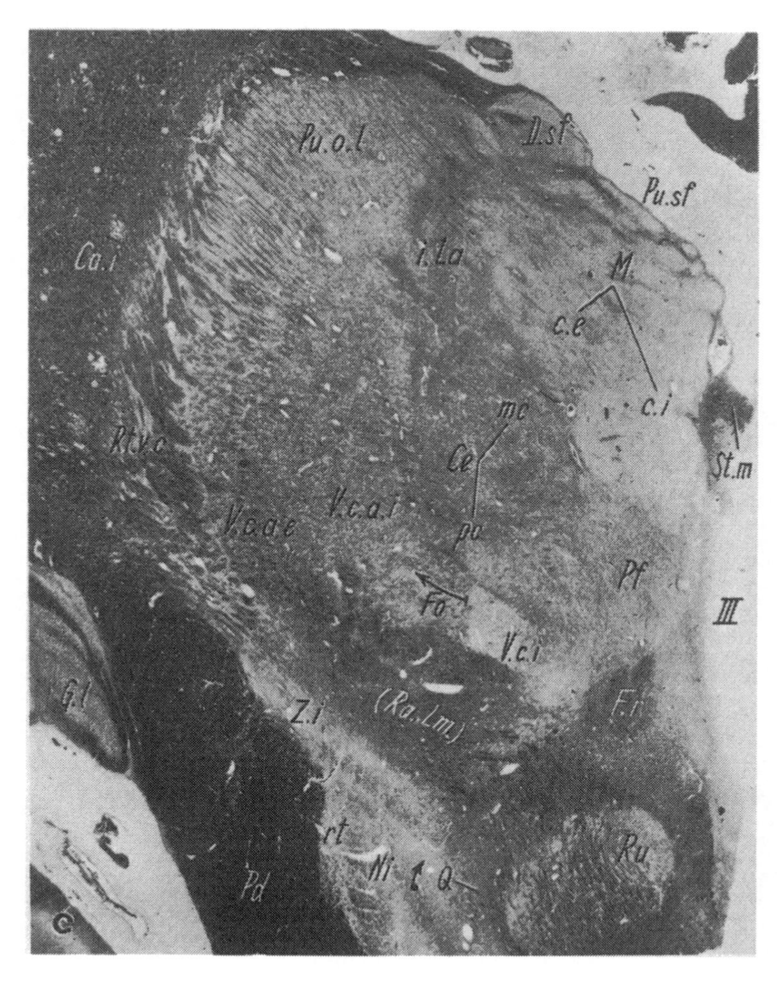

Fig. 78. (a) Unstained cross section through fixed brain in middle of thalamus in case No. 12. Left thalamus is occupied by a large hemorrhage. Therefore thalamus is enlarged, internal capsule is shifted laterally, and substantia nigra with peduncle (*Pd*) is shifted ventrally. Hematoma has thus penetrated region of sulcus hypothalamicus (Monroi) of 3rd ventricle. White matter of left hemisphere is also swollen. In right hemisphere, center of coagulation focus is situated in internal capsule between pallidum internum and V.o.p (*arrow*). Needle track can be recognized. Focus is separated from surroundings by fringe of softened tissue

(b) Macroscopic picture of left wall of 3rd ventricle in case No. 12. In middle of thalamus, hematoma expanded to subthalamus and 3rd ventricle, still partially covered by remnants of ependyma. No signs of hematocephalus can be seen 15 days after start of thalamic hemorrhage. Aqueduct (*Aq*) begins below posterior commissure (○).

Pi pineal gland; *St* stria medullaris; *X* anterior commissure

(c) Cross section, a few mm behind hemorrhage, shows swelling of all ventral thalamic structures: *centre médian* nucleus (*Ce*); parafascicular (*Pf*); ventrocaudalis anterior externus (*V.c.a.e*) and internus (*V.c.a.i*); ventrocaudalis internus (*V.c.i*); Radiatio lemniscalis (*Ra.Lm*); Fasciculus retroflexus (*F.r*). Fiber stain. ×3

to the V.o.a (Fig. 75c, d). It must be noted that bipolar individual stimuli caused questionable jerks of the left arm only when the strength of the stimuli was high (80–100 scale divisions). Stimuli at a rate of 4 cps also increased individual phases of the tremor and synchronized it only when the strength of the stimuli reached a higher level. Also, the unpleasant respiratory sensations and the short arrest of respiration in the inspiratory phase were in this case due to excitation of the fibers of the internal capsule next to the V.o.p. The acceleration of counting followed by the blocking of the stimulation effects, however closely resembled those from the V.o.p. The very unpleasant sensation of respiratory difficulty was the only unusual phenomenon. (We have fairly frequently observed arrest of breathing in inspiration when stimulating the pallidum at the rate of 50 cps.) The inactivation affected fibers of the internal capsule that run in a transverse plane 6–12 mm behind the interventricular foramen between the internal pallidum and the V.o.a. In addition, there was an incomplete interruption of the fibers of the ansa lenticularis (Fig. 75a, d) that run from the apex of the pallidum internum to bundle H_2. The coagulation also destroyed a part of the medial apex of the internal pallidum, and the lateral edge of V.o.a and V.o.p with the neighboring nuclei reticulati (Fig. 75a, d). The clinical effect of abolishment of rigidity in the left limbs for 3 years corresponded to the interruption of pallidothalamic fibers. Although the left-sided tremor was present to a slight degree on the operating table and up to the discharge 10 days later, it had disappeared completely at the first follow-up examination 3 months later and did not reappear in the next 3 years. This was due to inactivation of a large part of the V.o.p (Fig. 75d).

The effects of stimulation in the left V.o.p during the second operation included intensification of the tremor and its synchronization with the stimulating frequency at 4 and 8 cps as well as blocking of the tremor at the rate of 50 cps. In this case the effects were related to an area of V.o.p tissue 1 mm dorsal to the baseline and beyond. The slight flexion of the contralateral arm could be interpreted as being due to a spread of the stimulus into neighboring fibers (Fig. 127c) of internal capsule, since the autopsy showed them to be not further away than 3 mm. The stimuli at a rate of 50 cps caused a blocking of counting that obviously also could have originated in the V.o.p substrate.

Coagulation of the left thalamus led to a large hemorrhage (Figs. 78a, b; 127), which was responsible for (1) the impairment of consciousness that lasted until death and (2) the right-sided hemiparesis. Such a large thalamic hemorrhage presumes damage to a medium-sized arteriole. The point

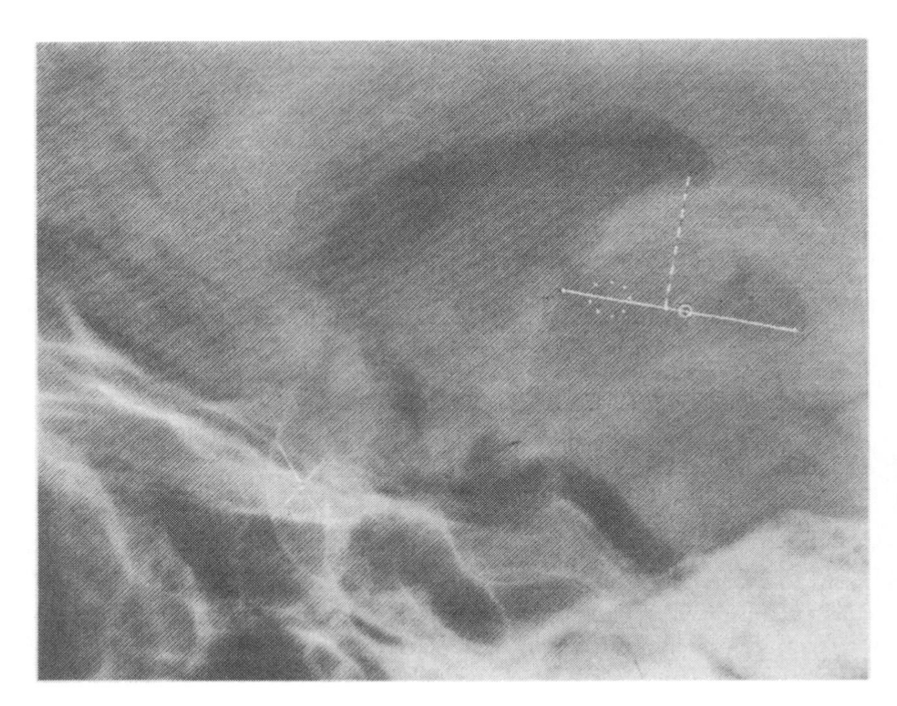

Fig. 79. Detail from lateral pneumoencephalogram of case No. 12 before 1st operation. In 3rd ventricle is shadow of large massa intermedia (within *dots*). Both anterior horns are dilated, but each to a different extent. Height of thalamus, in midst of baseline, is hyphenated. Circle in baseline is individual target V.o.a/p

of the electrode obviously had been lying in the neighborhood of such an arteriole. Since the current dropped at the first coagulation, it must be assumed that the temperatures around the electrode were higher than intended. In addition, coagulated red and black tissue was stuck to the point of the electrode. It is, therefore, probable that a vessel near the coagulating electrode was damaged at the first coagulation. At the two following coagulations it was necessary to apply 6.6 to 7 watts in order to achieve temperatures of 73° C and 70° C, respectively. This led to current of 200 and 240 mA at a very low voltage. The very low resistance of the tissue had apparently been caused by the blood outside the vessels.

The pressure of the blood outside the vessels was so considerable that the hemorrhage penetrated 10 mm to the wall of the third ventricle and broke into it (Fig. 78a, b). Further hemorrhages had occurred under the ependyma, below the interventricular foramen on both sides, but more rostrally right than left. The left fornix also sustained a small hemorrhage in the lateral ventricle. The multiple hemorrhages outside the field of operation were probably related to a general tendency to bleed, which could be attributed to the acute renal failure. The generalized clonic twitchings on the day before death could be regarded as a further sign of uremia. Further factors favoring the cerebral hemorrhage were (1) the severe (Fig. 99) arteriosclerosis of the cerebral vessels and (2) the displacement of brain matter in the area of the diencephalon and the interventricular foramen. This displacement must have particularly impeded the circulation in the anterior chorioid artery. The areas below the interventricular foramen, which showed subependymal hemorrhages on both sides, and the fornix in the lateral ventricle belonged to the distribution area of this artery.

The hemorrhage in the left thalamus destroyed the following thalamic structures (Fig. 127a, c): the ventral half of thalamus excepting the ventrocaudal nuclei, most medial structures including lamella medialis, parafascicularis, *centre médian* have lost all cells (Fig. 127a, c) and the zona incerta up to its midbrain extension is destroyed. The nucleus reticulatus thalami lateral to the V.o.a and a thin border area of the internal capsule also were victims of the hemorrhage. A tremor at rest in the right arm occurred again in the semiconscious patient despite the large extension of the thalamic destruction with paresis. The tremor could not be explained by the destruction of the ventral nuclei. It is possible that the additional lesion of the medial thalamus or of the subthalamic structures was responsible.

Despite the tracheostomy, the persistent impairment of consciousness and uremia due to hemorrhagic pyelitis led 15 days after the second operation to confluent purulent bronchopneumonia and to circulatory failure in the presence of myocarditis.

Case 13. F.L., aged 58; paralysis agitans; loss of consciousness caused by hemorrhagic infarct after thrombosis of right chorioid artery. First stereotaxic operation performed in left V.o.a (no. 1871); second, in right zona incerta below V.o.p (no. 2337). Death occurred Juli 21, 1964, 29 days after second operation, from pulmonary embolism.

History. Parkinson's syndrome developed at age 55 with tremor and rigidity of the right hand and foot. Eighteen months later the left side became involved to a lesser degree. All active movements were slowed down to such an extent that the patient had to have help for everyday activities and could be fed only chopped-up food. Medical treatment with trihexiphenidyl (artane), benztropine (cogentinol), and biperiden (Akineton) produced only moderate improvement of the symptoms. The patient herself, therefore, desired surgical treatment.

Clinical findings. At the first admission, rigidity and tremor of moderate degree were present in the right limbs. The right foot turned to the outside, there was adiadochokinesia, there was no marked akinesia, and there were hardly any vegetative disorders. Her speech could not be understood. As far as the patient was concerned, the tremor of the hand was the most important symptom.

The internal organs showed no abnormalities (BP 150/90 mm Hg). In the C.S.F., the Pandy test was positive. Helium encephalography was well tolerated; it showed a coarsening of the anterior horns and a dilatation of the lateral ventricles, particularly on the left side. The subarachnoidal space was increased in the right parietal area.

The *first stereotaxic operation* was performed November 23, 1962, in the left V.o.a. Stimulation with the electrode was carried out 1 mm medial and 1.3 mm rostral to the target point. Single bipolar stimuli caused no change in the marked tremor of the right arm. Stimuli at a rate of 4 cps produced jerks in rhythm with the stimulation. Stimuli at a rate of 8 cps first stopped the tremor in the right arm, then produced fast, irregular jerks of the right hand. Stimuli at a rate of 50 cps led to spontaneous speech, although she was asked not to speak, and to a stopping of counting. Stimulation of low intensity resulted in mydriasis, facial flushing, and upward jerking and flexion of the right arm.

Coagulation with a 4×2 mm thermoelectrode was performed at 70° C at the target point and in positions +2 and +4. Radiologically there was a medial displacement of 1 mm. The first coagulation (+2) led to stoppage of the tremor in the right arm and leg. After the coagulation in position +4, rigidity was abolished in the right arm (except in the wrist). After the last coagulation in position 0, tremor and almost all rigidity in the right wrist and ankle joint were abolished. A marked facial difference for expressive movements appeared on the right side with dilatation of the palpebral fissure.

Postoperative course. Upon discharge 8 days after the operation, rigidity and tremor on the right side had been completely abolished except in the foot. Speech had improved slightly; associated movements had returned on the right side. She could dress and feed herself. No mental abnormality was noted. Blood pressure had gone down to 120/80 mm Hg.

In the spring of 1963 tremor of the left hand and difficulty in walking became very noticeable. Upon readmission to the hospital on June 18, 1964, her condition was much improved in regard to everyday activities and to eating as compared to her condition before the first operation. She was unchanged mentally. The right limbs were free of rigidity and tremor, but on the left side, tremor of the arm, and moderate rigidity in the leg and slight rigidity in the arm were present. Severe adiadochokinesia was found on the left side. When walking, the patient dragged her left foot. Associated movements on the left side were much reduced. The blood pressure was 115/80 mm Hg, and had not, therefore, reached the peak measured before the operation (150/90 mm Hg).

The EEG showed a slowed-down α rhythm of 7–9 cps and on the left side, focal steep θ waves of 5–6 cps, indicating that the local changes due to the first operation had not diminished.

The *second stereotaxic operation* was performed June 22, 1964, in the right zona incerta with the starting coordinates 12, −5 (below the baseline), and 9. With the point of the central cannula, the target point was reached satisfactorily (1 mm ventral). The string electrode, which was insulated and which was inserted 6 mm dorsally and caudally and 30° medially, lay in the horizontal plane of the target point. When the string electrode was extruded (Fig. 80) without giving stimulance (so called "Setzeffekt"), it led to an interruption of the tremor that lasted for 6–7 min, and the tremor could be provoked either by emotion or by performing a calculation. During bipolar stimulation between the point of the insulated string electrode and the central cannula, the tremor, which in the meantime had reappeared in the left fingers, stopped when 17 volts were used. At 24 volts the tremor in the left hand and forearm increased. Stimuli of 12 volts at 4 cps caused irregular jerks in place of the tremor. From 8 volts on, stimuli at a rate of 50 cps blocked the tremor without the patient making any mistakes during serial subtraction. After this she made uncertain statements about "a feeling of swelling".

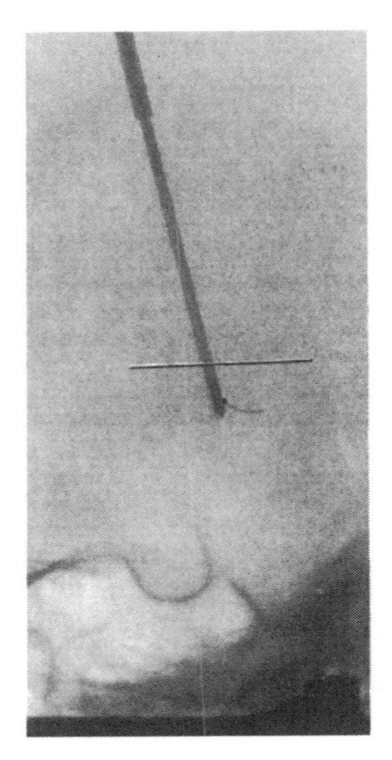

Fig. 80. X-ray check of position of protruded string electrode in target of zona incerta of case No. 13. Tip of electrode sheath, situated below V.o.p, has coordinates (in relation to baseline) 12 mm behind interventricular foramen and 5 mm below baseline. Baseline is drawn in afterward to demonstrate position of electrode and direction of protruded string electrode. Vertical and sagittal coordinates of target on operation day are transferred from pneumoencephalogram

Although at the beginning of the coagulation, tremor was not present and could not be evoked, and rigidity had already improved, the coagulation was done with an uninsulated string electrode. With reference to the model brain, the point of the string electrode lay 17 mm behind the interventricular foramen. The second coagulation with the string electrode was made 1 mm further dorsally. As a result, the rigidity of the left leg was influenced as well as it had been on the right side. For the third coagulation, the string electrode was moved 45° laterally and caudally. After this, the left-sided signs almost completely disappeared. There was no expressive weakness of the facial muscles, no spastic signs, and no increase in the speech disorder.

Postoperative course. Nothing special was noted apart from headaches on the first postoperative day. On the evening of the following day (June 23) she became somnolent and eventually lost clearness of consciousness. The pupils became small and did not react. Both arms were moved, the left less than the right. Since treatment with osmotic fluids did not improve consciousness, the surgical wound was inspected and the trephine hole enlarged. Neither an epidural nor a subdural hematoma was found, but the puncture of the ventricles showed increased pressure. After reducing the C.S.F. pressure, her condition improved; thus, a more extensive inspection was not made. During the night a tracheostomy was performed in order to improve ventilation. Only when painful stimuli were repeated could defense movements be evoked. On the right side these movements seemed coordinated while on the left, an extensor spasm occurred in the arm; however, the leg remained motionless. Extensor spasms were observed on both sides. Two weeks later no evidence of paralysis and no spastic signs were found. She opened her eyes, she reacted when spoken to, and she carried out simple instructions, but she did not respond verbally and she showed no spontaneity. A transitory rise of blood pressure became normal by the sixth postoperative day. Subsequently, bronchopneumonia developed insidiously and, at the same time, the cerebral state deteriorated. Brain stem automatisms occurred. Fourteen days after the operation the EEG showed a θ rhythm. There were short runs of high, steep δ waves in the frontotemporal region, particularly on the right side, with a frequency of 0.75–1.0 cps. In an arteriogram the right anterior chorioid artery could not be demonstrated. Three and a half weeks after the operation the patient suffered an acute severe pulmonary embolism with circulatory collapse. She no longer reacted to pain. The collapse could be ameliorated for a while, but the patient remained deeply unconscious. Her cardiac and circulatory states deteriorated again. Death occurred 29 days after the second operation, due to circulatory failure caused by a massive pulmonary embolism.

Anatomical findings in the brain. The frontal and parietal lobes showed definite cortical atrophy. The left lateral ventricle was considerably dilated. No pressure cone was found on the cerebellum, but there was a bulge measuring 35 × 35 mm on the right posterior frontal gyrus, corresponding to the dilated opening in the bone. The stimulating or the puncturing electrode passing the lateral ventricle had damaged a branch of the right choroidal artery (Fig. 81, white arrow). This damage caused softening of the dorsal thalamus, massa intermedia, and the anterior aspect of the wall of the third ventricle (Figs. 82a, b, 83a, b).

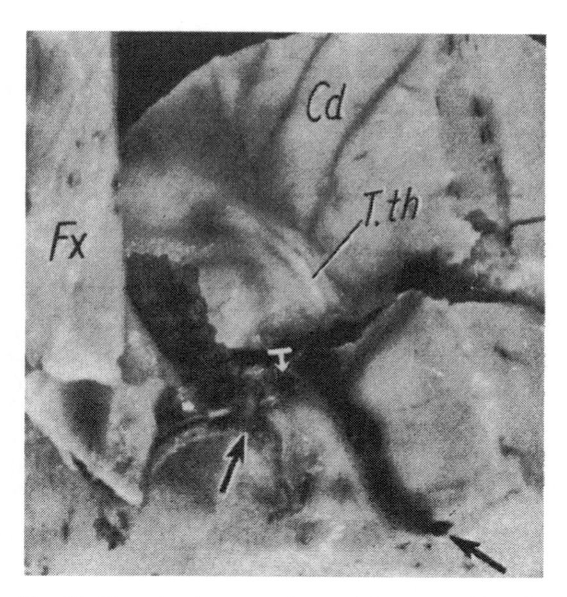

Fig. 81. Unstained aspect of ventricle in case No. 13. Contours of right thalamus, left of taenia thalami (*T.th*) are seen in cella media ventriculi lateralis, exposed by removal of ventricle roof. Behind section through the preparation, stereotaxic electrode tract (*white arrow*) is situated medial to branch of small vein. Lateral to it, small blood points are accumulated on branch of vena terminalis (*white arrow*). Medial to needle track, anterior choroidal artery is lowered and includes thalamic substance. This artery originated from dichotomy of choroidal artery (*black arrow with white border*). Posterior choroidal artery runs medially and caudally. In region of anterior choroidal artery (*arrow, white bordered*) is a malacia that destroyed dorsal wall of thalamus.
Black arrow another puncture;
Cd caudate nucleus; *Fx* fornix. × 2.5

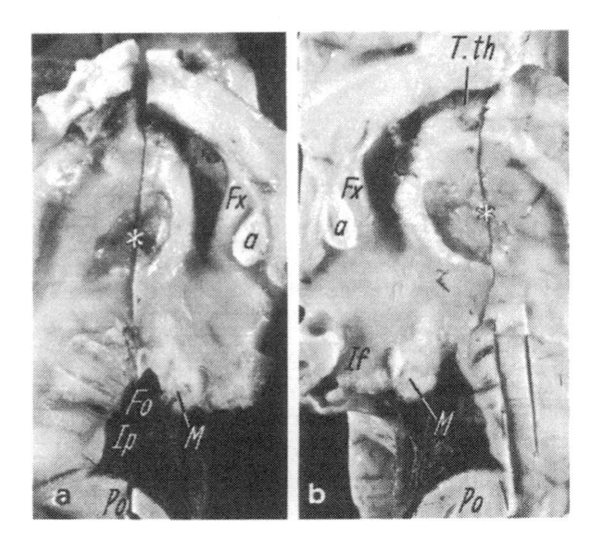

Fig. 82a and b. Medial aspect of left and right walls of 3rd ventricle in case No. 13. On left wall of 3rd ventricle, bean-sized wall deficiency is visible behind ascending part of stria medullaris thalami (*) with fleabite hemorrhage in its midst.
(b) In right lateral wall of thalamus is also bean-sized deficiency of ependyma with expanded softening below it with irregular border. These wall deficiencies are caused by occlusion of anterior choroidal arteries, which were touched by electrode track in lateral ventricle (Fig. 81). Massa intermedia, visible in x-rays, was destroyed by infarction

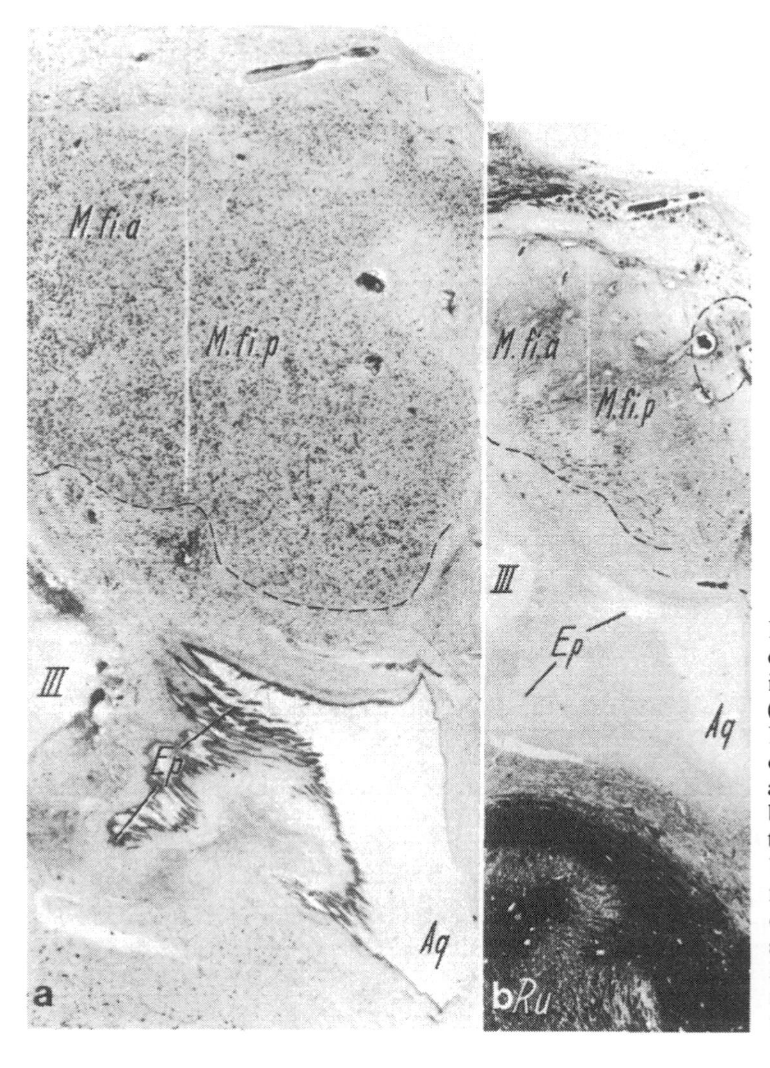

Fig. 83a and b. Sagittal section through caudal region of right thalamus. The thalamus is connected with midbrain by ependyma (*Ep*), in fiber- and cell staining of case No. 13. This section contains posterior part of 3rd ventricle before it is continued in aqueduct (*Aq*); thus 3rd ventricle is partially bridged by ependyma and subependymal tissue. In ependymal bridges between midbrain and thalamus, elongated infarction foci, same as those in subependymal part of medial thalamic nucleus, destroy commissural and paracentral nuclei partly. Hatched line indicates small necrosis due to hemorrhage. *Ru* red nucleus; above that, transition into aqueduct

On both sides of the substantia nigra, many black pigmented nerve cells were missing in a distribution that was typical for paralysis agitans. In the shrunken lateral edge of the main posterior section (Fig. 84), these cells were completely missing, but they had been spared in the medial groups containing smaller nerve cells. Spheroid inclusion bodies were found in the plasma of some damaged cells where they had replaced the melanin granules. Similar inclusions were found in the black pigment-containing cells of the locus caeruleus and the dorsal nucleus of the vagus without many cells having disappeared. These findings led to the conclusion that, in this case, the Parkinson syndrome had been caused by a genuine paralysis agitans. Many nerve cells in the basal nucleus contained an excess of lipofuscin. There was no clear loss of cells but some cells showed a glassy nerve cell change.

On both sides of the anterior thalamic part of the third ventricle, there were defects of the ependyma with an irregular yellowish-brown background and a dissolution of the massa intermedia (Fig. 82a, b). In addition, in the right wall of the third ventricle, there were shallow defects about 1 mm deep and the remains of subependymal hemorrhages caudal to the defects (Fig. 83a, b; 84).

The coagulated lesion in the left basal ganglia from the first stereotaxic operation had become partly a scar and partly a cyst. The lesion was angular in shape, with a vertical and a horizontal limb (Fig. 84). The vertical limb destroyed the V.o.p and parts of the nuclei lying above it and led to development of a scar in the lateral parts of the V.o.a and L.po, in the neighboring nucleus reticulatus and in the caudal V.o.i. The horizontal limb passed through the internal capsule at the level of the mamillothalamic bundle of Vicq d'Azyr from the base of the V.o.a to the medial apex of the pallidum. This limb formed a cyst that extended about 6 mm rostrocaudally. The tegmental field H of Forel showed considerable demyelination.

The lesion by coagulation in the right hemisphere had destroyed a large vertical segment in the V.o.a and V.o.p (Fig. 85) and a horizontal segment running in a caudal direction and ending 6 mm in front of the posterior commissure. The caudal segment again consisted of two parts (Fig. 86): (1) a coagulated area with coagulation

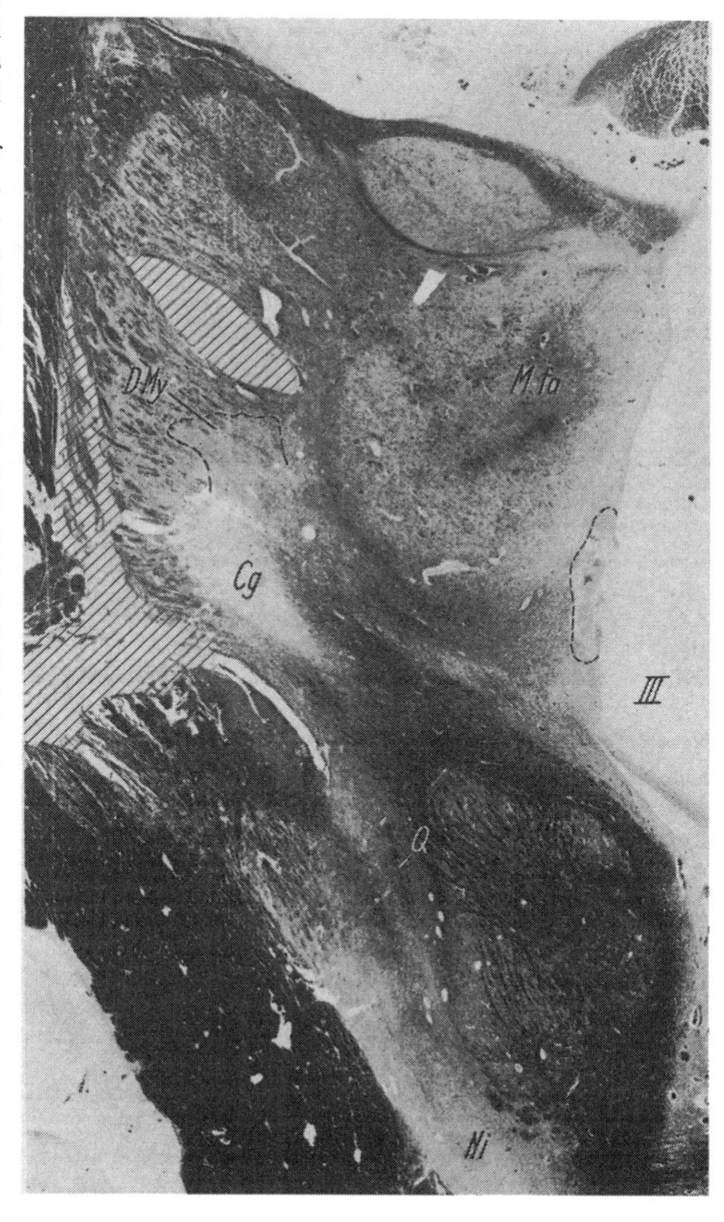

Fig. 84. Cross section through left thalamus midbrain region in case No. 13. Caudal pole of coagulation focus in V.o.a and V.o.p extended with ventral process into radiation of red nucleus and destroyed dentato-thalamic fibers. Above focus many efferent fiber bundles are demyelinated, but only in region of V.o.p. Remnant of softening focus in wall of 3rd ventricle is the level of paracentral nucleus. In substantia nigra (*Ni*) the lateral fibers are almost lost, and fiber accumulation dorsomedial to substantia nigra (*Q*) is thinned. Autolytic foci are cross-hatched. Fiber staining. × 5.5

necrosis but without dissolution of the structure of the tissue, going around the needle track and having a diameter of 3.4 × 2.7 mm; (2) a laterodorsal part, measuring at its greatest length 3.7 mm. While the lesion spared the internal capsule, or its transition to the cerebral peduncle, it destroyed caudolateral parts of the zona incerta (Fig. 86) (as intended; MUNDINGER, 1965), the V.im and the middle part of the tegmental area dorsal to the substantia nigra; it probably extended into the lateral parts of the prelemniscal radiation, but spared the Q-bundle of SANO (1911) (Fig. 85).

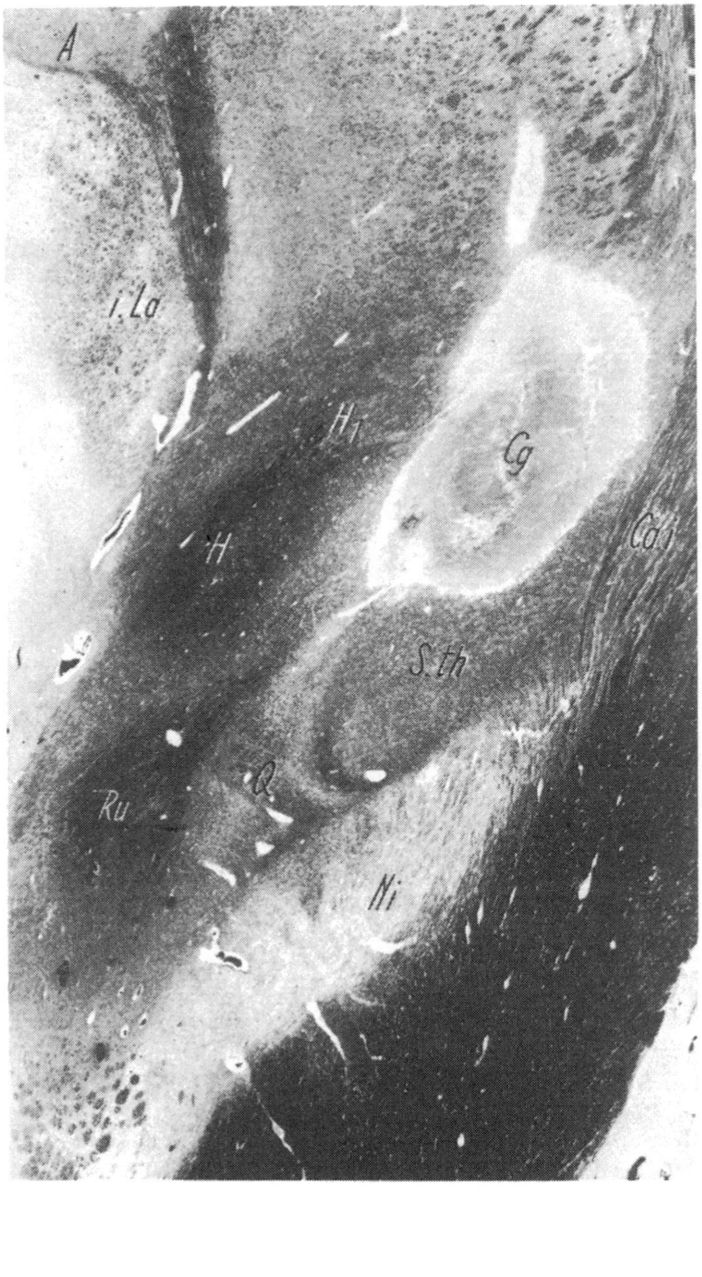

Clinical-anatomical correlations. According to the autopsy, concentric bipolar stimulation with a 5-mm difference in the right V.o.a had reached the nucleus (Fig. 85). A "Setzeffekt" with suppression of tremor was produced for 7 min only by introduction of the electrode. Current impulses of 4 cps made the tremor more rhythmic and slowed it down. Stimuli at a rate of 8 cps had caused the tremor to stop and provoked irregular twitches. These are the typical results of stimulating the oral ventral nuclei. Stimulation at a rate of 50 cps interrupted the requested speaking (counting) and evoked an urge to speak spontaneously even though he was told not to speak. This response, as well as the upward jerking of the contralateral arm, the turning of the eyes to the contralateral side, and the mydriasis, however, formed part of the effect of stimulating the supplementary motor area. (In the thalamus such results of stimulation can frequently be seen to be provoked in the area lying rostromedially to the V.o.a in lateropolar nucleus, see Fig. 61).

The three coagulations with the straight electrode at the target point up to position +4 led on the left side to complete destruction of the lateral V.o.a and large parts of the V.o.p (Fig. 84), while preserving the rostral V.o.i. During these coagulations, however, an infarction of the internal capsule developed up to a height of 3 mm, probably due to the thalamic coagulation causing blockage of a vessel that supplied parts of the internal capsule. Destruction of the left V.o.a, V.o.p, and a rostral part of the internal capsule had effected complete abolishment of the tremor and almost complete abolishment of rigidity with the exception of the ankle joint. Postoperatively, a facial mimic weakness with dilated palpebral fissure was observed and even the associated movements of the right side reappeared. There were no paralyses or abnormal reflexes in the right limbs. Preoperatively, the active movements were so badly im-

Fig. 85. Cross section through right anterior thalamus of case No. 13. Coagulation focus in base of V.o.a/p and in zona incerta produced by string electrode in zona incerta resembles rostrally that of the straight electrode. It is separated from surrounding by white zone of demarcation. Focus invades dorsal edge of subthalamic nucleus (*S.th*). Above focus, electrode track can be seen as 0.5-mm wide substance deficiency. String electrode protruded from focus caudally into zona incerta. Above anterior pole of red nucleus (*Ru*) with rubrothalamic fiber radiation, field H is met by section from which only a few H_1 bundles leave dorsolaterally. Fiber stain. ×6.7

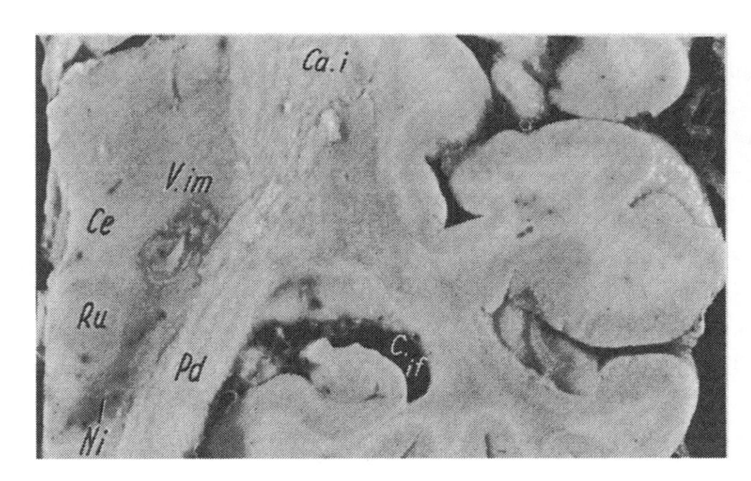

Fig. 86. Unstained cross section through caudal thalamus of case No. 13. Coagulation foci in medioventral zona incerta and in base of V.im are evident. Medial focus, which spares superior border of substantia nigra, shows in its midst electrode track of string electrode and in surroundings an oval focus of coagulation of 3.8 mm diameter. Dorsolateral to it, there is crumbled coagulation tissue produced by repositioning of string electrode. Centre median (*Ce*) is intact. Internal capsule is reached by border zone of coagulation but not damaged by it. ×5.6

paired that the patient had to rely on the help of others for everyday activities. After the operation she could again feed and dress herself. It follows that the lesion of the internal capsule on the left side lateral to the V.o.a had not impaired voluntary movements. The major part of the corticospinal systems in the posterior part of the internal capsule was intact; thus the patient was able to move the right side actively following abolishment of tremor and rigidity even the associated movements reappeared. The lesion in the left diencephalon had interrupted the pallidothalamic systems twice: (1) where they pass through the internal capsule at the level of the zona incerta and (2) at the base of the V.o.a (Fig. 84) after the fibers of the lenticular fasciculus turn in Forel's field H. This also explained the exceptionally marked demyelination of field H. After the operation, speech, which had been changed by the Parkinsonism, was improved but certainly to a lesser extent than the limbs. According to the somatotopic organization, the improvement could be related to the fact that the internal parts of the oral ventral nuclei were not destroyed. It is possible that supranuclear efferent fibers of the apparatus for articulation were interrupted in the damaged parts of the internal capsule. In this way the capsular part of the coagulation lesion may, perhaps, have caused impairment of the supranuclear regulation of articulation by interruption of the efferent fibers while the thalamic part in the V.o.a at the same time caused an improvement.

A part of the zona incerta, 12–18 mm behind the interventricular foramen and 5 mm basal to baseline, was reached in right hemisphere (Fig. 86) with the string electrode. Even before bipolar stimulation, the mechanical movement of the string electrode in the zona incerta caused an interruption of the tremor for 6–7 min and a reduction of rigidity. The recurrent tremor was blocked by stimuli at 4 and 8 cps, which then caused irregular twitches. Stimuli at a rate of 50 cps caused a blocking of the tremor as soon as 8 volts had been reached, without interfering with speaking or subtracting. At the same time the patient had an uncertain feeling of "swelling" on the left side. Unipolar stimulation in the zona incerta ventral to the V.o.p (12 mm behind the interventricular foramen) led to coarse jerks, a slow tremor, flexion of the left arm, and increased inbreathing—but no feelings of anxiety.

Stimulation in the longitudinal direction of the zona incerta produced results similar to those produced by stimulation in the V.o.a, which was reached by the electrode point (Fig. 85). This finding favored the assumption that the same neuronal substrate was excited in both places. The substrate probably consisted of the dentato-thalamic fibers, which, leaving the rostral pole of the red nucleus, passed through the zona incerta and entered the base of V.im and V.o.p. However, when establishing localizing relationships, it must be remembered that the sheath of the string electrode reaches the pallidothalamic fiber bundles rostrally. It seems more likely, therefore, to correlate the flexion of the arm and the increased inspiration evoked by unipolar stimulation below the oral ventral nucleus with stimulation of the pallidothalamic fibers, rather than with stimulation of the dentato-thalamic fibers (Fig. 85).

The act of pushing the string electrode (caliber, 0.4 mm) through the zona incerta for 6 mm was followed by a cessation of the tremor for 6–7 min and by a persistent reduction of rigidity. This could be explained by the mechanical action on the tissue, which blocked conduction in the fibers of the neighborhood and which probably was partly due to a serous exudate of the surrounding capillaries. This phenomenon corresponds to the *Setz-Effekt* of HESS in his animal experiments and also agrees in this case with the duration of 7 min. Introducing an electrode into the thalamic nuclear areas produces this phenomenon only rarely and then in a weak form; the difference is due to the concentration of the neuronal elements in the fiber bundles.

The tremor, which had reappeared for a time, did not recur after several stimulations with frequencies of up to 50 cps, nor could it be provoked again. It was therefore not possible to say anything in this case about the immediate effect of the coagulation on the tremor, since it had already disappeared before, as described previously. The left-sided rigidity had already diminished before the coagulation. After the first coagulation in the same position (30° in a medial direction), the rigidity diminished further. After the third coagulation (Fig. 86), performed 45° in a lateral direction, the rigidity of the left side was completely abolished, without the appearance of additional disorders of articulation, emotional facial weakness, spastic signs, or psychologic changes. This coagulation destroyed lateral parts of the zona incerta (Fig. 86) in a cylinder going in a longitudinal direction. At its point this cylinder had a diameter of 3.4 × 2.7 mm, approximately the same size as a V.o.p coagulation.

The histopathologic findings were interesting. On the cut surface of the tissue was a needle track (Fig. 86) that looked like a point (diameter of 0.4 mm). Around the track was a circular coagulation necrosis without signs of liquefaction, surrounded in turn by a small destroyed zone bordering on the normal tissue. In the latter zone the lesion resembled those that had been produced by the thermoelectrode but its measurements were much smaller. However, the part of the lesion near the central cannula was much more extensive (Fig. 85).

Because the combined coagulations in the base of the oral ventral nuclei and in the zona incerta had abolished tremor and rigidity, they probably had destroyed by coagulation the same substrates as the coagulations restricted to the base of the oral ventral nuclei – the afferent fibers of the oral ventral nuclei of the thalamus. Based on this one case it is impossible to determine whether the destruction of efferent fibers of the zona incerta itself or of other interrupted neighboring structures contributed to the effect, particularly because in this patient, the result could be observed without complications only during the first $1^1/_2$ days. A facial palsy of expressive movements, which occurs frequently after coagulations of the V.o.p had not been caused even temporarily by this lesion.

The second stereotaxic operation was performed without any complications. A rapidly increasing severe impairment of consciousness occurred $1^1/_2$ days later, followed soon after by a reduction in movements of the left arm. The use of osmo-onco-therapeutic measures did not produce a change in the level of consciousness – in fact the patient became comatose, with small nonreacting pupils. Inspection of the surgical wound 36 h after the coagulation excluded the presence of a superficial hemorrhage. Puncture of the ventricles showed an increased C.S.F. pressure, which was relieved. Because this was followed by an improvement of the general condition, the depth of the brain was not explored further. Consciousness improved during the next two weeks. The patient reacted when spoken to, woke up but remained completely lacking in spontaneity, was akinetic, and did not speak. This behavior was in keeping with the EEG results – a θ rhythm (4–5 cps) and groups of steep frontotemporal δ waves particularly marked on the right side. Following the reduction of intercranial pressure, neurologically a reduction of activity of the left limbs and a transient extensor spasm of the left arm were observed.

The severe acute clouding of consciousness with increased intercranial pressure was not explained at post-mortem by a space-occupying hemorrhage in the basal ganglia. Instead, a very unusual lesion was found on both sides in the rostro-medial thalamus. It had destroyed the massa intermedia (Fig. 82), which could be seen in the x-ray; in the rostromedial thalamus there were areas of softenings on both sides about 1 mm deep (Figs. 82, 83). It is probable that in this location bilateral softenings with hemorrhagic edges occurred. The dark pigmentation below the defect of the ependyma of the right thalamus also favored such a hemorrhagic infarct of the rostral thalamus, while the third ventricle was not particularly dilated. The bilateral softening of the wall could have led to a temporary obstruction of the outflow of C.S.F., with increasing intercranial pressure to follow, particularly because the subependymal hemorrhagic lesion in the right thalamus extended almost as far as the opening of the aqueduct. At first, however, a rapidly increasing impairment of consciousness occurred in the patient about 30 h after the stereotaxic operation; this loss could not be caused by intercranial pressure, which occurred only later. The impairment of consciousness was probably explained by the bilateral softening of the massa intermedia and the medial thalamic areas. It is known that the hypnogenic zone of HESS (1944) and part of the unspecific projecting nuclei (MORISON *et al.*, 1941; DEMPSEY and MORISON, 1942a, b; MORISON and DEMPSEY, 1942, 1943; JASPER and DROOGLEVER-FORTUYN, 1946; HASSLER, 1949b, c) are found in this softened area. The interpretation that the impairment of consciousness was caused by the bilateral medial thalamic lesion, was further supported by the facts that after the relief of the intercranial pressure by puncture of the ventricle, a state of akinetic mutism remained and that the EEG was permanently severely disordered, as evidenced by θ and δ waves. The unspecific thalamic projection nuclei are responsible for regulating the rhythm of sleeping and waking. They probably are responsible for clearness of consciousness and for regulation of the basic rhythm of the EEG. Their destruction could have led to severe impairment of consciousness, with the corresponding EEG in this case as the only localizing sign.

Because of the severely impaired consciousness and a lack of spontaneity, bronchopneumonia developed; death was caused by a massive pulmonary embolism $3^1/_2$ weeks after the operation.

Case 14. H.P., aged 57; postencephalitic Parkinsonism with pallidotomy and striatal apoplexy. Death occurred 32 days after the stereotaxic operation in left pallidum internum (no. 411), performed July 11, 1958, from bronchopneumonia.

History. Illness started at age 52 following a severe excitement. It began with tremor, especially of the legs, and developed rapidly, so that he was dependent on other people's help from age 55 on.

Clinical findings. Severe rigidity and akinesia, right more than left, with approximately symmetrical brisk reflexes. The legs were particularly badly affected: he walked with small steps; associated movements were missing. Weakness of the facial nerve was found on the right side. In the internal organs the following abnormalities were noted: pulmonary emphysema due to age, sclerosis of the aorta, and moderate cyanosis of the lips. In the EKG a right-sided damage to the auricle was suspected but no contraindication was found against surgical treatment. The finding in the C.S.F. was marginal. An encephalogram showed considerable dilatation of the internal and external spaces, particularly in the frontal region (Fig. 87a). Encephalography was tolerated without a reaction.

A *stereotaxic operation* was performed June 9, 1958, in the left internal pallidum. The blood pressure rose to 190/90 mm Hg even before the introduction of the electrode. Stimulating at the target point at a rate of 25 cps evoked a slight tremor of the right hand. Stimuli at 25 and 50 cps caused an inhibition of counting. Stimuli at 50 cps resulted also in an extension of the right arm. After the coagulations with the spark-gap diathermy apparatus, using an electrode that was bare at 4 or 2×1.1 mm $\varnothing$ in positions -2 to $+8$ (total of 11 coagulations), a definite reduction of the rigidity in the right arm and leg was found on examination, but the patient denied this result. Four further coagulations were then performed with the string electrode moved 4.5 mm rostrally, medially, and laterally, in positions 0, -2, and $+4$. After the 14th coagulation the patient became tired. After the 15th there was slight impairment of consciousness. After one of the coagulations with the string electrode, there was a definite loosening of the arm and better active mobility of it, while the right angle of the mouth and the right hand twitched. At the end of the operation there was no paralysis. When called to loudly, the patient reacted but definitely was tired.

Postoperative course. During the afternoon of the day of operation the patient could hardly be roused and had paresis of the right arm and leg with extensor response. On the day of the operation the C.S.F. was bloodstained; the next day it contained much blood. The C.S.F. pressure was 300 mm H_2O. On day 3 after the operation, a tracheostomy was performed and, because of suspicion of an intercerebral hemorrhage, a carotid angiography was performed on the left side (Fig. 87b, c). It showed severe arteriosclerosis of the cerebral vessels, with space occupation in the basal ganglia. Seventeen days after the operation the EEG was very abnormal. Its basic rhythm was 4 cps. It contained very slow δ waves from 0.75–1 cps on both sides, right more than left. The EEG indicated a gross organic lesion in the basal ganglia. Because of the deterioration of the general condition and the signs of increasing intracranial pressure, a cerebral puncture was performed at the site of the coagulation on day 24 after the operation. This produced no clinical improvement. On day 26 the C.S.F. pressure was normal and the C.S.F. was

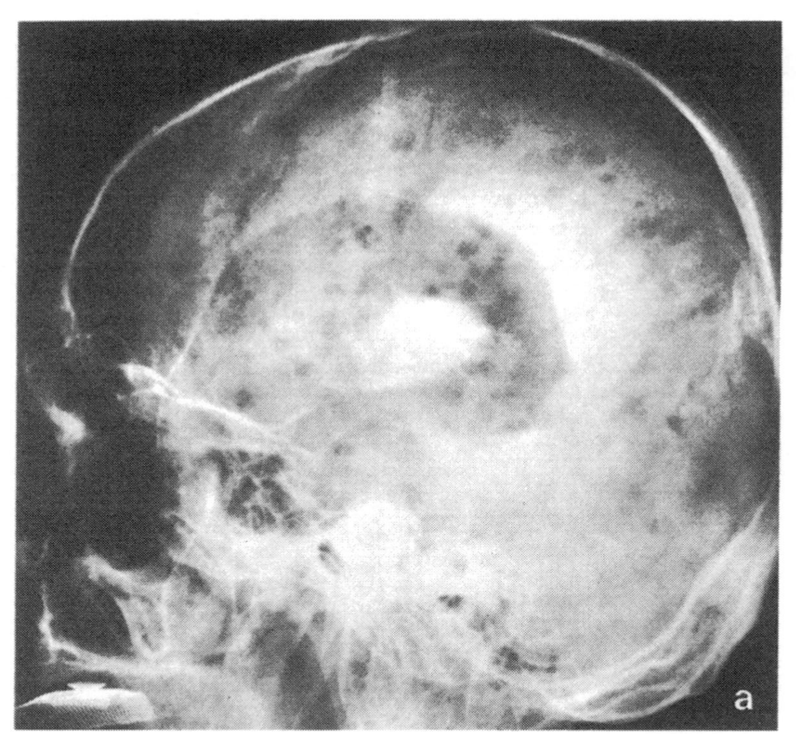

Fig. 87. (a) Lateral pneumoencephalogram of case No. 14. Marked oxycephaly with special density in region of coronal suture. Marked hydrocephalus internus in all regions of ventricular system especially in temporal inferior horn. Spotted appearance of subarachnoid spaces is expression of strong atrophy, particularly in frontal region with dilatation of sulci

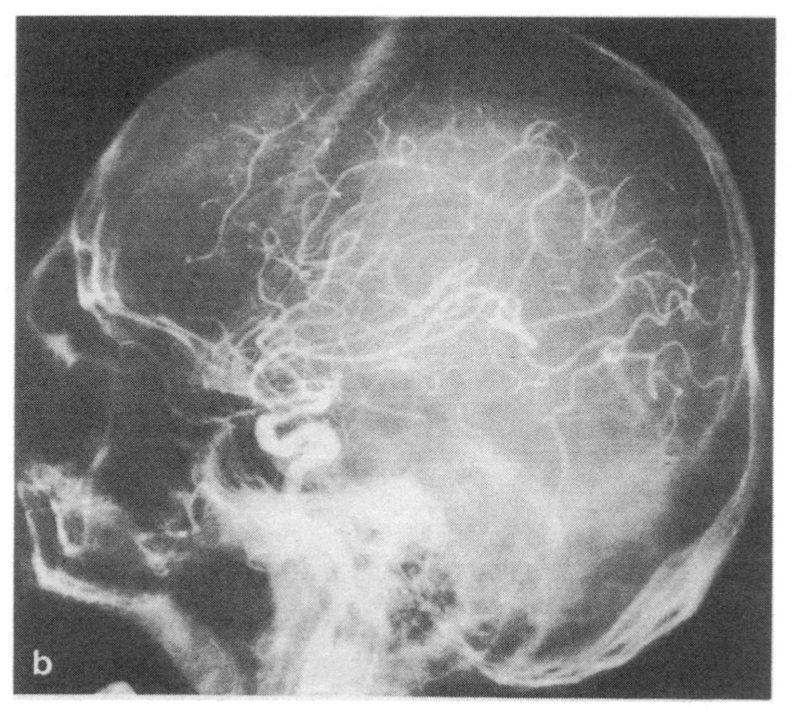

Fig. 87. (b) Left-sided arteriogram of common carotid artery in case No. 14 on day 3 after operation. Vascular system is relatively thin and wire-shaped. Internal carotid artery is greatly narrowed especially in last section of siphon; retarded, deficient filling of anterior cerebral artery. Most spectacular is arc-shaped shifting of deep branches of medial artery in direction of cranial base especially in middle 3rd of artery

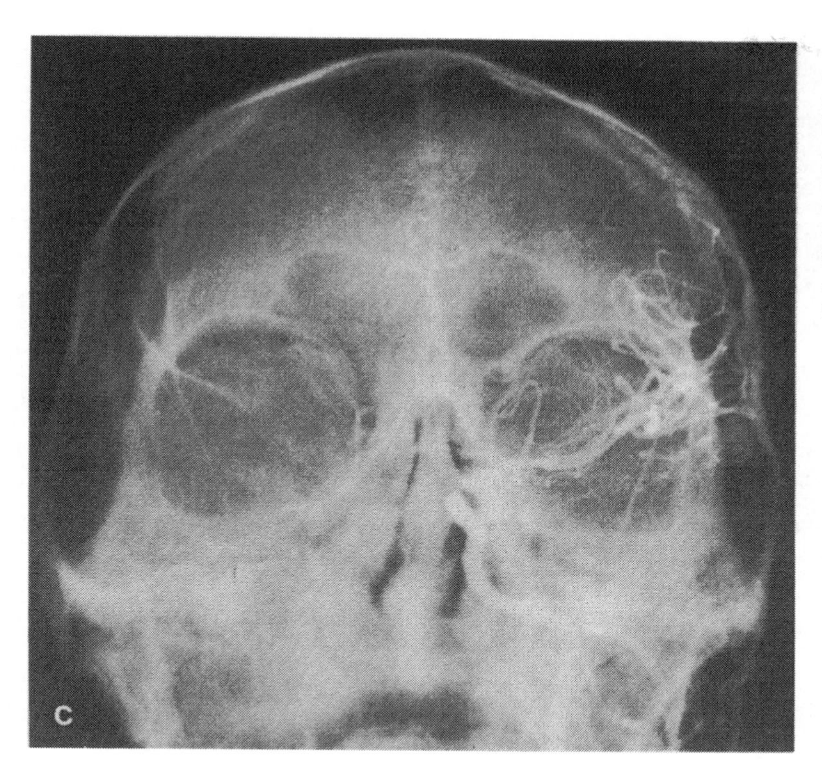

Fig. 87. (c) Anteroposterior arteriogram of carotid artery on day 3 after operation. Only small shifting of anterior cerebral artery with deficient filling. Signs of small depression of corpus callosum. Deep branches of medial cerebral artery show arch-shaped extension toward cranium. Lenticulostriate arteries are arch-shaped and shifted laterally and basally. Anterior artery is relatively thick in diameter and appears congested

clear. However, the deep impairment of consciousness remained. Consciousness did not become clearer despite intensive treatment of the circulation and the respiration. On the contrary, fever developed, respiration, and pulse became more rapid, the amplitude of the blood pressure diminished, and death occurred on day 32 after the operation due to circulatory failure.

Post-mortem of the body (Prof. Dr. H. NOETZEL) showed severe leukocytic bronchopneumonia in both lower lobes of the lung that began to form abscesses; dilatation and congestion of both cardiac ventricles were found together with a stenosing coronary sclerosis restricted to the beginning of the arteries. Also present were sclerosis of the aorta, which was partly ulcerated, and sclerosis of arteries and arterioles of the kidneys.

Anatomical findings in the brain. The sulci were smooth and there was a pressure cone on the cerebellar tonsils. The anterior part of the left hemisphere was somewhat wider than the right. In the left first frontal gyrus the needle track could be seen with marked dilatation of the sulcus. Severe arteriosclerosis with calcification and dilatation of the middle cerebral artery (Fig. 88a) were evident. The left anterior horn was almost completely filled by a massive hemorrhage coming from the head of the striatum. The septum was pushed to the right; the hemorrhage had entered the needle track and dilated it considerably. The hemorrhage of the left basal ganglia reached its greatest size at the level of the chiasma. There it went as far as the claustrum, and extended a bit beyond the claustrum into the white matter (Fig. 88a). The right anterior horn and lateral ventricle were also surrounded by small hemorrhagic lesions resembling purpura, and there were also larger hemorrhages into the white matter of the third frontal gyrus. The internal capsule and the pallidum showed necrotic and hemorrhagic areas in a plane about 12 mm behind the interventricular foramen (Fig. 88b).

The lateral ventricles and the third ventricle were lined with fibrinous membranes originating from the hemorrhage. At the trigonum no change was seen, but the posterior part of the third ventricle was still markedly dilated. Small petechiae were also found in the brachium conjunctivum and its surroundings.

With the exception of its cauda, the left caudatum and the left rostral putamen were completely destroyed. Except for its caudal parts, the pallidum was slightly damaged on the left side. The anterior limb of the internal capsule was completely, and the rostral third of the posterior limb partly, destroyed. The hemorrhagic lesion had severed the

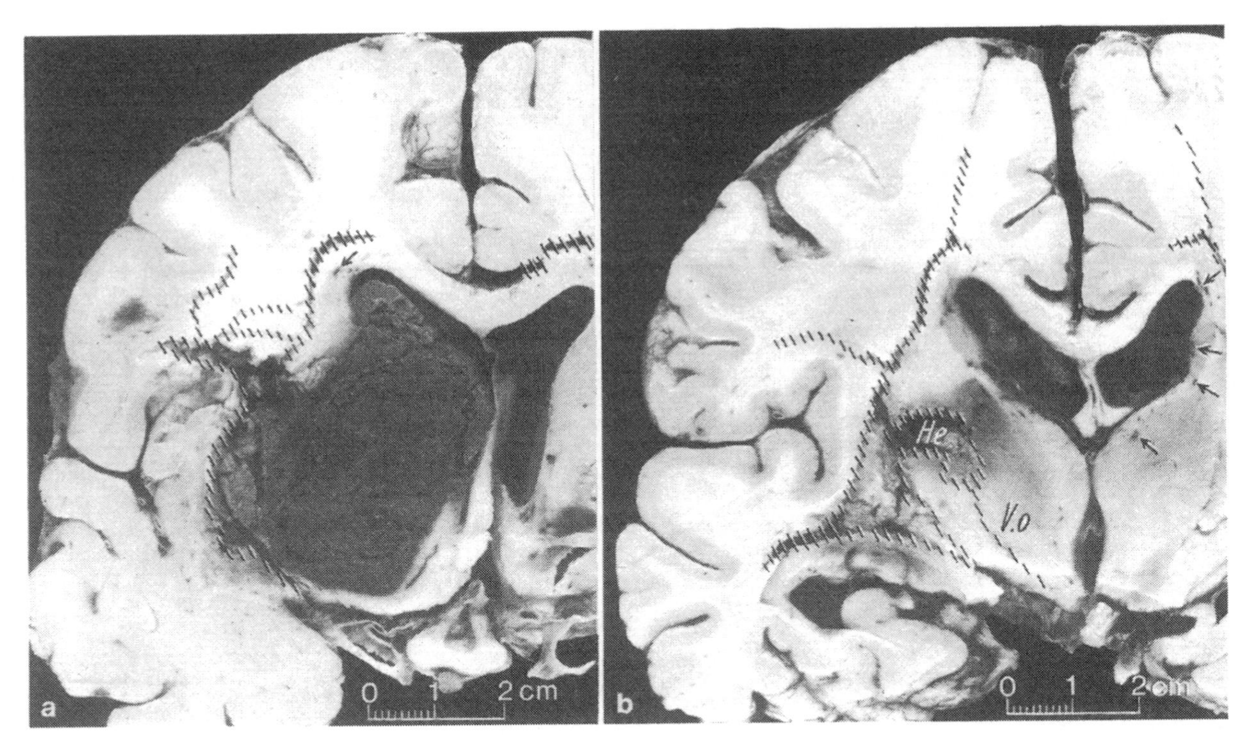

Fig. 88. (a) Cross section through unstained brain in case No. 14 at level of interventricular foramen. Basal ganglia are almost completely replaced by large blood coagulum that also fills left anterior horn. Hematoma is restricted to left ventricle by septum pellucidum, so right anterior horn is spared. Hemorrhage is surrounded laterally in region of putamen and external capsule by secondary infarction that is penetrated by small hemorrhages that continue above insula into white matter of forebrain. Small shifting of masses in white matter of inferior frontal gyrus. (| | | | |) = autolytic breakage of white matter.

(b) Unstained cross section through brain of case No. 14, about 15 mm further caudally than Figure 88a. Remnant of hemorrhage has expanded into internal capsule between dorsal pulvinar and thalamus. In this section, only dorsolateral parts of thalamus are infarcted. Small malacia exist in region of putamen and pallidum without mass shifting of thalami. Some disintegration of tissue is due to autolytic changes. The medial tip of right pulvinar exhibits small pea-sized softening focus with hemorrhage. *He* caudal extension of hematoma in internal capsule; → = small perivenous hemorrhages; (| | | | |) = autolytic breakage of white matter

whole prefrontal brain from its subcortical connections (Fig. 88a). The gray matter of the septum, the fundus caudati, the anterior commissure, and the basal nucleus were preserved on the left side as well as the caudal half of the left thalamus (Fig. 88b). The hemorrhagic lesion extended rostrally through the internal capsule into the anterior pole of the thalamus. The substantia nigra showed severe depigmentation on both sides, particularly in the caudal segments.

Clinical-anatomical correlations. This patient's Parkinson syndrome was probably postencephalitic in origin. In addition, he suffered from severe arteriosclerosis of the cerebral vessels. This showed itself in the angiogram as a severe narrowing of the first part of the middle cerebral artery. It corresponded anatomically to severe narrowing of the carotids and of the middle cerebral artery, with thickening of the walls and plaques in the intima. Because of the large massive hemorrhage in the basal ganglia, it could not be determined with certainty how accurately the stereotaxis reached its target. The radiologic check showed that the target point was reached with a slight deviation of 1 mm toward the dorsal side. The results of stimulation could, therefore, be regarded as coming from the pallidum internum, which was the intended target.

Tremor in the contralateral hand evoked by stimulating at a rate of 25 cps and inhibition of counting by stimulating at a rate of 50 cps were the results of stimulating the pallidum. It is possible that the extension of the contralateral arm was caused by the relatively powerful stimulating currents at 50 cps spreading into the neighboring internal capsule. The reduction of the rigidity by 70% in the contralateral limbs was probably achieved by inactivating the pallidum. The patient's denial of an improvement of mobility made us perform further coagulations with the string electrode in the neighborhood. His severe tiredness with following impairment of consciousness and the paresis of the contralateral arm were obviously the first signs of damage to brain substance by the hemorrhage into the basal ganglia beyond the coagulation focus.

Seventeen days after the operation the α rhythm in the EEG had been replaced by θ waves at 4 cps with bilateral, interspersed, very slow frontotemporal δ waves. This pattern corresponded to the severe impairment of consciousness of the patient, from which one could still recognize a change between sleep and waking. Because of the severing of all connections between the left orbital and prefrontal lobes on the one hand and the basal ganglia on the other, the frontal lobe on the left side lacked all direct neuronal excitation from the depth. This condition correlated with the very slow δ waves in the frontotemporal regions.

Since the middle and caudal thalamus was well preserved even on the left side, the absence of the parieto-occipital α rhythm on both sides could not be explained by an interruption of the thalamic connections with the parietal and occipital lobes. This case supported the idea that the α rhythm is initiated and regulated by the rostral parts of the basal ganglia (pallidum, putamen, caudatum, and rostral thalamus), which had in this case been destroyed in the left hemisphere.

By its extent, most of the large basal ganglia hemorrhage was similar to a striatal apoplexy. Lateral to the massive hemorrhage there was a hemorrhagic softening of the putamen, the internal capsule, and part of the white matter of the hemisphere over the insula. The coagulations with the straight and with the string electrode probably damaged smaller blood vessels in the pallidum. Until this time, the damage of a medium-sized vessel had not been demonstrated at autopsy. It seemed probable that the extension following the pattern of a striatal apoplexy was partly due to the severe arteriosclerosis of the basal cerebral vessels. At the post-mortem, the arteriosclerosis proved to be even more severe (Fig. 88a) than had been expected from the arteriogram.

The displacement of the middle cerebral artery and of the vessels of the basal ganglia in an arc convex toward the ventral side in the arteriogram 18 days after the operation corresponded to the ventral and lateral outlines of the hemorrhage found at autopsy. Consequent upon the lasting severe impairment of consciousness, the patient died from a bilateral purulent and, to some extent, abscess-forming bronchopneumonia 32 days after the operation despite intensive treatment of the respiration and circulation, and antibiotic medication.

Case 15. B.L., aged 62; postencephalitic Parkinsonism, low level of posterior commissure. Death occurred 52 h after stereotaxic operation in right V.o.a/p (no. 805) from acute cardiac failure with bilateral bronchopneumonia (gunshot wound of the lung with scar).

History. In 1918, at age 21, the patient was taken ill with Spanish influenza. Because of several wounds suffered in the First World War—gunshot wound of the lung and an operated empyema of the pleura on the right side—he became unable to work at age 45. Parkinson signs started at age 55 with tremor in the right hand; this was followed by rigidity. From age 59 on, his signs increased more rapidly so that he became completely dependent upon others.

Clinical findings. Very advanced Parkinson syndrome existed with severe akinesia, rigidity, and tremor, which affected the right side more, particularly the leg. Paresis of binocular movements was present. Mentally he was considerably slowed down and older than his age. In contrast to the C.S.F. taken occipitally, the lumbar C.S.F. showed increase of protein, slight pleocytosis, and curves resembling those in meningitis. Spinal arachnitis was suspected. The findings of the Medizinische Poliklinik (Prof. SARRE) included, in the internal organs, shrinking of the right half of the thorax with thickening of the pleura following a gunshot wound and empyema; calcium deposits and emphysema with the diaphragm standing high were observed radiologically. Slight cyanosis of the lips and sclerosis of the aorta were also present. The EKG indicated slight abnormality of conduction in the right ventricle. The blood pressure was 160/95; sedimentation rate and blood picture, N.A.D. An encephalogram showed extreme atrophy of the brain (Fig. 89a, b).

A *stereotaxic operation* was performed October 26, 1959, in the right V.o.a/p. According to a radiologic check, the target point was reached without deviation. Stimulating at low frequency resulted in stopping the tremor and jerks synchronous with the stimulation. Stimuli at the rate of 50 cps imme-

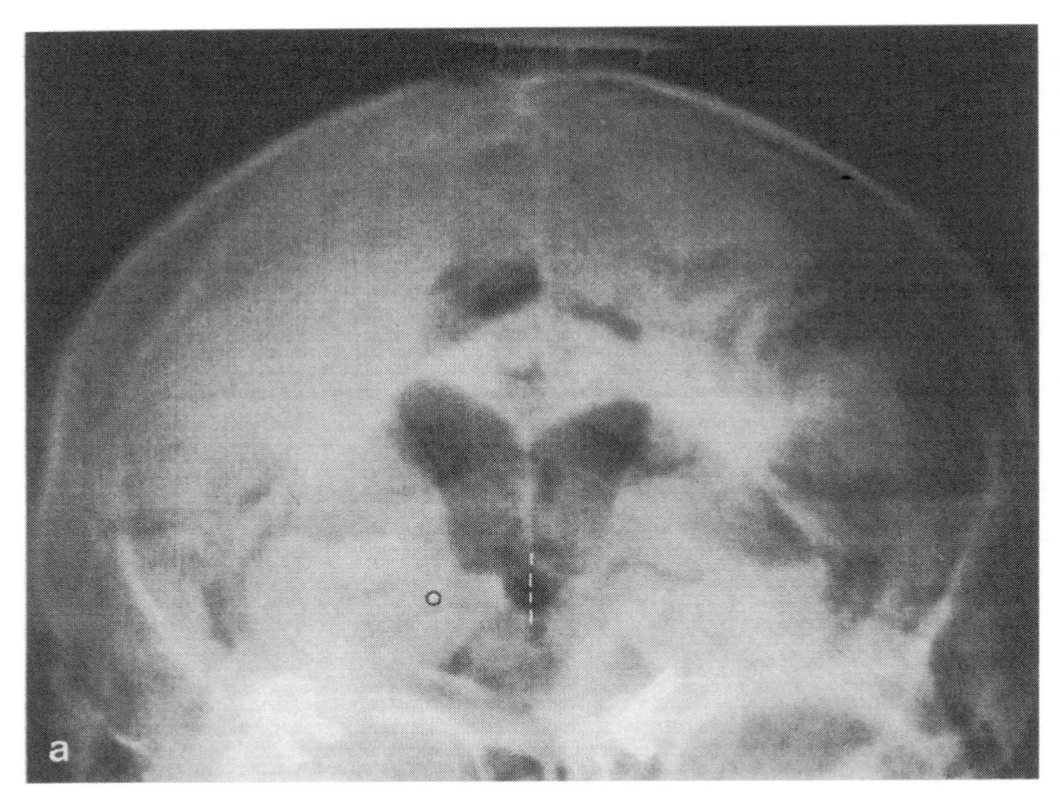

Fig. 89. (a) Anterior-posterior pneumoencephalogram of case No. 15. Small difference between both side ventricles with stronger dilatation of right dorsolaterally. Thalamic part of 3rd ventricle is round and slightly dilated. Marked coarsening of subarachnoidal spaces in left and right hemispheres. Sulcus cinguli and depth of Sylvian fissure are especially enlarged. Target of right V.o.a/p enclosed by a circle.

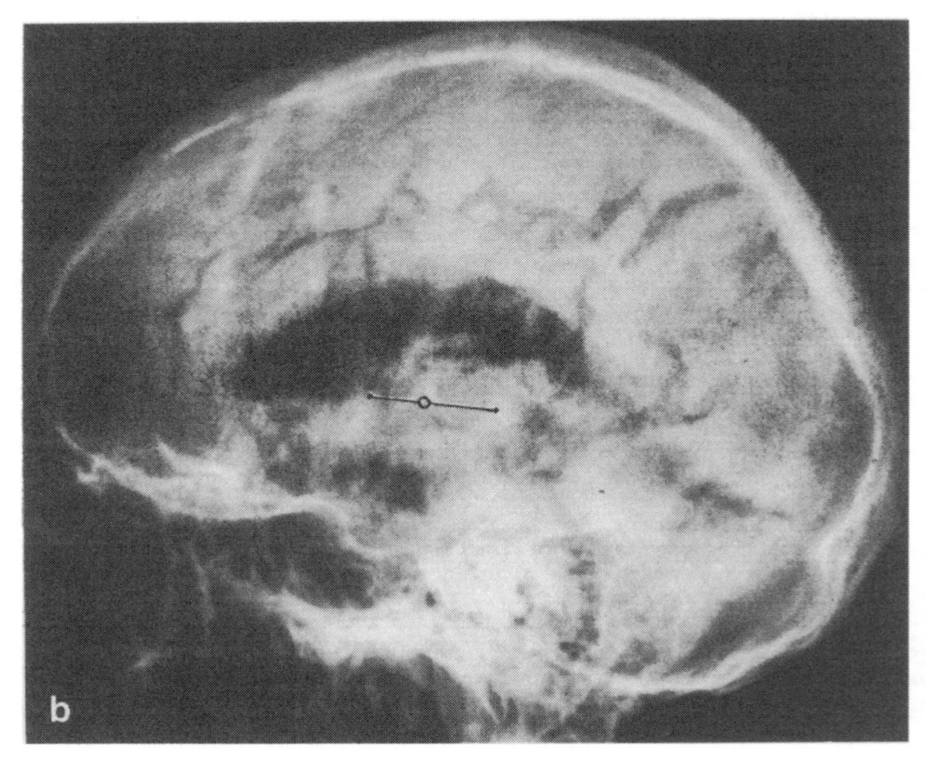

(b) Lateral pneumoencephalogram of same case with baseline and target V.o.a/p drawn in. Good filling of ventricles, dilated subarachnoidal spaces, particularly in medial surface

diately interrupted the active alternating movements and the counting. They led to the patient looking upward to the left, to mydriasis, and to blinking. Coagulation was performed from −2 to +5 with the Wyss coagulator and the straight electrode measuring 4×2 mm. During the first coagulation the current dropped after 7 sec from 125 to 65 mA. A drop also occurred in the following three coagulations. After the coagulation in position +4, the patient became more somnolent, shut his eyes, and could hardly be made to reply or to react. At the end of the operation the rigidity had been favorably influenced on the left, but the effect on the tremor could not be judged because of the somnolence. There was considerable mimical and voluntary paresis of the left facial nerve, which started after the coagulation in position +5 (4th coagulation).

After the operation the patient remained tired but was not confused. On the day following the operation the temperature rose to 39.2° C with slight rigidity of the neck and somnolence. The lumbar C.S.F. was slightly bloodstained without the pressure being increased (165 mm H_2O). Following treatment with antibiotics the temperature dropped and consciousness became a little clearer.

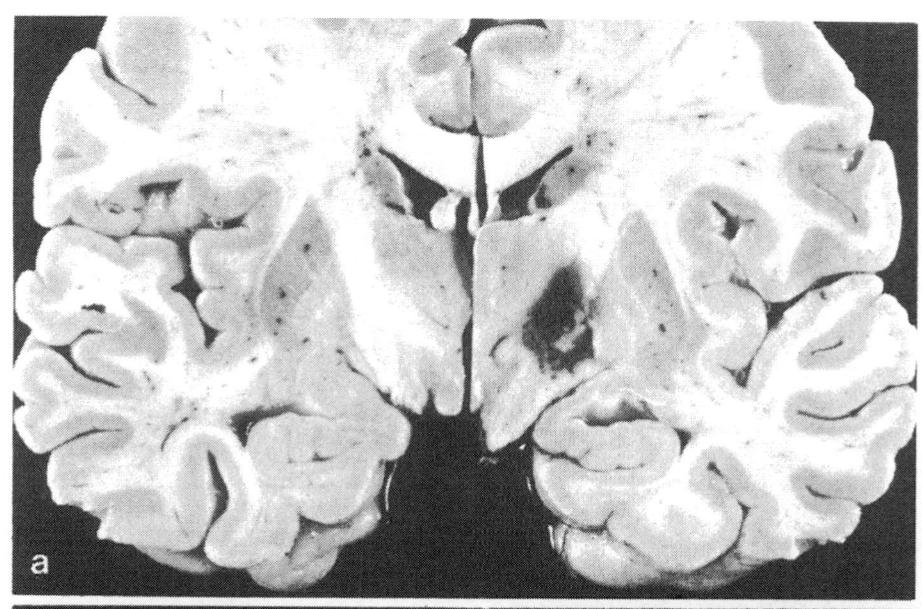

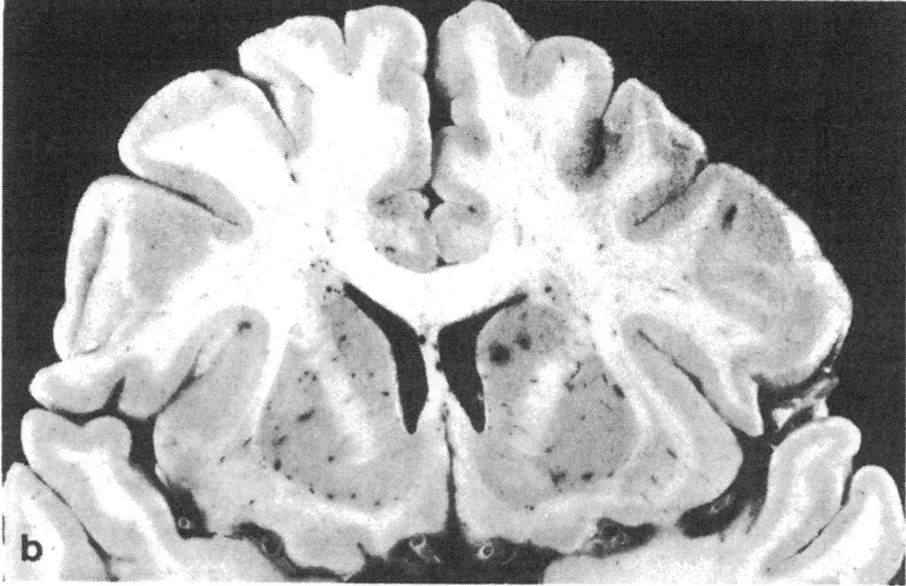

Fig. 90. (a) Cross section through unstained brain of case No. 15. In right thalamus at level of V.o.a, fresh coagulation focus invaded by hemorrhages is present 52 h after operation. Focus extends across basal margin of V.o.p into internal capsule and probably also has destroyed frontal pole of subthalamic nucleus and zona incerta at this level. Focus is continued by hemorrhagic necrotic route ventromedially into transition zone of internal capsule and cerebral peduncle, which extends in an arch to medial tip of internal pallidum. In comparison to right basal ganglia, left basal ganglia are increased in size 52 h after operation by marked swelling. Millet-sized hemorrhage in surroundings of terminal vein in medial border of left caudate nucleus.
(b) Cross section through frontal lobe of case No. 15. 52 h after operation. There is considerable swelling of right caudate nucleus and adjacent part of internal capsule caused by passage of electrode through nucleus. Subdural hematoma has flattened and broadened right frontal convexity. Arteriosclerosis of basal vessels

When the patient was placed upright in a chair in order to avoid pulmonary congestion, he suddenly collapsed and became pulseless. He was put to bed; because of respiratory arrest an endotracheal intubation with artificial respiration was started immediately. However, the heart did not start again and the patient died, 52 h after the operation.

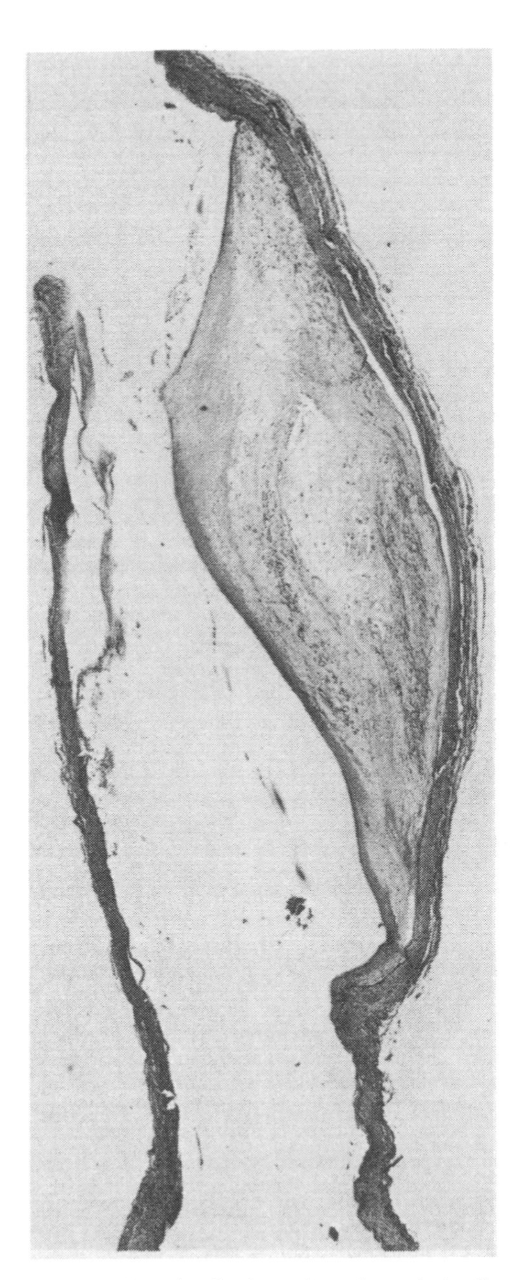

Fig. 91. Longitudinal section through basilar artery of case No. 15. Thick atheromatous plaque is embedded in intimal layer. Transformation of tissue in lipoid spots occurs in middle of plaques. Lumen of artery is reduced to $^1/_3$. ×11

Post-mortem of the body (Prof. Dr. H. Noetzel). The cause of death was circulatory failure because of confluent bronchopneumonia in both lower lobes with scar formation in the right lung (gunshot wound of lung). Ascending pyelonephritis with small abcesses in the kidneys was found as well as severe arteriosclerosis of aorta, coronaries, and cerebral vessels, and renal arteriosclerosis.

Anatomical findings in the brain. A flat (0.8 cm) subdural hematoma over the convexity of the right frontal lobe, a small subarachnoid hemorrhage, and a cortical hemorrhagic necrosis the size of a bean in the first frontal gyrus were found. Because of the severe arteriosclerosis, the basal vessels formed rigid tubes with enlarged diameters (Figs. 90b and 91). A hemorrhage arose from the needle track in the white matter of the frontal lobe and in the caudatum, with swelling in the neighborhood (Fig. 90b). The substantia nigra had shrunk and lost pigmentation with dense gliosis on both sides in its dorsal portions, right more than left, without the typical idiopathic pattern. In the locus caeruleus, there were many large cytoplasmic inclusion bodies without lamination. This led only to a slight loss of nerve cells despite dense gliosis (Fig. 92).

At the boundary between the right anterior thalamus and the internal capsule the coagulated area had a diameter of 8 mm. It lay dorsally and ventrally to hemorrhages so that the total lesion was 16×10 mm. The hemorrhage extended ventrally to the coagulated area up to the apex of the internal pallidum (Fig. 90a). The swelling of the thalamus led to narrowing of the third ventricle. The lesion destroyed the following structures: the lateral edge of the V.o.a with the neighboring nucleus reticulatus, the V.o.p with nucleus reticulatus, and the internal capsule between V.o.a and pallidum, including the apex of the internal pallidum. Below the thalamus, the bundles H_2 and H_1, the zona incerta, and part of the nucleus subthalamicus were involved.

Clinical-anatomical correlations. The severe postencephalitic Parkinsonism was due to the extensive loss of cells in both substantiae nigrae without the typical idiopathic pattern; the extreme rigidity of the left leg corresponded to the complete loss of the dorsolateral cell groups of the right substantia nigra. The "Spanish" influenza from 1918 resulted in a postencephalitic Parkinsonism after a latency of 34 years.

Due to the low position of the posterior commissure the electrode point was situated 3.0 mm too

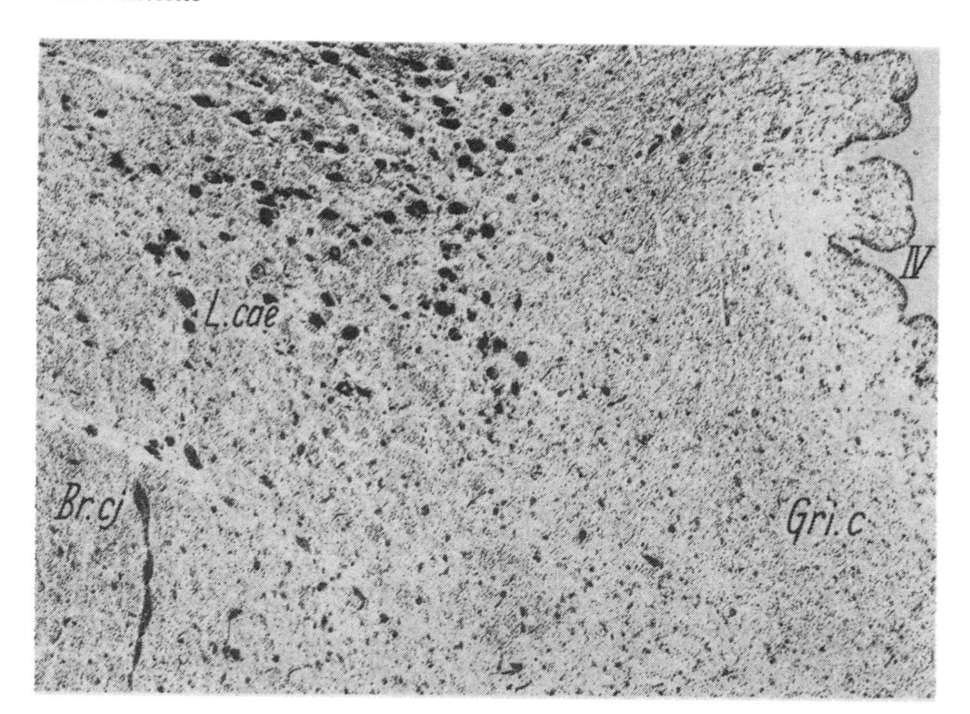

Fig. 92. Section of pontine tegmentum of case No. 15. Locus caeruleus (*L.cae*) is involved and has lost many nerve cells, replaced by dense gliosis. Also in central gray matter of 4th ventricle (*Gri.c*), the glia nuclei are increased.
Br.cj brachium conjunctivum. Nissl stain. ×60

far basally. According to the autopsy findings, the nucleus reticulatus was stimulated at the boundary between V.o.a and the internal capsule. The blocking of the tremor and the jerks at the same rate as the stimulation were probably due to excitation of the fibers entering the internal capsule. On the other hand, the cessation of counting, looking toward the other side, and mydriasis are phenomena that have always been observed when the V.o.a or the pallidum internum is stimulated without involving the fibers in the internal capsule. Blinking during stimulation of the V.o.a at a rate of 50 cps was an unusual occurrence. It is probably the result of excitation of supranuclear pathways in the internal capsule for muscles that shut the eyes rather than a special tendency in a postencephalitic Parkinson syndrome. Considerable swelling of the thalamus occurred 52 h after the coagulation, whereby the right wall of the third ventricle coincided with the midline (Fig. 90a). Taking the 4-mm displacement of the wall into account, the track was 11 mm from the original position of the ventricle wall, as originally intended. In a plane corresponding to the target point, the middle of the track probably lay inside the V.o.a.

The positive influence on the rigidity (by about 70%) after the coagulation correlated with a lesion of the lateral edge of the V.o.a, the neighboring nucleus reticulatus, and the inner capsule. That the rigidity was not completely abolished may have been due to the incomplete severing of the base of the V.o.a and the bundle H_1 or to the simultaneous damage to larger parts of the internal capsule. The interruption of the internal capsule at the height of the V.o.a showed itself clinically by facial weakness that was both expressive and voluntary. The somnolence and reduced reactivity of the patient, which appeared during the coagulation and lasted until death, may have been related to the 8-mm wide subdural hematoma or to the considerable destruction by hemorrhage and swelling thalamic structures.

The hemorrhage inside the altered zone and around the coagulated area was remarkable. The coagulation with the Wyss coagulator, with its pure frequency current of 0.5 MHz, led at the first coagulation with 115 mA, after 7 sec, to a drop of the current to 65 mA. Because of this, the coagulation was stopped. Apparently the increased temperature around the electrode caused such damage to the vessels that diapedesis occurred from several small vessels. The change of resistance in the hemorrhagic tissues again led to a drop of current during further coagulations; this, in turn, caused new vascular damage. Since only small vessels were affected, there was no massive hemorrhage; rather, there was bloody infiltration of the coagulated area and its surroundings. The altered zone determined the extent of the bleeding into the tissue. Death was caused by acute right cardiac failure (congested liver and cardiac ventricles full of blood) in the presence of bronchopneumonia in both lower lobes in a patient in whom the alveolar surface had been considerably reduced by emphysema and by the scar formation in lung and pleura following a gunshot wound and empyema.

Case 16. D.M., aged 59; postencephalitic Parkinsonism; coagulation of V.o.a/p and internal capsule; paranoid–hallucinatory psychosis. Death occurred 5 years after stereotaxic operation in right V.o.a/p (no. 603) from cardiac and circulatory failure.

History. In 1920, at age 15, she had influenza. Asthma bronchiale developed. At age 48, she developed trembling of the left arm, which gradually extended to the left leg. She dragged the left leg. For a few years a tremor in the right limbs had also been present. She suffered from stabbing pains in the left arm. She was slowed down but was able to look after herself completely. In the last few years, she developed an additional disorder, i.e., when she read a line, she felt as if perception suddenly stopped.

Clinical findings. The tremor of arm and leg was severe on the left but moderate on the right. Rigidity in leg and arm was of medium severity. The toes of the left foot were curled. When walking, the associated movements on the left side were lacking. There existed a general akinesia (Fig. 93a, b). Severe hyperhidrosis, slight reddening of the skin, and disordered peripheral circulation were found. She had slight pulmonary emphysema, early sclerosis of the aorta, and only moderate dilatation of the heart, with a BP of 130/85 mm Hg. Encephalography was well tolerated. It showed marked dilatation of the external C.S.F. spaces, which were shaped like plaques, mainly in the central parietal regions (Fig. 94b). The internal hydrocephalus was less; the waist of the basal ganglia was smoothed out (Fig. 94a). The C.S.F. was normal. The EEG showed an α rhythm of 9–10 cps.

Mental state. A report by her brother was very helpful. He stated spontaneously that the asthma had been caused psychologically for a number of years. Even as a small child she had suffered from anxiety complexes, with severe trembling of the whole body, which could not be explained and which could hardly be influenced. She was afraid of the other people living in the house; she watched anxiously any movement or statement by other people. As a result, she was indecisive and physically ill. In regard to the tremor, a change of surroundings had produced a surprising improvement. The brother compared her inner life to a "mental heap of broken china". During the interview her paranoic attitude was clearly evident (Fig. 93a, b).

A *stereotaxic operation* was performed December 23, 1959, in the right V.o.a/p. Stimulation was carried out in position −2. Single stimuli led to synchronous twitchings of the left hand. Stimuli at a rate of 4 cps caused the tremor to become slower and more rhythmic. Stimuli at a rate of 8 cps enhanced the tremor and could provoke it at times when it had stopped. At some stimulations she cried out that she had a feeling that there was a current running through her whole body. She let her left arm fall and laughed about it.

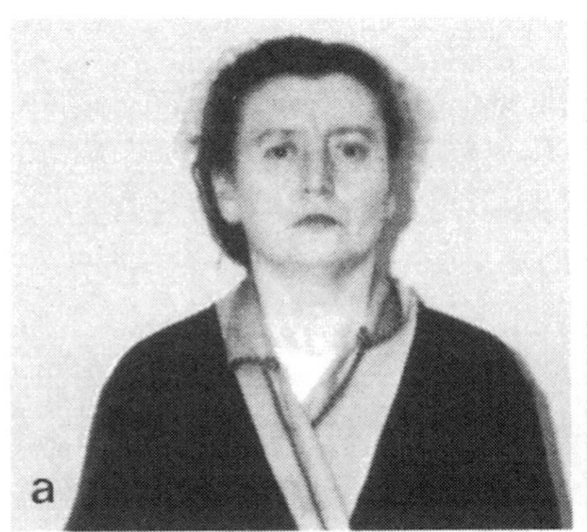

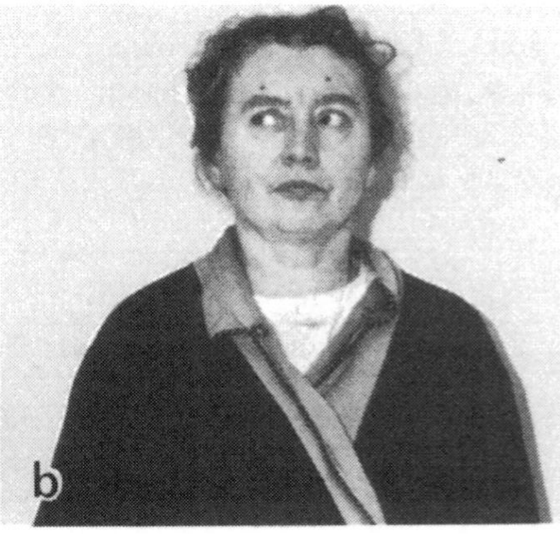

Fig. 93. (a) Patient No. 16, facing forward with strained expression.
(b) Same patient looking to right side with paranoid facial expression

Stimuli at higher frequencies regularly provoked laughter that stopped as soon as stimulation ceased. Alternating movements in the left arm were stopped during stimulation at a rate of 25 cps. She looked toward the left, and counting was inhibited. She could not state a cause for the laughter during stimulation, nor did she know what she had laughed about.

Using the spark-gap diathermy apparatus and a straight bare electrode of 4 mm, coagulations of 10 sec duration were carried out in positions −4 to +5. The tremor in the left arm was reduced only after the ninth coagulation. Three more coagulations in positions 0 to +3 were carried out with the string electrode, which had been moved 6 and 5 mm in the frontal, medial, and basal directions. These lasted 25 sec each. During the first of these coagulations the patient spontaneously talked of yellow fireworks in front of her eyes. The spontaneous tremor in the left hand was no longer present at the end of the operation but could still be provoked. Power in the left hand was slightly reduced. Tremor of the left leg had disappeared, but tremor continued in the right leg.

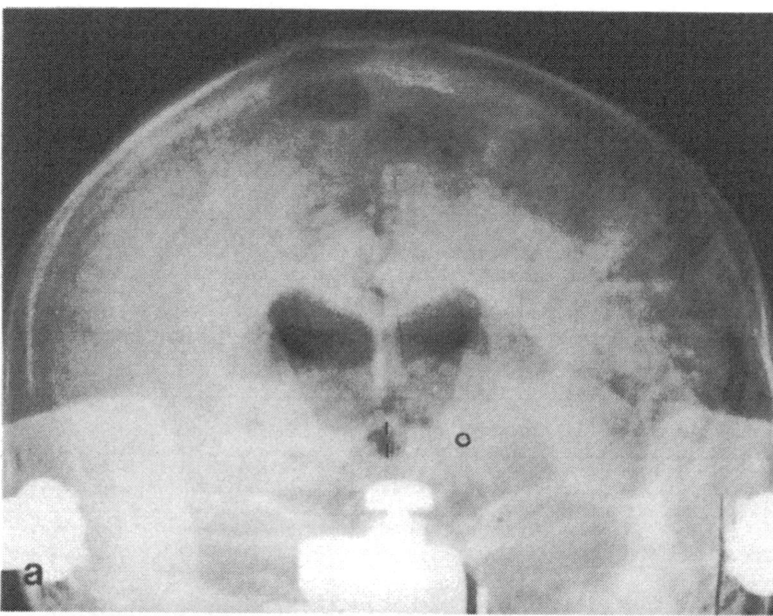

Fig. 94. (a) Anterior-posterior pneumoencephalogram of case No. 16. Moderate enlargement of lateral ventricles, left more than right, with slightly concave contour of left thalamus. Enlargement of subarachnoidal spaces with large blocks on right side, but only small dilatation of 3rd ventricle. Individually determined target is marked on right side.

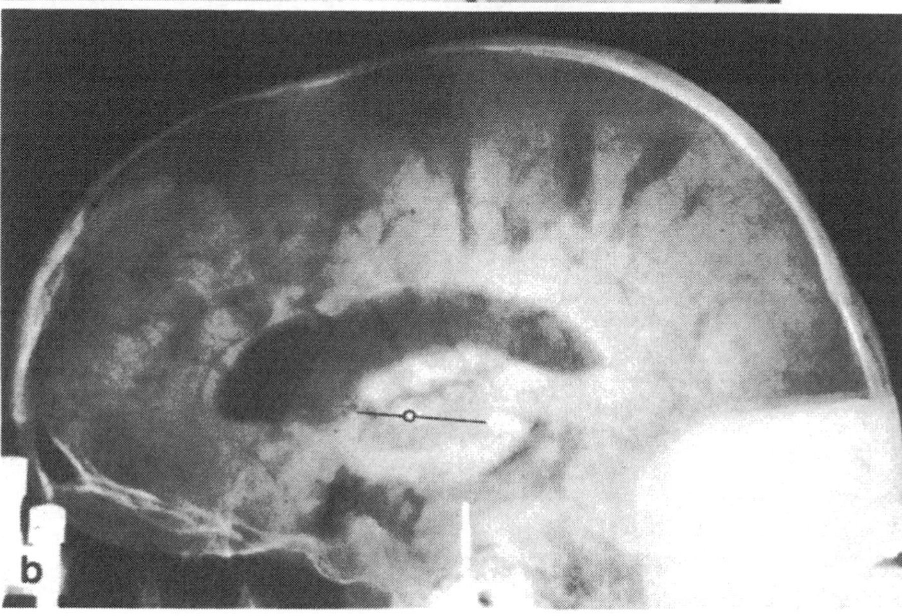

(b) Lateral pneumoencephalogram of case No. 16, with moderate dilatation of both ventricles. Massa intermedia is apparent in slightly dilated 3rd ventricle. Target V.o.a is drawn in, in anterior third of baseline of thalamus in middle of massa intermedia projection. Dilatation of both cisterna ambiens and of basal cisterns. Strongly dilated subarachnoidal spaces especially in parietal and frontal regions and two large almond-sized paramedian plaques of air

Postoperative course. The improvement of the tremor in the left limbs continued. A moderate organic psychosyndrome disappeared quickly. Seven days after the operation the EEG showed that the α rhythm on the right side had been slowed to 5–8 cps and also indicated single short groups of δ waves in the right frontal region extending into the frontotemporal region. These were clearly increased during hyperventilation. At the time of discharge, 8 days after the operation, the tremor occurred only occasionally and to a reduced degree.

At the follow-up examination 15 months after the operation she stated that she had resumed work, part-time, as a teacher of needlework. Her mobility was said to have been much improved, but she stated that she had choked several times when swallowing and had spasms around the mouth and frequent gastric and intestinal disorders. She was easily exhausted and sweated on slight provocation. She had to force herself mentally to do everything and had no initiative or spontaneity. The rigidity had disappeared in the left limbs but a slight tremor was still present on the left side. When walking, she leaned forward. In February 1961, she had to stop working and was pensioned off.

She wrote a letter 18 months after the operation stating that for a few months she had had a shuffling gait, that she walked with small steps and then could not stop (propulsion). She was no longer able to look after herself and had, therefore, entered another neurologic hospital. She asked whether she could be helped by a second operation.

In November 1961 (about 2 years after the operation) she was admitted to the Neurological University Hospital Würzburg (Dir., Prof. Dr. Scheller). The Parkinson symptoms and the coarse slow tremor affected mainly the right side of the body, which had previously been only minimally involved and which had not been treated surgically. Propulsion and retropulsion were so severe that she could no longer walk on her own. A change in her mental state was noted; she exhibited behavior of a paranoid, hallucinatory type, but since clinical psychiatric picture was equivocal, the diagnosis of a schizophrenic illness was avoided. Episodes occurred in which she was euphoric, overactive, and full of peculiar ideas that sometimes reminded one of delusions. She also talked of visual appearances and voices, but everything seemed to originate in a very vivid fantasy that was unrestrained because of lack of criticism. These episodes disappeared when she received Dominal (Prothipendyl), 140 mg/day.

Her condition deteriorated to the point that, nearly $2^1/_2$ years after the operation, she was admitted into a geriatric home. Because of increasing mental changes she could not be kept there and was transferred more than two years later (August 1964) to the Lohr psychiatric institution. On admission about 5 years after the operation, the diagnosis was of a Parkinson syndrome with personality change, irritability, excitability, and emotional instability. She explained the difficulties in the old people's home by the fact that she was continually stared at. Her complaints about that led to her transfer to this hospital. Though she was correctly oriented, she described her experiences circumstantially and very emotionally, particularly the acoustic and visual hallucinations. She reported that voices of relatives and friends came out of the air and frightened her, although they sometimes amused her. Her siblings were said to have climbed into the room and to have constructed cupboards in the cellar. Some apparitions had been up to mischief at night and had caused noises. Voices, which often disguised themselves, had claimed to be the rulers of the world and had mocked her. She became increasingly restless and, eventually, confused. She died after a rise in temperature from pneumonia on October 25, 1964, aged exactly 59 years, nearly 5 years after the stereotaxic operation.

Post-mortem[5] *of the body* showed pulmonary emphysema and valvular heart disease. The cause of death was cardiac and circulatory failure.

Anatomical findings in the brain. Thickening of the meninges over the left central region and slight external atrophy of both hemispheres at their convexities, particularly in the frontal region, were observed. An area of yellowish-red discoloration 25×15 mm, with a central cortical necrosis the size of a pea, was found over the right second frontal gyrus. The area merged into a funnel-shaped lesion that extended into the anterior horn (Fig. 100). Its diameter was maximally 8 mm, minimally 4 mm. Immediately below the cortex the funnel was lined by some hard yellowish-reddish tissue. Histologic examination showed that the funnel-shaped necrosis was a lesion, lined with glia

[5] We thank Prof. Dr. G. Peters, former Director of the Max-Planck-Institut für Psychiatrie in Munich for kindly supplying us with the post-mortem material.

and connective tissue, caused by the displacement of minute particles of dura into the depth with a severe reaction as to a foreign body. Beside the dura the funnel-shaped necrosis contained some amorphous material (Fig. 95).

The electrode entered the thalamus immediately rostral to the terminal vein, injuring it (Fig. 129). The track of the electrode went through the nucleus reticulatus, and the neighboring part of the lateropolaris at the edge of the internal capsule in a ventrocaudal direction. A cystic lesion, 6.5 mm high and 2.6 mm in saggital direction, developed in the oral ventral nuclei (Fig. 95). The lesion destroyed the V.o.a completely, the V.o.p to its base (Figs. 95, 129), and the nucleus reticulatus on the lateral and rostral sides. Around this lesion were considerable growth of glia and degeneration of fibers and cells. A narrow band of internal capsule was also involved. About half of the oral dorsal nuclei were destroyed (see Figs. 95, 129).

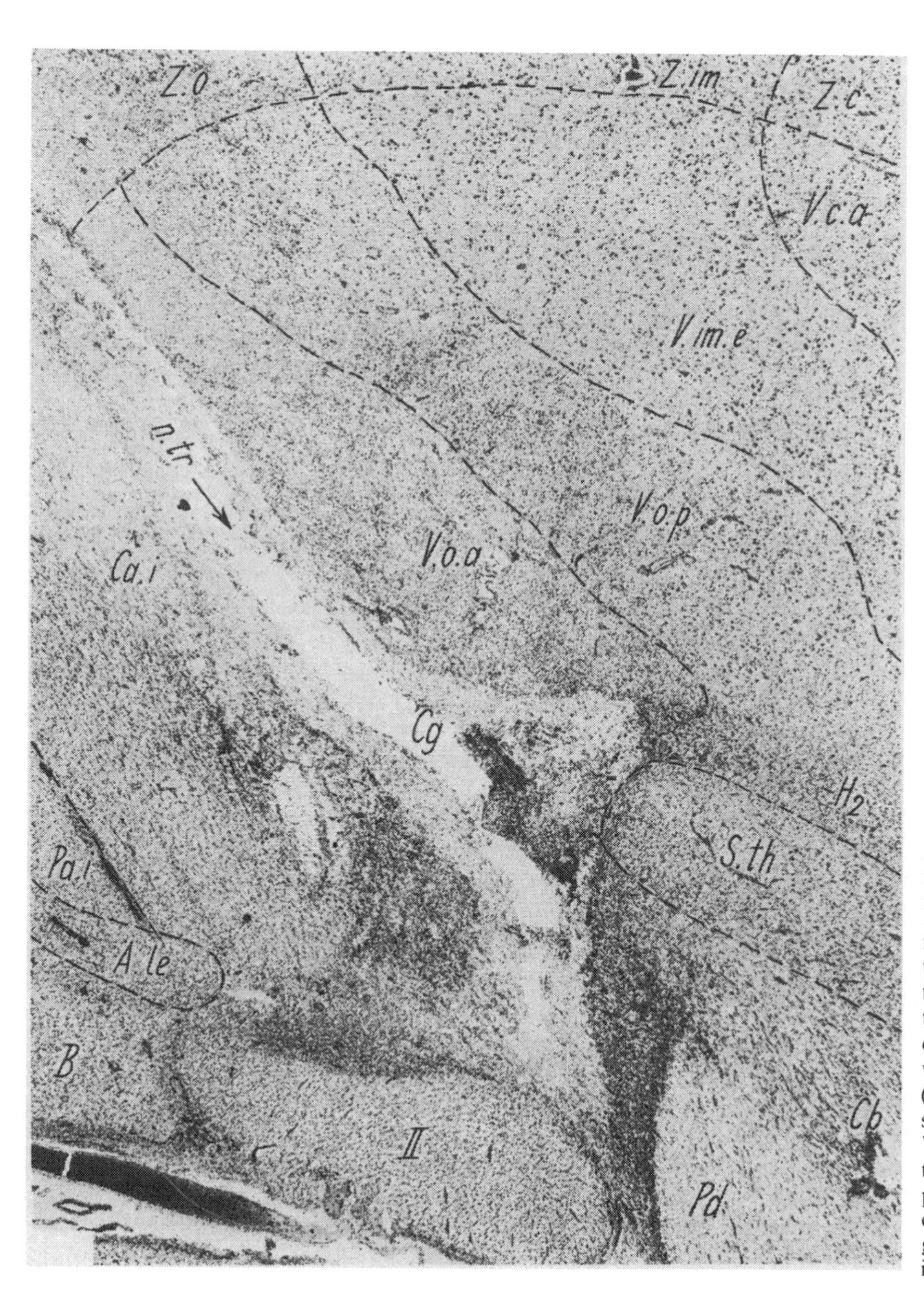

Fig. 95. Adjacent section to Figures 100 and 129 in Nissl stain at higher magnification (×15). Coagulation zone around needle track (*n.tr*) is partially transformed in elongated cyst, which is in connection with coagulation focus (*Cg*). Note complete destruction of nerve cells in V.o.a nucleus and in subthalamic nucleus (*S.th*) and preservation of some nerve cells of V.o.p nucleus with gliosis; no nerve cell loss in V.im.e

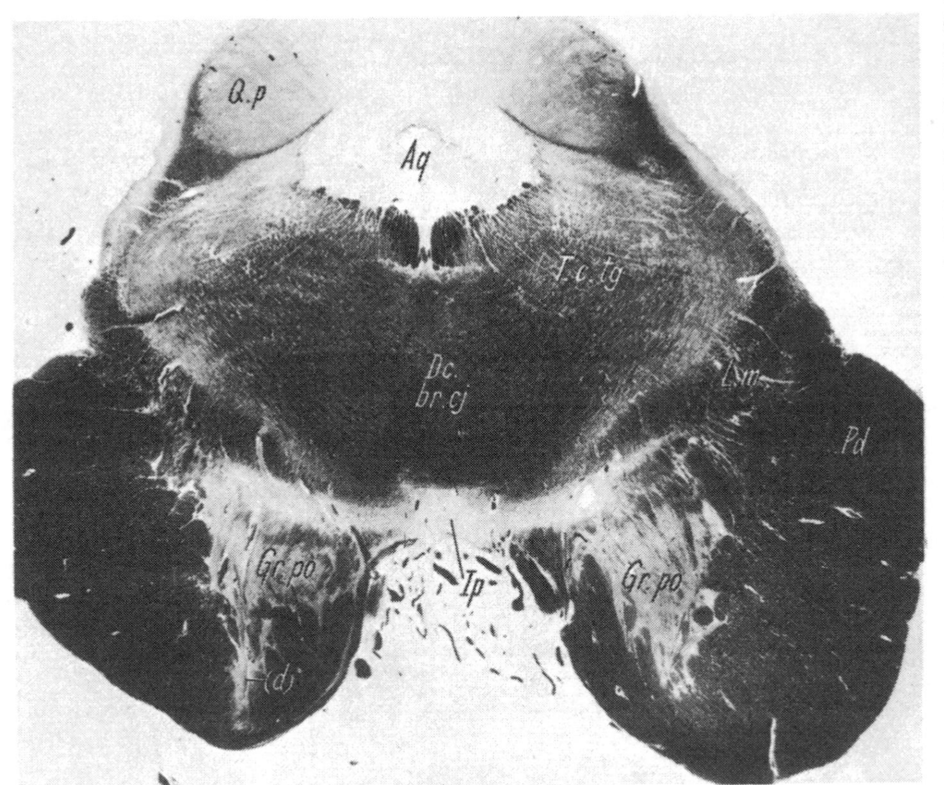

Fig. 96. (a) Cross section (reversed!) through caudal midbrain of case No. 16. Below decussation of brachia conjunctiva (*Dc.br.cj*), caudomedial expansions of substantia nigra can be seen on both sides of interpeduncular ganglion (*Ip*). On right side is degenerated strip (*d*) in transition zone of peduncle to pontine gray matter (*Gr.po*), which contains the capsular lesion. Fiber stain. ×4.8

From this cystic lesion, connective tissue growth extended through the rostral poles of the nucleus subthalamicus and through the internal capsule up to the caudodorsal edge of the optic tract (Figs. 95, 129). The latter also had small areas of degeneration, with increased growth of glia on its dorsal and lateral edges. The peduncle, which forms the caudal continuation of the internal capsule, was interrupted (Figs. 95, 129) in the sagittal plane of the lateral border of the substantia nigra—about 14 mm to the side of the midline. Bundle H_2 had been completely destroyed above the nucleus subthalamicus and replaced (Fig. 95) by growth of glia. Consequently there was marked atrophy of the pallidum internum (Fig. 129), with severe retrograde degeneration of cells. The track of the string electrode could be seen by the growth of connective tissue; at the base of the nucleus subthalamicus and below it in the peduncle one could also see a degenerative band with increased growth of connective tissue (Fig. 95). An individual abnormality was the broadening of the layer of fibers in the dorsal part of the zona incerta, particularly in the prerubral tract, which showed almost no degeneration. This was probably an aberrant bundle of fibers in this area.

An extensive loss of cells from the middle, lateral, and dorsal cell groups was noted in the substantia nigra compacta, replaced by a dense astro cytic proliferation (Figs. 96a, b). In a few cells that remained, there were unstained gelatinous cutouts the stiff surface of which are covered with a few melanin grains (ALZHEIMER tangles non impregnated). In the medial cell groups of the substantia nigra however, less severe cell losses on both sides. The marked density of glia proliferation in the right posterolateral group continues as a strand of gliosis (Fig. 96b) to the degenerated (cortico-nigral) fibers in the pes pontis. There were also some nerve cells deteriorated in the locus caeruleus with intraplasmic inclusion of non impregnated tangles (Fig. 97).

Clinical-anatomical correlations. According to the distribution of the loss (Fig. 96b) of, and the changes in cells, with astroglia proliferation this was a typical postencephalitic Parkinsonism after an influenza at age 15. The Parkinson symptoms started at age 48 with a trembling of the left hand. The fact that the symptoms were much more marked on the left side fits with the more marked loss of cells in the right substantia nigra in which the lateral and dorsal cell groups had almost

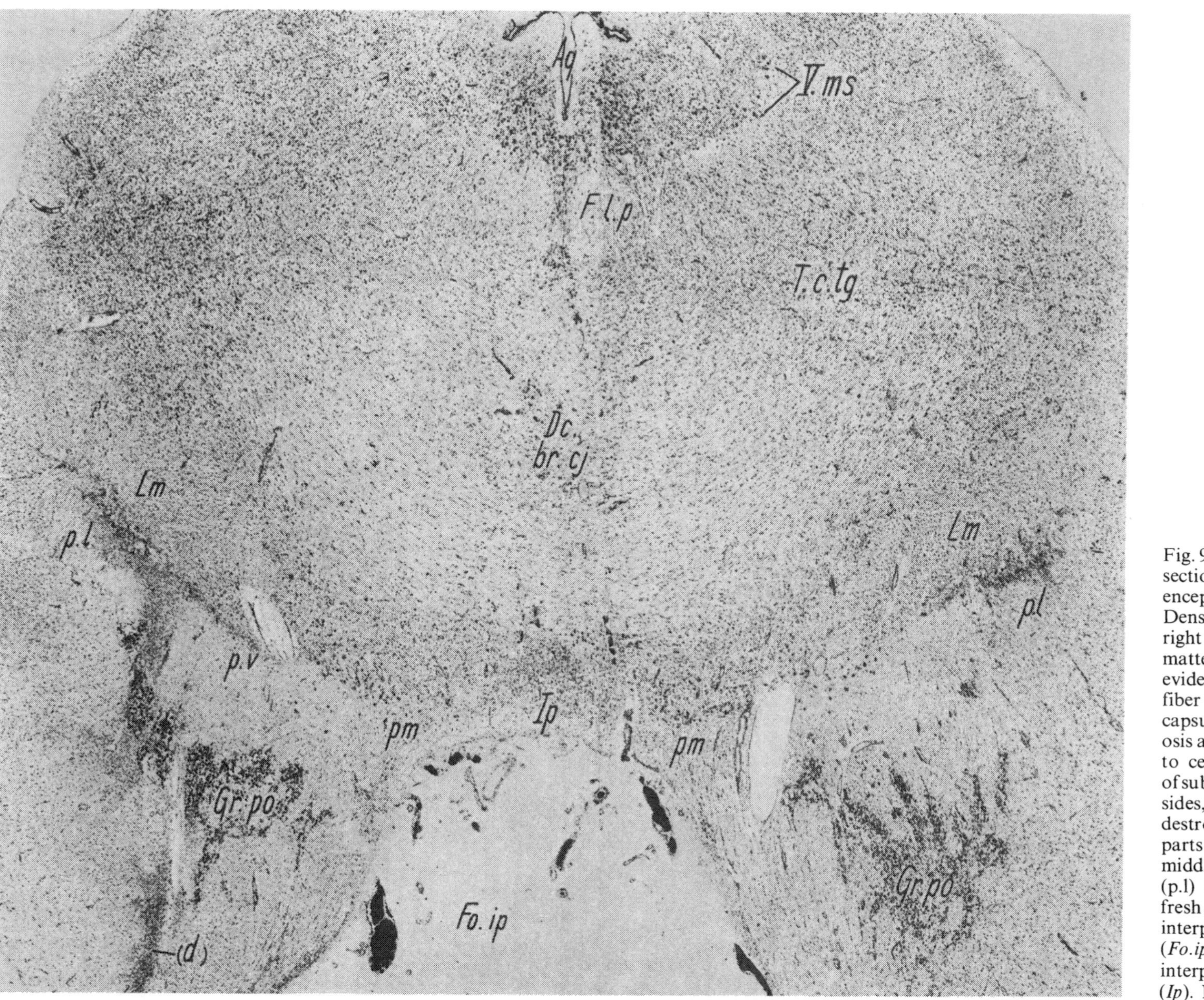

Fig. 96. (b) Neighboring section (reversed!) of postencephalitic case No. 16. Dense glia accumulation in right anterior pontine gray matter and in peduncle is evident due to descending fiber degeneration from capsular lesion. Further gliosis above it (*p.l, p.v*) is due to cell loss in cell groups of substantia nigra. On both sides, nigra cells are mostly destroyed also in medial parts (*pm*). On left side, middle group of nigra cells (p.l) is replaced by dense fresh gliosis. The roof of the interpeduncular foramen (*Fo.ip*) is occupied by the interpeduncular nucleus (*Ip*). Nissl stain. ×11.5

completely disappeared. A stereotaxic operation on the right thalamus was necessary at age 53 because of the rather rapid progression of the illness. The effects of stimulating 1.5 mm rostrodorsally to the target point (changes of the rhythm of the tremor by different frequencies of stimulation and irregular twitchings) were caused by the excitation of the V.o.a (Fig. 95). As the autopsy showed, the two stimulating poles, which were 5 mm apart, had been located mainly in the V.o.a (Fig. 129). The spread of current to i.La was responsible for smiling and laughing without motivation, which could be reproduced very clearly by stimulating at rates of 8, 25, and 50 cps (Fig. 130). At stimulation with higher frequencies her eyes turned upward to the left, she interrupted her active alternating movements, and stopped counting backward. She explained the interruptions of active movements caused essentially by stimulating the V.o as something hindering her that came from the upper left region.

The coagulations with straight electrodes using the spark-gap diathermy instrument had destroyed the V.o.a completely and severely damaged the V.o.p, including the nucleus reticulatus polaris rostral (Figs. 95, 129) to these nuclei. The improvement of rigidity and tremor except in the foot was explained by this. The three additional coagulations with the string electrode resulted in a reduction of the tremor of the foot, which was, however, accompanied by a mainly proximal paresis of the left arm and leg. This situation was due to destruction of part of the internal capsule at its transition to the peduncle immediately behind the optic tract. The involvement of the edge of the optic tract (Figs. 95, 129) explained her noting yellow fireworks at the first coagulation with the string electrode. This phenomenon is obviously a stimulating effect of heat and electricity produced by the high-frequency current, which did not consist of pure sine waves and which contained harmonics. Only 5% of the fibers of the optic tract were damaged,

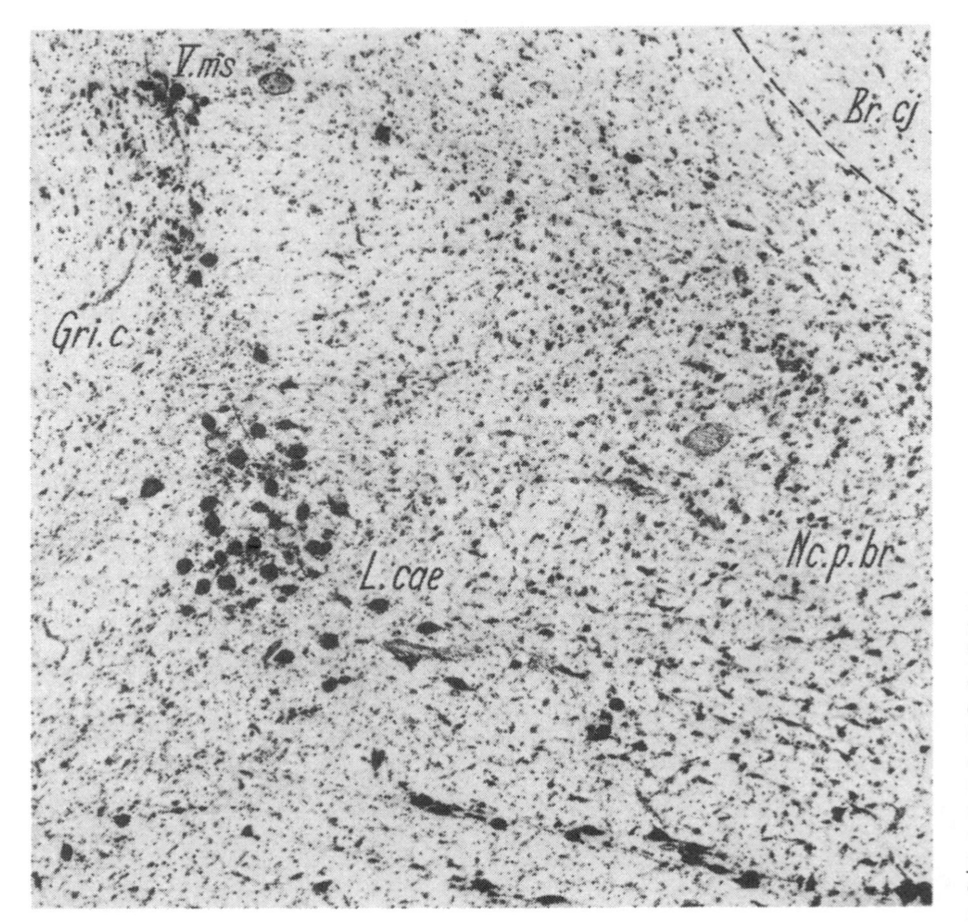

Fig. 97. Cross section through pons of case No. 16 shows cell loss and cell degeneration in locus caeruleus (*L.cae*), which becomes more severe in dorsal part where it contacts mesencephalic root of trigeminus (*V.ms*). Medioventral to brachium conjunctivum (*Br.cj*), peribrachial nucleus (*Nc.p.br*) is not involved. ×41

and there was no evidence of clinical loss. Postoperatively, the rigidity was permanently abolished, while the tremor of the left limbs did not disappear due to preserved cells in V.o.p (Fig. 95), it increased later. The fact that the tremor suddenly stopped during further damage of the internal capsule could be explained by the additional reduction of the tonic pyramidal influence on the innervation of the limbs. As the paresis decreased, the tremor increased to the same extent. This was comparable to the observations after pyramidotomy, i.e., the severing of the pyramidal tract in the spinal cord in the peduncle or in the internal capsule. (PUTNAM, 1938, 1940; EBIN, 1949; WALKER, 1949a, b; GUIOT and PECKER, 1949).

It was difficult to give a localizing explanation for the dysphagia and spasms around the mouth that appeared sometime after the operation. It is improbable that these symptoms were due to the primary illness, the postenc. Parkinsonism, since they were not present before the operation and the medial cell groups of the substantia nigra showed only moderate cell loss, although these cell groups belong somatotopically to the "bulbar" muscles. Through the demonstrated lesion of the corticospinal systems in the medial pedunculus, the supranuclear representation of the bulbar muscles had been damaged. The post-mortem showed in this case a cortical necrosis that was not seen in other cases to the same extent. It continued in the shape of a funnel through the white matter into the anterior horn. The ventricular system had, therefore, communicated with the external C.S.F. spaces. The fact that particles of the dura as well as amorphous material had been carried below the cortex explained the dilatation of the electrode track. The sharply defined funnel-shaped necrosis was caused by the exceptionally strong reaction to foreign body particles.

Although the patient had briefly been reemployed as a needlework instructress, her psychological symptoms became more conspicuous about 2 years after the stereotaxic operation. A paranoid hallucinatory psychosis developed that made her retirement necessary. Her personality had already shown deviate behavior patterns before the appearance of Parkinson's syndrome. This was the reason psychiatrists for some time avoided labeling the paraphrenic psychosis as such. Five years after the operation the patient died in a psychiatric hospital of bronchopneumonia with cardiac and circulatory failure after having developed considerable marasmus through her psychosis.

Case 17. O.A., aged 65; postencephalitic Parkinsonism with pallidotomy; seizures with loss of consciousness. Death occurred 7 days after stereotaxic operation in left pallidum internum (no. 1859) from hypostatic pneumonia.

History. The patient was buried by a shell explosion during the First World War. At age 22 (1919) he suffered for 6 months from an epidemic encephalitis and was at first unconscious for 4 weeks. At age 27, he suddenly developed a tremor of the right arm. From this a complete Parkinson syndrome with disorders of gait and speech developed gradually, and eventually he had to give up his work. At age 45, he suffered two short syncopal attacks during which he fell and each time broke his arm. From age 58 on, the Parkinsonism became much worse. It was complicated by a simultaneous chronic gastritis.

Clinical findings. The patient was practically helpless. His food had to be cut up, he had to be fed, he could not dress or shave himself, and he needed help to get up or to sit down. All signs were much more marked right than left. He complained about occasional spasms in the mouth and throat and a feeling of spasm in equinovarus posture occurred. Rigidity was more marked than the tremor; the leg was more affected than the arm. The right hand was deformed. Tremors of the right arm and of the tongue, with occasional spasms in the floor of the mouth, were present. The tendon reflexes were exaggerated on both sides. Trömner reflex was positive on the left but Babinski's sign was negative. A slight facial paresis was present on the right side. Sweating had increased, the face was shiny, and hypersalivation had been common since his 58th year. He walked with small steps, slowly, and without associated movements. Speech was almost unintelligible. Mentally he was slowed down, depressive, and irritable. His memory also was impaired. In the internal organs the blood pressure had risen to 160–200/115 mm Hg; the EKG indicated impairment of the coronary circulation, particularly in the left ventricle, left-sided decompensation, and pulmonary emphysema. Encephalography with helium was well tolerated. It showed a dilatation of the basal cisterns, external atrophy, and slight dilatation of the internal spaces, particularly of the third ventricle. C.S.F. findings bordered on normality.

A *stereotaxic operation* was performed November 10, 1962, in the left pallidum internum. Single stimuli 1 mm above the target point accentuated individual tremor phases on the right side and produced twitchings in the right side of the face. Stimuli at a rate of 4 cps slowed down the tremor of the right hand and evoked synchronous jerks of the entire right side of the body. Stimuli at 8 cps accelerated the tremor somewhat. There was a short phase of interference between tremor and the rhythm of stimulation, followed by severe twitching of the face. Stimuli at 25 cps stopped the tremor and led to a tetanic contraction of the right arm and the right side of the face. Stimuli at 50 cps reduced the amplitude of the tremor while accelerating it, and caused dilatation of the palpebral fissure, mydriasis, and turning of the eyes to the right. During stimulation at a rate of 50 cps, the rigidity in the right arm was considerably increased. During the stimulation in position +5 at a rate of 4 cps, synchronous jerks of the entire right side of the body were observed. At the time, he stated that he saw blue dots to the right side of the fixation point. Stimulating at a rate of 50 cps in the same position led to tetanic contractions of the right side of the body.

The first coagulation was performed with a 4×2 mm thermorod electrode in position +11.7. Immediately at the beginning of the coagulation a contraction of the right limbs occurred. This was regarded as due to an involvement of the internal capsule, and the coagulation was stopped immediately. A radiologic examination showed that the point of the electrode was 11.7 mm too far caudally, because the difference in length between the special stimulation electrode and the coagulation electrode had not been considered. After correcting the position of the electrode, five coagulations were performed in positions 0 to +6 at 70° C. The rigidity in the right arm was diminished after the first coagulation, and the right limbs were more mobile. After the coagulation in position +2, a facial difference of the right side and an extensor response on the right side were noted. Further improvement in the rigidity occurred after the third coagulation (+5).

After the left coagulation in position +6 the rigidity in the right arm was diminished by at least 60%. Paresis of the right foot and an extensor response on the right were present; the knee gave way when the patient walked to his bed; and facial paresis for voluntary and expressive movements was present but no marked impairment of consciousness was seen. A few hours after the operation the patient was very sleepy although correctly oriented. On the following day, about 18 h after the operation, the patient suffered three attacks in which he lost consciousness, and his head and eyes turned to the left, but there were no tonic or clonic convulsions of the muscles. After this he had a right-sided hemiparesis. The next day he was incontinent once, without any noticeable impairment of consciousness. His breathing was at times rapid, going up to 36 per minute. After

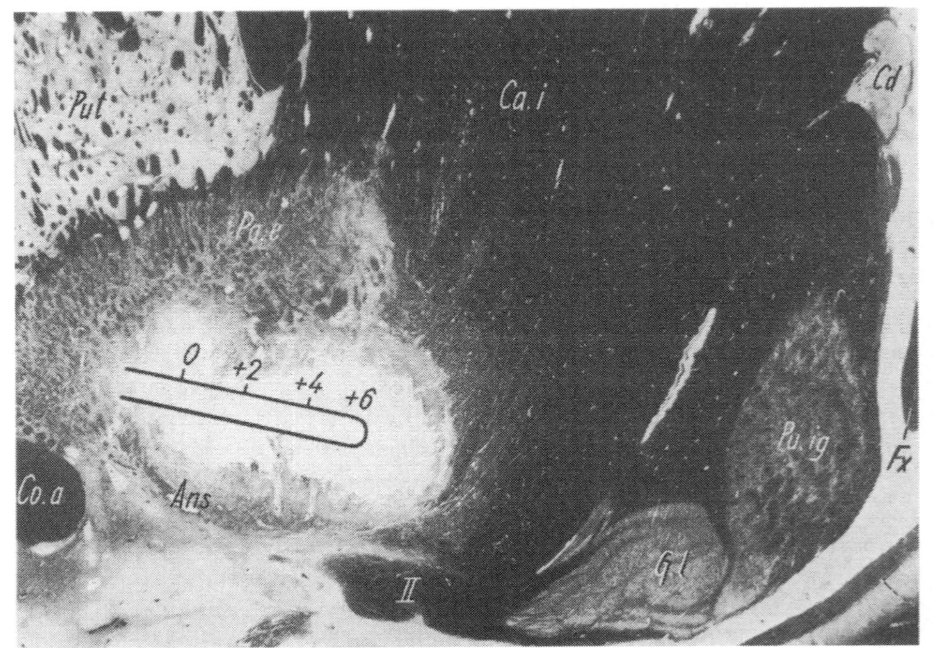

Fig. 98. Coagulation focus of case No. 17, situated in pallidum. In sagittal section, needle track actual in outer segment of pallidum (*Pa.e*) is seen on left lateral border of picture. Internal segment of pallidum is completely destroyed, but ventral layer of ansa lenticularis (*Ans*) is preserved. In focus, electrode positions 0, +2, +4, +6 are indicated. Caudal tip of focus is situated in cerebral peduncle in transition zone to internal capsule. Focus overlaps optic tract in small shell-shaped region. Fiber stain. ×4.
Co.a anterior commissure in base of external segment of pallidum; G.l first cut of lateral geniculate. *Pu.ig*= pulvinar intergeniculate

the attacks the blood pressure dropped to 120/80. His temperature was about 38° C. He reacted to being spoken to up to a day before his death. During the last day, his speech and consciousness were acutely impaired, and his breathing rose to 44 per minute. In order to improve pulmonary ventilation a tracheostomy was performed. Despite treatment for pneumonia, the patient died on day 7 after the operation, with signs of circulatory failure. Post-mortem of the body was not permitted.

Anatomical findings in the brain. The right hemisphere was slightly broader than the left; the arachnoid was thickened near the midline. Over the left frontal pole there were several recent hemorrhagic, superficial, cortical necroses of thumbnail size. Over the entire convexity there was marked atrophy of the gyri, which was less on the base. Arteriosclerotic changes were found in the branches of the vertebral, basilar, and carotid arteries. Severe cell loss was noted in both substantia nigra with involvement of medial cell groups.

On a sagittal section about 20 mm from the midline a coagulation destroyed the pallidum internum almost completely (Fig. 98). The external pallidum was well preserved except for its caudal border area. The center of the area contained a well-demarcated coagulation, while its edge was separated from the surrounding tissue by a liquefied necrosis. The lesion extended only about 1 mm into the internal capsule. Although the area of coagulation was close to it (ca. 1.2 mm), the optic tract was not affected at the base. Of the pallidum internum, 90% was inactivated; of the pallidum externum, 15%; of the internal capsule (over the pallidum internum) and the rostral peduncle, each 5% (Fig. 98). The basal nucleus had also been partially affected by the lesion (20%).

Clinical-anatomical correlations. In this case of postencephalitic Parkinsonism caused by a severe epidemic encephalitis suffered in 1919, the right side was much more affected than the left. In addition, there were severe vegetative symptoms, dysphagia, and speech disorders. The signs were due to a typical process leading to disappearance of cells from the medial substantia nigra. The cause of the two "apoplexies" at age 45, with two fractures of the arm, remains unclear all the more since the patient probably suffered from cardiac syncope with systolic blood pressure values between 160 and 200 mm Hg.

The stimulating electrode had reached the center of the pallidum internum. The effects of stimulation – accentuating individual phases of tremor, slowing the tremor and making it more rhythmic with stimuli of 4 cps, and accelerating it with stimuli at a rate of 8 cps – were due to excitation of the pallidum. Also explained by excitation of the pallidum internum were the turning of the eyes to the right, the increase in rigidity, and dilatation of the palpebral fissure, and the mydriasis when

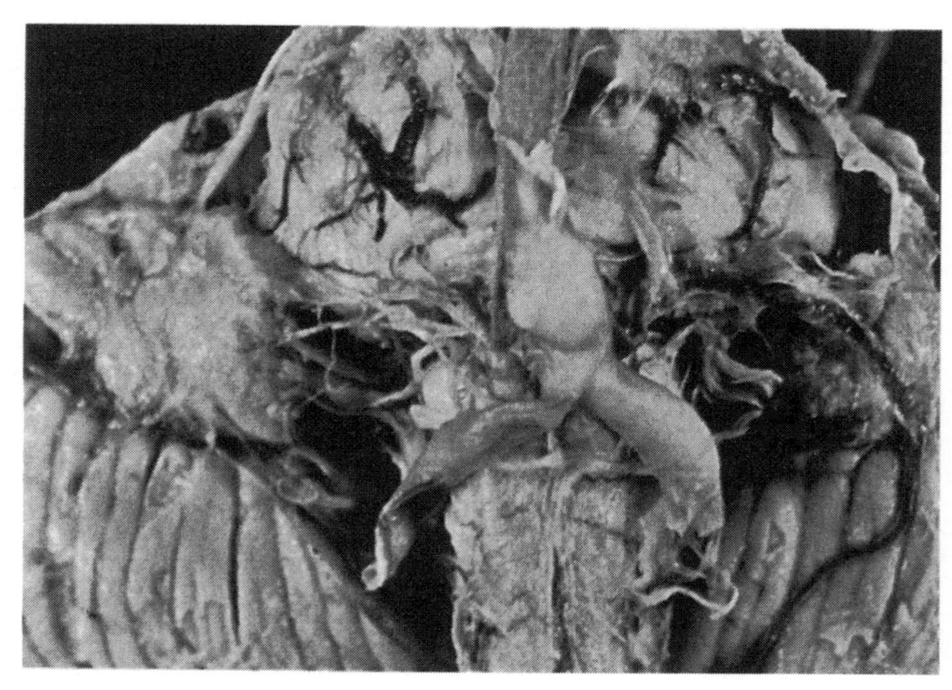

Fig. 99. View of basis of medulla oblongata and pons of case No. 12. Marked dilatation of left vertebral artery with thickening of its wall. In caudal part of basilar artery, large yellow plaque with sharp boundary toward front where basilar artery appears to have a thin wall. ×1.8

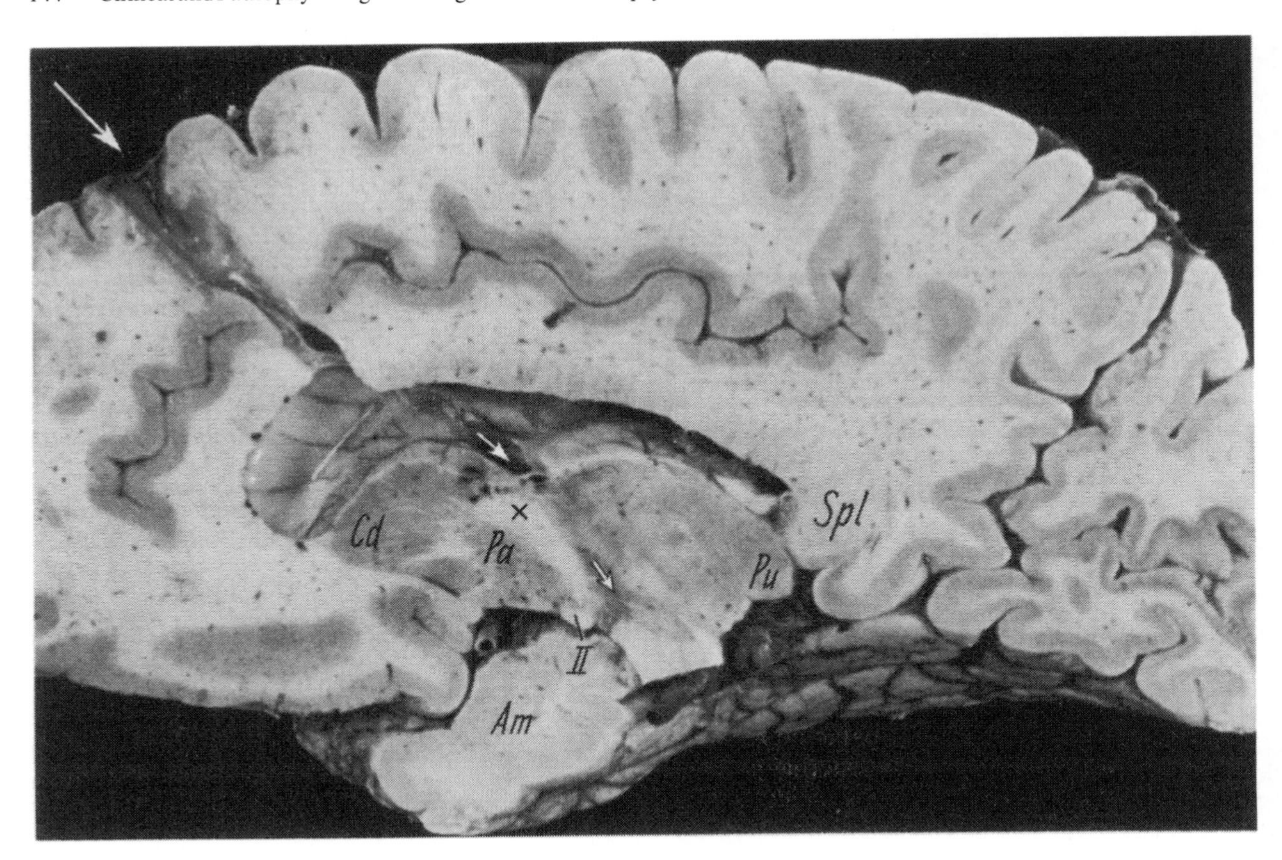

Fig. 100. Sagittal section through right hemisphere of case No. 16 in color. Needle track in white matter of right frontal lobe is surrounded by brownish-yellow necrotic tissue. Electrode has been introduced through this track. Focus is continued into lateral ventricle, where bleeding in ependyma marks path of electrode (*arrow*). In region of diencephalon, needle track penetrates caudal surface of internal capsule and rostral thalamus. Slender scarred necrotic zone behind optic tract (*II*) produced by side electrode. (We are indebted to Prof. PETERS, Munich, for this picture)

stimulating at 50 cps. The tetanic contractures of the entire right side including the facial muscles at position +5 was probably explained by the spread of the stimulating current to the pyramidal tract in the internal capsule, which was only 3–4 mm distant. When stimulating at 4 cps in this caudal position, blue points appeared in the right visual field; this was probably due to excitation of the optic tract, which was also only 3.5 mm away. It is very rare for stimulation of the optic nerve to evoke blue photomas.

An important factor was the unintentional use of a thermoelectrode that was 10 mm longer than the stimulating electrode. In this particular case, this led to the development of heat in the caudal part of the internal capsule (Fig. 98). Although the high-frequency current had a pure frequency, it provoked a stimulating effect, namely, a contraction in the contralateral side. The current was interrupted after about 2 sec. The high-frequency current of pure sine waves does not as a rule act as a stimulus, but its direct action on the very excitable pyramidal fibers had a strong stimulatory effect.

Among the five coagulations at 70° C in positions 0 to +6 the +5 was the most effective. They produced a lesion of 12.5 × 7 mm with layer formation (Fig. 98). The radii of such coagulations, therefore, were from 3.5 mm to a maximum of 4 mm. Through the flat approach the pallidum internum was almost completely destroyed in its whole rostrocaudal extent. The inactivation of the pallidum internum was followed by a 60% improvement of the rigidity and the tremor was also obviously improved after the operation. The fact that the high-frequency current switched off after a few seconds led to circumscribed damage of areas in the internal capsule 17 mm behind the interven-

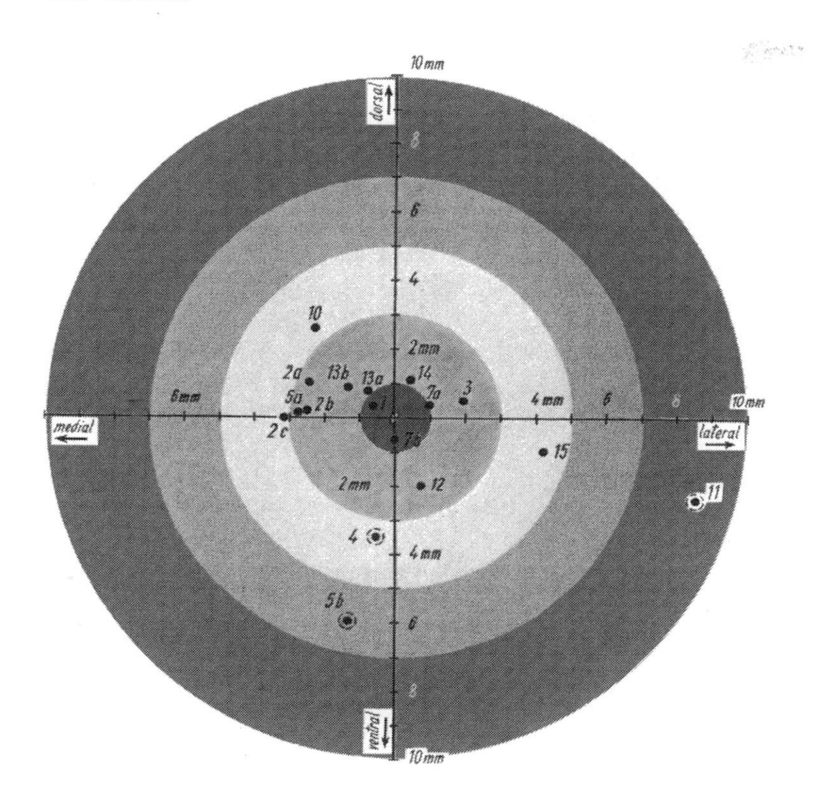

Fig. 101. Target disc for 17 hemispheres of Parkinson patients demonstrates aiming errors of all cases. Average localization error is 1.23 ± 1.48 mm for sagittal coordinate (not included in target), 1.64 ± 1.20 mm for frontal coordinate and 1.10 ± 0.92 mm for vertical coordinate. Numbers show sequence of operations. Cases No. 4, 5 b, and 11 (*broken line*) are not included in statistical calculation because of errors that would have been avoidable. (From HASSLER *et al.*, 1971)

tricular foramen and produced a paresis of the right foot with an extensor response on the right side.

The series of three attacks 18 h after the operation with unconsciousness, deviation of gaze to the left, a subsequent right-sided hemiplegia, and grossly impaired consciousness thereafter were probably not due to epileptic discharges; most likely they were of cardiac origin and were evoked by Adams-Stokes attacks. It was remarkable that the blood pressure after these attacks and the operation had gone down from values between 160–200 mm Hg systolic to an average of 120 mm Hg systolic. On the day after the attacks, an increase of the respiration and pulse rate were first signs of the development of pneumonia. With the reduced blood pressure, the previously existing emphysema, the damage to the coronaries, and the left-sided heart failure, a severe hypostatic pneumonia quickly developed. Despite tracheostomy and respiratory and circulatory treatment, the patient died from pneumonia on day 7 after the operation.

IV. Correlations

1. Remarks on the Accuracy of Reaching the Target

Since SPIEGEL *et al.* (1947), several stereotaxic instruments have been constructed that enable one to reach target points in the depth of the brain exactly with a probe, if the points have been marked on lateral and anteroposterior x-ray pictures (SPIEGEL *et al.*, 1947; SPIEGEL and WYCIS, 1952, 1954, 1961a, b, 1962; LEKSELL, 1949, 1957; TALAIRACH *et al.*, 1949; RIECHERT and WOLFF, 1951a, b; RIECHERT and MUNDINGER, 1956, 1959; GUIOT and BRION, 1952; NARABAYASHI and OKUMA, 1953).

The accuracy of reaching a radiologically determined target point is ± 0.5 mm, as shown by radiologic checks in 850 stereotaxic operations (MUNDINGER and UHL, 1965); but this accuracy does not tell us definitely whether the aimed-at structure of the brain has been reached exactly, anatomically speaking. The majority of subcortical structures cannot be shown in the radiologic picture. We are therefore forced to use several points of reference that have been made visible with negative or positive contrast media in the encephalogram. Our method for determining the cerebral target point consists of constructing a three-dimensional system of coordinates with the help of these cerebral reference points (HASSLER and RIECHERT, 1954a, 1955; MUNDINGER and RIECHERT, 1961, 1963; RIECHERT and MUNDINGER, 1956, 1959). The distance of the target point from these cerebral reference points is determined in relation to the system of coordinates in a model brain. The individual variability of diencephalic structures has been described by SPIEGEL and WYCIS (1952), TALAIRACH *et al.* (1949, 1957), BRIERLEY and BECK (1959), AMADOR *et al.* (1959), VAN BUREN and MACCUBBIN (1962), VAN BUREN and BORKE (1972), and ANDREWS and WATKINS (1969). Our method eliminates this variability to a great extent by comparing numerically the same reference distance in one patient's brain and in one model brain. This procedure avoids the use of an average value of a target structure (HASSLER and RIECHERT, 1954a), which does not take into account the individual variability of anatomical structures. In order to reduce to a tolerable degree target errors caused by individual anatomical variability, we compare three distances that are at right angles to each other, measured in the encephalogram of the individual patient, exclusively to the corresponding three distances in an individual cadaver brain (model brain) (Fig. 143a) or in a brain from an atlas (Fig. 144–163). The coordinates of the target point determined in the model brain are corrected numerically by the quotients obtained from the three distances and are entered on the encephalogram. With this method it is even possible to compensate for changes in the position of a target point brought about by abnormal changes in the structure of a patient's brain caused by the basic illness and which can be seen in the encephalogram. For this reason, we have chosen the series of frontal sections (see Figs. 144–163b) of a patient suffering from Parkinson's disease. We have shown that using a target point corrected for individual deviations produces a much smaller standard deviation from the individual target point than using the outer bounderies of the nuclei and determining their deviation (MUNDINGER and ZINSSER, 1965; MUNDINGER and POTTHOFF, 1960, 1961; MUNDINGER and UHL, 1965; MUNDINGER *et al.*, 1970). In order to reduce the uncertainty of the localization, we always employ a stimulating check before inactivating a nuclear area (RIECHERT and HASSLER, 1953; HASSLER and RIECHERT, 1954a, b, 1959, 1961; HASSLER, 1955b, 1957b, 1961, 1965, 1966b, 1968, 1970). This method improves the accuracy of reaching the target (HASSLER and RIECHERT, 1954a, b, 1957, 1961; MUNDINGER and RIECHERT, 1963; MUNDINGER, 1966).

This procedure, however, cannot provide absolute proof that the calculated and aimed-at part of the brain has been reached anatomically. The large number of therapeutic successes in more than 3000 operations on cases of Parkinson's disease demonstrate that the accuracy of the method is at least functionally adequate in the vast majority of cases for the therapeutic result. However, the ultimate certainty of showing the individual accuracy of the method can be proved only by post-mortem investigation. The following summarized findings serve this purpose (Fig. 101).

Post-mortem proof of the anatomical accuracy of our technique and method of localization has already been shown for one case of fornicotomy (HASSLER and RIECHERT, 1957), for thalamic operations for pain (HASSLER and RIECHERT, 1959), and for a few cases of Parkinson's disease (HASSLER, 1961, 1966b, 1969; HASSLER and RIECHERT, 1961; HASSLER *et al.*, 1965, 1967, 1969a, b, c, 1970).

2. Comparative, Radiologic, and Anatomical Evaluation of the Cerebral Reference Lines

As stated previously, the target points are localized with the help of reference lines measured in the encephalogram. A comparison of the measurement of the individual reference distances in each patient in the encephalogram and in the brain is made possible by the post-mortem material. The reference lines are defined as follows (Fig. 143a, b):

1. Sagittal reference line or "baseline" stretches from the posterior and lower edge of the interventricular foramen to the anterior edge of the posterior commissure. These are also linked in a slightly bent curve by the sulcus hypothalamicus. Although

Table 3. Intracerebral lines of reference comparing data from pneumoencephalogram and anatomical preparation

Case No.	Name			Lines of reference					
				Sagittal		Vertical		Frontal	
				Enceph.	Anat.	Enceph.	Anat.	Enceph.	Anat.
1	S.G.	I. Op.	rt.	23.6	20.5	14.4	14.5	19.7	21.5
			lt.	23.6	18.5	14.4	14.5		
		Hem.-Width	rt.					67.3	61.8
			lt.					67.3	65.6
		III. Ventricle (Co.p-plane)						9.9	7.0
2	F.J.	I. Op.	lt.	25.0		15.4		18.6	
		II. Op. a)	lt.	23.0		14.4	12.5	19.4	18.0
		b)						63.8	61.0
		III. Op.	rt.	24.3	17.0	15.25	12.5	19.6	18.0
		Hem.-Width	rt.					70.4	61.0
		III. Ventricle						12.7	9.4
3	R.L.	I. Op.	rt.	25.0	23.5	12.7	13.0	17.1	20.0
			lt.	Brain embedded					
		Hem.-Width	rt.					66.0	67.0
			lt.					66.3	67.3
4	H.J.	I. Op.	rt.	21.4	22.5	12.0	14.0	21.9	19.5
			lt.	21.4	23.5	12.0	13.5	21.6	20.5
		Hem.-Width	rt.					63.2	62.0
			lt.					64.8	66.5
5.	T.W.	I. Op.	rt.	25.1		14.6		20.9	19.2
			lt.	Brain embedded					
		II. Op.	rt.	25.5		14.7		22.0	21.0
			lt.	Brain embedded					
		Hem.-Width	rt.					65.7	65.0
			lt.					65.7	66.5
		III. Ventricle						13.8	7.0
6	P.Th.	I. Op.	rt.	28.4 unsuitable		14.8 unsuitable		22.1 unsuitable	
		III. Ventricle						11.5	8.0

Table 3 (continued)

Case No.	Name			Lines of reference					
				Sagittal		Vertical		Frontal	
				Enceph.	Anat.	Enceph.	Anat.	Enceph.	Anat.
7	C.Z.	I. Op.	rt.	Brain embedded				16.6	
			lt.	26.4		14.0		17.2	
		II. Op.	rt.	25.5		14.9		19.3	21.3
		Hem.-Width	rt.					75.0	71.5
		III. Op.	lt.	27.4		12.1			
		Hem.-Width	lt.					68.5	69.0
		III. Ventricle						22.0	21.9
8	S.H.	I. Op.	rt.	28.2	26.2	14.1	13.0	24.5	27.0
			lt.	28.2	28.5	14.1	21.0	25.2	21.0
		Hem.-Width	rt.					69.5	68.0
			lt.					71.8	72.0
9	A.I.	I. Op.	rt.	23.9		14.8		14.75	
		II. Op.	rt.	23.1	22.8	14.8	23.0	14.73	16.5
		III. Ventricle						17.4	20.5
		Hem.-Width	rt.					69.5	72.0
			lt.					66.3	64.0
10	B.W.	I. Op.	rt.					15.5	14.5
			lt.	23.6		13.6		21.0	18.5
		Hem.-Width	rt.					68.5	70.0
			lt.					67.0	68.5
11	B.H.	I. Op.	rt.	26.6	23.0	14.2	14.0	21.1	20.0
			lt.	26.6	23.5	14.2	14.5	21.4	19.0
		Hem.-Width	rt.					68.5	67.5
			lt.					67.6	63.0
		III. Ventricle						14.1	5.6
12	R.B.J.	I. Op.	rt.	26.4	23.5	15.0	19.0	17.7	18.0
							18.0 (above Co.p)		
		Hem.-Width	rt.					66.5	65.0
		II. Op.	lt.	26.4	24.0	15.0	22.0	17.7	19.0
		Hem.-Width	lt.					66.5	68.0
13	F.L.	I. Op.	lt.	24.2	19.5	15.3	15.5	19.8	20.5
		Hem.-Width	lt.					68.6	66.0
		II. Op.	rt.	24.2	20.0	15.3		19.8	21.5
		Hem.-Width	rt.					64.7	63.5
		III. Ventricle						6.28	5.2
14	H.P.	I. Op.	rt.		19.0		16.5	20.1	18.0
			lt.	23.0	18.0	14.1	14.0	19.2	19.5
		Hem.-Width	rt.					65.8	65.5
			lt.					64.5	63.0
		III. Ventricle						8.5	5.3
15	B.L.	I. Op.	rt.	25.6		14.5	Brain embedded	18.5	16.5
		Hem.-Width	rt.					66.0	66.0
			lt.					69.7	65.0
		III. Ventricle						9.0	4.3
16	D.M.	I. Op.	lt.	25.2	20.3	11.9	13.0	19.5	21.3
							(7 mm posterior to FMo)		
		Hem.-Width	rt.					68.4	68.0
			lt.					71.2	68.0
		III. Ventricle						7.2	
17	O.A.	I. Op.	lt.	25.2	21.5	13.8	14.5		
		Hem.-Width	lt.					65.5	

the thalamic structures continue much further occipitally, the "baseline" is a measure of the length of the thalamus.

2. The vertical reference line is the height of the upper border of the thalamus over the middle of the baseline, and it is determined by drawing a vertical line from the middle of the baseline to the lower border of the lateral ventricle.

3. The frontal reference line corresponds to the distance of the external border of the caudatum from the midline in the frontal plane of the posterior commissure. In cases of pathologic dilatation of the ventricular system, the maximum width of the hemisphere in the horizontal plane of the temporal lobe is considered the frontal reference straight line. This is the distance from the midline of the third ventricle to the tabula interna. Recently, 1.5 mm from each side have been subtracted from this for the dura and the subarachnoid space.

In Table 3 the three reference lines have been put together for the individual cases, and the values obtained by encephalography and anatomically have been put together. Only the measured values, after correction for x-ray distortion, have been compared. In cases of massive hemorrhage it is no longer possible to determine the measures reliably, which restricts the possible evaluation. The values have been placed into three groups: (1) hemispheres without stereotaxic interference; (2) hemispheres with stereotaxic interference; and (3) hemispheres with massive hemorrhages following stereotaxic interference. In addition to the frontal reference line, from the midline to the lateral wall of the lateral ventricle roughly in the Co.p level, in most cases also the maximal width of the hemisphere is measured in the x-ray, from the midline of the III. ventricle to the tabula interna of the temporal bone minus 1.5 mm, and in the postmortem of the individual case.

The statistical results, including mean values, standard deviation, and assessment of significance (after Student) of the calculations for errors of the median, are contained in Table 4.

Table 4. Mean values of the reference lines [mm]

Reference line	Group classification of hemisphere[a]	Pneumo-encephalography $\bar{x} \pm s$	Autopsied [mm] $\bar{x} \pm s$	Significance p		n
Sagittal	1	24.6 ± 2.21	22.8 ± 2.32	< 0.10		8
	2	24.8 ± 1.47	20.9 ± 3.25	< 0.0005		16
	3	25.4 ± 2.37	23.3 ± 4.15	< 0.30		4
					Total	28
Vertical	1	13.7 ± 0.71	14.2 ± 0.73	< 0.10		9
	2	14.2 ± 1.08	13.9 ± 3.99	< 0.40		16
	3	14.5 ± 0.47	20.0 ± 4.08	< 0.15		4
					Total	29
Frontal (width of lateral ventricle)	1	19.5 ± 2.64	19.9 ± 3.57	< 0.45		10
	2	20.1 ± 1.54	19.7 ± 1.70	< 0.25		15
	3	19.2 ± 4.38	19.0 ± 1.87	< 0.475		4
					Total	29
Frontal (width of hemisphere)	1	67.7 ± 2.41	65.5 ± 3.07	< 0.05		9
	2	67.1 ± 2.78	65.8 ± 2.79	< 0.10		16
	3	67.8 ± 10.59	69.4 ± 3.20	< 0.40		4
					Total	29
Frontal (width of third ventricle)	1	8.3 ± 1.20	3.8 ± 0.64	< 0.005		2
	2	9.5 ± 2.97	5.5 ± 1.82	< 0.0025		11
	3	10.5 ± 2.40	5.7 ± 2.41	< 0.025		3
					Total	16

[a] 1 = hemispheres without coagulations
2 = hemispheres with coagulations
3 = hemispheres with complications

[b] statistically significant difference (Student)

2.1. Sagittal Reference Line or Baseline

In normal hemispheres (group 1) of our material, this line is 24.6 ± 2.21 mm long, as determined from the encephalogram and 22.8 ± 2.32 mm long, according to anatomical investigations.[1] The difference between the baselines as determined anatomically and radiologically has a probability of only ($p < 0.1$) 90%. With a larger number of cases, however, it might become significant. A possible explanation of the difference is the enlargement of the interventricular foramen due to artifacts produced when fixing the anatomical preparation. Considering individual cases, it appears that the baseline is often thought to be longer on the encephalogram than in the anatomical preparation. One can take this fact into account by choosing the end point for the baseline in the interventricular foramen as far as possible in an occipital direction.

In the hemispheres with coagulation (group 2), there is a highly significant difference between measurements determined from the encephalogram and from the anatomical preparation. The post-mortems show that the shrinking of the sagittal reference line is due to the shrinking and scar formation produced by the coagulated lesion in the diencephalon. One case was not included in the statistical data—Case no. 2, in which the baseline in the encephalogram measured 23.0 mm, while in the anatomical preparation obtained 782 days after the operation its length was only 12.1 mm. The particularly marked shortening of the baseline in the anatomical preparation resulted because the thalamus, pallidum, and the internal capsule between them were coagulated in three extensive operations therefore, the entire length of the diencephalon was considerably shortened (see Table 3).

After massive hemorrhages in one hemisphere (group 3), there is no statistically significant difference ($p < 0.3$) between the preoperative encephalographic measurements and the anatomical values. The end points lie in the midline of the brain and have, therefore, been kept in place by the unaffected hemisphere.

A statistically significant difference between encephalographic and anatomical values is found only for the sagittal reference lines in the cases without hemorrhages. Here the values in the anatomical preparation are, on the average, smaller. This difference is obviously due to the sagittal shrinking caused by the scar from the stereotaxic operation.

[1] According to STAUFFER *et al.* (1953), this line measures on average 24.8 mm (22.5 to 28.5 mm).

2.2. Vertical Reference Line

Radiologically, this line exhibits the smallest differences from case to case (see also MUNDINGER and POTTHOFF, 1960, 1961; MUNDINGER and ZINSSER, 1965). Comparison of anatomical and radiologic measurements does not show a statistically significant difference, either. Except in cases of massive hemorrhage, the height of the thalamus varies very little. The anatomical height of a noncoagulated thalamus (group 1) is 14.2 ± 0.73 mm, as compared to a radiologically determined average of 13.7 ± 0.71 mm. This difference is not significant ($p < 0.1$).

In group 2, there was no significant difference between the anatomical height of the thalamus and the encephalographic measure. In group 3, we found an enlargement of the average height of the thalamus by 5.5 mm. However, the standard deviation is so great that the test for significance in these four cases does not show a difference. On the other hand, it may be pointed out that a hemorrhage in the thalamus or in the basal ganglia does increase the height of the thalamus considerably in individual cases (see Table 4).

2.3. Reference Line in the Frontal Plane

For this line, three measures are used: (1) the width of the lateral ventricle roughly in the plane of the posterior commissure; (2) the maximal width of the hemisphere at the height of the first temporal gyrus; and (3) the width of the third ventricle.

In the unoperated hemispheres (group 1), the median of the width of the lateral ventricle, as determined by post-mortem, is 19.9 ± 3.57 mm. The median of the width of the ventricle measured on the x-ray is 19.5 ± 2.64 mm. The test for significance shows no difference, with a probability of 45%. Nor is there a statistically significant difference between the two values (anatomically 19.7 mm, encephalographically 20.1 mm) in the hemispheres with coagulation (group 2). It is also remarkable that in the cases with large hemorrhages there is no difference in the width of diencephalon (19.0 mm : 19.2 mm). The constancy of these measurements, even after operation in the thalamus, is related to the fact that the end points of the measured distances are a long way away from the coagulated structures or from those that have been altered by a hemorrhage (see Table 3).

The anatomical measurements for the maximal width of the hemisphere at the height of the first temporal gyrus in group 1 show a median of 65.5 ± 3.07 mm, while the median width of one hemisphere in the encephalogram is larger, i.e., 67.7 ± 2.41 mm. The test for significance shows in group 1 no statistically significant difference ($p < 0.1$). In group 2, the discrepancy between anatomical and radiologic measurements is not statistically significant. In group 3, the anatomical measurements in three of the four cases are slightly larger because of the massive hemorrhages, but this is not statistically significant (see Table 4).

The diameter of the third ventricle in the anatomical preparation is smaller in all three groups (1, 2, 3) to a highly significant degree. The median for the width of the third ventricle in the encephalograms from group 1 is 8.3 ± 1.2 mm, while in the anatomical preparation it is 3.8 ± 0.64 mm. In group 2 the anatomical value is 5.5 ± 1.82 mm, while the encephalographic value is 9.5 ± 2.97 mm. In the group with hemorrhages in the basal ganglia (3), the anatomical value is 5.7 ± 2.41 mm and that from the encephalogram 10.5 ± 2.4 mm. In contrast to other reference lines, there is some uncertainty in the anatomical measurements for this point; for technical reasons the width of the third ventricle in the anatomical preparations was not always measured at its largest diameter before it merges into the aqueduct, which was where the radiologic measurement was taken. The measured anatomical values are, therefore, too small in some cases.

When such large deviations occur between two different methods of measuring the same structure, the question can be raised whether the values are influenced in both cases by the technique or by the method of measuring. We pointed out previously that the anatomical values were obtained in the same plane in every case. This fact certainly has an influence on the size of the measurement. In addition, the anatomical measurement of the third ventricle is altered by the fact that the C.S.F. disappears from the third ventricle in the first two hours after death. The C.S.F. enters the tissues surrounding the third ventricle and they swell after death, narrowing the space of the third ventricle by this swelling. This explains why the third ventricle has a smaller diameter post-mortem.

On the other hand, a number of factors combine to produce an enlargement of the diameter of the third ventricle in the encephalogram: (1) the technique of fractionally increased pressure, (2) the later warming and expansion of the helium (which probably plays a small part, since the gas is warmed to body temperature before it is blown in), and (3) the extraction of liquid from the surrounding tissues. Acting as a foreign body, the gas induces increased production of C.S.F. As a rule, the a.p. picture is taken last, 30 min after filling has begun and some C.S.F. has been extracted, and is therefore particularly effective. During encephalography, all factors tend to increase dilatation, while post-mortem, all factors work toward narrowing the third ventricle. This situation is the cause of the great discrepancy between the anatomical and radiologic measurements. Assessing the changes in the width of the third ventricle acting in opposing directions, it seems that the encephalographic values for the third ventricle come nearer to its normal size during life.

In summary it can be said that apart from the width of the third ventricle, there are no statistically valid differences between the values for the reference lines obtained radiologically and those obtained anatomically. As is shown in particular by comparison of normal brains, there is justification for using the previously described reference lines for determining the target point. Individually, the widths of the hemispheres show only slight scatter, but in all three groups the radiologic measurement is always about 1.5 mm too large. In order to calculate the frontal coordinate in the individual case, it is, therefore, necessary to measure the width of the hemisphere from the midline to a point 1.5 mm inside the tabula interna in order to subtract the meninges and the subarachnoid space.

3. Anatomical Accuracy of Our Stereotaxic Procedure for Reaching the Target

This section is a discussion of how to determine the preciseness with which the target point, calculated from the encephalogram, coincides anatomi-

Table 5. Aberration and direction of aberration from the calculated target in anatomical preparation

Case No.	Target reached		Interval between operation and encephalography [days]	Aberrations			Cause of aberration
				Sagittal	Vertical	Frontal	
1	V.o.a	+	3	0	0.3 ↑	0.6 lat.	dilation of 3rd ventricle
2a	Pa.i lt.	+	0	0	3.1 ↑	3.5 med.	scarring after $3^1/_2$ years
2b	V.o.a lt.	+	0	0	2.0 ↓	0.0	scarring
2c	V.o.a. rt.	+	3	1.3 rostral	0.0	4.6 med.	dilation of 3rd ventricle
3	V.o.a/p rt.	+	9	2.7 rostral	0.5 ↑	1.9 lat.	none: (inaccurate localization of anatomical substrates!)
4	V.o.a lt.	+	10		3.5 ↓	0.6 med.	cannula too large (avoidable aberration)
5a	Pa.i rt.	(+) Pa.i	11	4.4 rostral	0.1 ↑	2.8 med.	dilation of 3rd ventricle (13.8:7.0)+scars (operation 19.5 months before)
5b	V.o.a lt.	– Pa.i	6		6.0 ↓	1.4 med.	reading error for coordinates by 5.0 mm; this target was reached exactly even with a 0.5 mm dilation
7a	V.o.a lt.	+	0		0.4 ↑	0.9 lat.	scarring after $3^1/_2$ years (incorrect determination of anatomical substrate target!)
7b	Pa.i rt.	+	0	0	0.6 ↓	0.0	cyst, therefore doubtful
10	V.o.p lt.	+	9		2.6 ↑	2.4 med.	none
11	V.o.a/p rt.	– (Pa.i)	6		2.3 ↓	8.0 lat.	shifting by parietal hemorrhage and oppression of corpus callosum
12	V.o.a rt.	+	3	0.1 rostral	2.0 ↓	0.6 lat.	too low position of posterior commissure (4 mm basal, correctable mistake)
13a	V.o.a/p lt.	+	4	1.0 rostral	0.8 ↑	0.9 med.	scar shrinkage $1^1/_2$ years after operation
13b	Z.i rt.	+	574	0	0.9 ↑	1.35 med.	none
15	V.o.a rt.	+	4		1.0 ↓	4.1 lat.	slight bleeding within coagulation area, frontal distance from midline 1.0 mm lat. (point too deep!)
16	V.o.a rt.	+	7	1.3	0.8 ↑	1.2 med.	anatomically exactly in the center
17	Pa.i lt.	+	8	0.3	1.1 ↑	0.3 lat.	anatomically in the center of the nucleus
19/13	16 + 1 (+) 2 –			1/19 9/18 $\bar{x}=8$	6 med. 3 dors. 6/18 3 mm $\bar{x}=1.$	4 med. 2 lat. 6 mm	

cally in fact with the place reached by the point of the electrode. As seen from the following survey, the cut could not always be placed in the previously calculated target point plane; the cut's locale only approximated the target point plane in the four cases that were not examined serially. The necessity of approximation leads to a discrepancy between the anatomically measured coordinates and the reached target point, dependent on the angle of the electrode in the three dimensions of space. For this reason, the previously calculated individual target point sometimes does not coincide exactly with the anatomical point. This is particularly the case for the coordinates in a sagittal direction, since in all but one case, frontal series of cuts were performed. Another source of error in the values for the coordinates of the target point measured anatomically is that the coagulation of thalamus or pallidum and following complications have changed the reference lines in their position or size. In addition, the brains were examined at different intervals after the coagulation and after different periods of formalin fixation. It is known that formalin alters the size of brain structures (see STEPHAN, 1956) according to the length of fixation (Table 5).

We compensated for the deviation of the anatomical plane of a cut from the radiologic target point plane by preparing a drawing. On graph paper, we drew in a proportion of 10:1 the baseline and the actual angle of the electrode to the baseline (if the target point lies on the baseline) or the same angle in parallel displacement (see Fig. 144). From this drawing, the point where the electrode should have met the anatomical plane can be determined. The point in the target plane, which has been reached by the point of the electrode, is found by entering the anatomically determined electrode track in the transverse plane and drawing the position of the electrode as a parallel line through the track. Where the parallel line passes through the target point plane is the spot where the point of the electrode has in reality been lying. For this special evaluation we have omitted from the table and the discussion the autopsies of cases with hemorrhages in the basal ganglia because the accuracy of reaching the target cannot be judged in these cases (see HASSLER et al., 1959).

Case 1. Coordinates of target point V.o.a: sagittal →10.0; vertical ↕0; frontal 11.1 toward the left. According to the anatomical preparation, the electrode reached the target, the V.o.a. In the plane of the target point (10.0), the location of the point of the electrode lay anatomically 0.3 mm dorsal to the calculated target point (Fig. 34a). This deviation lay in the range of radiologic error since the plane of the target point lay at a certain distance from the central ray, which was projected onto the center of the base ring. At a maximum distance of 3 m between focus and film (as used by us), this error had a value of 0.28 mm (MUNDINGER and UHL, 1965, 1967).

In the frontal cut the aimed-at structure had also been reached despite the swelling on the 23rd day, according to the autopsy. In the graphic reconstruction of the position of the electrode, the factual lateral coordinates deviated laterally by 0.6 mm. For this coordinate the radiologic error measured 0.52 mm.

Case $2_{\ell 1}$. Coordinates of the target point pallidum internum: →5.0, ↓4.1; 16.5 left. At the first operation the target was the pallidum internum. According to the anatomical finding, it had been reached by the point of the electrode. The reconstruction of the target point from the anatomical plane of the cut 7.5 mm behind the interventricular foramen showed a deviation of the target point in the target plane 5.0 of 3.1 mm in the dorsal direction. This considerable deviation is in contrast to the proved displacement of the 2-mm long silver clip that had been left in the coagulated area. Eleven months after first operation (Fig. 40a), this clip had already moved 2–3 mm in a basal direction. This shift (Fig. 132a, b) led us to doubt whether we had localized the puncture track correctly $3^1/_2$ years after the first operation in the very scarred basal ganglia of the left hemisphere. In the frontal plane (5.0 mm behind the interventricular foramen) the anatomical target point lay 3.5 mm medial to the calculated target point. This deviation was probably due to a lateral displacement of the basal ganglia during the first 3 days after encephalography, which had not disappeared at the time of the operation. It is known that air in the ventricles leads to dilatation because of the drying out of the surrounding brain tissues (so-called 24-hour encephalogram, after BRONISCH, 1951, 1952). During the stereotaxic operation, the frontal coordinate was determined from the midline of the skull after the third ventricle had filled again with CSF. Therefore, the lateral displacement of the wall of the third ventricle because of the drying-out process could neither be recognized nor taken into account. (See Fig. 132a–c.)

This situation occurs only if the operation is performed less than 5 days after encephalography. After that time, the third ventricle resumes its normal size.

At the second intervention $2_{\ell 2}$ (coordinates of the target point in the V.o.a/p: →12.8; ↕0; 11.0 left), the V.o.a was reached by the electrode as intended at the boundary with the V.o.p. Anatomically a 2.0 mm deviation in a basal direction was found, but in the frontal and sagittal planes the calculated target point had been reached exactly. The vertical deflection was due to scar formation in the two coagulated areas (Figs. 133b, 132a–c).

At the third intervention in the right hemisphere 3 years later, we reached the base of the target structure with the stimulating electrode with the coordinates in the V.o.a/p →12.3, ↕0, and 10.6 right. According to the autopsy, the point of the electrode (see Fig. 44) deviated 4.6 mm medially and 1.3 mm rostrally from the calculated target point. The dilatation of the ventricle after putting in air may explain the medial deviations in both hemispheres. The drying effect of the air in the third ventricle on the surrounding structures of the diencephalon leads to a displacement of the thalamus, and consequently of the target point, in a lateral direction. Because the third ventricle becomes filled with CSF., this change cannot be recognized on the day of the operation and therefore cannot be taken into account when the frontal coordinate is measured from the midline. If the operation takes place a few days after encephalography, at a time when the lateral displacement of the thalamic structures has not been compensated for, the target point will be assumed to be too far in a medial direction on the day of the operation.

In **case 3,** coordinates of the target point V.o.a/p: →11.7; ↕0; 9.6 right, the base of V.o.a was anatomically reached by the electrode. The autopsy showed that there was a 2.7 mm rostral deviation of the point reached, a 1.9 mm lateral deviation, and a vertical deviation of only 0.5 mm dorsally. The electrode reached the basal nuclear area of the V.o.a instead of the V.o.p. It is possible that the deviation in a rostral direction was due to the difficulty incurred in determining anatomically the zero point of the coagulation. The 1.9 mm lateral deviation had already been observed on the control x-ray. This relatively slight deviation was not corrected because we did not want to damage the brain by a new puncture in the immediate neighborhood of the first puncture track (Fig. 50a).

In **case 4** (postencephalitic Parkinsonism), the predetermined target point in the V.o.a (coordinates: →9.5; ↕0; 10.6 left) had not been anatomically reached by the electrode. Its point when stimulating lay below the base of this nucleus in the border of peduncle (Fig. 52). The x-ray check of the electrode point had already shown a ventral deviation of 4.5 mm. From the x-ray (Fig. 51), this deviation could be explained by the fact that the electrode guide had a larger-than-usual internal diameter. In this case, an additional electrode of a larger diameter was introduced in order to record from depth in the caudatum. When reconstructing the position of the electrode in the plane of the target point at autopsy, it was found that the point actually reached was 1.0 mm higher than could be assumed from the x-ray photo (Fig. 51), or 3.5 mm below the target point. The point at which the stimulation took place was in the frontal plane, 0.6 mm medially from the aimed-at point. This last deviation lay within the limits of error of the radiologic technique and of the stereotaxic instrument.

In **case 5** the aim of the first operation (5_r) was the pallidum internum (coordinates: ↔5.4; ↓3.1; 18.3 right) but according to the autopsy the electrode reached the pallidum externum. In the vertical direction there was almost no deviation (0.1 mm). In the sagittal direction the deviation was 4.4 mm rostrally, but according to the autopsy the electrode lay 2.8 mm medial to the target point. The shortening of the two coordinates was due exclusively to the strong pull of the scar in the basal ganglia. The fact that the destruction of the pallidum externum (Fig. 55a) was greater than that of the aimed-at pallidum internum—in other words, that the electrode during the operation was lying too far laterally—could be related to the marked dilatation (Fig. 54b, d) of the third ventricle during encephalography (13.8 mm as opposed to the autopsy finding of 7.0 mm). The dilatation had disappeared 11 days later, at the time of the operation. The marked scarring of the coagulated area later displaced the basal ganglia medially in the direction of the third ventricle so that the frontal coordinate was even anatomically displaced medially.

The aim of the second operation (5_l) was to operate on the V.o.a (coordinates →10.9; ↕0; 12.4 left) because of an error in reading the vertical coordinate in the picture at the operation. In fact, a point was aimed at that was 5.5 mm too far ventrally (Fig. 56a, b). The middle of the puncture track reached in the target plane (10.9 mm behind the interventricular foramen), a point that deviated from the point aimed at (reading error) vertically by 5.0 mm (Fig. 123a). A frontal deviation of 1.4 mm medially was also noted. In other words, the point of the needle almost accurately reached the target point, which had been altered through the error in reading. This fault could not be discovered from the radiologic check of the position of the needle because the numerical value that had been determined by the error in reading had, in fact, been reached. Autopsy showed that the point reached was 6 mm ventral, and 1.4 mm medial, to the target point. Despite the mislocation of the coagulation focus, it interrupted the dentato-thalamic fibers, the degeneration of which is indicated by the gliosis in the terminal nucleus V.o.p (see Fig. 123b).

In **case 6,** it was not possible at the post-mortem to determine the accuracy with any certainty because of the massive hemorrhage. Despite this uncertainty, we were able to ascertain that the aimed-at V.o was in fact reached because the hemorrhage had mainly destroyed the ventral nuclei of the rostral thalamus and extended only slightly into the neighboring areas (see Fig. 58a).

Case 7. The first operation (7_l) in the V.o.a (coordinates: →11.2; ↑1.8; 9.1 left) had reached this nucleus $3^{1}/_{2}$ years before the patient's death. Scar formation developed with narrowing of the thalamus (see Fig. 63). The center of the scar lay 0.4 mm dorsal and 0.9 mm medial to the target point. The anatomical accuracy was, therefore, good in view of the fact that the measurements of the thalamus were changed by the scarring.

At the second stereotaxic intervention (7_r) in the right pallidum (coordinates: →4.1; ↓0.5; 15.5 right) 19 months before the patient's death, the pallidum internum had been anatomically reached. As far as could be ascertained in the presence of a cyst through softening, there was a slight deviation ventrally of 0.6 mm (Fig. 64).

The third operation (7_{l2}) was performed 45 days before death. The pallidum internum (coordinates: →4.6; ↓4.3; 19.2 left) had been reached by the coagulation (see Fig. 62). Because of a hemorrhage into the pallidum and caudatum, the exact size of the coagulated lesion could not be measured on the anatomical preparation. The ventral-most edge of the lesion lay 1.7 mm from the electrode axis and 3.2 mm medial to it. In view of the hemorrhage that had destroyed the pallidum completely, this deviation did not indicate a wrong localization (see Fig. 118).

Cases 8 and 9 could not be correlated to the x-ray because of the massive hemorrhages into the basal ganglia.

In **case 10**, the reconstruction indicated that the target point (V.o.a, coordinates: →10.1; ↓1.5; 11.4 left) had been reached anatomically in the plane of the target point. Dorsally, the electrode point deviated by 2.6 mm; medially, by 2.4 mm. These deviations were caused by the swelling of the thalamus due to the hemorrhage into the frontal lobe and the operation for it (Fig. 71).

In **case 11** the target was a point in the V.o.a/p (cerebral coordinates: →12.4; ↑1.1; 13.8 right). Anatomically, the coagulation (Fig. 119) was found in the pallidum internum. This deviation could not be blamed on the method of aiming, on the apparatus, or on the method of localization. Rather, it resulted from a large hemorrhage that occurred in the right parietal and occipital lobes sometime during the period between encephalography and the stereotaxic operation. Although hardly noticeable clinically, it led through swelling to a displacement of the basal ganglia into the other half of the skull (see Fig. 73a, b). At encephalography, the right wall of the third ventricle was 3.5 mm to the right of the middle of the skull, but anatomically it was found 4.5 mm across the midline in the left half of the skull. The frontal coordinate, which had been correctly determined from the encephalogram before the occurrence of the hemorrhage, was used on the day of the operation, but at that time the third ventricle was no longer filled with air. The center point of the coagulated area was anatomically 16.5 mm away from the midline of the skull. According to the reconstruction, the needle track had reached the frontal plane (10 mm behind the interventricular foramen) at a distance of 14.0 mm from the wall of the third ventricle. The actual distance from the middle of the skull was 17.5 mm. The remaining 3.5 mm corresponded exactly to the distance of the wall of the third ventricle from the middle of the skull. The center of the lesion lay anatomically 2.3 mm on the basal side of the target point. The vertical difference was explained by the hemorrhage above the corpus callosum, pressing it down. The determination of the target point was, therefore, not wrong, within the limitations of the apparatus and radiology. The deviation resulted from the displacement of cerebral structures by the parietal hemorrhage (Fig. 73a) that occurred in the period between encephalography and operation. The effects of stimulation and of coagulation did not differ very much from the aimed-at V.o.a, since the pallidum internum (the neurons of which were superposed on the V.o.a) had been reached and stimulated (see Fig. 119).

In **case 12** (V.o.a/p, coordinates: →11.4; ↕0; 10.0 right) the middle of the coagulated area was anatomically 0.1 mm rostrally, 0.6 mm laterally, and 2 mm below the target point. The basal deviation was caused by the posterior commissure being 4 mm too low relative to the neighboring structures of the thalamus. The displacement in the plane of the halved baseline where the lesion lay was only half the size (2 mm). In the coagulated area in the target plane →11.3 mm (see Fig. 75d), there were two lesions; the medial one was in the nucleus reticulatus next to the V.o.a and measured 10.6 mm from the wall of the third ventricle and 2 mm below the baseline, while the lateral one was found at the dorsal edge of the pallidum internum with the centerpoint 13.3 mm from the wall of the ventricle and 4 mm below the baseline. These two lesions resulted because, besides the five coagulations with the straight thermoelectrode from the target point to +5 ventrally, seven more were carried out later with the string electrode 6 mm to the lateral side. The string electrode had been moved from the position's target point, +2 and +4 ventral to the target point. The string electrode had been turned three times in a lateral direction by 90°, which meant that its point had reached 6 mm lateral to the track of the central cannula. The lateral edge of the coagulation (Fig. 75d) in a plane 7 mm behind the foramen extended to 7 mm lateral to the first track. The deviation in a ventral (basal) direction was explained by our having performed four coagulations with angles of 30°, 35°, and 40° in a laterobasal direction: the lesions in the pallidum internum and the internal capsule were caused by the string electrode. When looking at the ventricles, it was noticeable that the posterior commissure was about 4 mm lower than usual in relation to the neighboring thalamic structures. This deviation had shown itself in the steep course of the stria medullaris thalami, descending at about 55°. The position of the two coagulated lesions was, therefore, in agreement with the position of the rod electrode and the bent-string electrode during the operation.

A massive thalamic hemorrhage occurred during the stereotaxic operation performed on the left side (2nd operation) (12_ℓ) three years later. In the plane 13 mm behind the foramen, the center of the hemorrhage was in the baseline and 13 mm left (Fig. 127a, c). The center of the coagulated lesion coincided well with the point determined by coordinates and reached radiologically (Figs. 74; 78a).

In **case 13** the target point at the first left-sided operation (13_ℓ) lay in the V.o.a (coordinates: →10.7; ↕0; 11.2). The point reached by the electrode coincided with the target point in all three coordinates with deviations of 1.0 mm rostrally, 0.8 mm dorsally, and 0.9 mm medially. The shrinking of the thalamus after 18 months was responsible for these small deviations (see Fig. 84).

At the second operation (13_r) a target point was chosen in the zona incerta (coordinates →13.4; ↓5.7; 9.2 right). According to the x-ray check, the point was reached by the string electrode with a deviation averaging 0.7 mm. The point of the anatomically determined lesion deviated by 0.9 mm in a dorsal, and 1.4 mm in a medial direction (Fig. 86). The edge of the aimed-at structure had been reached by the point of the coagulated area. Further rostrally, the zona incerta and the fibers going through it had also been destroyed by the coagulation. Anatomically, coagulating with the string electrode in a position 30°–45° laterally, caused a lesion that was located further laterally than the primary lesion, and that reached into the nucleus reticulatus thalami and the neighboring internal capsule (see Figs. 85, 86).

In **case 14,** the target point and the puncture track were not measured anatomically because of the massive hemorrhage. The radiologic check of the position of the electrode showed a deviation of 0.3 mm rostrally and medially and 1 mm dorsally as compared with the target point determined in the encephalogram.

Case 15. The target point in the V.o.a (coordinates →10.4; ↕0; 10.4 right) had been reached anatomically with a lateral deviation of 4.1 mm and a ventral deviation of 1.0 mm (Fig. 90a). The ventral deviation was caused by the posterior commissure lying too far basally relative to the thalamic structures. The lateral deviation from the wall of the third ventricle occurred because the ventricle was narrowed by the hemorrhage, with subsequent swelling of the thalamus, with the result that the ventricular wall, in the plane of the target point, coincided anatomically with the midline of the brain. When the lateral coordinate was determined from the midline, the deviation measured only 1.1 mm. Radiologically (Fig. 89a), the third ventricle had a width of 9 mm, anatomically (Fig. 90a) of 3 mm, even in that part where the hemorrhagic swelling of the thalamus was insignificant. This situation alone would force the needle track to lie in the anatomical structure 3.0 mm further laterally than intended. Therefore, 3 mm of the 4 mm deviation resulted because, upon determination of the target point, the width of the 3rd ventricle was assumed, from the encephalogram, to be larger than it actually was.

Case 16. The target point was the V.o.a (coordinates: →8.8; ↓0.8; 9.8 right). Anatomically, the track and the coagulated area (Figs. 95, 129) had reached the V.o.a exactly, and the nucleus had been coagulated down to its base. The radiologic check of the electrode position showed that the point of the electrode deviated from the target point by 1.3 mm rostrally, by 0.8 mm dorsally, and by 1.2 mm medially. Severe retrograde cell loss in front of and behind the coagulation (Figs. 95, 129) was also evident.

Case 17. The target point was in the pallidum internum (coordinates: →5.9; ↓3.0; 16.1 left); the target structure had been reached exactly and almost completely destroyed. The actual location of the point of the electrode was 0.3 mm rostral, 1.1 mm dorsal, and 0.3 mm lateral to the calculated target. A 1.0 mm dorsal deviation of the electrode point in the x-ray picture corresponded to the anatomical findings (Fig. 98).

3.1. Discussion of Accuracy

Of the 17 cases that could be evaluated, the calculated, aimed-at structure was, therefore, reached in 14 (Fig. 101). In one intervention in the pallidum internum, the point of the electrode was positioned lateral to the lamella pallidi interna in the pallidum externum (5_r). In two other operations (5_ℓ, 11) the target point was not reached by the point of the electrode. These deviations were not due to the method but to causes that will be discussed below. In each case, the aimed-at structure was defined by its cytoarchitectonic and myeloarchitectonic characteristics. In case 11, the point of the electrode reached the pallidum internum instead of the V.o.a because the diencephalic structures had been displaced by a hemorrhage into the white matter of the parietal lobe that occurred between encephalography and the operation 7 days later (Figs. 73a, b; 119, 101).

In 3900 subcortical stereotaxic interventions, this was the only time that we observed this phenomenon. In the other case (5_ℓ), the 5.5 mm basal deviation of the point of the electrode was due to a mistake in reading off the vertical coordinate in the x-ray. The reading error coincides exactly with the observed deviation (Fig. 56a; 101).

In case 5_r the third ventricle was unusually dilated at the encephalography (Fig. 54a) 11 days before the operation, as compared to the post-mortem finding, which explained the wrong localization of the point of the electrode laterally into the pallidum externum. According to the post-mortem measurement, the electrode point had been displaced 2.7 mm medially because of the marked formation of scar tissue in the basal ganglia during the 19 months after the operation (Fig. 55a).

In all other interventions, the architectonically defined substrate was reached. In case 15, the 4-mm lateral displacement of the electrode point was caused by a swelling of the thalamus at the coagulation lesion that had afterward increased the distance between the wall of the ventricle and the target point. Without hemorrhages, the real deviation would have been only 1.0 mm in a lateral direction (Fig. 101).

In case 12 the posterior commissure was located low in relation to the thalamic structures. As a result, our baseline and, therefore, the point of the electrode, were located 3 mm too far basally (Fig. 75d). A similar displacement in a ventral direction because of the low positioning of a posterior commissure, though to a lesser extent, was present in case 15. In case 14, a relatively large guiding cannula was used in order to record from the basal ganglia. The thinner electrode, therefore, was permitted to move too freely and was responsible for a ventral displacement of 4.5 mm. This error can be avoided by using the correct cannula.

Scar formation in the basal ganglia (see Fig. 132a, b) over years renders numeric anatomical determination of the coordinates inaccurate, which explains smaller deviations of the electrode point in our cases. The uncertainty of determining anatomically a localization of the electrode point is responsible for the frequent measurement of a vertical deviation of up to 2 mm, although the electrode reached the target structure. The anatomical determination of the vertical coordinate is also made difficult by the fact that, through the forward movement of the electrode for further coagulations, the first localization of the electrode point often can no longer be reconstructed. In three cases (1, 2_r, 12) it is possible that a medial displacement of the localization of the point of the electrode in the anatomical preparation was caused by the fact that the dilatation of the third ventricle a few days after encephalography had not yet returned to normal on the day of the operation.

The errors of localization in our technique made it advisable to analyze accurately many factors, utilizing data from as many cases as possible. Analysis of a large number of cases also makes visible avoidable errors. The cases in which post-mortems were possible were, as a rule, not representative of all the cases; in fact, they constituted a rather unfavorable selection (Fig. 101).

Some of the errors noted can be avoided by using very precise instruments that allow the accurate guiding of the electrode (case 4). In such a complicated stereotaxic method, it is also necessary to check the calculations and readings (5_l) twice. In order to avoid hemorrhages in the coagulation area (15), the tissue should be damaged as little as possible. This can be achieved by using a device that measures the temperature of coagulation.

To a certain extent, errors in localization that are due to anomalies of those brain structures seen in the x-ray picture and used in determining the target point could be avoided. One such anomaly is the abnormally low position of the posterior commissure relative to the neighboring thalamic structures (12 and 15), as described previously by Brierley and Beck (1959). In such cases it is possible to compensate empirically for major deviations by measuring the vertical distance from the posterior commissure to the outline of the thalamus. In order to avoid deviations, one should not take the entire lateral coordinate from the edge of the ventricle if the diencephalic structures, as viewed in the antero-posterior encephalograms, are much contracted (5_r, 2_r).

It is possible that some of the numerically stated errors of localization were not real. In patients who survived for a long time, it was no longer possible to determine accurately the location of the electrode because of changes of the anatomical substrate through scar or cyst formation. However, these errors have not been corrected for in the following ascertainment of our accuracy of hitting the target (see Fig. 101).

It is not possible to avoid all errors of localization caused by individual variations of the brain structures aimed for, although our method of localization for the most part takes into account or compensates for the numerical deviation between the model brain and that of the patient.

Utilizing the data obtained from the foregoing post-mortem findings, it is possible to express numerically the accuracy of reaching the target when the avoidable errors of localization are excluded but not those caused by scar formation. According to our post-mortem findings, the median total error of accuracy in reaching the predetermined target was 1.23 ± 1.48 mm in the vertical, 1.68 ± 1.20 mm in the frontal, and 1.10 ± 0.92 mm in the sagittal coordinates (Fig. 101).

The median error of the stereotaxic instrument including the radiologic error if used correctly was about ± 0.5 mm (Mundinger and Uhl, 1965). In view of the post-mortem findings, we are now able to state the median total error of localization, which contains in addition the error due to the instrument and radiologic technique. One can assume that the total error of localization would be smaller if it were determined from cases other than those selected by mortality. It is remarkable that the median error, which contains instrument, x-ray, and wrong-use errors, is similar to that obtained from 1837 stereotaxic operations that were not checked anatomically (Mundinger *et al.*, 1970).

4. Correlation of the Effects of Stimulation During the Operation with the Anatomical Substrates

4.1. Method of Stimulation

Since 1950, when we performed the first stereotaxic operations for extrapyramidal movement disorders, we have carried out stimulations at the target point in order to have a physiologic check of the localization of the electrode in view of individual variations in anatomical structure. Contrary to previous expectations, we found that stimulations of the oral ventral nuclei of the thalamus do not produce sensory effects. Recording the effects of stimulation has been an integral part of the stereotaxic operations performed with our method since 1950 (RIECHERT and HASSLER, 1953; HASSLER, 1953b, 1961, 1966; HASSLER and RIECHERT, 1954a, b, 1961; HASSLER *et al.*, 1960; KROTZ, 1962; MUNDINGER and RIECHERT, 1963, 1966; MUNDINGER, 1966; VAN MANEN, 1967). In the majority of cases, potentials have been recorded from depth with multiple electrodes and with scalp electrodes when stimulating in the subcortical ganglia (JUNG *et al.*, 1950, 1951; JUNG and RIECHERT, 1952, 1955; RIECHERT and SCHWARZ, 1952; RIECHERT and UMBACH, 1955; UMBACH, 1957; GANGLBERGER, 1959b, 1961b, c, 1962a; HASSLER *et al.*, 1960; UMBACH, 1964, 1966a, b; UMBACH and EHRHARDT, 1965a, b). In some cases, the motor effects of stimulating inside the subcortical ganglia have been recorded electromyographically (GANGLBERGER and BRUNZEMA, 1962; GANGLBERGER, 1963a, b; STRUPPLER and STRUPPLER, 1962; KRIENITZ, 1961).

For the bipolar stimulations in the subcortical ganglia, a coaxial electrode is used with a distance between poles of either 5 or 3 mm. Initially we used a thyratron instrument; this was subsequently replaced by one with time control that produced sawtooth and triangular stimuli, designed by TÖNNIES. The most frequently used stimulus is a rectangular impulse of 1 msec duration. The usual frequencies of stimulation are 1, 4, 8, 25, and 50 cps. Higher frequencies are avoided because of the danger of epileptic discharges. In the last 15 years the voltage of stimulation has been measured with an oscillograph. This voltage varies from 2 to 30 volts with intensity values of 0.3–1.2 mA. The stimuli are applied to conscious patients who have received no premedication, which enabled us to note the subjective effects during and after stimulation. The patients are lying on their backs and can move freely except that their heads are fixed through the basal ring of the stereotaxic instrument. Apart from the motor effects and the experiences during the stimulation, we also observe or register the vegetative effects: frequency of respiration, intravascular or external registration of the blood pressure (SCHMIDT, 1965; SCHMIDT and KANIAK, 1960; SCHMIDT and UMBACH, 1961), registration of the pulse wave (HÖHNE, 1964), and changes in the hematopoietic systems (MUNDINGER and SCHOLLER, 1956a, b).

Several publications by other authors describe the effects of stimulation of subcortical structures during stereotaxic brain operations (SPIEGEL and WYCIS, 1952, 1962; GUIOT and BRION, 1953; TALAIRACH *et al.*, 1952; RIBSTEIN, 1960; SHEER, 1961; BERTRAND, 1966; and others).

In the following discussion, only those effects of stimulation are used for localization, for which the stimulated substrate has been ascertained anatomically post-mortem. However, the structure surrounding the point of the electrode cannot be regarded as the only reacting structure, since the stimulation was performed with a distance between poles of 5 mm and because sometimes relatively high voltages were used for the stimulation. A spread of the stimulating current to structures lying further is therefore possible, particularly when higher frequencies are used. This applies especially to the internal capsule, which has a lower threshold for stimulation than the nuclei because of the closeness of the neuronal elements.

The evaluation of the effects of stimulation based on the records is done separately according to the anatomically determined structure reached by the stimulating electrode. The five regularly used frequencies of stimulation (1, 4, 8, 25, and 50 cps) are also considered in terms of their effects.

In this monograph a total of 162 effects of stimulation in seventeen patients with 25 operations and with 27 target points are considered. Twenty-two factors of function have been recorded (see Table 6a, b). Such factors include effects of stimulation that either cause a change of the symptoms of the illness, e.g., influencing the tremor, or are independent of the symptoms of the illness, e.g., twitches synchronous with the stimuli or changes of active movements. Many effects of stimulation on the limbs affect particularly the contralateral side of the body, although the ipsilateral limbs can react in the same way when stimuli of greater strength are employed.

Table 6a. Correlation of effects of electrical stimulation with anatomical substrates[a] (162 effects of stimulation in 17 patients)

Numbers of stimulated substrate targets in right hemisphere or in **left hemisphere** (boldface)		Number of stimulation effects with different parameters	Increasing tremor	Acceleration of tremor	Slowing of tremor	Blocking of tremor
Pa.e	1	5	No. 5 (4 cps, 8 cps)			No. 5 (50 cps)
	0	**0**				
Total	1	5	40% ⟨0%⟩			20% ⟨50%⟩
Pa.i	4	13	No. 7 (8 cps); No. 9 (1 cps, 25 cps); No. 11 (4 cps, 8 cps, 50 cps); **No. 17 (1 cps)**	**No. 8 (25 cps, 50 cps);** No. 9 (8 cps); **No. 17 (50 cps)**	No. 9 (4 cps); **No. 17 (4 cps, 8 cps)**	**No. 2 (25 cps);** No. 7 (50 cps); **No. 8 (1 cps);** **No. 14 (4 cps, 8 cps);** **No. 17 (25 cps)**
	4	**15**				
Total	8	28	20% ⟨14.3%⟩	14% ⟨21.3%⟩	8.5% ⟨0%⟩	17% ⟨21.3%⟩
V.o.a	4	26	No. 6 (4 cps, 8 cps); No. 9 (4 cps, 8 cps); No. 12 (4 cps); No. 16 (8 cps)	**No. 4 (8 cps);** No. 9 (8 cps)	**No. 4 (4 cps);** No. 9 (4 cps); No. 12 (4 cps); No. 16 (4 cps)	No. 9 (50 cps); **No. 13 (8 cps)**
	4	**12**				
Total	8	38	13.3% ⟨0%⟩	4.5% ⟨0%⟩	8.9% ⟨0%⟩	4.5% ⟨5.6⟩
V.o.p	2	6	**No. 2 (4 cps, 8 cps);** No. 2 (1 cps, 4 cps); **No. 12 (4 cps, 8 cps);**	**No. 2 (25 cps);** **No. 10 (4 cps);** **No. 12 (8 cps);**	No. 2 (4 cps); **No. 12 (4 cps)**	**No. 2 (25 cps);** No. 3 (4 cps, 8 cps); **No. 10 (50 cps)** **No. 12 (50 cps)**
	3	**14**				
Total	5	20	20% ⟨0%⟩	12% ⟨10%⟩	8% ⟨0%⟩	20% ⟨30%⟩
V.o.i	1	3				
	1	**5**			**No. 7 (8 cps)**	**No. 7 (1 cps, 8 cps)**
Total	2	8			10% ⟨0%⟩	20% ⟨0%⟩
Ca.i	1	7	No. 15 (4 cps, 8 cps)			
	0	**0**				
Total	1	7	40% ⟨0%⟩			
Ped.c.	0	0	**No. 5 (1 cps)**	**No. 5 (8 cps)**	**No. 5 (4 cps)**	
	1	**4**	20%	20%	20%	
Total	1	4	⟨0%⟩	⟨0%⟩	0%	
Zi	1	4	No. 13 (4 cps)			No. 13 (8 cps, 50 cps)
	0	**0**	20%			40%
Total	1	4	⟨0%⟩			50%
Hemisphere						
right	14	63	19	2	5	7
left	**13**	**51**	**6**	**8**	**6**	**11**
Total	27	114	25	10	11	18
Influencing of tremor 65 Preferential rate of stimulation			38.5% (4 cps, 8 cps) >	15.4% (8 cps) >	16.9% (4 cps > 8 cps)	27.7% (50 cps, 25 cps) > 8 cps

[a] The columns of different stimulation effects contain the case number (No.) and, in parentheses ( ), the stimulation frequencies in cps at which the special stimulation effect appeared. The percentage number without brackets refers to the stimulation effects produced by the listed stimulation frequencies. The percentage in angular brackets ⟨ ⟩ represents the stimulation effects observed only with 25 cps or 50 cps, respectively. Boldface indicates the left hemisphere.

Jerks at rate of stimulation	Evocation of tremor	Slowing of alternating movements until blocked	Interruption of counting	Miscounting	Acceleration of counting	Supplementary area effects
		No. 5 (50 cps) 20% ⟨50%⟩		No. 5 (50 cps) 20% ⟨50%⟩		
No. 17 (4 cps) 2.9% ⟨0%⟩		**No. 8 (50 cps);** No. 9 (25 cps); No. 11 (50 cps) 8.5% ⟨21.3%⟩	**No. 8 (50 cps);** No. 9 (25 cps); No. 11 (50 cps); **No. 14 (50 cps)** 11.4% ⟨26.6%⟩			
No. 1 (4 cps, 8 cps); No. 6 (1 cps, 4 cps, 8 cps); No. 12 (1 cps); **No. 13 (4 cps, 8 cps);** No. 16 (1 cps) 20% ⟨0%⟩	No. 16 (8 cps) 2.2% ⟨0%⟩	**No. 1 (50 cps);** No. 6 (25 cps, 50 cps); No. 16 (25 cps) 8.9% ⟨22.2%⟩	**No. 1 (50 cps);** No. 9 (50 cps); **No. 10 (50 cps);** **No. 4 (50 cps);** No. 12 (50 cps); No. 16 (25 cps) 13.4% ⟨33.3%⟩	**No. 4 (50 cps)** 2.2% ⟨5.6%⟩	No. 12 (50cps) 2.2% ⟨5.6%⟩	No. 6 (50 cps); **No. 13 (50 cps)** 4.5% ⟨11.1%⟩
No. 3 (4 cps); **No. 10 (4 cps);** **No. 12 (1 cps)** 16% ⟨0%⟩			**No. 12 (50 cps)** 4% 10%			
						No. 2 (8 cps, 25 cps, 50 cps); **No. 7 (25 cps, 50 cps)** 50% ⟨100%⟩
No. 15 (1 cps, 4 cps, 8 cps) 60% 0%		No. 15 (50 cps) 20% ⟨25%⟩	No. 15 (50 cps) 20% ⟨50%⟩			
			No. 5 (50 cps) 20% ⟨50%⟩			
No. 13 (4 cps) 20% 0%						
10 **7** 17	1 **0** 1	7 **2** 9	6 **7** 13	1 **1** 2	1 **0** 1	4 **3** 7
(4 cps, 8 cps) > (1 cps)	1.5% (8 cps) >	(5 cps) > (25 cps)	(50 cps ≫ (25 cps)	(50 cps)		(50 cps > 25 cps)

Table 6b. Correlation of effects of electrical stimulation with anatomical substrates[a] (162 effects of stimulation in 17 patients)

Numbers of stimulated substrate targets in right hemisphere or in **left hemisphere** (boldface)		Number of positive stimulation effects with different parameters	Tetanic contraction	Turning the eyes	Opening the eyes	Binocular raising of the eyes
Pa.i	2 **3**	2 **5**	No. 14 (50 cps) **No. 17 (25 cps)**	**No. 8 (50 cps)** **No. 19 (50 cps)**	**No. 17 (50 cps)**	
Total	5	7	5.7% ⟨14.3%⟩	5.7% ⟨14.3%⟩	2.9% ⟨7.1%⟩	
V.o.a	5 **3**	17 **6**	No. 6 (25 cps, 50 cps) **No. 1 (50 cps)**	No. 2 (25 cps, 50 cps) No. 12 (50 cps)	No. 6 (50 cps) **No. 1 (50 cps)**	No. 9 (50 cps) No. 16 (25 cps)
Total	9	23	6.7% ⟨16.7%⟩	6.7% ⟨16.7%⟩	4.4% ⟨11.1%⟩	4.4% ⟨11.1%⟩
V.o.p	0 **2**	0 **3**	No. 10 (50 cps) **No. 12 (50 cps)**			
Total	2	3	8% ⟨20%⟩			
V.o.i	1 **0**	4 **0**				
Total	1	4				
Ca.i	1 **0**	4 **0**	No. 15 (25 cps) 20%	No. 15 (50 cps) 20%		
Total	1	4	⟨50%⟩	⟨50%⟩		
Ped.c	0 **1**	0 **7**				
Total	1	7				
Hemisphere right **left**	 9 **9**	 27 **21**	 3 **5**	 4 **2**	 1 **2**	 2 **0**
Total	18	48	8	6	3	2
Preferential rate of stimulation			(50 cps) > (25 cps)	(50 cps) > (25 cps)	(50 cps)	(50 cps)

[a] The columns of different stimulation effects contain the case number (No.) and, in parentheses ( ), the stimulation frequencies in cps at which the special stimulation effect appeared. The percentage number without brackets refers to the stimulation effects produced by the listed stimulation frequencies. The percentage in angular brackets ⟨ ⟩ represents the stimulation effects observed only with 25 cps or 50 cps, respectively. Boldface indicates the left hemisphere.

4.2. Increasing the Tremor

The most frequent effect of stimulation of V.o.p, V.o.a or pallidum internum is a change in the tremor, which can be facilitated by assuming a definite posture before the stimulation starts. The prevalence of this effect of stimulation is due to the fact that stimulation is performed in those substrata that belong to the mechanism causing the tremor. The most frequent change of the tremor, an increase in amplitude (Fig. 102a–c), occurs in 39% of all changes of tremor. Single stimuli increase those phases of the tremor that coincide with the stimulation. Increases of amplitude occur usually with stimulation at rates of 4 and 8 cps (see Fig. 43a). With higher-frequency stimulations (25–50 cps) the increase of tremor frequently merges into a blocking of it. Low-frequency stimu-

Blinking	Stimulation of the tractus opticus (photism)	Mydriasis	Inspiratory apnea	Reddening of the face	Feeling of anxiety	Laughter
	No. 17 (50 cps) 5 mm target 2.8% ⟨7.1%⟩	No. 17 (50 cps) 2.9% ⟨7.1%⟩				
No. 9 (50 cps) 2.2% ⟨5.6%⟩		No. 2 (25 cps, 50 cps) **No. 4 (50 cps)** No. 12 (50 cps) **No. 13 (50 cps)** 11% ⟨27.0%⟩	**No. 1 (50 cps)** No. 12 (50 cps) 4.4% ⟨11.1%⟩	**No. 13 (50 cps)** 2.2% ⟨5.6%⟩	No. 12 (50 cps) 2.2% ⟨5.6%⟩	No. 16 (8 cps, 50 cps, 25 cps) 6.7% ⟨16.7%⟩
		No. 12 (50 cps) 4% ⟨10%⟩				
				No. 2 (25 cps, 50 cps) 20% ⟨50%⟩		No. 2 (25 cps, 50 cps) 20% ⟨50%⟩
No. 15 (50 cps) 20% ⟨50%⟩	No. 15 (25 cps) 20% ⟨50%⟩					
		No. 5 (8 cps, 25 cps, 50 cps) 60% ⟨100%⟩	**No. 5 (8 cps, 25 cps, 50 cps)** 60% ⟨100%⟩		**No. 5 (50 cps)** 20% ⟨50%⟩	
2 **0** 2	1 **1** 2	4 **6** 10	1 **4** 5	2 **1** 3	1 **1** 2	5 **0** 5
(50 cps)		(50 cps)	(50 cps)	(50 cps)	(50 cps)	(25 cps + 50 cps)

lation of the corticospinal systems in the internal capsule or the peduncle increases the tremor. It is improbable, however, that the increase of tremor by stimuli inside the pallidum and in the oral ventral nuclei acts via the corticospinal fibers directly. If this were the case, then spread of current would produce the same effects of stimulation from very different parts of the internal capsule, which are 7–8 mm apart in a sagittal direction. Stimulating the V.o.p (see Fig. 44), especially at rates of 4 and 8 cps, and the pallidum internum (see Fig. 98; 119), at all frequencies, results with equal frequency in an increase of tremor. More rarely, one obtains an increase of tremor by stimulating the V.o.a at rates of 4 and 8 cps (see Fig. 45c).

Among the autopsy cases, all frequencies of stimulation – though primarily rates of 4–8 cps – caused an increase of tremor. The latter occurred three times as frequently from the nondominant as from the dominant hemisphere.

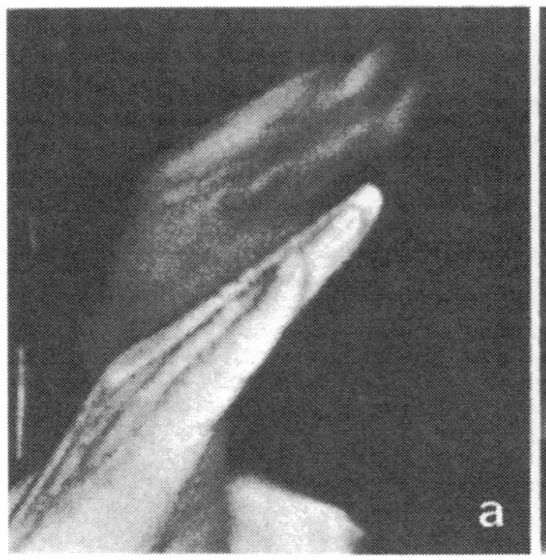

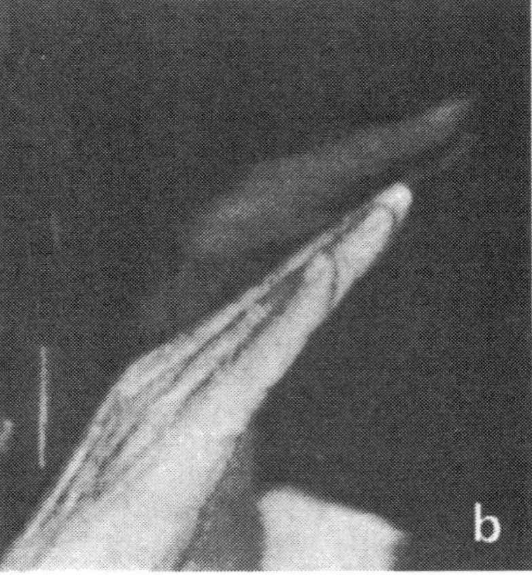

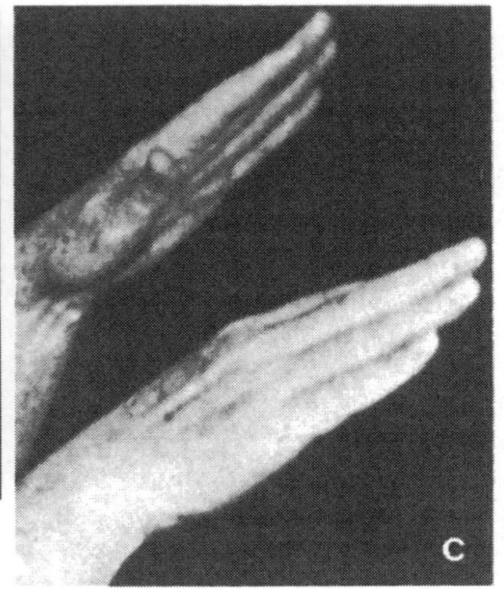

Fig. 102a–c. In Parkinson case with severe tremor at rest, more prevalent in left than right hand, (a) before, and (b) immediately after coagulation of right V.o.p and V.o.a. (c) Some minutes after coagulation, left no longer shakes. Before operation, shaking was so strong that despite recording frequency of 16 pictures/sec, the hand was blurred. (After HASSLER *et al.*, 1961)

4.3. Blocking of Tremor

Of the influences on tremor, blocking occurs in 28% of all instances, particularly with higher frequencies of stimulation (25 and 50 cps), but also sometimes with stimuli at 8 cps. It is also possible for the tremor to disappear or to be hidden by jerks or contractions of antagonistic muscles in response to stimuli. However, stoppage of the tremor through single stimuli in the pallidum as well as in the V.o is always due to acting on the central pacemaker of the tremor without tetanic contraction of the muscles. It is remarkable that blocking of tremor occurred four times as frequently from the pallidum internum and externum (see Fig. 55a) and nearly five times as frequently from the V.o.p as from the V.o.a (Fig. 126). In the one case in which the autopsy findings indicated that the zona incerta had been stimulated, blocking the tremor followed stimuli at 8 as well as at 50 cps (Fig. 86). Because these dentato-thalamic fibers are bundled, this blockade belongs to the V.o.p effect. The blockade of the tremor during stimulation of pallidum or V.o.p is due not to a lack of motor impulses, but to a lack of their synchronization.

In contrast to the increase of tremor, blocking the tremor followed twice as frequently upon stimulation of the pallidum and V.o.p of the dominant than of the nondominant hemisphere.

4.4. Slowing of Rhythm of Tremor

In one third of the cases, stimulation made the tremor more rhythmic. Usually, stimuli at 4 cps slow down the tremor (Fig. 126). Rarely, stimuli at 8 cps slow down the tremor, probably because of an interference between the rhythms of tremor and stimulation. Slowing the tremor follows more frequently stimulation of the V.o.a and pallidum (Fig. 98) than of the V.o.p. Neither hemisphere is preferred. Slowing of tremor shows no preference for the dominant hemisphere; it seems to be due to an increased synchronization of motor impulses reaching the agonists and antagonists.

4.5. Acceleration of Tremor

As a rule, stimuli at 8 cps accelerate the tremor. In contrast, stimuli at 25 and 50 cps produce this effect only rarely and only from the pallidum. A marked difference was observed between stimulation of the V.o.p and the V.o.a on the one hand and stimulation of the pallidum on the other. Higher rates of stimulation (25 and 50 cps) from the pallidum internum accelerate the tremor. In only one case did the autopsy show definitely that stimulation occurred in the rostral peduncle. Here stimulation at 8 cps also produced acceleration of the tremor. From this summary, we obtained the surprising result that in contrast to the increase of the tremor, acceleration of the tremor occurs four times as frequently from the dominant than from the nondominant hemisphere (see Fig. 126).

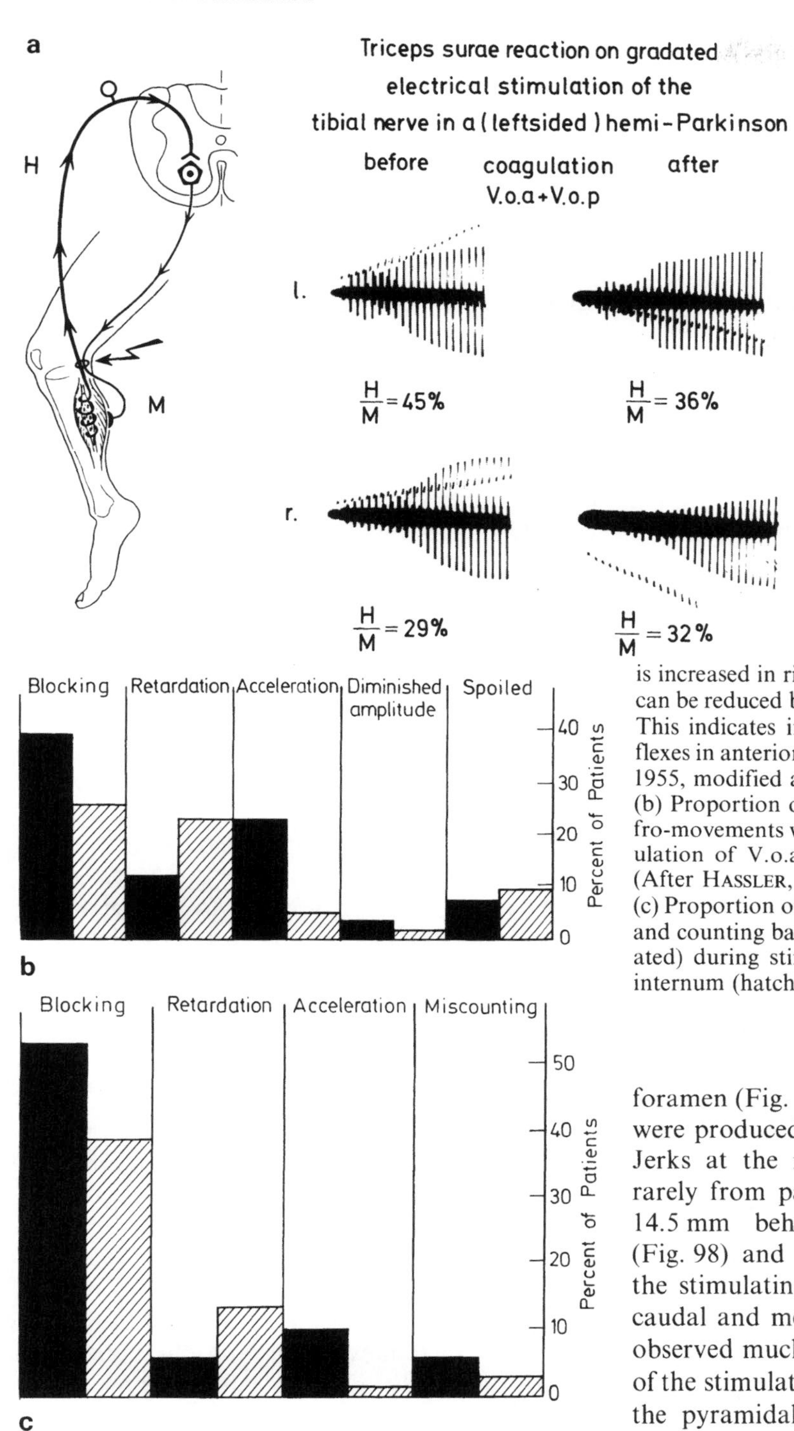

Fig. 103. (a) Response records of right and left triceps muscles in hemi-Parkinson patient to electrical stimulation in back of knee before and after right-sided coagulation of thalamic nuclei, V.o.a and V.o.p, which relieved left-sided rigidity. In response to weakest electrical stimuli, applied first, motor fibers of tibial nerve are excited directly and produce early electrical response that increases with increased electrical stimuli. First electrical stimulus also elicits reflex response that passes through reflex arch. This H reflex increases with further enhancement of electrical stimuli so that after 8th stimulus, only H reflexes occur in response to further increase in electrical stimuli applied to tibial nerve. Ratio of H reflexes to direct motor responses (M) was 45% in rigid left leg and 29% in normal leg before operation. After operation for left side, H:M ratio in right leg is almost the same but is decreased from 45 to 36% in left leg. H:M ratio is increased in rigid extremities of Parkinson patients and can be reduced by relief of rigidity by V.o.a thalamotomy. This indicates increased excitability for tonic stretch reflexes in anterior horn in rigid condition. (After KRIENITZ, 1955, modified after HASSLER, 1972)
(b) Proportion of patients (in %) in which active to- and fro-movements were altered in different ways during stimulation of V.o.a (black) and pallidum (hatch); n = 400. (After HASSLER, 1966)
(c) Proportion of patients in which the faculty of counting and counting backwards was impaired (slowed or accelerated) during stimulation of V.o.a (black) and pallidum internum (hatch); n = 400. (After HASSLER, 1966)

4.6. Jerks at Rate of Stimulation

They occur in the contralateral limbs and side of the face in response to low-frequency stimuli (1, 4, 8 cps). In one case (15) in which autopsy findings confirmed that the internal capsule had been stimulated about 10 mm behind the interventricular foramen (Fig. 90a), jerks at the rate of stimulation were produced with all stimuli at 1, 4, and 8 cps. Jerks at the rate of stimulation were produced rarely from pallidum internum (17) in the plane 14.5 mm behind the interventricular foramen (Fig. 98) and 2.5 mm below the baseline. When the stimulating electrode was pushed 5 mm in a caudal and mediobasal direction, such jerks were observed much more frequently because the point of the stimulating electrode had made contact with the pyramidal part of the internal capsule. We have no doubts in ascribing this effect to a stimulation of the pyramidal tract fibers. If the internal capsule is stimulated in a frontal plane just in front of the posterior commissure, then jerks synchronous with the stimuli can be obtained using much weaker currents. These effects are due to stimulation of the thicker fibers in the pyramidal tract in the caudal part of the internal capsule's posterior limb.

Jerks in rhythm with the stimulation (Table 6a) are observed in one-third and one-fourth of the cases from the V.o.a and V.o.p, respectively. In some of these cases, the effects of stimulation are again possibly due to the spreading of current into the neighboring, thicker fibers of the pyramidal tract. Stimulation of the internal capsule in case 2 showed that in this plane, the threshold for stimulation of the internal capsule is lower than the average threshold of the neighboring V.o.a or V.o.p (Fig. 43a).

If in the same location in the internal capsule, stimuli are applied at rates of 25–50 cps, tetanic contraction of the contralateral limbs instead of jerks at the rate of stimulation occurs in 50% of the stimuli. The absence of the tetanic contractions in the remaining 50% is related to the fact that sufficiently strong currents were not always applied, since the occurrence of jerks in rhythm with the stimulus of a tetanic contraction depends on the frequency of stimulation used. The jerks and tetanic contractions do not occur only in patients suffering from Parkinson's disease. Jerks synchronous with the stimulus evoked from the V.o.a or the V.o.p consist of contractions of different, even antagonistic, muscle groups. For instance, consecutive stimuli can make the hand move in different directions. This modification does not occur when the caudal part of the internal capsule is stimulated. Then, repeated stimuli always produce brisk contractions in the same muscle groups, similar to the jerks evoked by direct galvanic stimulation of the muscles.

Finding latency differences by electromyographic methods enabled GANGLBERGER (1963b) to differentiate between the effects of stimulating the oral ventral nuclei and the pyramidal tract. For the pathophysiologic interpretation of the brisk jerks in rhythm with the stimuli, it is significant that the electrode is regularly found near that part of the pyramidal tract in the internal capsule that contains thick fibers. Descending impulses are caused by stimulation in the thick fibers of the pyramidal tract; the large number of simultaneously excited fibers leads to a strong contraction of one and the same muscle, the flexors being more frequently affected. Stimulation of the V.o.p and V.o.a results in less-brisk and less-regular contractions of different, even antagonistic, muscle groups, in rhythm with the stimulation. These contractions are obviously due to excitation of the ascending thalamocortical fibers going to the motor cortex, since the projections of V.o.p and V.o.a to areas 4 γ and 6a α have been strongly indicated. The motor neurons are affected only via the cells forming the origin of the pyramidal tract fibers (see Fig. 124a, c). Between these cells and the fibers of thalamic origin, there lies at least one and usually several intermediary neurons, which explains the less-brisk jerk. In addition, the excitability of the cells from which the pyramidal tract originates varies considerably with time and is for a while refractory after a successful single stimulus. A volley of afferent impulses produced by stimulation of the thalamic nuclei V.o.p or V.o.a will, therefore, cause the contraction of less-coordinated and locally not identical motor units. In this way, the stimulation effect of the thalamic nuclei differs from that of the efferent pyramidal tract in the internal capsule. For this reason, it is not always the same muscles that respond with a contraction to repeated stimulation in the V.o.p or V.o.a; therefore, the movements evoked by their stimulation can change rapidly.

4.7. Change in Speed of Movement

The speed of active alternating movements of the arms is tested during stimulation. Active alternating movements are slowed down by higher-frequency stimulation (25 and 50 cps) in the pallidum internum (22%), pallidum externum (50%), or the V.o.a (22%) (Fig. 103b). This localization has been ascertained by post-mortem investigation. In contrast, active alternating movements are never slowed down by stimulation of the V.o.p. In cases in which the site of stimulation was confirmed by autopsy, an acceleration of active alternating movements in response to 50 cps stimulation of V.o.a/p (Fig. 34b, 35) was present, but it is known from the results of stimulation not checked post mortem that stimulation of the V.o.p usually accelerates active alternating movements. On the other hand, these movements are slowed down by stimulation of the pallidum internum (KROTZ, 1962; GUIOT *et al.*, 1962b, c; HASSLER, 1961, 1965, 1966b). An anatomically proved stimulation in the anterior third of the internal capsule at a rate of 50 cps also shows a slowing of active alternating movements (Table 6a).

Large clinical series show that acceleration of active movements can indeed be caused, but only by stimulation (25 and 50 cps) in the V.o.p. This effect is not disproved by the five cases which were negative as far as stimulation of the V.o.p is concerned and which were examined post mortem, particularly since a slowing-down was never caused. It is also possible that in some of these cases, which were distributed over many years, no active alternating movements have been ordered during the stimulation.

Stimulation of the pallidothalamocortical system activates the premotor cortical area 6a α which probably facilitates initiation of the active alternating movements. The tonic stretch reflexes and tonic motor neurons are involved in this system and cause tonic contractions of the muscles. The latter interfere with the rapid alternation of agonists and antagonists, which is interpreted clinically as the speed of movement. Increased excitation of the pallidothalamocortical system for the tonic contractions of muscles slows in this way the speed of movements (see Figs. 103b; 129).

The opposite effect is achieved by stimulation of the V.o.p. The excitation of the dentato-thalamocortical system is increased to area 4 γ, the area from which the most rapidly conducting pyramidal tract fibers originate. An increase in the speed of active alternating movements results in about 40% of the cases. This acceleration of active movements is specific for V.o.p-stimulation.

However, only V.o.p-stimulation at higher frequencies is effective. The most rapidly conducting pyramidal fibers, which are influenced by these afferent systems, cause contraction of the phasic motor neurons. They activate the part of the motor units that contracts rapidly. Rapid contractions of muscles and the speed of movement are improved by the increased excitation of the V.o.p-area 4 γ system, where the thickest pyramidal fibers originate.

4.8. Change in Counting

Counting forward or backward on instruction is altered mainly by stimulation at a rate of 50 cps. The most frequent alteration is the interruption of counting by stimuli at higher frequencies (24%). Of the different structures stimulated, stimulation in the V.o.a resulted in interruption of counting in 33%, and in the pallidum internum, in 26% of the cases; stimulation of the V.o.p resulted in interruption of counting in only 10% of the stimulations. Similarly, interruption of counting was obtained in both cases in which the stimulation was demonstrated to have been in the internal capsule or in the rostral peduncle (5_ℓ). In this series, miscounting occurred once by stimulation of the pallidum internum (50 cps) (Figs. 55a; 103c).

The *interruption of counting* observed only with stimulation at higher frequencies has been evoked six times as frequently from the pallidum (Fig. 119) and the V.o.a as from the V.o.p. Here, also, there is apparently a localizing relationship to the pallidothalamocortical system and its efferent fibers in Capsula interna (Fig. 75d) or in peduncle (Fig. 56a). Increasing the excitation of this system by stimulating the pallidum or the V.o.a increases the tonic slow contractions and thereby impedes the rapid contractions necessary for speech. A change in the psychomotor activity caused by stimulating the system of the pallidum may also interfere with articulation (Table 6a).

Accelerated counting was observed once, in the case in which stimulation of the V.o.a/p (50 cps) was ascertained by autopsy (Fig. 75d). The same mechanisms that have been attributed to the acceleration of movements can be adduced for the pathophysiologic interpretation of acceleration of counting. In only two cases of miscounting, did the post-mortem verify stimulation of the V.o.a (Fig. 52) or pallidum internum (Fig. 55a). In this case, one has to consider the unspecific activation of the cerebrum by the pallidum. This phenomenon could have impaired the concentration necessary for counting backward on instruction. In the large clinical series in which locus of stimulation was not checked anatomically, acceleration of counting was evoked five times more frequently from V.o.a than Pallid (Fig. 103a) and particularly through stimulating the V.o.p, similarly to the acceleration of alternating movements. Acceleration of counting had the same pathophysiologic explanation as the acceleration of movement. Slowing of counting with early stopping of the counting is clinically infrequent, probably due to the same mechanism as that responsible for miscounting—psychomotor interference of concentration (Table 6a).

4.9. Ocular Effects

Many ocular effects of stimulation have been observed. They would probably be less frequent if the head was not fixed during the stereotaxic operation, because under these circumstances turning the eyes takes the place of turning the head.

Turning the eyes to the opposite side, sometimes obliquely upward (Fig. 104), occurs frequently

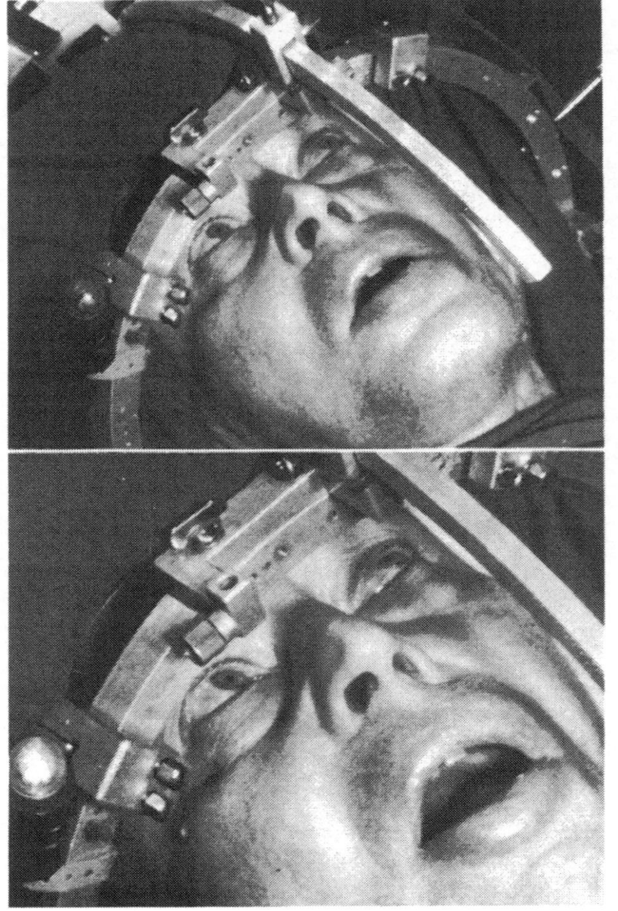

Fig. 104. First picture: Parkinson patient before, and second picture: during stimulation in right zona incerta, 50 cps (6 V), which produces deviation of binocular eye movements to contralateral side and upward, without mydriasis, and opening of mouth

(i.e., in 27.8%) with stimulation at higher frequencies of the V.o.a – i.e., in 27.8%. In contrast, this movement occurs only in 14.3% when the pallidum, and never when the V.o.p, is stimulated. Contraversive turning of the eyes can be evoked from the rostral part of the posterior limb of the internal capsule by stimulating at a rate of 50 cps (Table 6b).

Opening of the eyes (Fig. 105) is an effect that arises only with stimuli of 50 cps from V.o.a, zona incerta, and pallidum. It is usually linked to mydriasis; however, not every mydriasis occurs in conjunction with a widening of the palpebral fissures or with opening of the eyes. It is remarkable that in these cases, no effect on the eyes has been obtained by proved stimulation of the V.o.p.

Stimulating the V.o.p at a rate of 50 cps caused a slight dilatation of the pupils only once. In this case (17) stimulation of the caudal pallidum internum in a plane 10 mm behind the interventricular foramen at a rate of 50 cps caused the patient to see "blue points" in the contralateral visual field. This effect of stimulation can probably be ascribed to irradiation of the optic tract, for the coagulation of the pallidum in the caudal plane went beyond the basal edge of the pallidum in a ventral direction and partially destroyed the basal nucleus, which lies immediately above the optic tract.

Binocular turning to the opposite side in response to stimulation with higher frequencies is particularly common during stereotaxic operations – it replaces turning of the head or trunk, since the head is kept in a fixed position. In unrestrained animals, turning of the head and body occurs following circumscribed stimulation of the efferent systems of the pallidum and the pallidothalamic fibers (Montanelli and Hassler, 1962). During stereotaxic operations, turning of the eyes when the head is fixed is the analogous result. In those cases that were not checked anatomically, stimulation in the V.o.a frequently evoked a similar response. Functionally, this proves the presence of a continuation of the system from the pallidum into the V.o.a. The pathway from the V.o.a for the binocular turning goes to the adversive areas of the cerebral cortex.

Blinking has been observed once in a proved 50-cps stimulation of the V.o.a (9) (in Fig. 106, of zona incerta) and once during stimulation of the anterior third of the internal capsule (15). It appears that blinking is not caused by excitation of thalamic or pallidum structures; rather, it appears to be an effect of excitation of the rostral third of the internal capsule through which the efferent pathways for bilateral closing of the eyes also pass. Flickering of eyelids combined with olfactory hallucinations has been observed by Nashold and Wilson (1970).

On the other hand, *opening of the eyes,* usually combined with mydriasis, is the result of stimulating the pallidum and the pallidothalamic structures, especially since it is not caused by stimulating the V.o.p. The pallidothalamocortical systems exert an unspecific influence on the activity of most cortical areas. An increase in excitation is expressed by increased activity and by interested turning of the body, head, or eyes. Opening of the eyes with mydriasis belongs to the adversive movements.

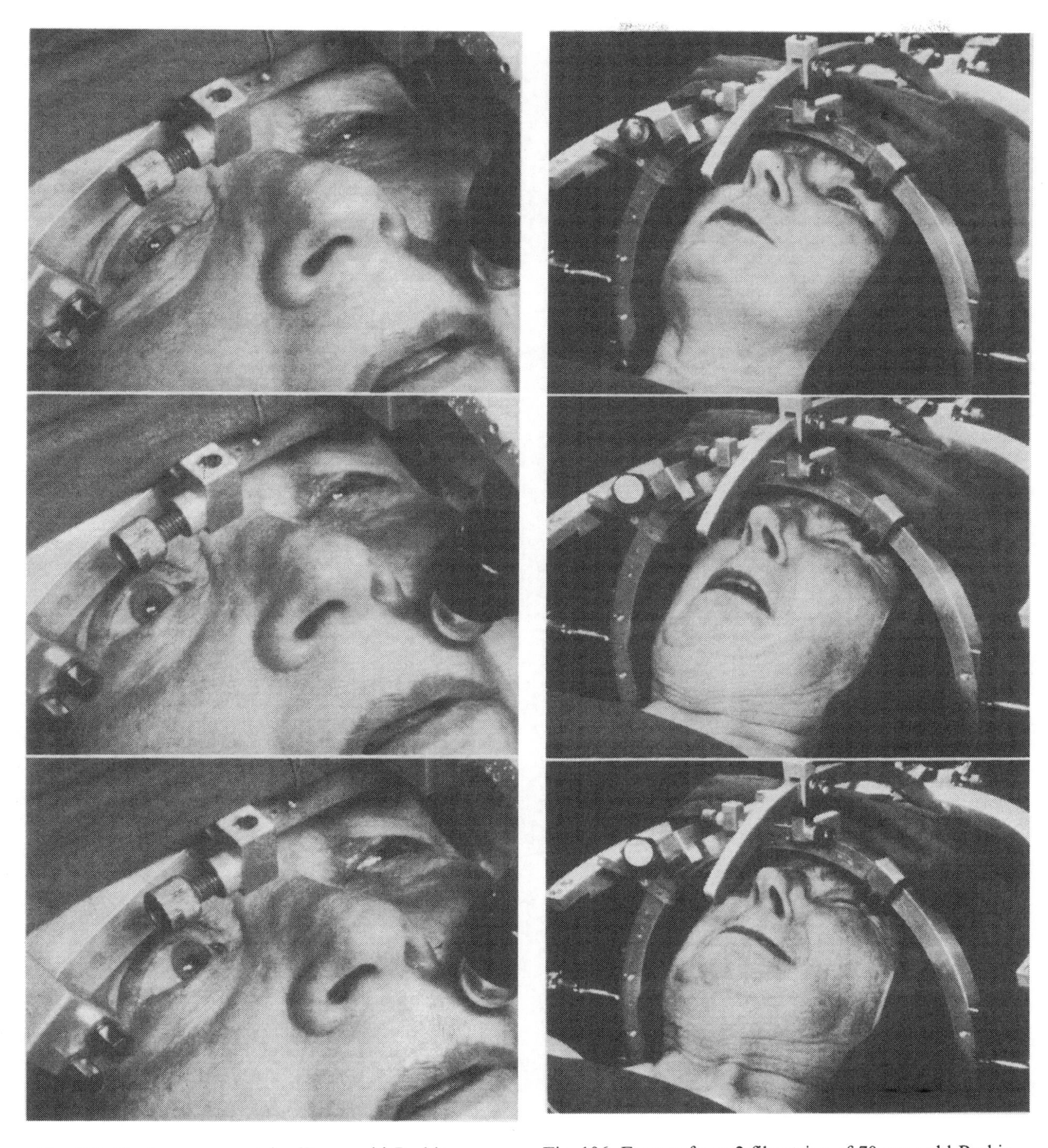

Fig. 105. Ocular movement in 67-year-old Parkinson patient, due to 50 cps stimulation of left zona incerta region. First picture: before stimulation. Second picture: movement of right eyeball to midline, dilation of right palpebral fissure and mydriasis. Third picture: 2 sec later, deviation of right eye to left and dilation of pupil, more pronounced. Also, left eye is now deviated to left side.

Fig. 106. Frames from 2 filmstrips of 70-year-old Parkinson patient. Stimulation in right zona incerta region. Stimulation frequency: 50 cps. First picture: before stimulation. Second picture: 2 sec after start of stimulation, strong contraction of all left-sided facial muscles with lid closure and mouth opening. Third picture: 4 sec later, stimulation effect is even stronger and moves left corner of mouth lateralward. Left eye is more tightly closed. Ipsilaterally, stimulation effect is much weaker

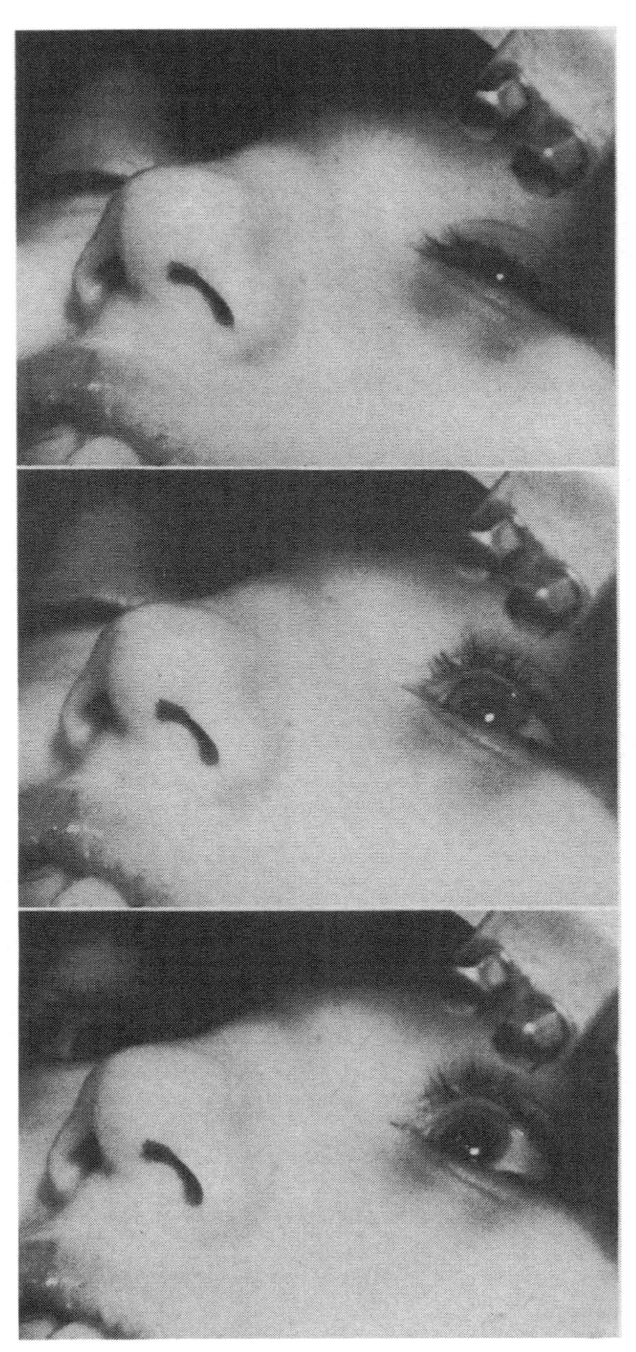

Fig. 107. Pupil dilation as stimulation effect of zona incerta in case of double athetosis. First picture: before stimulation. Second picture: 2 sec after stimulation, widening of palpebral fissure and beginning mydriasis. Third picture: 2 sec later, palpebral fissure is maximally opened with maximum dilation of pupil; eye is directed forward

Mydriasis was observed in one-fifth of the stimulations at higher frequencies. With one exception, the frequency of stimulation was 25–50 cps. Only once did stimulation of the rostral peduncle at 8 cps cause dilatation of the pupils. Mydriasis (Fig. 107) appears frequently when the zona incerta or V.o.a is stimulated (27%). It happens much less frequently and to a lesser extent when the pallidum internum and V.o.p are stimulated.

The preceding mechanisms are only partly responsible for isolated, usually symmetrical, mydriasis. To the effects provoked by pallidothalamocortical stimulation are added those that are due to direct excitation of vegetative centers in the thalamus. HESS (1948) obtained mydriasis in the cat from the medial thalamus, which includes the intralaminar nuclei.

4.10. Vegetative Effects

Other vegetative effects of stimulation include reddening of the face, inspiratory apnea, often combined with a feeling of anxiety located in the chest. *Reddening of the face* has been observed in only 11% of the cases during high-frequency (50 cps) stimulation. It is of note that reddening of the face occurred in two out of four anatomically proved stimulations in the V.o.i (2_l, 7_l) and in the V.o.a (5.6%). Stimulation of cerebral peduncle (Fig. 56a) elicited anxiety in the chest and respiratory arrest.

4.11. Psychological Effect of Stimulation

A feeling of *tightening in the chest* was evoked once by stimulation from the V.o.a (12) and once from the rostral peduncle (5_l). This effect of stimulation in the rostral peduncle is of importance because stimulation in this structure has only rarely been proved. The vegetative effects caused by stimulating the V.o.i are probably due to additional stimulation of the autonomously active intralaminar nuclei. The same has to be assumed for the effects of stimulating the V.o.a. The feeling of tightening, which has been evoked from the rostral peduncle, can be related to the corticofugal fibers passing through the peduncle on the medial side coming from the vegetative cortical systems in the orbital lobe and the frontal convexity.

Laughter occurred in one-fifth of the higher-frequency stimulations. Of the four stimulations of

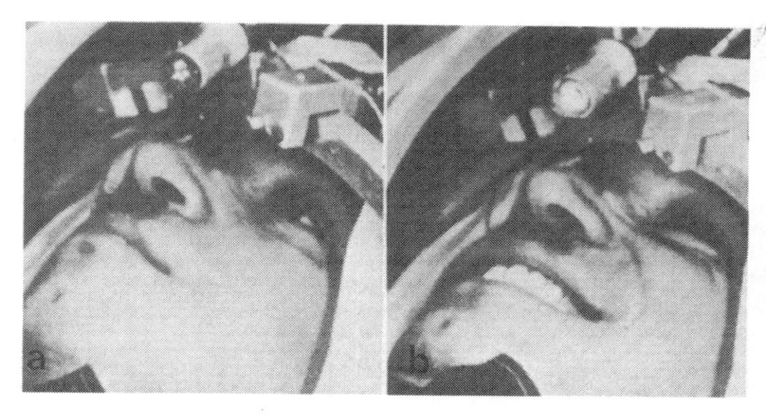

Fig. 108. (a) In patient affected by dystonia musculorum deformans, (b) 50 cps stimulation of right pallidum led to hearty ringing laughter, which could be reproduced any time. It occurred a few seconds after onset of stimulus even if patient was asked beforehand to refrain from laughing. During stimulation, everything he experienced seemed funny to him. (After HASSLER, 1961)

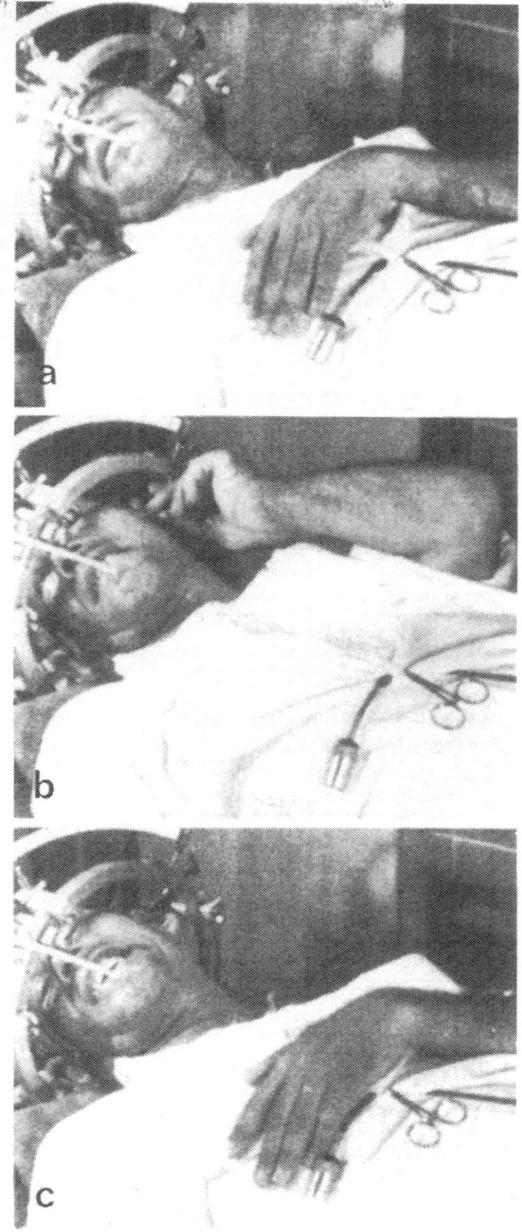

Fig. 109a–c. Unanesthetized patient, affected by postencephalitic torsion dystonia and Parkinsonism before (a) and during (b) 50 cps stimulation of right V.o.i near lateral part of intralaminar nucleus, shows prompt raising of left arm and grasping of left ear lobe. Repetition of stimulus (c) caused true laughing effect with appropriate facial expressions. (After HASSLER, 1961)

the V.o.i, two led to marked laughter (Figs. 108, 109). Laughter also occurred in 17% of the 50 cps-frequency stimulations from the V.o.a restricted to stimulation (Fig. 129). These patients were awake, had no cause for laughter, and could not explain it. The neurophysiologic interpretation is given on pp. 231–232.

Stimulating the *supplementary motor area* on the medial surface of the hemisphere results in a combination of vegetative, motor, and psychomotor symptoms as described by PENFIELD and WELCH in 1951. In our autopsied cases, these symptoms occurred in 13% of all higher-frequency stimulations in the thalamus. The patients jerked the contralateral arm upward or touched either their face or the basal ring of the stereotaxic instrument (Figs. 110a–c, 109b). At the same time they turned their eyes to the opposite side, mydriasis occurred, and they uttered more or less unintelligible sounds. These responses have their origin ros-

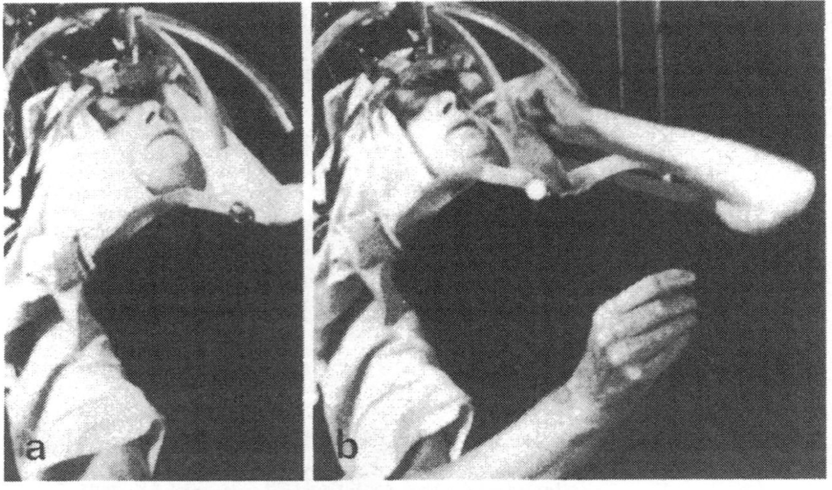

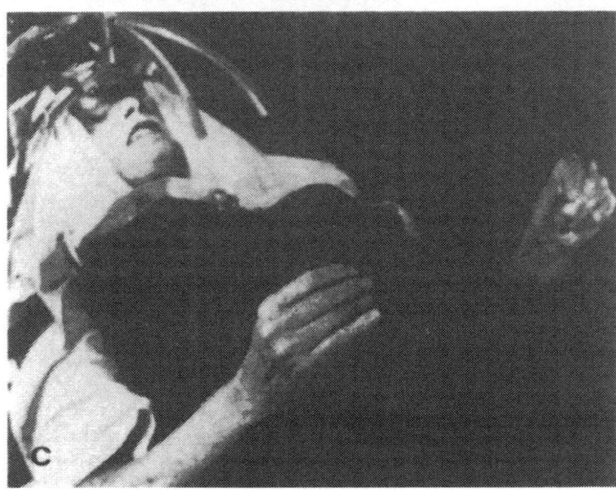

Fig. 110a–c. Parkinson patient before (a), during (b) 50 cps stimulation of right V.o.a and after (c) stimulation. At onset of stimulation, patient quickly raised left arm and pointed to left eye with index finger, turned eyes to opposite side, and exhibited urge to utter unintelligible words, which outlasted stimulation for some seconds. (HASSLER, 1961)

tromedial to the V.o.a (Fig. 45a) and especially close to the V.o.i, since results of this type have been proved from post-mortem examinations only in these two locations. From the V.o.i, the effects from the supplementary area occur only through stimulating at higher frequencies. Since the distance between the stimulating poles measures 5–7 mm, the proximal pole is usually further rostral in L.po (Fig. 61). Therefore, the stimulating current affects not only the V.o.i and V.o.a, but also the more rostrally lying lateropolar nuclei. It is probable that from these a projection goes to the supplementary area.

5. Correlation of the Electrophysiologic Findings with the Anatomical Substrates

The recording in man of electric activity from diencephalic structures and the study of the electroencephalographically recorded changes of these structures in response to electrical stimulation were first necessitated by stereotaxic operations. Very different patterns of potentials were recorded from nuclei and fibers of the diencephalon. The EEG potentials continue independently of the activity of the cortex for a long time after experimental decortication (SPIEGEL and WYCIS, 1950). The hope that electrophysiologic criteria would be of help in localizing certain nuclei in the thalamus or the basal ganglia was not confirmed by the use of macroelectrodes. GUIOT *et al.* (1962a, b, c) and ALBE-FESSARD *et al.* (1962) found, when introducing electrodes sagittally in a frontal direction, that they could determine a definite boundary within the thalamus. This boundary corresponded to a line between the (magnocellular) intermediate ventral nuclei (V.im) and the posterior part of the oral ventral nuclei (V.o.p), due to the considerably smaller size of the nerve cells in the V.o.p as compared to those in the sensory (V.c) and intermediate nuclei (V.im).

It is possible to record evoked potentials from circumscribed areas by combining depth electrodes from some diencephalic structures with physiologic stimulation of the skin or with electrical stimulation of other cerebral structures (NARABAYSHI, 1968, 1972; HAIDER *et al.*, 1972; STRUPPLER *et al.*, 1972). In one of our first stereotaxic thalamotomies in man in 1952, we obtained a circumscribed cortical response in a posterior central region when stimulating the sensory nuclei of the thalamus. This result was published by JUNG (1954). ALBE-FESSARD *et al.* (1962), GUIOT *et al.* (1962a, b, c), ARFEL (1962) and other colleagues as well as BERTRAND and JASPER (1965; JASPER and BERTRAND, 1964, 1966) and TASKER *et al.* (1972) recorded electric activity from certain thalamic nuclei with electrodes of different sizes. In response to tactile stimulation of different parts of the contralateral arm, these authors obtained potential changes in somatotopically circumscribed areas of the sensory thalamic nuclei. Several maps and anatomical correlations regarding the sensory thalamic nuclei have been published (BATES, 1969; TASKER et al., 1972).

In two of our patients, post-mortem analysis showed that there was a good localization of the stimulation in the V.o.a (2 and 9), and we obtained cortical responses to the different stimulations of the V.o.a. Single stimuli led to precentral biphasic responses in the EEG; in the secondary complex a slow wave occurred that had a latency to the peak of the wave of 220–250 msec. Stimulation in the V.o.a at rates of 4–8 cps led to responses resembling a recruiting response with waxing and waning (Fig. 66). Stimulation at a rate of 25 cps did not show definite changes while it was in progress. However, after the end of stimulation, a few slow, not strictly localized waves appeared.

The responses to stimulation in the V.o.a are less clearly localized in a cortical area than are those to stimulation of the sensory systems. There is a tendency for responses to extend in a frontocaudal direction and also to involve the precentral area of the other hemisphere. (See also NISHIMOTO and MATSUMOTO, 1970.) This tendency can be explained by the fact that the pallidothalamocortical systems contain part of those systems that cause an unspecific influence of the cortex. These can become effective through the frontal cortical regions and can spread quickly bilaterally through the commissure systems of the pallidum.

This effect of stimulating the pallidothalamic systems on recruitings in the motor and premotor area has also been observed in experimental animals, even when microelectrodes (field potentials) in steps of 250 μ (DIECKMANN and SASAKI, 1970) are used. Among the cases reported here, only one exhibited phase reversal (Fig. 66) so we were unable to verify by autopsy the location of stimulation in V.o.a due to a hemorrhage. GANGLBERGER (1962a, b) demonstrated phase reversal in a large number of stimulations in our stereotaxic cases (HASSLER *et al.*, 1965). Since post-mortem findings have confirmed the accuracy of reaching the target with our stereotaxic method of localization, the assumption is justified that in the cases reported by GANGLBERGER (1959b, 1961a, b, 1962a) the V.o.a and the pallidum internum have, in fact, been stimulated as intended. These findings agree completely with the investigations carried out quite independently regarding the fiber connections of the pallidocortical systems and their electrophysiological criteria. These cortical responses can, therefore, be regarded as specific for the pallidothalamocortical systems (see Figs. 138; 66).

The inactivation of the same pallidothalamocortical system in the V.o.a or in the pallidum internum shows itself postoperatively in the EEG in a precentral δ focus. In two cases, this finding could be related to the V.o coagulations, which were proved by autopsy (cases 1 and 2; see Figs. 33b; 45b). GANGLBERGER'S (1959a, 1962a, b) EEG findings have also been reported by other authors (R. HESS, 1961; LIM *et al.*, 1964) before and after coagulation of the V.o nucleus. This local change a week after the coagulation of the pallidum internum and the V.o.a was observed in 35%; and in 15%, it even affected the other side. In addition, it was found that in 35% of the patients with coagulation of V.o or the pallidum, the α rhythm was considerably slowed down, almost always on both sides, while a moderate slowing was found in 40%. In case 1, this finding was, for the first time, correlated with the coagulation of the V.o and proved by autopsy. These changes frequently disappear after 6–12 weeks. According to SPIEGEL and WYCIS (1962), the focus is usually frontal or frontotemporal, while GANGLBERGER (1967a) stresses the precentral accentuation. In case 1, we observed for the first time convulsive

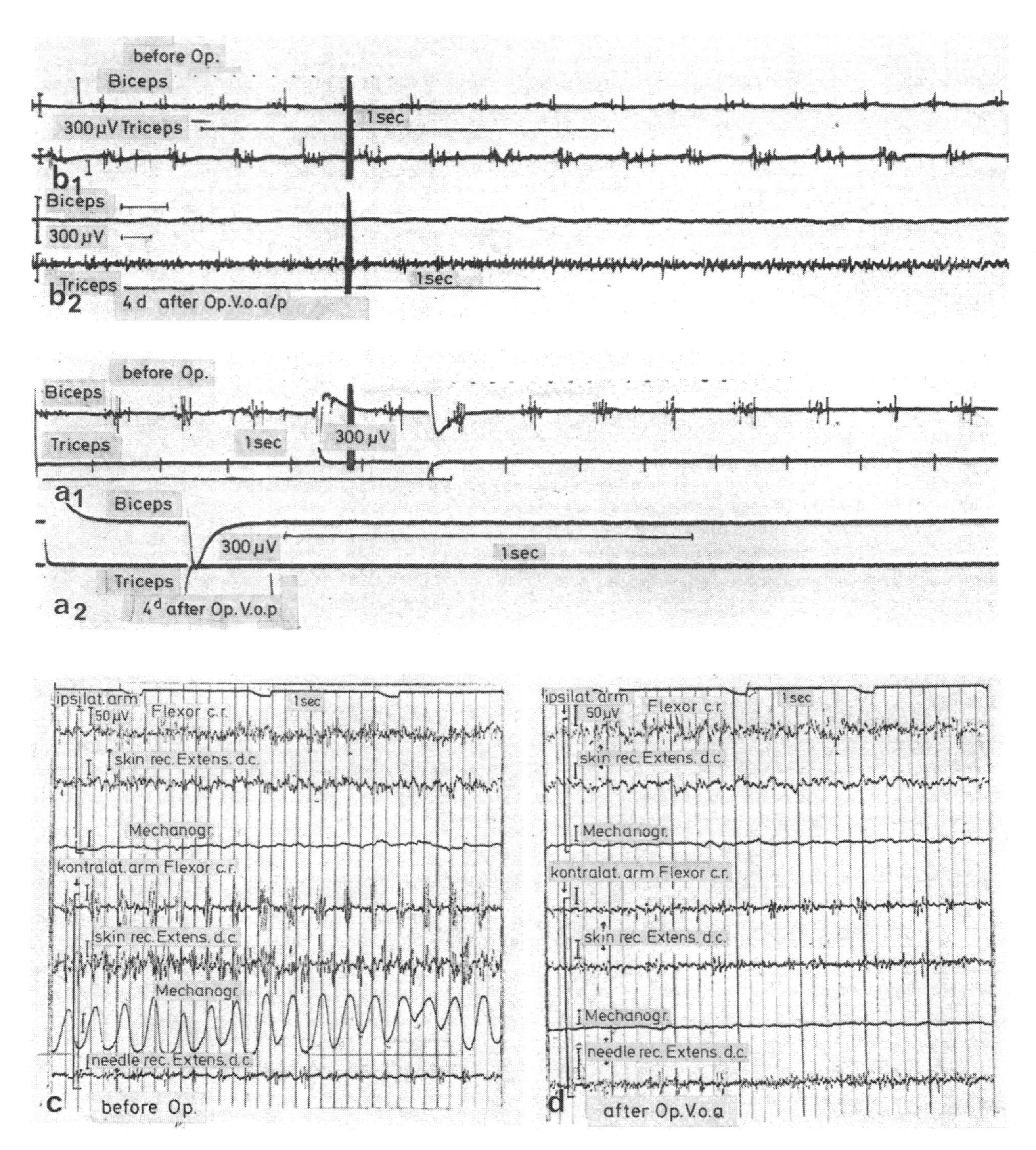

Fig. 111. (a–l) Electromyogram of left biceps and triceps muscles before operation in Parkinson patient, age 53, with predominantly left-sided tremor and rigidity. Rhythmic activity in biceps.
(a-2) Same recording 4 days after coagulation of right V.o.a that abolished tremor and rigidity in left arm. Rhythmic activity at rest in biceps is no longer present.
(b-1) Preoperative electromyogram of left biceps and triceps muscles during active extension of left arm in same case as in Figure 111a. Regular rhythmic activity in biceps, (sometimes also in triceps) in biceps always 40 msec earlier than in triceps.
(b-2) Same recordings under active extension of left arm 4 days after coagulation of right V.o.a/p that abolished tremor and rigidity in left arm. During active extension, electrical activity in biceps muscles has completely disappeared in electromyogram, whereas it is always present in triceps but without rhythmicity. (From GANGLBERGER and BRUNZEMA, 1962).

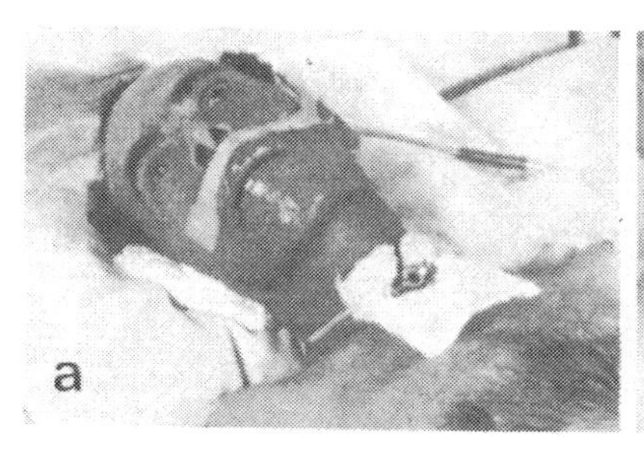

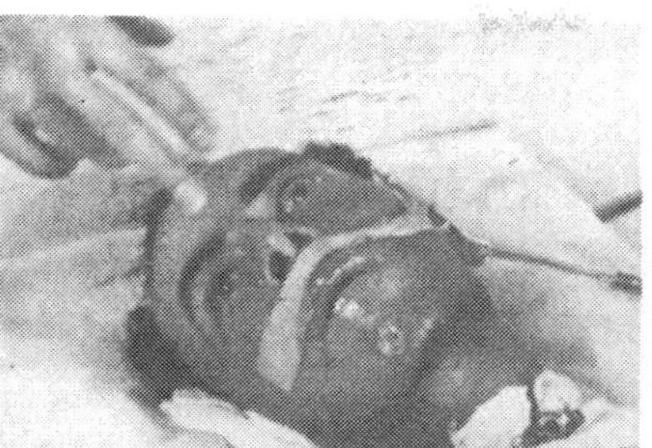

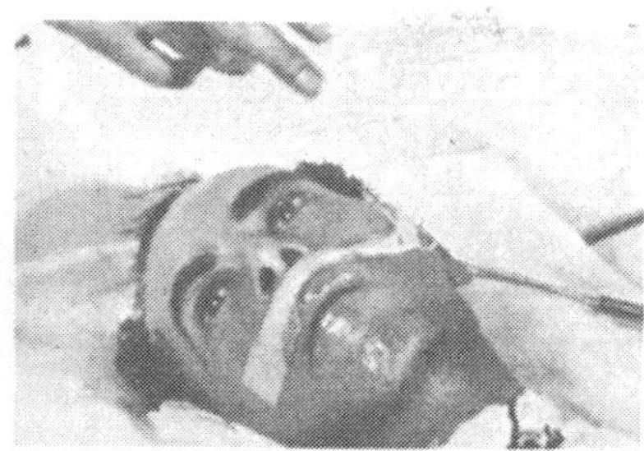

Fig. 112. (a) Individual frames from filmstrip of patient No. 6. After operation, patient had an apallic syndrome without any conscious reaction to external stimuli and negativistic episodes with eyeclosure. Tracheostomy performed some weeks after operation. Patient is lying with divergent open eyes, completely akinetic, similar to vigil coma. He is able to follow finger in front of his eyes only for a few seconds. He has hyperhidrosis and greasy face. He exhibits no reaction to visual or painful stimuli (UMBACH and RIECHERT, 1963)

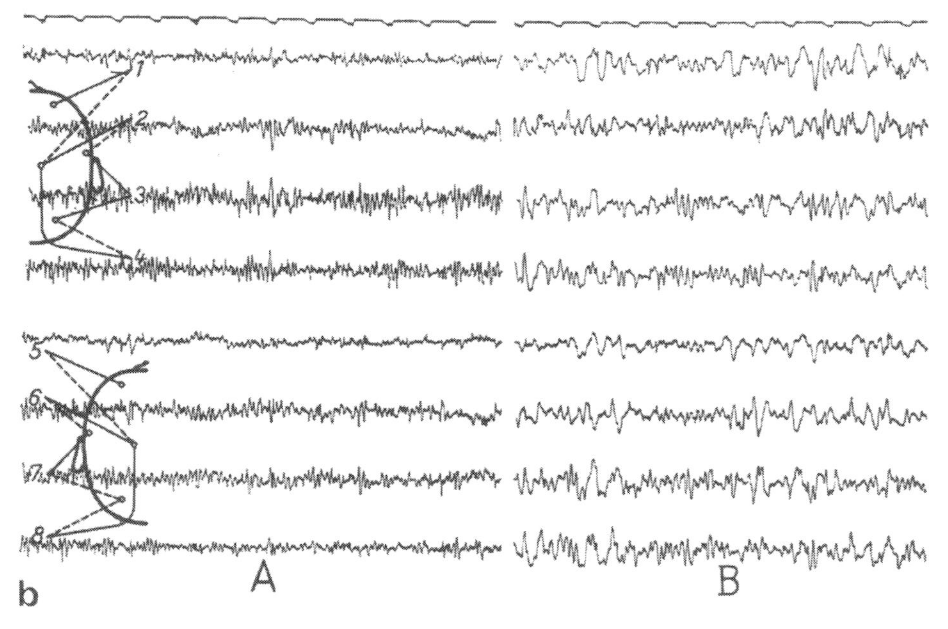

(b) EEG of same Parkinson syndrome patient with postoperative vigil coma. (A) Before stereotaxic intervention, patient shows slowed α rhythm with frequency of 8–9/sec. (B) 8 weeks after operation in vigil coma due to damage of both hypothalami and thalamic hemorrhage. Meanwhile akinetic mutistic condition was ameliorated. EEG shows, left more than right, slowing of basic rhythm in frontal recordings: 2/sec δ waves, θ waves predominate, and α rhythm is slowed to 7/sec (UMBACH and RIECHERT, 1963)

◀ (c) Electromyogram and mechanogram of both arms of 64-year-old Parkinson patient before operation. Long tremor at rest in mechanogram of right arm, with 5 cps frequency with waxing and waning of amplitude. Correspondingly marked rhythmic discharges in right flexor carpi radialis (*c.r.*) muscle that are more rhythmic than in extensor digitorum communis (*d.c.*) muscle. Mechanogram of right arm shows slight oscillations but no distinct alternating tremor. Discharges in electromyogram of flexor carpi radialis muscle and extensor digitorum communis muscle.

(d) Same recording 4 days after coagulation of left V.o.a/p that led to clinical abolition of tremor in right arm whereas rigidity was not completely abolished. Thus, mechanogram shows only nonrhythmic oscillations, now smaller right than left. EMG of right flexor carpi radialis muscle shows less activity than in left, in which 5 cps frequency is barely detectable. (After GANGLBERGER and BRUNZEMA, 1962)

potentials after the coagulation of the V.o, although it must be admitted that the patient suffered from a mitral stenosis, damage to the myocardium, and multiple pulmonary embolisms.

The main point of this monograph is the correlation of localization of lesion and electrodes, as ascertained by autopsy, with clinical findings. Therefore, the electrophysiology of the stereotaxic operations for Parkinson's disease is dealt with only from the viewpoint of the recordings that were made of the cases analyzed here. A monograph of UMBACH (1966a) and in particular the papers of GANGLBERGER (1961a, b; 1962a; GANGLBERGER and HAIDER, 1969) have dealt with the electrophysiology of our stereotaxic material, which contains several thousand cases.

6. Correlation of Coagulations of the Anatomical Structures with Functional and Therapeutic Effectiveness

6.1. Physical Parameters and Size of Coagulated Area

In addition to the exact localization of the structure to be inactivated, the size of the lesion is decisive for the therapeutic result. Shape and size of the lesion depend on the method employed for the destruction of tissue.

6.1.1. Methods of Inactivation. Physiologists investigated this problem thoroughly as early as 1907–1908. At that time, inactivation by electrolysis (ROUSSY and MOSINGER, 1934) was compared to the effects on structure of high-frequency current (HORSLEY and CLARKE, 1908). In 1952, CARPENTER and WHITTIER concluded that a linear relationship between intensity of current and duration of its application, on the one hand, and size of the lesion, on the other, could most easily be obtained by coagulation with a high-frequency current. WYSS (1945) and HUNSPERGER and WYSS (1953) further developed the technique of coagulation with high-frequency current by producing a pure sine wave high-frequency current. According to WYSS, the size of the lesion can be predicted from the intensity of the current (mA) applied, if the tissue structure is the same.

Repetition of the coagulation with the same parameters does not lead to a coagulated area of the same size. In high-frequency coagulation, the amount of heat produced in the tissue is decisive for the inactivation (MUNDINGER *et al.*, 1960; MUNDINGER and RIECHERT, 1961, 1963, 1966); it is therefore necessary to measure the temperature reached at the location of the structure to be inactivated.

Another method of stereotaxic inactivation in man is the setting of a mechanical lesion with a wire loop, as in leucotomy (OBRADOR, 1947; LAPRAS, 1960), or with stylets that can move sideways and are turned in the tissue (BROWN and HENRY, 1934, 1935; C. BERTRAND, 1958; HOUSEPIAN and POOL, 1960). Injections with oil and wax (NARABAYASHI, 1952) also act mechanically; the addition of procaine is of only minor significance. SPIEGEL and WYCIS (1949) and COOPER (1955) used alcohol as a chemical agent to achieve destruction, following Kirschner's inactivation of the Gasserian ganglion. COOPER (1955) used ethyl cellulose dissolved in alcohol (etopalin) in order to reduce diffusion into the tissue. Because the damage to the tissue caused by the spreading of alcohol cannot be controlled (WHITE *et al.*, 1960; GILDENBERG, 1960; ROSENSCHON and WECHSLER, 1964), most surgeons have since abandoned the injection of chemical agents in stereotaxic operations. For the inactivation of subcortical structures, irradiation by radioactive isotopes can be used. This has been done by implantation of radioactive gold (^{198}Au) by TALAIRACH and TOURNOUX (1955), of ^{109}Pd by MULLAN *et al.* (1959), of ^{192}Ir (Fig. 113) by MUNDINGER (1958, 1963b) and of ^{90}Y by RASMUSSEN *et al.* (1953). The desired therapeutic effect can be expected only when the accumulation dose has been reached, which depends on the half-life of the individual isotope. Only then will an adequate distribution have been achieved by the energy emitted as β or γ rays. In addition, it is not possible to stop the effect of radiation once the isotope has been implanted and the dose has been applied. LEKSELL *et al.* (1960) and KJELLBERG and PRESTON (1961) destroyed deep structures by bombarding them with heavy particles (protons, deuterons) that had been highly accelerated in a synchrocyclotron. For this purpose, KJELLBERG and PRESTON (1961) used the Bragg-peak, which occurs toward the end of the ionization curve. Recently, LEKSELL (1971) reported a preference, for practical clinical use, for cross-fire irradiation from multiple-focused ^{60}Co sources. It is also possible to produce relatively circumscribed lesions by ultrasound (FRY *et al.*, 1954; MEYERS *et al.*, 1959), but in this method as with the bombardment with heavy par-

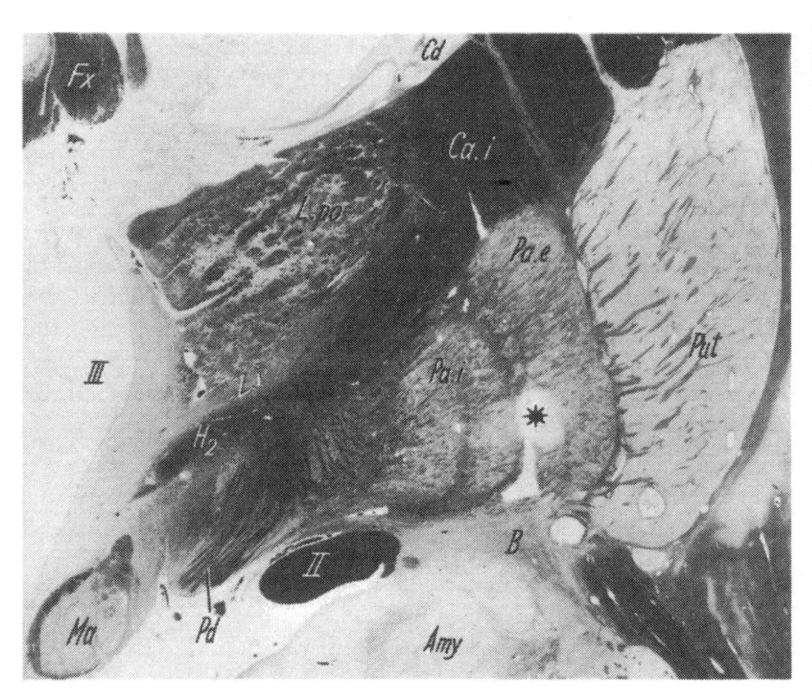

Fig. 113. Cross section of basal ganglia in Parkinson patient in whom radioactive ^{192}Ir wire was permanently implanted in pallidum. Track of this wire (*) is seen immediately beside lamella pallidi interna. Area surrounding wire track shows discrete demyelination. Fiber stain. ×3.2

ticles, it is difficult to localize and limit the lesion. In addition, ultrasound needs a bilateral bony decompression (craniotomy) in order to make the ultrasound act only on the subcortical target point.

Inactivation by freezing through cooling to below −43° C has been done experimentally by von Baumgarten and Kapp (personal communication; 1950/1951) (see also Balthasar, 1957) and clinically by Mark *et al.* (1961) and Cooper (1961). Through the considerable difference in temperature between the point of the probe through which the cooling agent flows and the surrounding tissue, one obtains first a freezing of the water content of the tissue with toxic concentration of the electrolytes and a freezing of the cell membrane with subsequent crystallization and destruction of the protein molecules (see also Riechert *et al.*, 1964). The formation of ice in the tissues around the point of the probe explains the dangerous hemorrhages and the marked edema surrounding the necrosis caused by the cold. Because the configuration of the cryolesions is not regular (Coe and Ommaya, 1964), many authors prefer not to use cryolesions for the therapeutic inactivation of subcortical structures.

6.1.2. Used Methods of High-Frequency Coagulation. We have used high-frequency coagulation since 1949, though the technique has developed considerably in the course of time. From 1950 to 1958, we performed coagulations in 475 stereotaxic operations using a high-frequency current of 1.2 MHz, produced by a spark-gap diathermy instrument (made by F. Hüttinger, Freiburg). This instrument, which had unfiltered high-frequency currents, sometimes led to uncontrolled, undesirable paroxysms of high temperatures (up to 1000° C). Such temperatures caused severe damage to the tissue, particularly the blood vessels. For this reason we switched in 1958 to the valve oscillator constructed by Wyss (1945). It produces a high-frequency current with a pure sine wave at 0.5 MHz. With this instrument, we performed an additional 202 stereotaxic subcortical interventions. Even so, the proportion of dangerous local complications was not considerably reduced (Mundinger and Riechert, 1961, 1963). Since 1959, therefore, we have used a thermoprobe and a high-frequency oscillator that delivers a pure frequency of 0.29 MHz, which we ourselves developed (Mundinger *et al.*, 1960; instrument made by F.L. Fischer, Freiburg). The use of this instrument, in which the temperature is checked by the thermoprobe (Figs. 114, 115b) and a definite dose delivered, has resulted in a considerable reduction in vascular complications following electrocoagulation in subcortical stereotaxic interventions. We have since performed more than 3000 stereotaxic interventions with this instrument in patients suffering from Parkinson's disease.

6.1.3. Anatomical Determination of the Size of Coagulation and the Consequent Alterations. Among the cases of Parkinson's disease on which we have performed post-mortems, coagulation lesions were produced with all three high-frequency instruments. Of the patients in whom the coagulation was performed with the spark-gap diathermy instrument, 6 out of 475 (1.27%) died from subcortical hemorrhages. When the Wyss coagulator was used, 2 out of 202 patients (1%) died. Among the patients in whom the coagulation was performed with the temperature check, one died from subcortical hemorrhages (0.05%). All interventions refer to patients with Parkinson syndrome.

The post-mortem analysis includes an investigation of the size of the coagulated lesions produced by the different methods of destruction, taking into account the other parameters of the coagulation. We made post-mortem examinations of nine interventions in which the coagulations were performed with the spark-gap diathermy instrument. Of these, three had massive hemorrhages into the basal ganglia. Of the seven operations with the Wyss coagulator, two had massive hemorrhages, in one case combined with a traumatic subdural hematoma. Of the nine operations examined by autopsy, and in which the high-frequency coagulation with temperature check was used, there was only one with a massive hemorrhage that led to the patient's death. Despite the small numbers of post-mortems, it is evident that the number of massive hemorrhages has decreased since the coagulations have been performed with the temperature check (see also mortality rate, p. 46). In the following evaluations of the size of the coagulation, the massive hemorrhages are not included because the coagulated lesions were rendered invisible in the hemorrhage.

Of the 9 cases receiving spark-gap diathermy coagulation and examined post-mortem, three exhibited massive hemorrhages and were therefore excluded from those cases in which the size of the coagulation as related to its parameters was measured. One of the hemorrhages ($7_{\ell 3}$) was not caused by the heat of the coagulation but by a mechanical lesion (Fig. 118), since it appeared immediately after a mechanical severing above the substantia perforata anterior with the string electrode (Fig. 115b). As can be seen from the histologic picture, the hemorrhage originated in an artery above the substantia perforata. The hemorrhage destroyed completely the pallidum and formed a narrow track through the internal capsule into the caudate nucleus, where it continued to spread out. The other two massive hemorrhages (14 and 8) were attributed to arterial damage caused by the coagulation, although the vessel affected could not be determined anatomically with exactitude (Fig. 88a).

(a) The **Coagulation With Spark-Gap Diathermy** led to scar formation of the softened substrate in five cases ($2_{\ell 1}+2_{\ell 2}$, 5_r, $7_{\ell 1}$, 7_r). The period of survival after these operations ranged from 760 to 1,296 days. In three cases, the scar formation (Figs. 41b, 43a) was so extensive that no cavity was left and the topographic relationships of the remaining surrounding structures had been altered. In one case (2_ℓ), there was only a scar in the shape of a puncture, but in the second case ($7_{\ell 1}$) the volume amounted to $6 \times 0.5 \times 8$ mm = ca. 24 mm^3. In the third case ($2_{\ell 2}$) the volume was $7 \times 4 \times 6$ mm = 168 mm^3. In this case ($2_{\ell 2}$), a mechanical lesion with the string electrode was added in order to unite the thalamic lesion with part of the pallidum. Apart from the additional possibility of complications by hemorrhage, the mechanical lesions certainly contributed to the increased scar formation.

A small *cyst* (Fig. 55a) remained 605 days after another intervention (5_r). The neighborhood of this cyst was scarred and had led to a considerably diminished volume of the pallidum. The volume was $6 \times 4 \times 12$ mm = ca. 288 mm^3. 530 days after coagulation with the spark-gap diathermy instrument (7_r), a large cystic lesion remained in the pallidum which, from the ventral side, included the internal capsule. The cyst extended vertically for 13 mm, and the size of the lesion was $13 \times 7 \times 11$ mm = ca. 1001 mm^3. The size cannot be explained solely by the two coagulations at the target point with the bare electrode surface measuring 1 mm or 3 mm × 1.1 mm in diameter and a duration of the coagulation of 7–10 sec. The fact that the whole cross section of the internal capsule was included in the lesion suggests occlusion of an arteriole supplying this part of the internal capsule (Figs. 63, 64). It is probable that such an event had also enlarged the already large cyst in the pallidum beyond the coagulation by spark-gap diathermy. The subsequent defective scar formation in the neighborhood of the cyst also favors additional softening that damaged the glia to the same extent as the nervous elements. In only one case (11) could the fresh area of coagulation produced by the spark-gap diathermy (Fig. 73a) instrument be measured exactly at the post-mortem. The greatest length of the lesion, which was 11 mm after 7 days, corresponds to the distance of 8 mm, which the electrode measuring 2 × 1.1 mm in diameter had been moved. The diameter of the cylindrical lesion measured 6.5 × 7 mm. Its volume was ca. 500 mm^3. The lesion extended radially for about 3 mm from the electrode, which had a diameter of 1.1 mm, and when the duration of coagulation was 10 sec. The volume of the individual coagulation could not be determined from the lesion because six consecutive coagulations were set in the direction of the electrode.

(b) In the seven **Coagulated Lesions Made with the Wyss Coagulator,** massive hemorrhages were found twice at autopsy. One was so extensive that neither diencephalic structures nor the remains of the two previous coagulated lesions were recognizable (9_r, 6_ℓ). At the first coagulation with the electrode measuring 4 × 2 mm, a rapid drop of current from 110 to 50 mA occurred; the coagulation was therefore stopped after 5 sec. The other massive hemorrhage occurred in an atypical case (6) of Parkinsonism in which a spongioblastoma of the diencephalon was found that grew into the third ventricle and led to an internal hydrocephalus of the ventricles (Fig. 58a). Below the massive hemorrhage in the right thalamus, a previously coagulated lesion, 8 × 9 × 9 mm = ca. 647 mm^3, was at least partially recognizable. The hemorrhage was obviously due to a vascular lesion. At the first coagulation the current dropped from 120 to 60 mA. The thalamic vessel affected appears not to have been large, so that the hemorrhage destroyed the thalamus only partly (Fig. 58a). The patient, therefore, survived the incident by 82 days.

One of our autopsy cases (16) died $5^1/_2$ years after coagulation with the Wyss coagulator (Figs. 95, 129). A cystic lesion measuring 3.5 × 2.6 mm was found at the base of the oral ventral nuclei. The coagulation was performed with an electrode whose bare point measured 4 × 2 mm. In this case, too, a decrease of current occurred during several coagulations. The electrode was moved 9 mm, and the lesion, extending from the base of the thalamus up into the oral dorsal nuclei, measured 6.5 × 2.6 × 11 mm = ca. 185 mm^3. Individual coagulations lasted 5–10 sec apiece. This case indicates that it is also possible to obtain a cystic lesion with the Wyss coagulator, at least when a drop in current has occurred. During the $5^1/_2$ years since the coagulation, this cystic lesion shrank similarly to the coagulated lesions achieved by the spark-gap diathermy instrument. A considerable shrinking of the diameter was also caused with the string electrode, but the extent could no longer be determined. The string electrode was pushed out from position +3 in the basal, ventral, and lateral directions, and went through the subthalamic nucleus and the peduncle as far as the optic tract. In the optic tract only 5% of the fibers were damaged. The coagulated lesion from the string electrode was ca. 3,5 × 21 × 6 mm = 441 mm^3. The coagulation lasted 24 sec.

In the remaining three cases treated with the Wyss coagulator, the autopsies showed recent coagulated lesions. In case 15, hemorrhages occurred dorsally and ventrally to the coagulated lesion. These were probably related to the drop in current from 115 to 64 mA. Because of this, the coagulation was stopped after 7 sec. The bare part of the electrode measured 4 mm; it was moved 7 mm. The coagulated area had a length of 16 mm: It should only have had a length of 13 mm (4 + 7 + 2 mm). The diameter of the lesion was 8–10 mm, also larger than average. The total volume was, therefore, ca. 1440 mm^3 despite the relatively short periods of coagulation of 5–20 sec. In two other interventions with the Wyss coagulator drops in current were avoided, and, in keeping with this, there were no hemorrhages in the area surrounding the lesion. In case (2_r), the 4 × 2 mm electrode was moved 2 mm; the periods of coagulation lasted 20 sec each; the size of the lesion (Fig. 44) was 4 × 7 × 9 mm, making a total of 252 mm^3. In case 5_ℓ the electrode was moved 5 mm deeper and there were three coagulations of 30 sec duration each (Figs. 56a, 123a). The same electrode was used. The coagulated area was 9.5 × 9.5 × 11 mm, for a volume of ca. 993 mm^3. The extension of the duration of the coagulation by 50%, together with the fact that the electrode was moved 3 mm more, caused a lengthening of the lesion by only 1 mm but an enlargement of the diameter by about 4 mm. It obviously follows that the duration of the coagulation is of decisive importance for the enlargement of a coagulation cylinder. In case 6, in which a hemorrhage occurred, the diameter of the coagulated lesion was 1 mm less because of the shorter time of coagulation despite a more extended movement and a larger number of coagulations (4 coagulations). The survival times of both these cases were identical—21 days. The small width of the coagulation cylinder in case 16 (Fig. 129) ($5^1/_2$ years survival) is explained by the fact that although the number of coagulations was larger, the duration of each was not extended beyond 10 sec.

(c) Only one massive hemorrhage in the thalamus was found among the 9 coagulated lesions set with **Our Own Thermocontrolled Coagulator** (MUNDINGER *et al.*, 1960). The patient survived for 15 days (12_t). At autopsy, it was no longer possible to determine the size of the coagulated area (Fig. 78a). In this case also, a drop in current occurred after a few seconds during the first coagulation. For the following two coagulations unusually high currents were needed, i.e., 200 and 235 mA, or twice the normal strength, in order to reach a temperature of 70° C at the point of the electrode. This situation correlates with the presence of vascular damage. The high current with a power of 6.6 or 7.0 watts suggests that the point of the electrode was surrounded by a liquid with low electrical resistance, which was, in fact, confirmed by the measurement of 170 ohms at the second and 125 ohms at the third coagulation. The values that can be read from our coagulating instrument during the performance of the coagulation permit us, therefore, as was confirmed in this case by the post-mortem findings, to conclude that in certain cases there is a collection of fluid around the electrode point.

In the same case (12_t) a lesion was formed in the right internal capsule (Figs. 75a, d and 77a) which, after 1,106 days, measured $7 \times 7.5 \times 9$ mm, equalling ca. 470 mm^3 when the electrode had been moved 5 mm. The lesion had become smaller by the pull of the scar. Extension of the lesion into the pallidum internum was caused by the coagulations with the string electrode, which had been moved 6 mm laterally and occipitolaterally.

Another patient (13_t) survived the temperature-controlled coagulation by 595 days. The electrode was bare for 4×2 mm at its point, and was moved 4 mm. Three coagulations of 30 sec duration each were performed. The lesion was cystic and measured $8.5 \times 4 \times 13$ mm = ca. 440 mm^3 (Fig. 84).

There were three cases (3, 4, 17) with short survival periods of 5–7 days. In these, temperature-controlled coagulations with electrode movements of 4–6 mm produced lesions that varied in size between 290 and 650 mm^3 (Figs. 52, 98). The smallest lesion occurred with the smallest movement of the electrode, 4 mm. In this case (4) three coagulations lasting 20 sec each were performed. The length of the coagulated cylinder was 8.3 mm; its

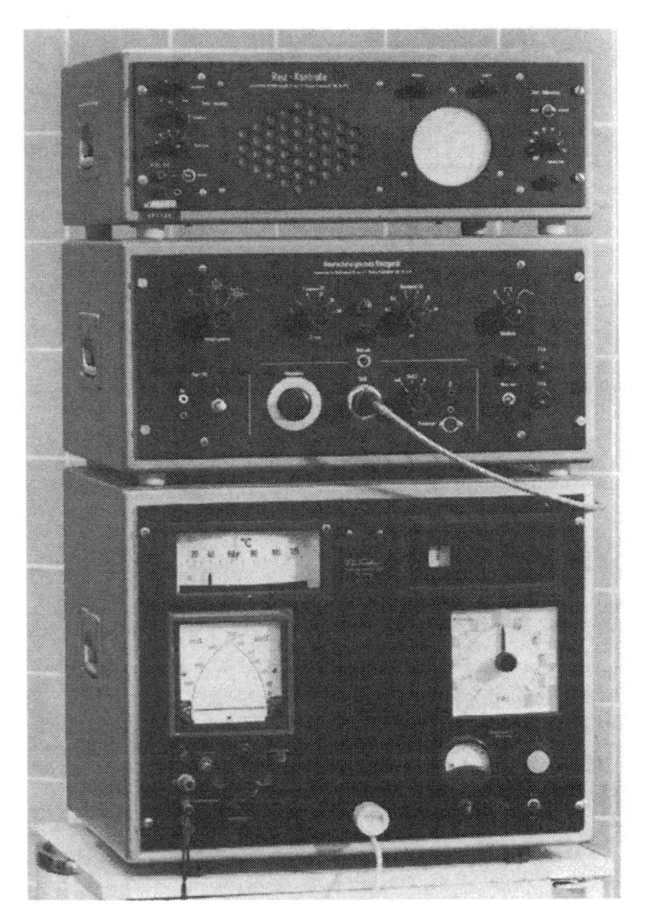

Fig. 114. Temperature-controlled high-frequency coagulator of MUNDINGER *et al.* (1964), used to produce stereotaxic brain coagulations. Temperature required for controlled high-frequency coagulation in range of 70° C is measured by thermal couple included in electrode tip and can be read on control instruments. When predetermined temperature is exceeded, electric timer automatically starts. During coagulation, milliamps and voltage of high-frequency current, as well as temperature, can be read and regulated

diameter was 6.5 and 7 mm (see Fig. 52). Next in size came a lesion produced by movement of the electrode for 6 mm and with coagulations lasting 30 and 25 sec at a temperature of 70° C (17). The length of this lesion was 12.5 mm, which corresponds exactly to the distance moved plus the length of the bare electrode (4 mm), also considering the lesion around the point of the electrode. The diameter of the coagulated cylinder measured 7 mm; the volume was ca. 612 mm^3. The lesion was located in the pallidum. It had damaged the internal capsule beyond it by 5%. The largest (Fig. 50a) lesion (3) had a volume of 650 mm^3. It was caused

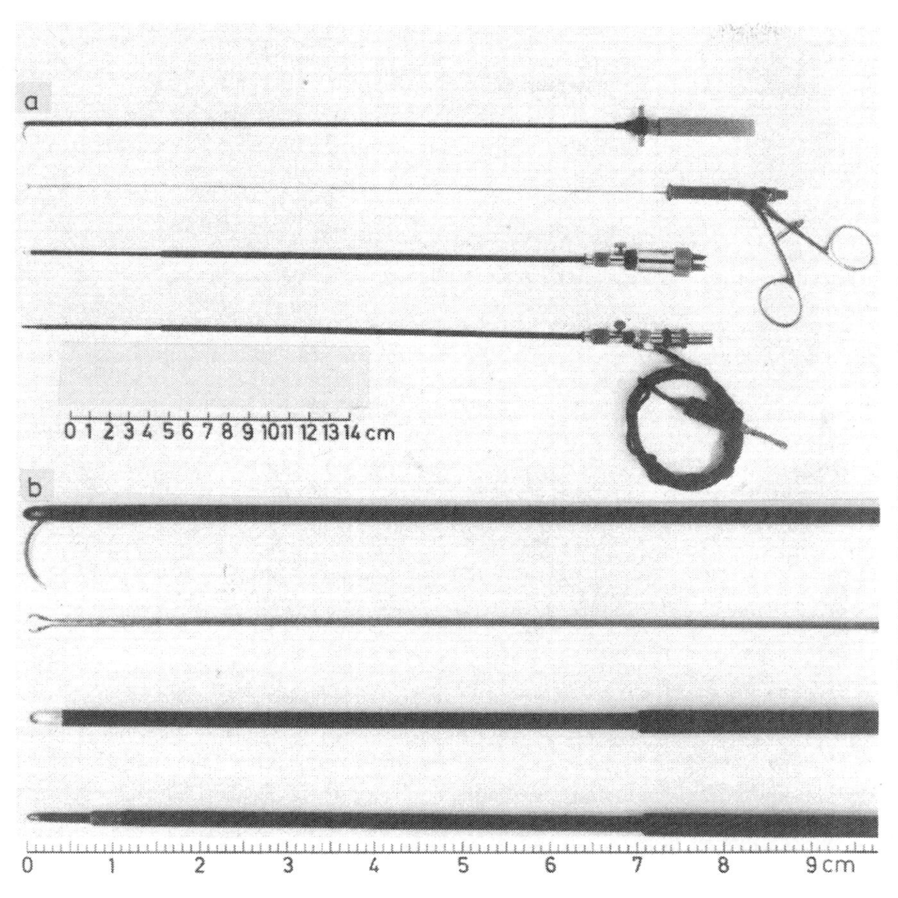

Fig. 115a and b. Set of 4 different types of electrodes for high-frequency lesions: (a) full electrode; (b) magnification of tips. First row: string electrode insulated except for 1 mm at tip can be replaced by uninsulated electrode with same shape. Both electrodes can produce round or sausage-like lesions in directions other than that of electrode sheath. Second row: microforceps for 1.1 mm diameter brain tissue biopsy of 3 mg tissue weight. Third row: thermo-controlled coagulation electrode with gold tip of 4 mm length and 2 mm diameter. Last row: straight electrode with diameter of 1.1 mm and uninsulated tip of 1 mm for bipolar stimulation. One pole is uninsulated tip, other is uninsulated ring at end of sheath (MUNDINGER and RIECHERT, 1966)

by a movement of the electrode of 5 mm and by coagulations lasting 25 and 30 sec at a temperature of 70° and a current of 85–105 mA. The length of the lesion, 12 mm, was slightly longer (1 mm) than expected. The diameter varied between 8.5 and 11 mm. The greater-than-expected size of the lesion is explained by the fact that the lesion extended from the V.o.a into the internal capsule and damaged 60% of it in addition to the V.o.a. The fibrous structure in the inner capsule offered higher resistance, which led to a higher local temperature in the field of the high-frequency current.

In case 1, surviving 23 days, the electrode was moved 5 mm and a temperature of 70° C was applied for 30 sec. The resulting lesion was ca. 704 mm^3, i.e., 11 mm long, 8.5 mm wide, and 8 mm in diameter. The lesion lay in the oral ventral nuclei; while its length was as expected, its width was greater than expected, because the coagulation involved the internal capsule, of which 15% lateral to the V.o.a had been destroyed (Fig. 34a, b).

These observations indicate that the temperature-controlled, measured coagulation with a high-frequency current of pure sine waves produced lesions whose size at post-mortem corresponds fairly accurately to the intended size. When the temperature is controlled, the size of the lesion depends mainly on the length of the movement of the electrode and the duration of the coagulation. This fact can be shown, although less definitely, in the coagulations with the Wyss instrument. It appears to be important that the lesion be widened asymmetrically as soon as the coagulation reaches a bundle of fibers, usually the internal capsule. A bundle of fibers offers higher electrical resistance, and since higher resistance at the same energy level caused an increase in temperature, the asymmetrical widening of the coagulated lesions in the direction of the internal capsule is due to fibrous tissue (Fig. 34b). HUNSPERGER and WYSS (1953) described this phenomenon in animal experiments. They found a linear dependence of the diameter of the coagulated cylinder upon the strength of the high-frequency current, but a nonlinear enlargement of the lesion in the direction of fibrous structures, if these were involved.

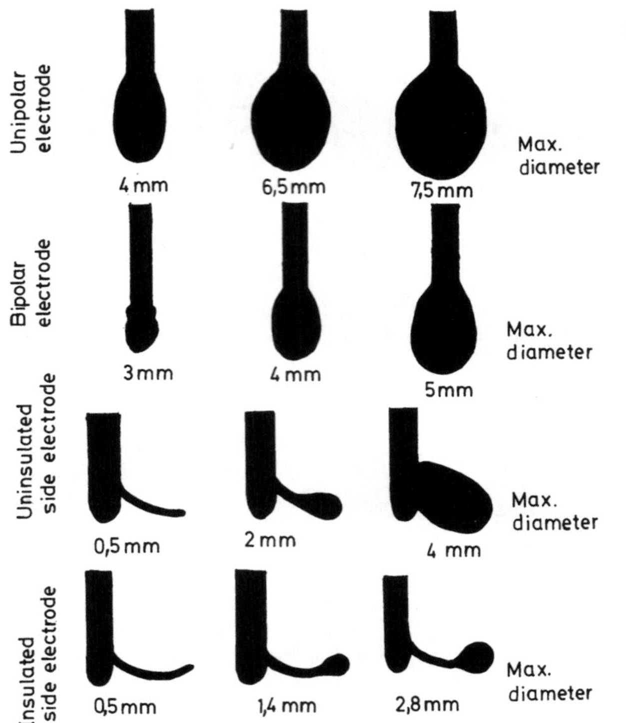

Fig. 116. Different forms and sizes of denatured protein in egg albumin solution, produced by various electrode types. Unipolar thermoelectrode, diameter 2 mm, and 4-mm long uninsulated tip, produces spherical focus of 4, 6.5, and 7.5 mm, depending on time at fixed temperature of 73° C. Bipolar thermo-electrode, diameter 2 mm, produces drop-shaped focus of 3.4 or 5 mm diameter, depending on time. Uninsulated tip (0.5 mm diameter) of string electrode in 3rd row of Figure 115b produces sausage-shaped coagulations of varying lengths, i.e., 0.5, 2, and 4 mm, which taper from tip to shaft. Insulated side electrode (2-mm uninsulated tip) produces the smallest coagulations of the 4th row.

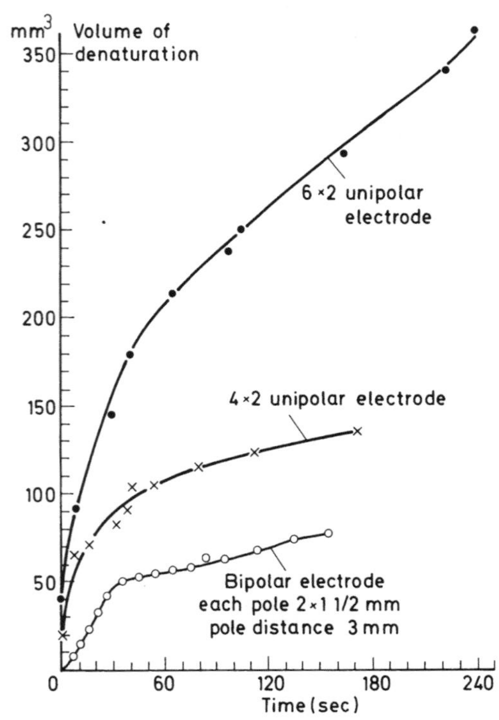

Fig. 117. Diagram of denaturation volumes produced in egg albumin solution by various types of electrodes at predetermined temperature of 73° C and depending on time. Steepest slope of curve (volume) represents unipolar type of thermo-controlled electrode with uninsulated gold tip of 6×2 mm. Curve shows no constant level after 240 sec in protein solution. Unipolar type of thermo-controlled electrode, with uninsulated gold tip of 4×2 mm, causes coagulation volume that reaches constant level after 3 min. Bipolar type of thermocontrolled electrode, with 2×1.5 mm uninsulated tip, with distance of 3 mm between two poles, reaches constant level after 40 sec. (From MUNDINGER and RIECHERT, 1966)

Some of our autopsy findings confirm the dependence of the size of the lesion upon the duration of coagulation. When rod electrodes (Fig. 115a, b) are regulated to heat in chicken albumen to 73° C (Figs. 116, 117), an almost linear relationship between the volume of the coagulation and its duration can be observed in the first 30–40 sec. After that, the increase in volume of the coagulation with increasing duration is diminished (MUNDINGER and RIECHERT, 1966; MUNDINGER *et al.*, 1960). With an electrode with a larger surface (6×2 mm), the volume of the coagulation is less dependent on time (Fig. 117).

The factors affecting the volume of coagulation have been investigated by DIECKMANN *et al.* (1966) in 82 experiments on cats, rabbits, macaque monkeys, and one sheep. With a thermoprobe measuring 4×2 mm and a regulated temperature of 70°, the following median values for the volume of the lesions were found:

At 10 sec duration, 114 mm³
At 15 sec duration, 159 mm³
At 20 sec duration, 186 mm³
At 25 sec duration, 251 mm³
At 30 sec duration, 223 mm³

These figures show that in cerebral tissue with equal parameters, the volume of the coagulation increases linearly up to 25 sec, but not beyond. In the cerebral tissue, the cut-off point occurs about 10 sec before it occurs in chicken albumen. The variable size of the lesion depends on the cerebral structures affected. Lesions occurring in structures with myelinated fibers are larger than those occurring in nuclear areas and at the borders with the C.S.F. spaces (Fig. 34b).

6.2. Correlation of Parkinson Symptoms with the Coagulated Structures

This section is a discussion of which structures must be inactivated in order to relieve specific symptoms. For this purpose we verified the specific site of inactivation and measured the size of the lesions as percentage of the nuclear or fibrous structures involved in the serial investigation of seventeen cases of Parkinson's disease. The improvement of individual symptoms has then been viewed in relation to the substrate inactivated and the size of the area inactivated.

The procedure in eleven cases was to cut parts of the hemispheres containing the diencephalon into slices of 20 μm thickness. In these series, the exact size of the lesion was reconstructed, including the extension produced by coagulation with the string electrode (Fig. 115b). We did not attempt reconstruction in cases where a long postoperative survival period caused such marked scar formation that the original size of the lesion could no longer be determined. Measurement of the inactivated area is particularly difficult in the internal capsule and the peduncle, since a cross section of an inactivated fibrous structure corresponds to the number of interrupted fibers only if the lesion is at a right angle to the direction of the fibers. Measurement of the cross section of such fiber structures has so far not been performed in this material because it requires a specific technique of dissection, particularly in the internal capsule. However, in four cases the corticospinal fibers on the cross section of the pyramid in the medulla oblongata were counted and compared to those on the normal (noncoagulated) side. Table 7a, b shows in detail the percentage of the individual structures inactivated in individual cases.

In the following discussion, the post-mortem findings regarding inactivation are correlated with the postoperative alteration of individual Parkinson symptoms and with the additional symptoms that appeared after the operations. In order to determine the degree of correlation, we have correlated for every symptom the percentage of improvement with the percentage of inactivation of individual substrates (Table 7a, b).

The degree of correspondence is divided into five groups: very good, good, medium, slight, and absent. The degree of correlation is represented in the following histograms by the increasing darkness of shading; white represents an absence of correspondence, i.e., the substrate was not destroyed, though a symptom improved or—the opposite—the substrate was destroyed, but the symptom did not improve. A $k>$ means that the therapeutic effect on the symptoms was clinically greater than expected from the anatomical lesion. An $a>$ signifies that the effect of treatment was less than expected considering the size of the anatomical lesion.

6.3. Correlation of Rigidity

6.3.1. Thalamotomy. A greater or lesser degree of correlation was present in all 14 cases between improvement of rigidity and inactivation of V.o.a or H_1, as shown in histograms (Fig. 120) and in Table 7. Admittedly in three cases (3, 5_ℓ, 12_r), the proportion was only 30, 40, and 100%, respectively. On the other hand, in 10 out of 14 cases with improvement of rigidity, there was no destruction of the H_2; in 9 cases there was no lesion of the internal capsule next to the V.o.a, and in 10 cases, no lesion of the internal capsule next to the V.o.p. Therefore, obviously, the pallidothalamic system is decisive for the improvement of rigidity (Fig. 120).

In case 12_r the rigidity was influenced 100% without a lesion of the V.o.a. Only 30% of the bundle H_1 was interrupted, but in the internal capsule, which lies between pallidum internum and V.o.p and through which the pallidothalamic fibers (Fig. 75d) pass, 80% of its fibers were destroyed. In addition, 30% of the pallidum internum was damaged. In this case, too, therefore, there was good correlation between improvement of rigidity and the lesion of the pallidothalamic fiber system.

In case 13_r, H_1 was not destroyed. However, the good (80%) influence on rigidity is in keeping with inactivation because 50% of the terminal nucleus of H_1, i.e., V.o.a (Fig. 85), was destroyed and 20% of the corresponding part of the internal capsule next to the V.o.p was also destroyed. In cases 4 and 15, the improvement of rigidity, 75% and 70%, respectively, exceeds the extent of inactivation in the V.o.a and H_1, whose respective sizes were only 30% and 60%, and 40% and 50%. But in both cases, all fibers of H_1 which always passed through this H_2 bundle are completely destroyed in the H_2 bundle (Fig. 120a). Therefore the pallidothalamic system was completely interrupted, which explains the good effect on rigidity. Further effective factors in case 15 were the 30% interruption of the internal capsule next to the V.o.a and the 10% destruction of the pallidum internum (Fig. 90a).

Table 7a. Inactivated structures in percent (cases 1–5)

Case No.	1	2			3	4	5	
	lt.	lt.	lt.	rt.	rt.	lt.	rt.	lt.
Therapeutic effect	rig. (arm, leg) 100%, tremor 100% (except emotionally	rig. (arm, leg) 90%, tremor 0%	tremor at rest 100%	rig. 100%, tremor 50%	tremor (leg) 100%, tremor (arm) 90%, rig. 40%	rig. (arm) 85%, rig. (leg) 70%, tremor (arm > leg) 60%	rig. 70%, tremor 100%	rig. 30%, tremor 10%
Side effects			persistent fac. paresis	severe organic psychosyndrome		aphonia	proximal paresis (arm), confusional state, experience of the double	confusional state, choke when swallowing, articulatory disturbance, ballism
Inactivated structures	%	%	%	%	%	%	%	%
V.o.a (+Rt)	100 (100)		60 (30)	50 (0)	50	30		40 (70)
V.o.p (+Rt)	70 (0)		40 (0)	30 (0)	100 (50)	25		
V.o.i				20	60			
V.im								
V.c								
C.e								
La.m								
ma. thal.				20		70		
Z.i	15 (↓V.o.p)			10 (V.o.a)	10 (↓V.o.p)	100 (↓V.o.a)		50 (↓V.o.a)
H_1	100		50	100	100	60		100
H_2						100		
Br.cj			50	90	100	50		15
S.th				20				
L.po					60		10	75
Ca.i (L.po)		30					20	30
Ca.i (V.o.a)	15		30					
Ca.i (V.o.p)								
Ca.i (V.im)						50		30
Ca.i (V.c)								
Pd.m								
Pd.e								
D.o							50	50
D.im							75	
Pa.i		60					70	
L.Pa.i		70					5	
Pa.e		40						
Put			25 atroph.					
Ans.lent.						30		
Caud.								10
Hypoth.lat.								30
Tr. opticus								

Table 7a (continued, case 6–10)

Case No.	6	7			8	9		10
	rt.	lt.	rt.	lt.	lt.	lt. rt.	2 rt.	lt.
Therapeutic effect	rig. 80%, tremor 60% hemorrhage	rig. 70%, tremor 70%	rig. 40%, tremor 50%	rig. 100%, tremor 60% hemorrhage		rig. 100%, tremor 0%	tremor 70%	rig. (arm) 70%, rig. (leg) 50%, tremor (arm) 100% hemorrhage into the frontal lobe
Side effects	negativism, dysarthria		spast. paresis (arm), dysarthria	myoclonia, confusion, sopor, deviation to left, proximal paresis, amentia			hemi-paresis, somnolence	somnolence
Inactivated structures	%	%	%	%	%	%	%	%
V.o.a (+Rt)	70	80 (40)			100 (100)	80		60
V.o.p (+Rt)	50 (50)	40 (40)			100 (100)	50		75
V.o.i		80			100			
V.im					70 (100)			
V.c								
C.e					50			
La.m	80				90			
ma.thal.					25			
Z.i	70 (V.o.a)				100			
H_1	100	100			50			
H_2					100	100		
Br.cj	30	30			100			
S.th					100	30		25
L.po	50				100			
Ca.i (L.po)	100		40					
Ca.i (V.o.a)				70	100		50	
Ca.i (V.o.p)				70	100		50	
Ca.i (V.im)					100			
Ca.i (V.c)	30				70			
Pd.m					20			
Pd.e	30				100			
D.o	60		100		70			
D.im			80		100			
Pa.i			65		50			
L.Pa.i				100	100		100	
Pa.e				100	100		100	
Put				100	100		100	
Ans.lent.				100	100		70	
Caud.				75	80		40	
Hypoth.lat.					30			
Tr.opticus					20		30	

(+Rt) adjacent reticular zone included; ↓ ventral

Table 7b. Inactivated structures in percent (cases 11–17)

Case No.	11	12		13	
	rt.	rt.	lt.	lt.	rt.
Therapeutic effect	rig. (arm) 50%, rig. (leg) 100%, tremor 40%, subcortical hemorrhage in parietal and occipital lobe	rig. 100% tremor 50%	tremor 50% (temporary) hemorrhage	rig. 80% tremor 100%	rig. 100% tremor 100%
Side effects	spast. hemi-paresis lt.		hemiparesis rt. somnolence	mimic facial paresis rt.	severe mimic facial paresis lt. transitory unconsciousness
Inactivated structures	%	%	%	%	%
V.o.a (+Rt) V.o.p (+Rt) V.o.i		 25 (100)	100 (100) 100 (100) 100	50 80 (+90)	70 (30) 70
V.im (+Rt) V.c			100 (100) 20		20 (basis)
C.e La.m			50 100		
M ma.thal.			50		 50 (V.o.a)
Z.i			100 (V.o.a/p) (V.im 70)	50 (V.o.p)	80 (V.o.p, V.im)
H_1 H_2		30	100 100		100 100
Br.cj S.th L.po			100 100	100	100 20
Ca.i (L.po) Ca.i (V.o.a) Ca.i (V.o.p)	 10	 30 80	 15 15	 20	
Ca.i (V.im) Ca.i (V.c) Ca.i (Pa.i)			15		
Pd.m Pd.rostral Pd.e			10		
D.o D.im			50		
Pa.i L.Pa.i Pa.e	50 40 30	30			
Put Fornix Ans.lent. Caud.					
Hypoth.lat. Tr.opticus B.-Kern					

(+Rt) adjacent reticular zone included; ↓ ventral

14	15	16	17
lt.	rt.		
rig. 70% hemorrhage	rig. 70% tremor 50%	rig. 60% tremor 80%	rig. 60% tremor 40%
prox. paresis (arm) tiredness	facial paresis, somnolence	paresis (arm), choke when swallowing, spastic cramps of floor of mouth, no initiative	paresis (rt. foot), Babinski, facial paresis, deviation to left, head turning to left, 3 seizures
%		%	%
40	40 (100) 60 (100)	90 (100) 75 (100)	
	25	80 (rostral)	
	50 100	100 100	
100	20	20 (rostral) 40	
100 50 20	30 15	5 (rostral Thal.)	
			5
		60	
		50	
75 75	10		90 50
85 100			15
30 (caudal) 75		5	20

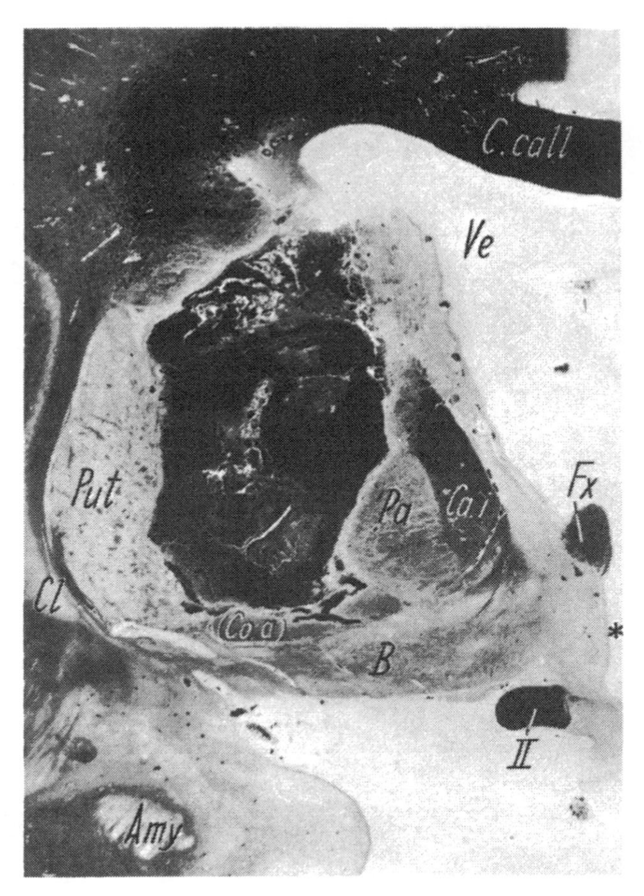

Fig. 118. Cross section through basal ganglia of case No. 7 at level of interventricular foramen. Irregularly limited hemorrhagic focus has destroyed outer part of pallidum externum and inner parts of putamen. Focus has penetrated internal capsule dorsally, destroyed lateral parts of head of caudate nucleus, and invaded lateral ventricle. Beside blood clot in ventral part, there is lighter structure with spike in dorsal direction, probably remnant of high-frequency coagulation. Border of focus on ventral side is almost horizontal and involves a small part of anterior commissure (*Co.a*) that is demyelinated. Anterior commissure is penetrated by cleft-shaped small hemorrhages especially in medial part. They are situated above anterior perforated substance penetrated by many ascending vessels. Cleft-shaped hemorrhages seem to have resulted from mechanical lesions by string electrode. Small artery supplying medial putamen and lateral pallidum is also apparently interrupted mechanically by string electrode. Fiber stain ×4

Twenty-five to 100% of the part of the brachium conjunctivum going to the thalamus was interrupted in 6 of 14 cases. One cannot, however, assume a causal relationship between this inactivation and the influence on rigidity because in all cases there was a corresponding interruption of the pallidothalamic system.

It is much more difficult to correlate slight therapeutic effects (Fig. 120b) to stereotaxic lesions than it is to interpret the therapeutic effects observed in cases with extensive or complete destruction of the V.o.a or H_1 (cases 3, 5_ℓ, 6, 7_ℓ, 16). In case 8, the left basal ganglia were almost completely destroyed by a massive hemorrhage after the stereotaxic intervention. The favorable effect on rigidity was only 70%, although the pallidothalamic system, pallidum externum, and internal capsule were completely destroyed by the hemorrphage. Case 8 seems to favor the view that there are systems that are antagonistic to the pallidothalamic system as regards effect on rigidity. The therapeutic effect on rigidity is less than expected if the systems in the basal ganglia, which will be discussed later, are destroyed in addition to the pallidothalamic system, which has an effect on rigidity. Case $7_{\ell 1}$ is not an exception. Here rigidity was 70% improved while 80% of the V.o.a and 100% of H_1 were destroyed. In all those cases in which the effect on rigidity was slight, 30–100% of the internal capsule next to the rostral pole (Fig. 118) of thalamus (L.po) was also destroyed. It is possible that the additional damage of a part of the internal capsule, through which fibers run from the rostral pallidum to the thalamus, had reduced the positive effect of the inactivation of the V.o.a and H_1. It should also be noted that the fibers from the supplementary motor area on the medial surface of the brain also pass through these rostral segments of the internal capsule immediately behind its genu. TRAVIS (1955a, b) produced an increase of tone in the direction of spasticity in monkeys as soon as the supplementary area in addition to the motor cortex was inactivated. For further explanations see p. 220.

In three of the cases (5_ℓ, 8, 16) in which the therapeutic effect of coagulating the V.o.a or H_1 was slight, 15–20% of the subthalamic nucleus (Figs. 123a, 129) was damaged. In two other cases (13_r, 15), however, the subthalamic nucleus was damaged to the same extent without impairment of the therapeutic effect on rigidity. Thus, partial damage of the subthalamic nucleus cannot be regarded as the explanation for an inadequate therapeutic effect (Figs. 56a, 85, 90a, 95, 129).

On the other hand, it is interesting that partial damage (15–20%) of the subthalamic nucleus has not evoked an even transient contralateral hyperkinesia or ballismus. This fact is in agreement with the experimental results of WHITTIER and METTLER (1949). They found in experimental monkeys that

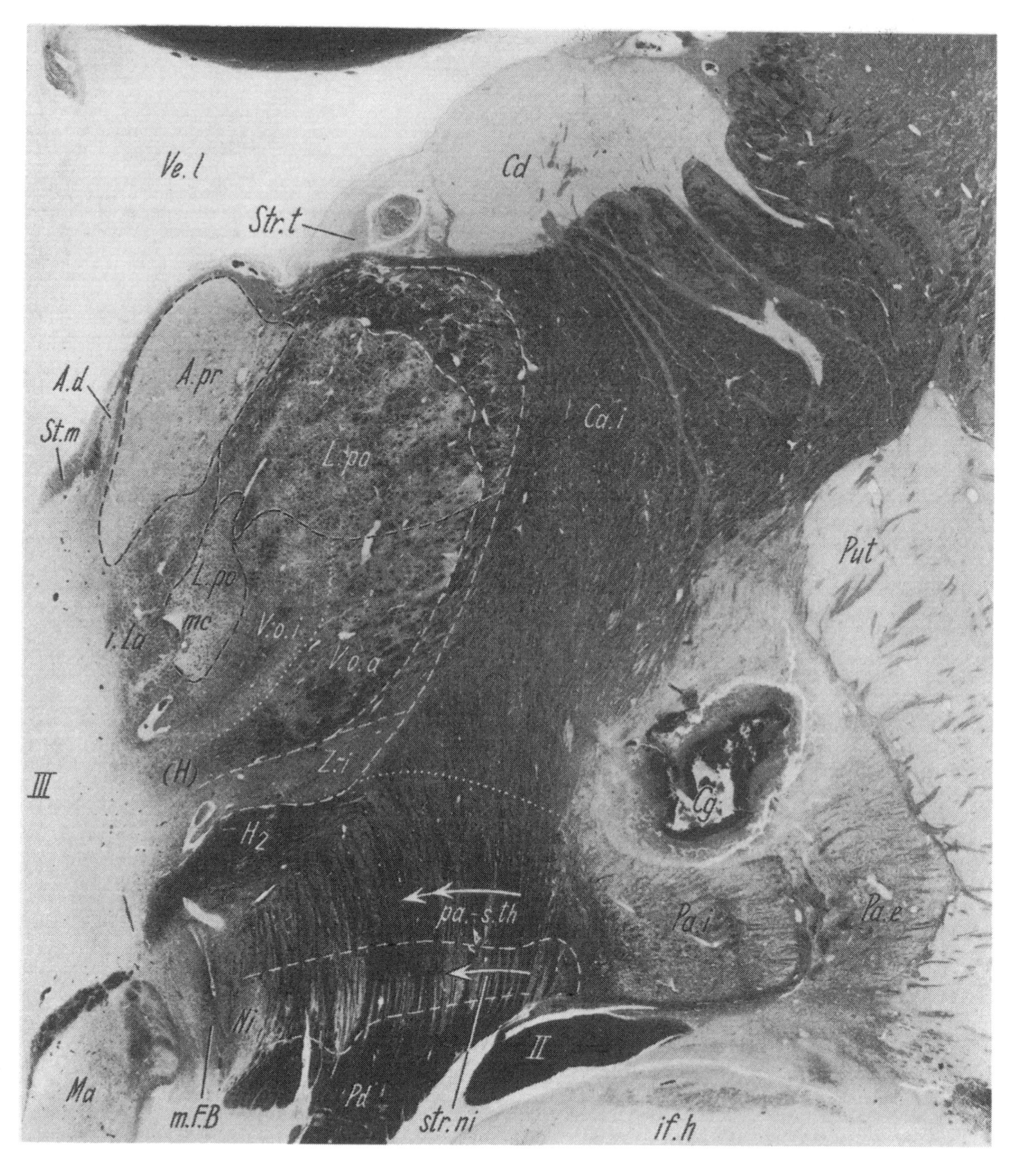

Fig. 119. Cross section through right thalamus of case No. 11, showing coagulation focus in dorsal pallidum. Coagulation focus target was intended to be right V.o.a. Right thalamus is shifted considerably into left half of cranium due to large swelling of right hemisphere caused by large parieto-occipital hemorrhage (Fig. 73a, b). Therefore, focus was placed in pallidum. Swelling of basal ganglia is revealed by paleness in fiber picture of structures in putamen, pallidum, and internal capsule, whereas cerebral peduncle and bundle H_2 have normal dark fiber staining. Note the complete loss of field H and bundle H_1 7 days after the coagulation. Striking staining difference in comb system of peduncle between dark pallido-subthalamic fibers (*pa.-s.th*) and very light strionigral fibers (*str.ni*). × 5

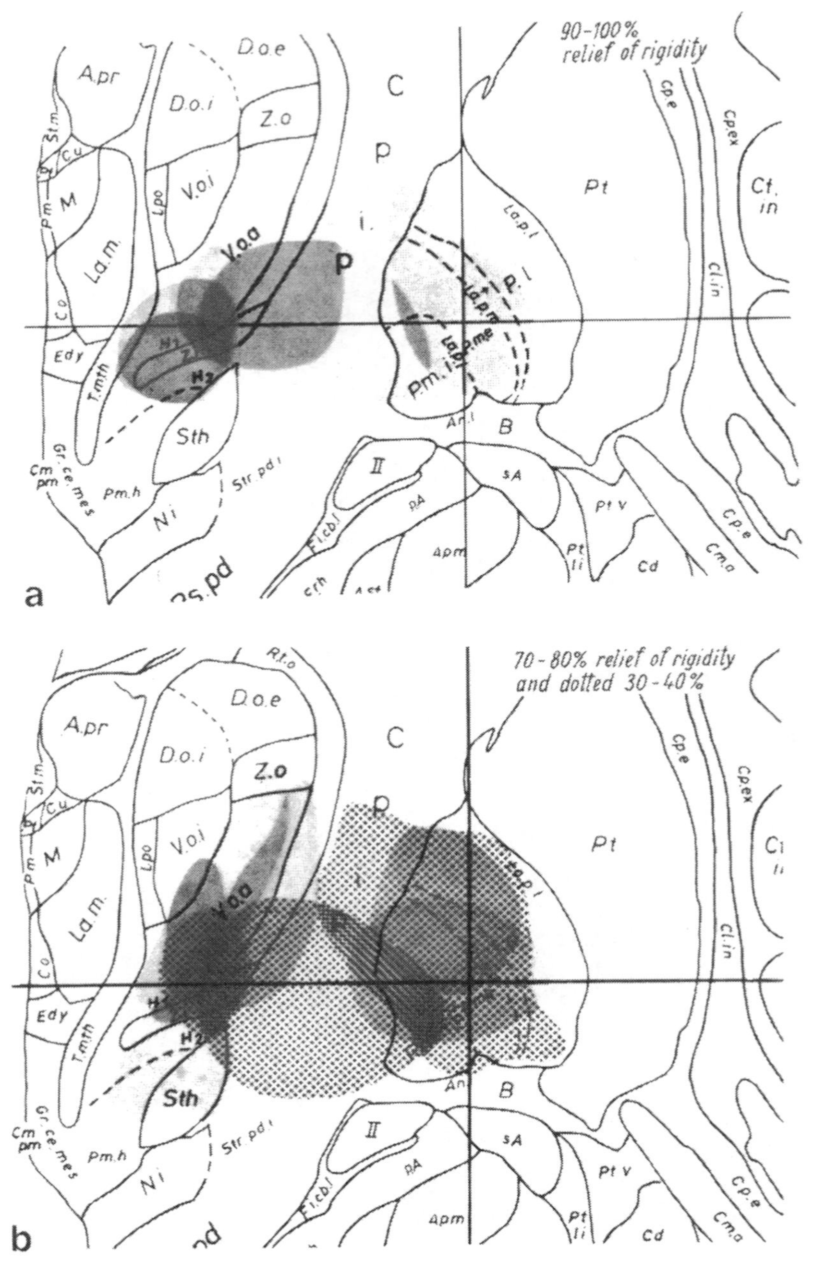

Fig. 120. (a) Diagram of transverse section of human diencephalon at level 10 mm caudal to anterior commissure (see Fig. 150a, b atlas section). Dark shadows represent extent of individual lesions in 5 autopsied Parkinson cases with 90–100% relief of rigidity. Foci extend from base of V.o.a nucleus through Forel's field H_2 through reticular nucleus of thalamus and narrow part of internal capsule, which connect pallidum internum to V.o.a nucleus. One focus is located in pallidum internum and lamella pallidi interna (Hassler *et al.*, 1965).
(b) Diagram of same cross section of human diencephalon at level 10 mm caudal to anterior commissure. Uniformly gray shadows indicate lesions in 6 cases of 70–80% improvement of rigidity. Dotted zones, involving large parts of internal capsule, represent 3 cases of 30–40% improvement of rigidity. These foci are concentrated in region of V.o.a and its immediate ventral and dorsal surroundings as well as in inner segment of pallidum, from which they extend dorsally in external segment and medially in that portion of internal capsule that connects pallidum with V.o.a (Hassler *et al.*, 1965)

choreiform hyperkinesia is produced only if the damage to the subthalamic nucleus exceeds 20%. Admittedly, it is possible that the absence of hyperkinesia after slight damage to the subthalamic nucleus is due to its prevention by the simultaneous far-reaching or complete interruption of the pallidofugal pathways in H_2. This statement should not create the impression that more extensive destruction of the main substrate of the nucleus subthalamicus during a stereotaxic intervention could not cause a contralateral ballistic hyperkinesia, as was shown by Hopf *et al.* (1968). Cases of postoperative ballismus without serial sections have been reported several times, e.g., by Dierssen *et al.* (1961) (with negative results) and Krayenbühl and Yarsagil (1961). Among cases that did not come to autopsy we also have observed postoperative ballismus only rarely. The order of magnitude is approximately 0.2% in about 3500 stereotaxic operations for extrapyramidal diseases.

Table 8 shows the correspondence between improvement of rigidity and the extent of pallidum destruction. Parallelism between improved rigidity and the extent of the coagulation of pallidum internum, externum, and lamella pallidi interna (Fig. 119; 60% in case 11) is very good in four cases and good in two. There are no cases without any correlation. Of the five cases in which there was no good correspondence, there were only two (7_r, 17) in which the effect on rigidity was less than expected. In four cases, the influence on rigidity was better than expected. This is adequately explained in cases 2_ℓ and 5_r by the simultaneous inactivation (Fig. 55a) of lamella pallidi interna and the pallidum externum, which belong to the same neuronal system. The insufficient improvement of rigidity in case 17 could be explained by the fact that only half of the lamella pallidi interna and one-sixth of the pallidum externum were destroyed. The inadequate effect on rigidity in case 7_r is probably connected with the fact that again the internal capsule had been damaged next to the rostral pole of the thalamus. On the other hand, in cases 2_ℓ, 7_ℓ, and 14 with similarly localized damage, the degree of pallidum inactivation corresponds completely to the degree of improvement of rigidity, but in these cases 40–100% of the pallidum, from which many fibers run to the nucleus lateropolaris, had been coagulated. In order to explain the effect on rigidity in contrast to the thalamotomy, it is not necessary to consider the simultaneous damage to the internal capsule in the plane of the V.o.a and in one case of the V.o.p. The damage to the internal capsule was completely absent in 2 of 7 cases (2_ℓ, 17), in which correlation was good. On the other hand, simultaneous damage to the internal capsule in the plane of the V.o.a does not interfere with the influence on rigidity that is caused by a lesion of the pallidum. The therapeutic effect was not impaired in cases 7_ℓ, 8, 14 (Fig. 118) in which 75% of the caudatum had been destroyed by hemorrhage. We cannot say anything about the effect on rigidity of a simultaneous lesion of the putamen, since in one case (8) with a large lesion of the putamen, the basal ganglia had at the same time been almost completely destroyed, while in the other case (5_r) only about 5% of the putamen had been damaged (Fig. 55a).

6.3.2. Significance of the Pallidothalamic Systems for the Production of Rigidity. Rigidity is a special form of increase in muscle tone that differs from the increase of tone observed in spastic paresis. The rigid increase of tone is caused by an increase of the tonic stretch reflexes. That the rigidity is due to reflexes is proved by the fact that severing the posterior roots of the rigid limb (POLLOCK and DAVIS, 1930b) at first leads to a complete loss of rigidity in that arm followed *a few days later* by an increase in tone. This delayed increase of tone, which depends on the position of the labyrinth and the head and which does not depend on segmental reflexes, is called *basic rigidity*, in contrast to the reflex component of rigidity, the *plastic muscle* tone. Two different forms of rigidity observed after decerebration in experimental animals correspond to these two kinds of rigidity. SHERRINGTON'S (1947) decerebrate rigidity is produced by a transsection between the collicles of the corpora quadrigemina toward the posterior hypothalamus. One obtains an increased muscle tone that depends on the reflex arc regulated by the γ fibers being intact (LEKSELL, 1945). POLLOCK and DAVIS' (1924, 1930a) anemic decerebration caused rigidity that is independent of the peripheral reflex arc, a so-called α rigidity. In anemic decerebrate rigidity, tying the two vertebral arteries causes a softening of the brainstem in the anterior part of the pons. Similar α rigidity can be achieved by a postnigral transsection of the brainstem in the pons (Fig. 27c), as evidenced in animal experiments by WAGNER (HASSLER, 1972).

While there is little doubt that the rigidity in Parkinson's disease contains a reflex component, there has been controversy during the last few years regarding production of this rigidity by mediation of the γ innervation (Figs. 103, 121). The fact that Parkinson-rigidity depends partly on the reflex arc being intact does not prove the dependence of the rigidity on the γ innervation. Contraction of the muscle can be caused via a reflex arc without its excitability being shifted by the γ innervation. Some work has been reported that seems to favor the view that the γ innervation is increased in rigidity (NARABAYSHI *et al.*, 1965; LANDAU, 1969), while other findings indicate that in rigidity the γ innervation either has been weakened or has disappeared (STEG, 1964, 1972; HASSLER, 1956a). BOYD (1962) showed that there is a separate innervation for the nuclear bag spindles (γ-I) and another for the nuclear chain spindles (γ-II). These have different thickness of fibers and different speeds of conduction. This finding clarifies the opposite views regarding the activity of the γ innervation in rigidity.

It is probable that the rapid γ-I innervation of the nuclear bag spindles is decreased in rigidity while the slower γ-II innervation of the nuclear chain spindles is increased. The increased γ-II innervation would lead to an increase of tonic stretch reflexes. It is known that the phasic stretch reflexes are not increased in Parkinson rigidity in contrast to tonic stretch reflexes. In addition, according to KRIENITZ (1955), electrically elicited stretch reflexes (H reflexes) are increased in Parkinson rigidity (Fig. 103), but not after V.o.a. coagulation, although the tendon reflexes are not enhanced and cannot be reinforced by the Jendrassik maneuvre (HASSLER, 1956).

The apparatus of the tonic stretch reflexes and that of the two different kinds of γ innervation are under several central influences. The best pathologic example for this is the Parkinson syndrome. In it the rigidity is caused by the loss of the dopaminergic nerve cells in the substantia nigra. When these are lost one finds an increase of the tonic stretch reflexes. It follows that it is a normal function of the substantia nigra to have an inhibitory effect on the tonic stretch reflexes. The rigidity in Parkinsonism depends on the degree of wakefulness of the patient. During sleep, the rigid increase of tone disappears in the same way as does the tremor. Thus, during wakefulness the arc of the tonic stretch reflexes is additionally under another influence that facilitates the tonic stretch reflexes. The previous operations against the Parkinson syndrome, pyramidotomy by PUTNAM (1940) and the extirpation of the motor cortex by BUCY (1938, 1940, 1942; BUCY and CASE, 1939), brought the first hint regarding the localization of this additional influence on the tonic stretch reflexes. Hence, the final pathway for the increase of the tonic stretch reflexes passes through the corticospinal systems of the pyramidal tract to the relays of the tonic stretch reflexes in the anterior horn of the spinal cord (see also RONDOT, 1965).

The areas in the motor cortex in which the pyramidal tract originates receive specific afferent fibers from special nuclei in the thalamus. The nucleus ventro-oralis anterior (V.o.a) is one of these nuclei. This nucleus and the inner part of the pallidum, which sends the greatest part of its efferent fibers to the V.o.a, are the two most important targets for abolishing rigidity (Fig. 120a, b).

The present evaluation of stereotaxic brain operations for the Parkinson syndrome has shown that in every case improvement of rigidity is linked to inactivation in the pallidum internum, in the V.o.a, or in the pallidothalamic fiber bundle (Fig. 128a) that joins these two nuclei, ending as Forel's bundle H_1. The cases discussed here which survived operations for 2 or more years demonstrate that inactivation of these pallidothalamic systems (Figs. 120a, b; 121) can influence rigidity permanently (HASSLER *et al.*, 1965). Long-term observations in a previous group of 1171 patients suffering from the Parkinson syndrome showed that a recurrence of the rigidity occurred only in 13.7% of the cases (MUNDINGER and RIECHERT, 1966). It is, therefore, obvious that the pallidothalamic neuronal systems are the most important neuronal systems for improvement or abolition of rigidity, though they may not be the only systems that influence rigidity. It can be assumed that the midbrain reticular activating system that controls the degree of wakefulness has an influence on the tonic stretch reflexes in the spinal cord.

We shall now discuss in what way the pallidothalamic systems influence the tonic stretch reflexes,

Fig. 121. Scheme of peripheral and central conduction mechanism of rigidity and of its stereotaxic relief. Tonic stretch reflexes are elicited from secondary endings of muscle spindle (*II*). Impulses reach spinal cord via II-afferents and are transmitted to special set of interneurons (*red*) that finally excite tonic motor neurons ($\alpha 2$; *red*). This reflex arc is antagonistic to, but not directly interrelated with, arc of monosynaptic reflex (*blue*) originating from anulospiral endings on muscle spindles that have direct contact with phasic motor neurons ($\alpha 1$) via I-A afferents. Both reflex arcs, however, are inhibited by I-B afferents (*green*), which originate from Golgi-tendon organs. — Parkinson rigidity is due (1) to loss of ascending inhibitory nigrostriatal (*green*) influence on pallido-reticular efferent neurons (*red*); (2) to loss of descending inhibitory nigro-reticulo-spinal influence (*green*) on tonic motor neurons ($\alpha 2$; *red*) of tonic stretch reflex either directly or indirectly via interneurons. If, in Parkinson syndrome, nerve cells of posterior portion of substantia nigra are deteriorated, then tonic stretch reflexes are no longer subjected to inhibitory action of substantia nigra, but only to facilitatory influences (1) of corticospinal fibers originating from area 6aα, (*purple*) (2) of pallido-reticulo-spinal pathway originating from pontine reticular nucleus (*Rt.po*). The thereby enhanced tonic stretch reflexes constitute Parkinson rigidity. To restore disturbed balance between inhibitory and facilitatory influences on interneurons of stretch reflexes in awake state, it would be necessary to restrict facilitatory influences from corticospinal and pallido-reticulo-spinal pathways. This has

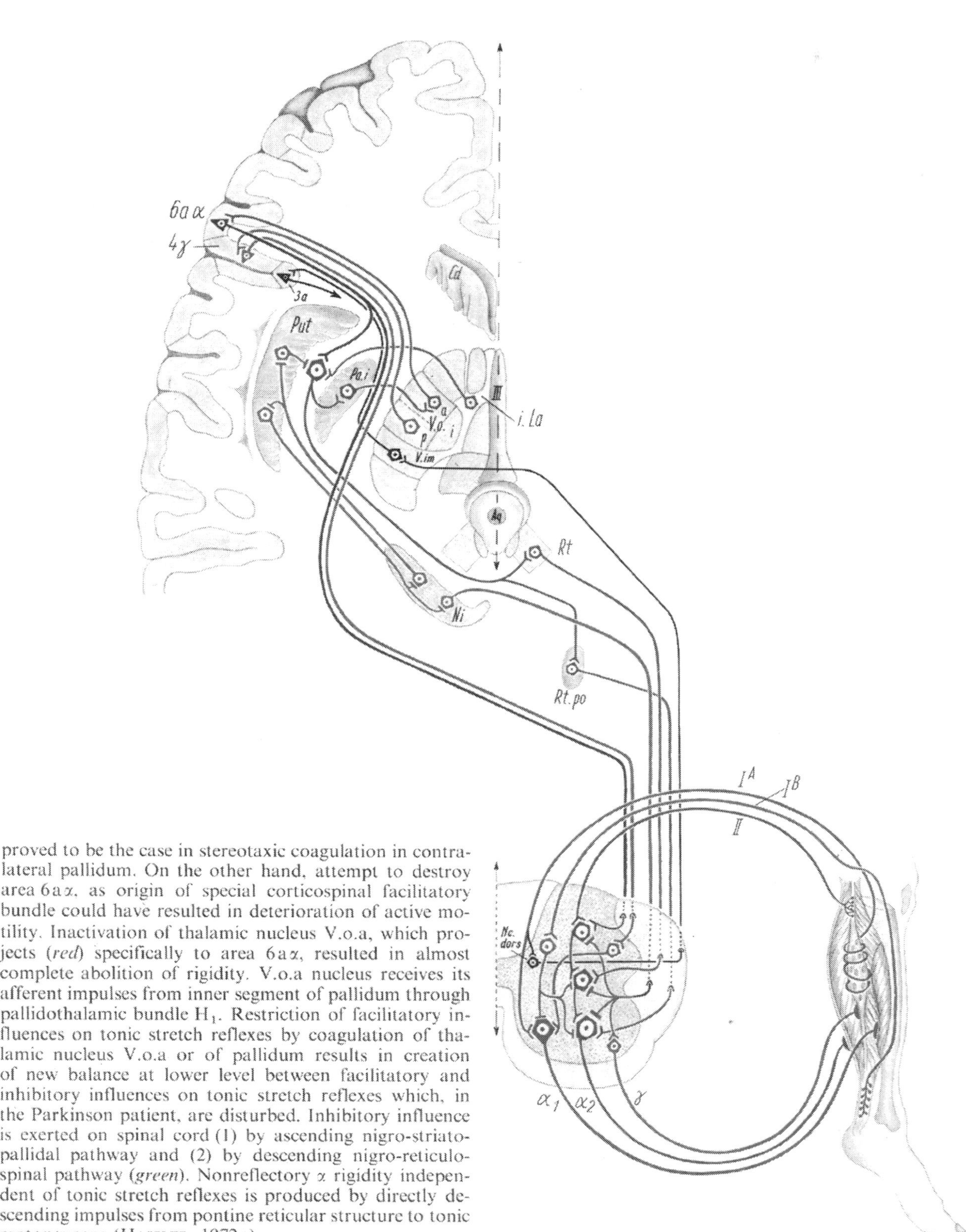

proved to be the case in stereotaxic coagulation in contralateral pallidum. On the other hand, attempt to destroy area 6aα, as origin of special corticospinal facilitatory bundle could have resulted in deterioration of active motility. Inactivation of thalamic nucleus V.o.a, which projects (*red*) specifically to area 6aα, resulted in almost complete abolition of rigidity. V.o.a nucleus receives its afferent impulses from inner segment of pallidum through pallidothalamic bundle H_1. Restriction of facilitatory influences on tonic stretch reflexes by coagulation of thalamic nucleus V.o.a or of pallidum results in creation of new balance at lower level between facilitatory and inhibitory influences on tonic stretch reflexes which, in the Parkinson patient, are disturbed. Inhibitory influence is exerted on spinal cord (1) by ascending nigro-striato-pallidal pathway and (2) by descending nigro-reticulo-spinal pathway (*green*). Nonreflectory α rigidity independent of tonic stretch reflexes is produced by directly descending impulses from pontine reticular structure to tonic motoneurons (HASSLER, 1972a) of which the rigidity consists. It has been shown anatomically that nearly 100% of the efferent fibers of the inner portion of the pallidum go into the thalamus (HASSLER, 1949a). They terminate in thalamic nuclei from which fibers do not descend but go to the cerebral cortex. The most important terminal nucleus of the pallidothalamic fibers, the V.o.a, projects onto the motor cortex, probably to area 6aα in man. A projection of the V.o.a or the pallidum internum to the precentral motor areas has also been shown by electrophysiological investigations. Consecutive recording has shown phase reversal of cortical responses (Fig. 30c) in precentral region to stimulation of the pallidum or the V.o.a (GANGLBERGER, 1961b, 1962a, 1964, 1970; GANGLBERGER and PRECHT, 1964; HAIDER *et al.*, 1972). It is frequently possible to demonstrate the presence of a focus high in the frontal areas in the first few weeks after the operation, which is explained by the inactivation of afferent fibers going to the motor cortex. This work shows, therefore, not only by anatomical methods but also by electrophysiologic (stimulating and inactivating) experiments, the projection of the thalamic ventral nuclei including V.o.a and V.o.p to the motor cortex. According to SHIMAZU *et al.* (1962) there is a directly descending pathway in addition to this link to the motor cortex, but the anatomical findings show this applies only to subthalamic systems for which there are also descending connections to the brainstem. There is unequivocal anatomical and electrophysiologic evidence that the ventral nuclei of the thalamus project directly to the cortex, particularly since all their nerve cells show retrograde degeneration (HASSLER, 1949) or labeling after a cortical ablation or horseradish peroxidase treatment (NAKANO and HASSLER, in preparation).

GUIOT (1958) and GILLINGHAM (1962); GILLINGHAM *et al.* (1960) pointed out that the best effects on rigidity of Parkinsonism can be achieved by placing a lesion in the internal capsule. The area involved is a small segment of the internal capsule lateral to the V.o.a and V.o.p in the horizontal plane of the subthalamus. In case 12_r a good effect on rigidity was achieved with a lesion of the internal capsule in the frontal plane of the V.o.p. However, the improvement was due not to an interruption of the descending capsule fibers but rather to the fact that at that height the internal capsule is crossed by pallidothalamic fibers (Figs. 75c, 119). These fibers could be demonstrated (HASSLER, 1964) in a human patient who had not had a stereotaxic operation but who suffered complete degeneration of the internal capsule following a striatum apoplexy (Fig. 128a). In this patient only the bundle of fibers that crosses the internal capsule from the pallidum to the thalamus was preserved. When the internal capsule is not destroyed, these fibers cannot be recognized because they are hidden by the longitudinal fibers. The area in question is intermediate between the caudal ansa lenticularis and Forel's bundle H_2 medial to the internal capsule.

This circumscribed segment of the internal capsule was also affected in cases 12_r and 15 (see Figs. 75a, 90a, 129) by a coagulated area from the medial side. In both cases the improvement of rigidity was better than could have been expected from the extent of destruction of H_1 or V.o.a. It follows that the effect on rigidity of a stereotaxic operation can be improved by interrupting the pallidothalamic fibers in this part of the internal capsule. However, this applies only to the circumscribed ventral segment of internal capsule lying laterally to the V.o.a/p. In several cases, improvement of rigidity was less than could have been expected from the amount of destruction in the V.o.a or the H_1. In these cases, in addition to the coagulation of the V.o.a, the internal capsule was interrupted further rostrally, lateral to the anterior V.o.a or laterally to the anterior thalamic pole (L.po) e.g., 5_l (Fig. 56a); 6 (Fig. 58a).

These data indicate, therefore, that simultaneous damage to the internal capsule next to the rostral thalamic pole should be avoided in the attempt to influence rigidity of Parkinsonism. During the stereotaxic treatment of athetosis, we also found that inactivation of the rostral segment of the internal capsule behind its genu leads at first to a reduction of mobile rigidity and of muscle tensions. A few weeks later, however, the muscle tensions recur, often to an increased degree.

Abolition of rigidity in Parkinsonism by interruption of the pallidothalamic system has been clinically confirmed from the large number of cases of Parkinsonism that have been operated on in many centers. Consequently, it can be concluded that, normally, the pallidothalamic system has an enhancing action on muscle tone, which is based on the tonic stretch reflexes (myotatic reflexes). This action of the pallidothalamic system goes via the motor cortex into which it projects through the V.o.a. At present, inactivation of this system has proved to be the best method for abolishing Parkinson rigidity on the contralateral side.

The effect on rigidity occurs immediately during coagulation of the pallidothalamic system. The effect is maintained even after the muscle hypotonia, which often appears during the operation, has again disappeared. Clinically, hypotonia is evidenced by the tendency the patients have in the first few days after the operation to deviate and to fall on the side of the body that has been influenced by the operation. This phenomenon is due to the reduced basic innervation of limbs and trunk. The lessened muscle tone is replaced by a normal muscle tone after an average of 2–3 weeks.

The findings in the post-mortem material have led to the following interpretation of the mechanism of rigidity. Rigidity is caused by the tonic stretch reflexes (myotatic reflexes). The loss of nigra nerve cells in Parkinson cases affects the relays of tonic stretch reflexes in the anterior horn *via* two possible pathways. The first path passes through the nigrostriatal fiber bundle, which exerts an inhibitory action *via* the dopamine transmitter on the striatal interneurons. The pallidum thereby loses some striatal inhibition, so that the facilitatory influence on tonic stretch reflexes prevails either through the descending or through the corticopetal pallidar pathway (V.o.a and its projection). The second pathway descends from the substantia nigra (Fig. 121) indirectly to the spinal cord. The tonic stretch reflexes are enhanced by the disappearance of an action of the substantia nigra on their relays in the anterior horn of the spinal cord. The substantia nigra normally has an inhibiting effect on the relays of the tonic stretch reflexes while the efferent corticospinal system (pyramidal tract) has an enhancing influence on the same relays. The enhancing influence is due to the afferent fibers of the corticospinal system from the pallidum which run through the pallidothalamic fibers and through the V.o.a. Coagulation in the pallidum internum in H_1 or in the V.o.a abolishes the enhancing corticospinal influence on the tonic stretch reflexes. In patients with rigidity, a new equilibrium at a lower level is created between the enhancing and inhibiting influences on the tonic stretch reflexes, and normal strength is regained, leading to abolition of rigidity (Fig. 121).

Physiologically, up to now, only enhancement of the tonic stretch reflexes through the corticospinal system has been proved by stimulating the V.o.a. In patients with Parkinson's disease, stimulating the V.o.a with higher frequencies (25 and 50 cps) before inactivating it, leads to an increase of resistance against passive movements that exceeds the rigidity of the patients. This interpretation of the central mechanism of rigidity is valid whether rigidity is due to an increase or a decrease of peripheral γ activity (Fig. 121).

The afferent fibers of the motor cortex, whose inactivation is therapeutically effective against rigidity, become parts of the internal capsule. It is, therefore, possible that additional lesions in individual parts of the internal capsule have an influence on rigidity, particularly if the long fibers from area $6a\alpha$ are affected. According to the general rules governing the course of cortico-fugal fibers, it can be assumed that these long fibers lie in the internal capsule immediately rostral to the pyramidal tract. However, this has not been demonstrated. Apart from the lesions discussed here, we have no information of additional lesions in the internal capsule having a diminishing influence on rigidity. In the two cases (12_r, 13_ℓ) in which rigidity was more improved than could have been expected after destruction of the V.o.a, H_1, or internal pallidum, the internal capsule was included in the lesion lateral to the V.o.p. The improved effect on rigidity can be explained in terms of the interruption of the pallidothalamic fibers which cross that part of the internal capsule. The therapeutically successful inactivations performed by Guiot (1958) and Gillingham *et al.* (1960) are located in this area.

In cases 5_ℓ, 3, 6, 8, and 16 the effect on rigidity was less than could have been expected after complete inactivation of H_1 or the V.o.a. In all these cases the longitudinal bundles of the internal capsule were interrupted next to the rostral thalamic pole by 50–75%; in case 16, 40% of the L.po was inactivated (Fig. 129). It is obvious that the complete abolition of rigidity is hampered by additional damage to the internal capsule or by damage to the corresponding thalamic nucleus, L.po. These parts of the internal capsule lie rostrally to the pallidothalamic fibers crossing the internal capsule. The anatomical as well as the physiologic findings by stimulation (see p. 40) indicate that the regio lateropolaris (L.po) contains the nucleus that projects into the supplementary area (Fig. 61). According to Travis (1955a, b), the supplementary motor area is that cortical region which in experiments on monkeys is responsible for the spastic increase of tone if the pyramidal tract has been severed at the same time. The efferent fibers of this area run through the internal capsule closely behind its genu (Bertrand, 1966). The influence of the supplementary motor area on the increase of the spastic tone makes it probable that an additional lesion of these fibers in the internal capsule or of the L.po reduces the abolition of rigidity because it is able on its own to cause an increase of tone.

6.4. Correlation of Tremor

Of the 23 operations in which the effect of the operation could be studied at autopsy, 16 were carried out against the tremor in the oral ventral nuclei of the thalamus and 7 in the pallidum. Since the inactivations affected two quite different neuronal systems, the thalamotomies and pallidotomies are discussed separately.

6.4.1. Thalamotomy. In all 16 thalamotomies the tremor was improved or abolished by coagulation of the oral ventral nuclei (Table 6). This was combined in 11 cases with an interruption of the dentato-thalamic fibers entering the V.o.p (Figs. 46, 47a).

Abolishment or almost complete relief from tremor (90–100%) was achieved in 6 cases after coagulation in the V.o.p (Fig. 122a). In case 3 this was achieved by complete destruction of the V.o.p (Fig. 50a). In 5 other cases (1, 2_{ℓ}, 10, 13_r, 13_{ℓ}), only 40–80% of the V.o.p had been coagulated (Fig. 122b). In both operations on case 13, there was in addition a complete severing of the dentato-thalamic fibers in the zona incerta before entering the V.o.p (see Figs. 84, 86). In this case, therefore, abolishment of the tremor was due to the complete

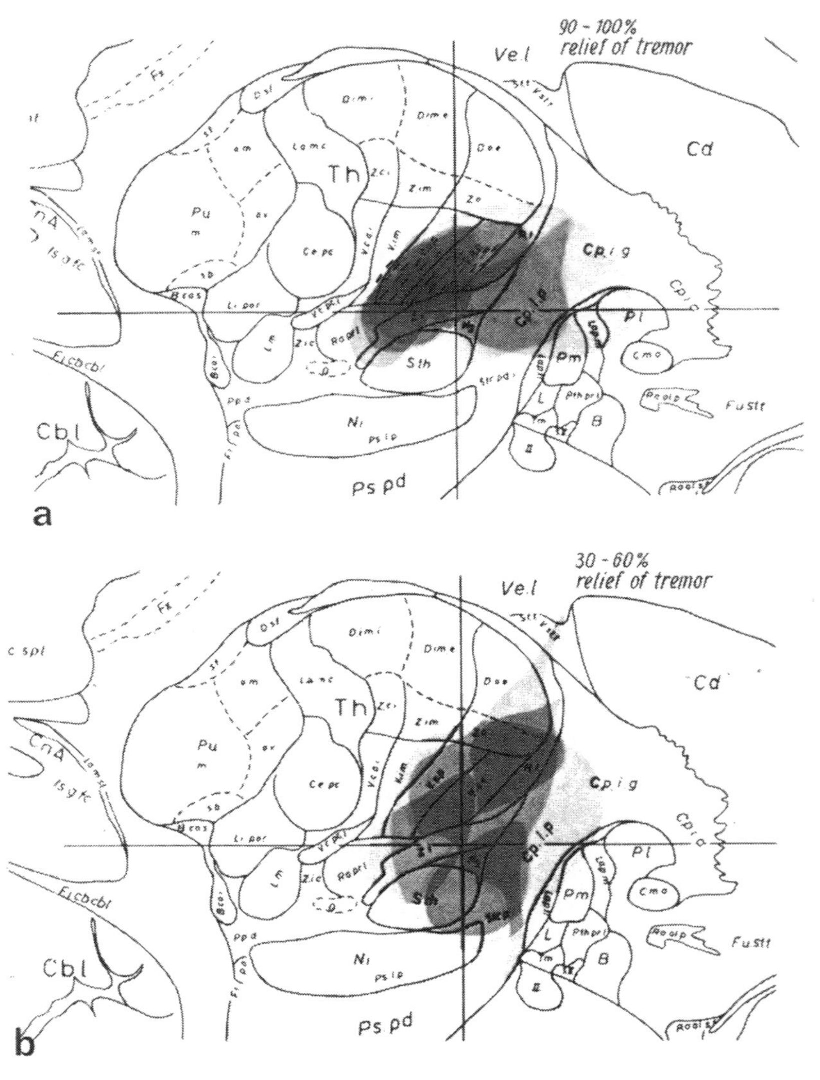

Fig. 122. (a) Diagram of sagittal section through human diencephalon 11 mm lateral to midline. Dark shadows represent extent of individual lesions in 9 autopsied Parkinson cases, which resulted in 90–100% relief of tremor. Most dense superposition of lesions is located in basal $^2/_3$ of V.o.p and to lesser extent in basal V.o.a. Ventro-intermedius nucleus (*V.im*) is minimally involved in only 3 cases with almost complete relief of tremor, whereas it is completely spared in all other cases. Rostral internal capsule is involved in only 2 Parkinson cases, with best results. (b) Same diagram of sagittal section with shading of extent of individual lesions in 7 Parkinson cases with 30–60% improvement of tremor. In 3 cases, V.im is also partially involved. Base of V.o.p is coagulated in only one of all cases with 30–60% relief of tremor. In 3 cases, base of V.o.p is spared by coagulation. Lesions in cases with 30–60% tremor relief are always more extensive and involve more adjacent structures, e.g., internal capsule and subthalamic nucleus, than in cases with 90–100% tremor relief (HASSLER *et al.*, 1970)

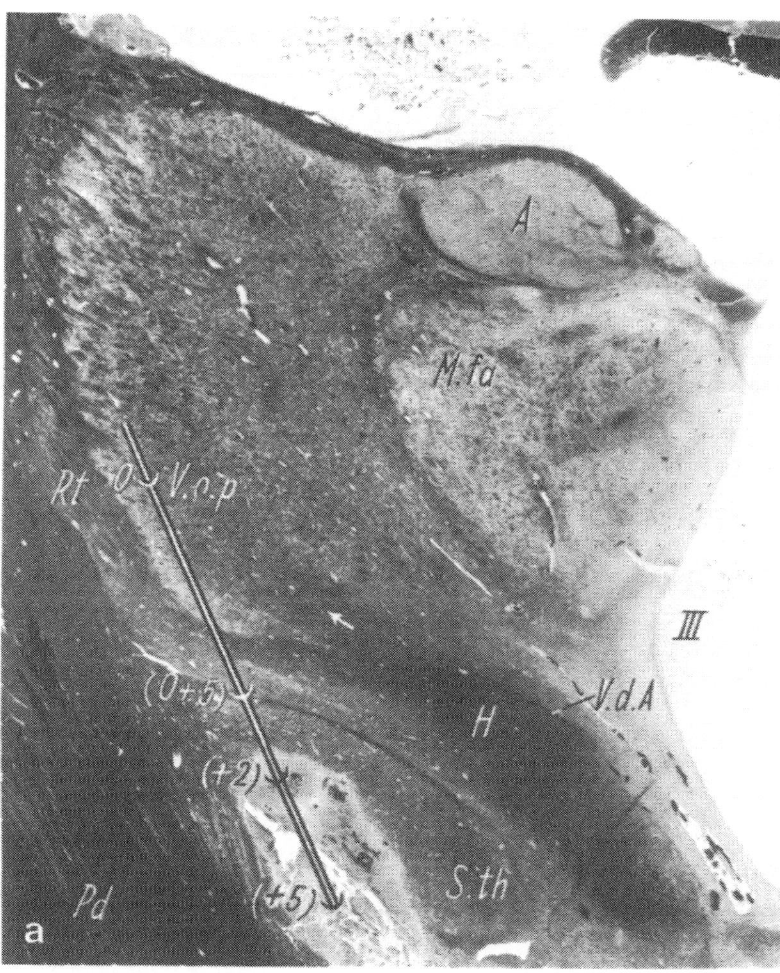

Fig. 123. (a) Cross section through left basal ganglia of case No. 5 showing coagulation focus within subthalamic nucleus (*S.th*) and cerebral peduncle (*Pd*). Dorsal border of subthalamic nucleus, zona incerta above it, nucleus V.o.p with entering dentato-thalamic fibers (*arrow*) are all preserved because coagulation was performed 5 mm lower than intended. Fiber stain. ×6.

(b) Coagulation focus, located too deeply (see Fig. 123a), has interrupted dentato-thalamic fibers at point of entry in base of thalamus. Thus, terminal nucleus V.o.p displays in contrast marked glia increase, perhaps combined with discrete nerve cell atrophy. Nissl stain. ×9

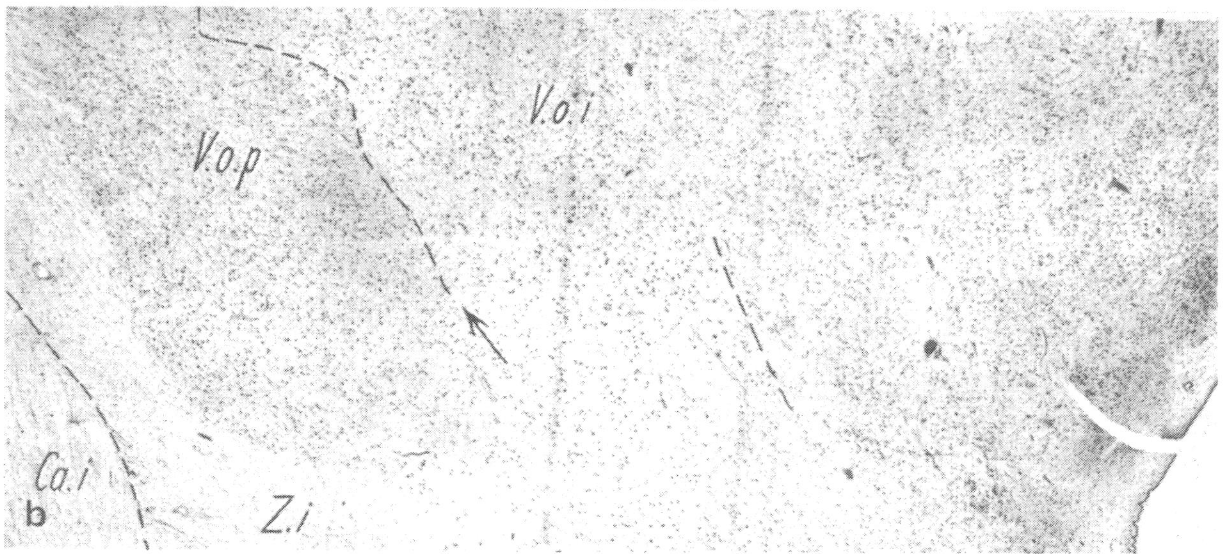

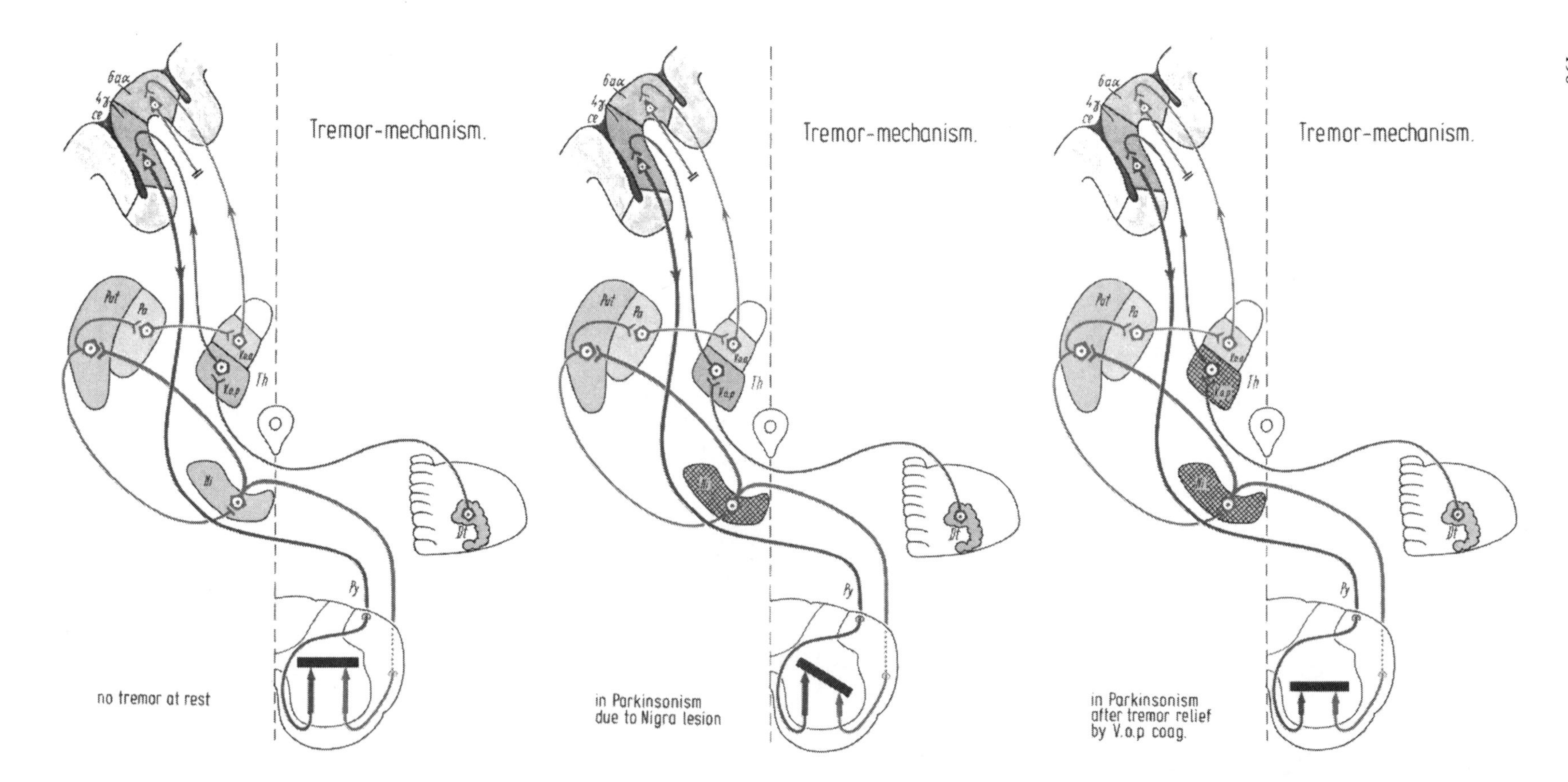

Fig. 124a–c. Schematic drawing of tremor mechanism:
(a) Under normal awake conditions, pyramidal system originating mainly in area 4γ (*red*) exerts strong synchronizing influence (*red arrow*) on peripheral motor apparatus in anterior horn, whereas strionigral system (*green*) exerts strong desynchronizing influence (*green arrow*)—these systems are normally in balance.
(b) In Parkinsonism, which is due to nigra cell loss (*Ni* crosshatched), desynchronizing influences are sharply reduced (*short green arrow*), thereby disturbing the balance so that synchronizing influences prevail (*long red arrow*). This imbalance is the neurophysiologic basis of tremor at rest.
(c) In Parkinsonism, balance can be regained by reduction of synchronizing influences exerted by pyramidal system (*red*) and their specific exciting afferents. Consequently, pyramidal synchronizing and strionigral desynchronizing influences in anterior horn regain balance on lower level and tremor at rest is relieved. This can best be performed by stereotaxic coagulation of specific exciting afferent pathway to pyramidal system, which comes from thalamic nucleus V.o.p (crosshatched), which is supplied by dentato-thalamic fibers

inactivation of the V.o.p system due to destruction of V.o.p itself or of its afferent fibers. The same applies to the left operation (Figs. 41 and 133a, b) of case 2. Only one-fourth of the dentato-thalamic fibers were affected in case 10. This, together with the inactivation of 70% of the V.o.p, contributed to the complete abolishment of the tremor.

Damage to the zona incerta and the dentato-thalamic fibers passing through it in case 1 was only slight, but the nucleus reticulatus, through which the efferent fibers from the V.o.p pass to the internal capsule, had been completely destroyed. In this case, too, the neuronal system of the V.o.p had been completely interrupted, although this time in its efferent part (Fig. 34b, c).

A striking relief of 80% of tremor occurred in two cases. In one (16) 75% of the V.o.p cells were destroyed (Fig. 95). In case 12_r, the tremor was improved by 80%, although only 25% of the V.o.p was inactivated. However, the adjacent nucleus reticulatus (Fig. 75a, d) was coagulated. It is probable that the additional damage to the efferent fibers of the V.o.p running lateralwards (Fig. 75d) and to the dentato-thalamic fibers (Fig. 77a) contributed to the good result against tremor.

A favorable influence on the tremor (30–70%) occurred in the remaining eight cases. In case 7_r the tremor was improved by 70% while only 40% of the V.o.p had been destroyed, and more than 50% of the nucleus reticulatus (Fig. 64). In this case 30% of the dentato-thalamic fibers were interrupted. The degree of coagulation of the V.o.p alone or together with the interruption of the dentato-thalamic fibers corresponds relatively well to the degree of improvement of the tremor even in cases 15, 6, 4, and 2_r. In case 5_ℓ the V.o.p had not been damaged, yet the tremor was improved by 30%. This can be explained partly by damage to the dentato-thalamic fibers in the zona incerta. The afferent fiber degeneration is evident from the glia proliferation almost without nerve cell degeneration in the V.o.p (Fig. 123b). Additionally, the V.o.a, its base, and the internal capsule lateral to the V.o.a were damaged 30–40%. As seen from the next case, a 75% inactivation of the internal capsule close behind the genu at the frontal level of the L.po cannot be regarded as the cause of a slightly improved tremor.

In case 12_ℓ the tremor was only 50% improved and the improvement did not last, even though the V.o.p and the adjacent nucleus reticulatus were completely destroyed; there was slight destruction on the lateral side of the internal capsule as well as an additional complete interruption of the dentato-thalamic fibers and large parts of the caudal zona incerta (Fig. 75d). The discrepancy between the only temporary improvement of the tremor and the fact that the V.o.p and its afferent fibers were completely destroyed can be explained only by assuming that some structures which inhibit tremor were inactivated. In case 12_ℓ a hemorrhage destroyed large parts of the rostral thalamus; the V.o.a, the L.po, and the V.im were destroyed completely (see Figs. 78a, 127a, c). These structures may contain a factor that is normally instrumental in inhibiting the tremor. The rostral thalamic pole (L.po) receives efferent fibers indirectly from the caudatum through the external part of the pallidum. These fibers cross the internal capsule closely behind the genu. They could transmit impulses inhibiting the tremor, particularly since damage to the substance of the caudatum and putamen in the shape of an *état précriblé* is the cause of the essential tremor (HASSLER, 1939). In case 5_ℓ the internal capsule was also destroyed lateral to the nucleus lateropolaris (Fig. 56b) and in this way reduced the expected improvement of the tremor, as was discussed previously. Although in case 9_r half of the V.o.p had been inactivated, a recurrence of tremor appeared just a few days later. Whether the recurrence of tremor was due to an additional lesion in the neighborhood of the coagulated area could not be ascertained, since 3 weeks later, following a second operation performed because of the recurrence of tremor, a massive hemorrhage occurred in the basal ganglia. The hemorrhage destroyed the first coagulated lesion and part of the internal capsule lateral to the L.po. After consciousness improved, tremor of the right arm at rest recurred on the third postoperative day, although the V.o.p. and the dentato-thalamic fibers were entirely destroyed. This is interesting because the patient had a hemiparesis on the right side caused by the extension of the thalamic hemorrhage into caudal parts of the internal capsule. Since PARKINSON (1817) it was generally accepted that a hemiplegia will abolish a previously existing tremor at rest. This did not occur in this case. The large area of destruction must have involved substrates in the left diencephalon that normally have an inhibiting effect on the tremor. These substrates will be discussed in more detail below.

6.4.2. Pallidotomy. In only two (5_r, 11) of the patients of Parkinson's disease who had a pallidotomy and came to post-mortem did the degree of improvement of the tremor correspond to the destruction of the pallidum (Figs. 55a, 119). The degree of improvement was 80% and 40%, respectively. Although the pallidum internum and externum as well as the lamella pallidi interna (see Figs. 62, 64) were almost completely inactivated in three cases (7_l, 7_r, 9_r), the tremor was not improved by more than 60–70%. Although the pallidum was completely destroyed, these cases showed no abolition of the tremor. Also, in case 17, the 90% destruction of the pallidum internum (see Fig. 98) caused an improvement of the tremor of only 40%. These autopsy findings agree with our clinical observations from large series (HASSLER and RIECHERT, 1954; MUNDINGER and RIECHERT, 1961, 1963, 1966). The clinical observations indicate that inactivation of the pallidum has much less effect on the tremor than inactivation of the V.o.p and its afferent fibers. In one case (7_l), even the complete destruction of the ansa lenticularis in addition to that of the pallidum (Fig. 62) did not abolish the tremor. The 40–60% inactivation of the pallidum internum and externum in case 2_l did not improve the tremor at all (Fig. 43a).

6.4.3. Influence of Different Parts of the Internal Capsule on the Effect on the Tremor. The dentato-thalamic fibers and their terminal nucleus, the V.o.p, form part of the afferent chain of neurons belonging to the primary motor area in the anterior central gyrus. The findings at autopsy show that the therapeutic influence on the tremor is correlated to inactivation of this afferent chain of neurons. It is, therefore, necessary to discuss how the altered conditions of excitation of the motor cortex following a coagulation of the V.o.p influence the peripheral motor apparatus. Of the fiber connections one must consider first the fibers that descend from the motor cortex and pass through the internal capsule. It is, therefore, possible that the favorable therapeutic influence of the coagulation of the V.o.p is impaired by the further interruption of these efferent fibers of the motor cortex. For this reason the change of the therapeutic effect on the tremor by lesions in different segments of the internal capsule will be discussed in terms of our material (Fig. 134).

In cases 3, 5_r, and 6, damage of 50–75% of the internal capsule at the level of the anterior thalamic pole or at the level of the V.o.a, did not have an unfavorable influence on the improvement of the tremor. It must be noted that in cases 7_l and 9_r the complete inactivation of the pallidum (see Figs. 62, 68a) and its afferent fibers had led to an improvement of the tremor of only 60–70%. Here 50–70% of the internal capsule had been interrupted at the level of the V.o.a and the anterior thalamic pole. But it is not likely that this alone explains the lack of correspondence with the effect on the tremor, since in both cases large parts of the putamen or the caudatum, respectively (75%), were also destroyed, and it is known that damage to these areas causes essential tremor or in Parkinson patients increased tremor amplitudes. Only in case 12_l did the additional lesion of the internal capsule (see Fig. 75a, d) impair the effect on the tremor of complete inactivation of the V.o.p and of the dentato-thalamic fibers. In this case 15% of the internal capsule had been destroyed lateral to the intermediate ventral nucleus (V.im.e).

We do not wish to attach too much importance to this single case, although generally the effect against the tremor requires the preservation of the efferent pathways of the motor cortex to the peripheral motor apparatus, which pass through this part of the internal capsule (Fig. 134). In previous operations by BUCY and CASE (1937, 1939) and PUTMAN (1940), the inactivation of efferent pathways from the motor cortex by extirpation of the motor cortex or pyramidotomy had a favorable effect on the tremor. The discrepancy between complete inactivation of the V.o.p and its efferent fibers and the only transient improvement of 50% of the tremor can be related to the complete destruction of L.po without, in this case, the internal capsule lateral to the L.po being involved. The L.po is a relay nucleus of the caudatothalamocortical system, which normally inhibits tremor. If the system is interrupted in the L.po, the mechanism inhibiting the tremor is reduced and the tremor is not abolished as quickly as expected from the inactivation of the V.o.p and its afferent fibers.

Further interruptions of fibers in the different part of the cerebral peduncle (Figs. 56a, 77a) had had no unfavorable influence on abolishment of the tremor in four cases (4, 5_l, 6, 12_r). This is to be expected, because prefrontal systems and only those motor systems that represent head and face areas, descend in the medial peduncle.

6.4.4. Significance of the Dentato-thalamocortical System for Mechanism of Tremor. The tremor at rest is characterized by alternating contractions in antagonistic muscles at a rate of 4–5.5 per sec; KLEIST (1908, 1918) called this tremor a tremor of the antagonists (*Antagonisten-Tremor*). The rhythm of the tremor is independent of an interruption of the peripheral reflex arc. In the classical case of POLLOCK and DAVIS (1930b), the trembling arm was de-afferentiated by severing the posterior roots. This did not abolish the tremor but rather increased and coarsened it. Blocking the afferent fibers to muscle spindles by injecting novocaine increases rather than stops the tremor in the injected muscle (WALSHE, 1924). Interruption of the proprioceptive reflex arc does not, therefore, cause inhibition but rather a coarsening of the tremor in the affected muscles. The intact arc of the proprioceptive reflex has an inhibiting effect on the tremor by maintaining the constant length of the muscle. Therefore, the tremor rhythm originates in the gray matter of the anterior horn, where, according to JUNG (1941), the pacemaker of the tremor rhythm is located, causing tremors of different frequencies in the upper and lower limbs of the same side of the body.

The tremor is not controlled from the periphery. The rhythm of the tremor is rather influenced by central systems (Fig. 124a–c). Anatomical investigations have shown that a tremor at rest always appears when there is a disappearance of nerve cells in the substantia nigra (tremor of Parkinsonism) or sieve-like lesions in the putamen and caudatum (essential tremor). Lesions of the nucleus dentatus or the brachium conjunctivum cause a different kind of tremor, namely the postural tremor, which is often associated with myoclonus. Neuronal losses in these strionigral systems can alter the processes of excitation in the anterior horn of the spinal cord in such a way that impulses for a tremor are sent out. From a neurophysiologic point of view, one is dealing here with the synchronization of excitations in the motor neurons which, under normal conditions during the waking state, are continually sent out in a desynchronized pattern. Weak impulses of low frequency are sent out from the normal motor neurons all the time when movement is neither intended nor visible. In tremor, these impulses are synchronized and therefore cause a visible shaking. When the strionigral systems are intact (Fig. 124a), they normally exercise a desynchronizing effect on the motor neurons. If in diseases affecting the strionigral systems the desynchronizing influence on the motor neurons is absent (Fig. 124b), these motor neurons are solely under the influence of other cerebral centers that favor synchronization. However, this mechanism increasing tremor is only active in the waking state. The resting tremor disappears after a short time when patients suffering from Parkinson's disease go to sleep. One must therefore conclude that the neuronal systems favoring tremor stop their activity during sleep.

The observation by PARKINSON in 1817 (republished in 1922) that after hemiplegia the tremor disappears on the hemiplegic side was the first hint as to which systems are involved. In 1937 BUCY (1938; BUCY and CASE, 1939) resected the motor cortex in order to influence the tremor and achieved a good and permanent effect on the tremor of the contralateral arm. This observation led PUTMAN in 1940 to sever the corticospinal system in the lateral pyramidal (corticospinal) tract in order to diminish or abolish the tremor on the same side. A "pyramidotomy" affects the tremor completely but only until active movement of the affected limbs returns. Normally excitation of the pyramidal tract increases the tremor at rest because it synchronizes the activity of the motor neurons.

The tremor as well as rigidity can be therapeutically influenced not only through the efferent corticospinal system but also through the afferent fibers coming from the thalamus. If the nucleus ventro-oralis posterior thalami (V.o.p), which projects into the primary motor cortical area, is stimulated with low frequencies in patients suffering from Parkinson's disease, the tremor rhythm on the opposite side is influenced according to these frequencies (Fig. 125, where because of the lack in these cases of an autoptic differentiation between V.o.a or V.o.p the stimulated thalamic structure was called V.o.a). Stimulations at rates lower than that of the tremor rhythm slow it down, while as a rule, frequencies above 6 cps accelerate it. Single stimuli interrupt the tremor rhythm for a short period. During single stimuli in the V.o.p, interference between the spontaneous rhythm of the tremor and the rhythm of stimulation is sometimes observed. This interference indicates that the tremor rhythm can be influenced by stimulation of the dentato-thalamic fibers running to the V.o.p or to a lesser degree also by the pallido-thalamic fibers running to the V.o.a. The nature of this influence going to the main area in which the pyramidal tract originates depends on the frequency of stimulation and is observed clinically as acceleration, retardation, or temporary interruption (Fig. 125a–i, 126).

Low frequency stimulation of V.o.a or V.o.p (Fig. 125a–c) have a restricted influence only on the tremor rhythm and amplitudes on the contra-

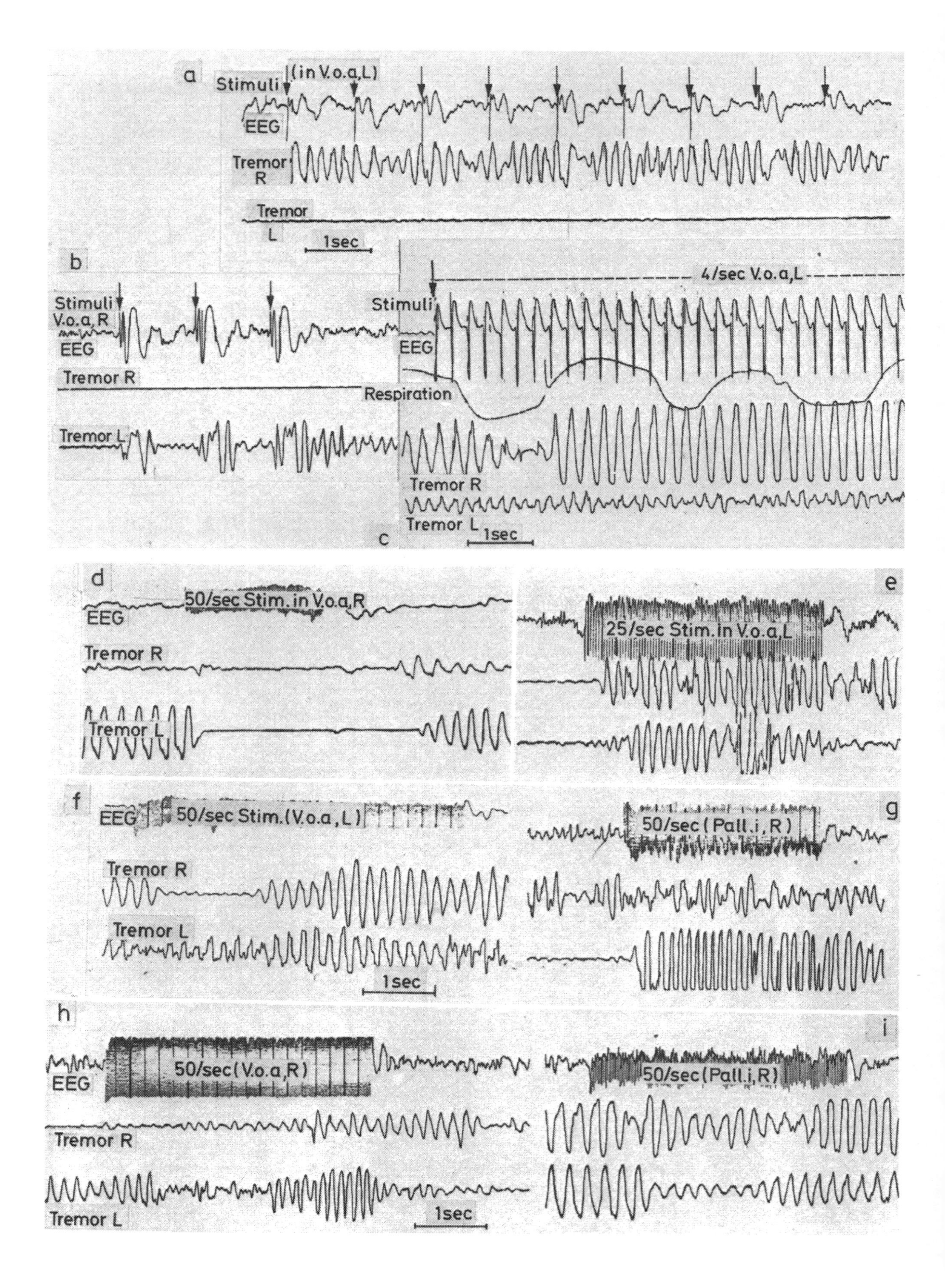

a
Stimuli
(in V.o.a, L)
EEG
Tremor R
Tremor L
1sec
b
Stimuli V.o.a, R
EEG
Tremor R
Tremor L
4/sec V.o.a, L
Stimuli
EEG
Respiration
Tremor R
Tremor L
c
1sec
d
EEG
50/sec Stim. in V.o.a, R
Tremor R
Tremor L
e
25/sec Stim. in V.o.a, L
f
EEG
50/sec Stim. (V.o.a, L)
Tremor R
Tremor L
1sec
g
50/sec (Pall.i, R)
h
EEG
50/sec (V.o.a, R)
Tremor R
Tremor L
1sec
i
50/sec (Pall.i, R)

lateral side. 50 cps (and less at 25 cps) stimulation of V.o.a or V.o.p, (see Fig. 125d, f, h) lead first to an interruption of the contralateral tremor and after 1–3 sec to increased amplitudes and a very regular tremor. On the ipsilateral side the stimulation produces, after a lesser inhibition, an increased regularity and a higher amplitude of the tremor. 50 cps Pallidum stimulation (Fig. 125g, i) has a lesser effect on ipsilateral tremor and increases the regularity of contralateral tremor. – With thalamic (V.o.a or V.o.p) stimulation there was a higher incidence of acceleration, reduced amplitudes and blockade of tremor rhythm, with pallidar stimulation a higher percentage of slowing tremor was observed (Fig. 126).

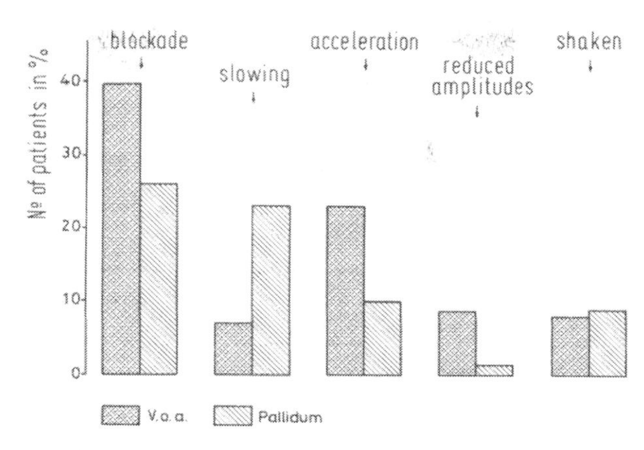

Fig. 126. Frequency of alterations of tremor rhythm in Parkinson patients with stimulation of V.o.a or pallidum, respectively. Since the stimulation focus was not checked by post mortem examination V.o.a stands for V.o.a or V.o.p

The post-mortem material contains essential evidence regarding the relationship of tremor to specific inactivated structures. In serial examinations of six operations that were followed by complete abolishment of tremor at rest, we found coagulations either in the V.o.p (Fig. 122a) or in the dentato-thalamic fibers, which end in the V.o.p. The volume of inactivated tissue in these thalamic structures corresponded largely to the degree of influence on the tremor. Of the ten cases in which the tremor was improved by 30–60%, in only one was improvement of tremor greater than could have been expected from the extent of inactivation of the V.o.p (Fig. 122b) or of the dentato-thalamic fibers. A special feature of one case (12_r) is that, in addition, 15% of the internal capsule lateral to the V.o.p was damaged. As a result the additional lesion of the pallidothalamic fibers running through the internal capsule improved the effect on the tremor (see Figs. 75a, 77a).

Inadequate improvement of the tremor despite inactivation of the V.o.p or its afferent fibers occurred in three cases (12_ℓ, 5_ℓ, 9_r). These cases also had, however, large lesions affecting the internal capsule next to the rostral thalamic pole (L.po).

◀ Fig. 125a–i. EEG recording and tremor registration during intraoperative stimulation of either V.o.a or pallidum internum (*Pall.i*).
(a) Single stimulus in left V.o.a produces alteration of tremor rhythm on right index finger so that this tremor rhythm becomes irregular (see mainly 4th, 5th, and 8th stimulus). Ipsilaterally no tremor was produced by this stimulus. At same time, three-phasic wave always appears in EEG.
(b) Single stimuli in right V.o.a elicit bursts of tremor in left index (see tremor registration *L* below) that outlast single stimuli in reduced manner.
(c) 4 cps stimulation in left V.o.a reduces tremor of contralateral index finger during 1st 3 sec, after which tremor exhibits increased amplitudes and regular 4 cps rhythm. On ipsilateral index finger (see tremor *L*), no alteration of tremor rhythm appears. In EEG, each single stimulus produces spike-wave complex.
(d) 50 cps stimulation in right V.o.a leads to cessation of tremor of left index finger after one sec, whereas tremor in right index finger is only reduced and irregular. Both effects outlast termination of stimulation for 2–3 sec. Afterward regular spontaneous tremor reappears on the left side, less on the right side.
(e) 25 cps stimulation in inner part of left V.o.a elicits tremor of contralateral index (2nd line) finger rather than of ipsilateral (3rd line). Contralateral tremor outlasts termination of stimulation.
(f) 50 cps stimulation in left V.o.a almost suppresses tremor of contralateral (right) index in 1st 2 sec. Continued stimulation enhances tremor to even higher amplitudes than before stimulation. Tremor of ipsilateral index finger becomes irregular during beginning of stimulation and is also followed by tremor of enhanced amplitude after 2 sec (see tremor *L* registration).
(g) 50 cps stimulation in right pallidum internum has no influence on tremor of right index finger whereas in left index finger they elicit regular 6 cps tremor with large amplitude and without latency.
(h) 50 cps stimulation in right V.o.a reduces tremor of left index finger after $^1/_2$ sec for duration of 2 sec. Afterward, tremor *L* with larger amplitude reappears without higher frequency and outlasts termination of stimulation, tremor *R* (ipsilateral index) appears in the last 1 sec of stimulation and outlasts.
(i) 50 cps stimulation in right pallidum does disturb regular tremor of left index finger only slightly and also noticeably reduces ipsilateral tremor (2nd line: Tremor *R*). (From Ganglberger, 1970)

The fibers from the striatum to the L.po, which cross the internal capsule, transmit, apparently, the impulses from the striatum that normally inhibit the tremor (see p. 199). Inactivation of these fibers (Fig. 122b) impairs the improvement of the tremor achieved by inactivating the V.o.p.

6.5. Correlation of Akinesia, Neglect, and Festination

6.5.1. Discussion of the Concept of Akinesia and Hypokinetic Signs. The assessment of akinetic disorders in the Parkinson syndrome and their alteration through stereotaxic interventions are particularly difficult. One reason is the conceptual difficulty of determining differences; a lack of movement can be due not only to rigidity of muscles but also to disappearance of automatisms and to disorders of initiative without even a functional disorder of the extrapyramidal system. If mental spontaneity and initiative are diminished, movements mediated through the extrapyramidal system will no longer be initiated. Extrapyramidal akinesia can be simulated by lack of initiative, i.e., of psychologically determined movements. In the Parkinson syndrome, akinesia involves different levels of complexity, from motor through psychomotor, up to the level of mental activities. It is possible, for instance, that the synergic movements of the arm are as disordered as the reactive movements, e.g., automatic changes of posture or movements to defend against or to avoid painful stimuli. It is also possible that in Parkinsonism the primary impulse for spontaneous mobility has disappeared as part of the bradyphrenia. No tests are known for measuring certain forms of akinesia. The difficulty of analyzing akinetic disorders must be kept in mind during the following discussion.

The multiple causes of an akinetic symptom in a patient suffering from Parkinson's disease are difficult to differentiate because rigidity also causes a diminution of automatic movements and changes of the spontaneous and reactive movements. There is no doubt that an independent pure akinesia exists without rigidity, which Bostroem (1922) called stiffness without rigidity (*rigorfreie Starre*), but it is extremely difficult in an individual case to determine the respective contributions made to an akinetic symptom by loss of synergic movement, by rigidity, or by pure akinesia. The *pulsions* form a particular group of deficiencies of movement characteristic of the Parkinson syndrome. At first glance, they appear to be due to an excess of movement, because going backward, forward, or sideways cannot be stopped deliberately. In fact, they are due to the absence of a normal inhibiton of automatic innervation (Fig. 137a, b). Going forward, sideways, or backward, which sufferers of Parkinson's disease also have difficulty in initiating, is an automatic process in normal people that continues without voluntary control once it has been deliberately set in motion. We believe that the different forms of pulsions result because patients suffering from Parkinson's syndrome are no longer able to brake intentionally the movement automatisms once they have been set in motion.

When looking for comparable motor phenomena in the Parkinson syndrome one finds in the excessively rapid speech a further example of the inability to brake automatic movements once they have begun. The speech in Parkinsonism is characterized, apart from the aphonia, by the monotonousness, and the inadequate articulation, by the fact that speaking gradually becomes increasingly rapid while the loudness of the sound diminishes. It is a known phenomenon that normal people speak more rapidly when the loudness diminishes, particularly when they whisper, and speak more slowly when loudness increases. One could talk of a disorder of the "gait of speaking" in connection with the loudness of the sound. The excessively rapid speech in Parkinsonism can then be interpreted by assuming that the rapid pace of the whisper, once it has started normally, can no longer be deliberately stopped by the patient. In addition, patients suffering from Parkinsonism have a reduced amplitude of respiration and disordered inspiration automatisms. They attempt to accommodate as many words as possible during the expiration. These changes of motility, which appear to be excessive phenomena, are in fact true signs of a loss of motility, because inhibitory processes are integral parts of every aimed motor action.

The handwriting of Parkinson patients reveals three types of disturbances: (1) disfigurement caused by shaking, so that all the letters are jagged (Fig. 136a); (2) distortion due to rapid exaggerated shaking, a result of the absence of normal inhibition of rapid *physiologic tremor*, which has a frequency of 16–20 cps (Fig. 136b); (3) progressive reduction in the size of letters during writing, so that while at first the letters are almost normal or only somewhat reduced in size, they become increasingly smaller until they can no longer be

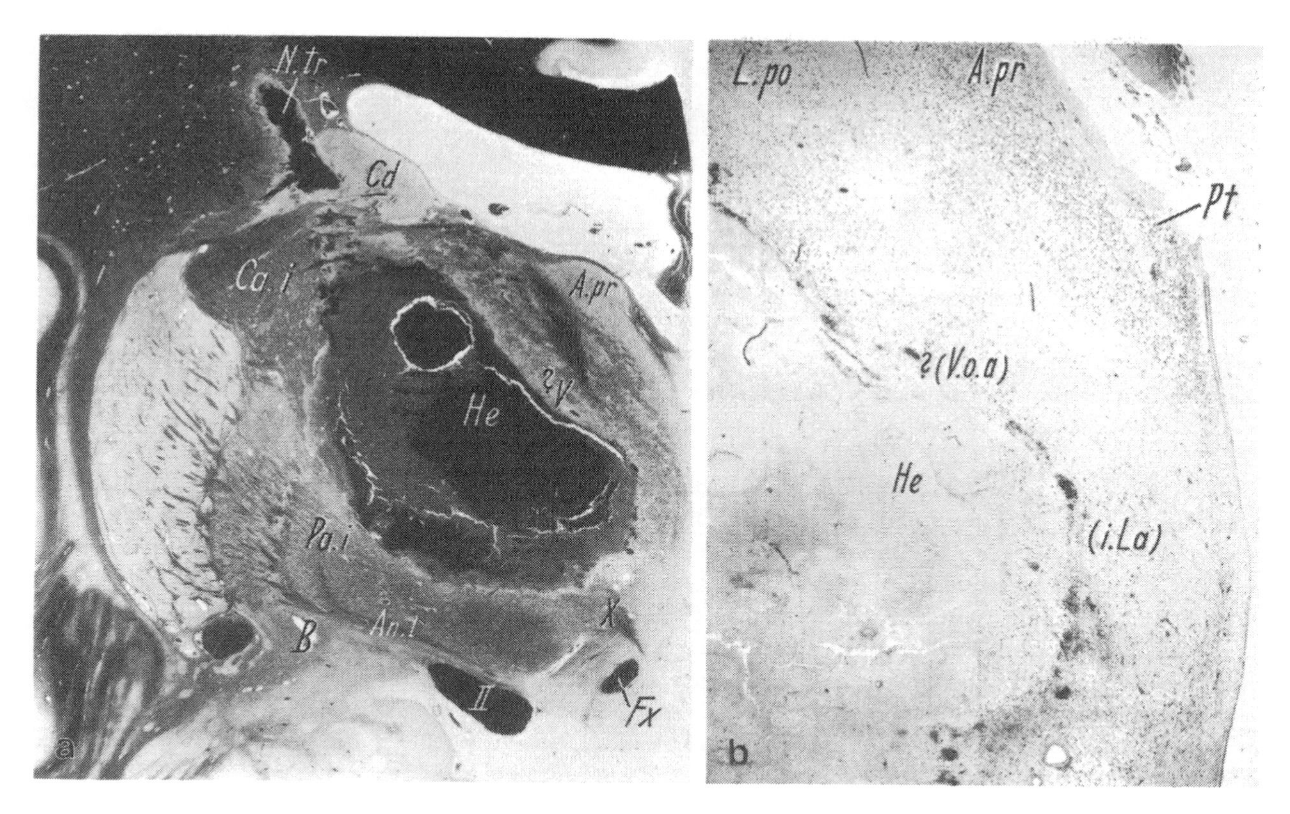

Fig. 127. (a) 15-day-old coagulation focus in left thalamus of case No. 12 is expanded by large hemorrhage (*He*). Focus occupies ventrolateral half of thalamus and destroys middle of internal capsule. V.o.a and Z.o nuclei of thalamus as well as larger parts of dorsal nuclei are destroyed by hematoma. Electrode track was dilated by pressure of hematoma during its passage through caudate nucleus, and white matter lateral to caudate nucleus is filled with blood coagula (*N.tr*). Dorsal delimitation is sharply defined, with no transition zone in fiber picture. 0.5-mm wide demarcation zone with demyelination surrounds hematoma ventrolaterally, toward inner section of pallidum.

(b) Adjacent section in Nissl stain demonstrates preserved cell content in anterior principal nucleus (*A.pr*), parataenialis (*Pt*) and anteromedial nucleus. Lamella medialis (*i.La*), situated medioventral to mamillothalamic tract, has almost completely lost its nerve cells 15 days after coagulation. At the same time, nerve cell loss is almost complete in first cut of commissural nucleus, both due to interruption of efferent fibers.

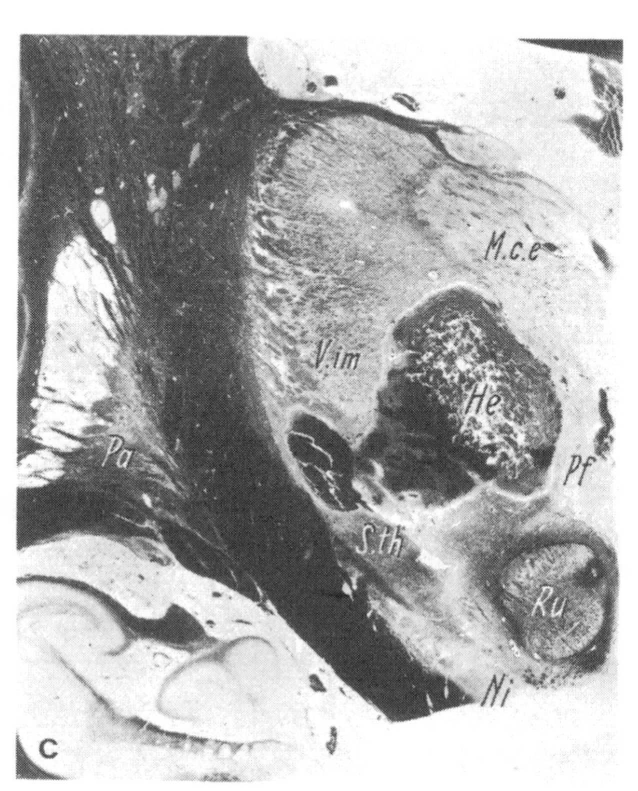

Fig. 127. (c) Cross section through thalamus at level of nucleus ventrointermedius (*V.im*). Focus in ventrointermedius region is divided into 3 parts. Medially, 12-mm long coagulation focus invaded by hemorrhage (*He*) is located in region of Ventrointermedius internus and *centre médian* nuclei. Laterobasally lies 7-mm long focus that has infarcted thalamic tissue with blood but has not destroyed it completely. Almond-shaped, 6-mm long focus filled with blood coagula occupies reticulate nucleus of thalamus immediately above subthalamic nucleus (*S.th*) and has more completely destroyed thalamic structures. Although combination of foci spares dorsal part of lamella medialis, this nucleus has lost almost all nerve cells, as seen in Nissl picture (Fig. 127b), probably due to interruption of efferent fibers of these nerve cells below lamella medialis

distinguished from a wavy line and are no longer recognizable. All three disturbances occur during the development of a Parkinson syndrome and do not depend on the side of the body affected. Deformation of the written letters due to shaking is the most common disturbance; it results in all the lines becoming jagged (Fig. 135a). This deformation is more clearly shown in Figure 136a where the obvious diminution in the size of the letters is not present. Both examples are taken from handwriting of Parkinson patients before stereotactic operations. After a left-sided coagulation of the pallido-thalamic neurons or their terminal nucleus V.o.a, of the dentato-thalamic neurons or their terminal nucleus V.o.p, the writing is less jagged because the tremor is reduced (Fig. 135b; also after ipsilateral pallidum coagulation in Fig. 136b). The letters are less disfigured because of shaking, but are more distorted because of a lack of coordination, which results in the letters and writing pressure being irregular (Fig. 135b). In addition the writing becomes smaller (Figs. 135b, 136b, c), even though tremor and rigidity have been improved. The third type of disturbance, micrographia in the strict sense, can worsen and increase even after therapeutic improvement of tremor and rigidity. Figure 135c contains an example in which, after an additional operation in the left hemisphere with almost complete destruction of the efferent pallidar neurons and some parts of the V.o.p, the micrographic alteration was abolished although the jaggedness of the straight lines and the deterioration of writing is increased. The micrographia worsens only as the Parkinsonian process progresses (Fig. 136c). The disinhibition of the rapid rhythm of *physiologic* tremor during writing is clearly demonstrated in Figure 136d, so that some type of involuntary increase in the speed of hand movements occurs in situations involving emotional reactions.

In view of the difficulties involved in quantifying the hypokinetic signs and in conceptually dividing them into different groups, pure psychomotor akinesia in the sense of lack of initiative and slowing down of thinking, i.e., bradyphrenia, is not included in the following discussion. Bradyphrenia involves not only the slowing of all mental events and lack of initiative but also difficulty in making voluntary decisions and, particularly, in translating the intention to make movements into action. This is one of the most agonizing symptoms of a severe Parkinson syndrome. Analysis of the individual phenomena that constitute bradyphrenia has not been carried out in the clinical reports of the majority of these cases. We therefore avoid discussing them.

In contrast to what can be done regarding rigidity and tremor, only limited statements can be made regarding the akinesia and its treatment by stereotaxic operations in the Parkinson syndrome. For theoretical reasons alone, one could hardly expect an improvement of akinesia by a stereotaxic operation. This was confirmed in the first few series of operations. We therefore did not operate later on cases with predominant or virtually exclusive akinesia. In many clinical disorders, it is difficult or, for practical purposes, impossible to estimate the respective contributions of rigidity and true akinesia to the motor loss forming a lack of movement. What is needed is a concept for defining these motor defects in the Parkinson syndrome without separating their different pathophysiologic components from pure akinesia. For this purpose, we should like to introduce the concept of "hypokinetic signs".

The following disorders comprise the hypokinetic signs in the Parkinson syndrome and have been considered in a schema for examination (Riechert, 1973): disorders of posture, general reduction of movement, difficulty in turning in bed, difficulty in dressing, need for help on the lavatory, inhibition of starting, akathisia, walking with small steps, propulsion, retropulsion, lateropulsion, loss of synergic movements, lack of expressive movements, dysphagia, aphonia, monotony of speech, reduction of breathing movements, dysdiadochokinesia (shaving, brushing and combing hair, and buttoning), disorders of writing, and micrographia.

6.5.2. Correlations of Hypokinetic Signs. The following post-mortem material is from ten patients in whom the hypokinetic signs had to be assessed after the operation. The following hypokinetic signs were assessed in these cases (Table 8; see p. 209): hypomimia, aphonia, monotony of speech, reduction of breathing movements, dysdiadochokinesia, micrographia, akinesia in everyday activities (e.g. dressing, buttoning, shaving, and using cutlery), inhibition in starting, akathisia, disorder of gait usual in Parkinsonism, loss of synergic movements, bent posture with pains in the back and joints, and propulsion, retropulsion, and lateropulsion.

Hypomimia was not improved in any of them at first, regardless of whether the operation was directed at the pallidum or at the V.o.p and V.o.a.

In contrast, the other hypokinetic signs were favorably influenced. Inactivation of the pallidum internum and particularly at the base of the V.o.a produces in the first week a contralateral reduction of purely expressive facial movements irrespective of what disease is present. As a rule the reduction of expressive facial movements disappears within a few weeks, but even when these movements have been reduced on the contralateral side immediately after the operation, they can, after a few months, become considerably more improved on both sides than they were before the operation. The reason for this is the double representation of expressive facial movements in the substantia nigra and in the pallidothalamic system. The irreversible loss of the cell groups of posterior nigra has two effects: one descending in form of diminished supranuclear inducing of expressive facial movements and one ascending in form of diminished inhibition of the pallido-thalamo-cortical pathways for facial movements. These movements are caused by the excessive tonic impulses characteristic in Parkinson patients. The reduction of expressive movements due to stereotaxic lesion of this pallido-thalamo-cortical system itself can be very quickly compensated for, so that only the releasing effect of the stereotaxic lesion remains, leading to an improvement of facial expression.

Neither the *aphonia* nor the monotony of speech was improved in these cases through the operation. Because this agrees with the general clinical experience, our patients, as well as those of the majority of surgical teams, are informed of the uncertainty of improving speech disorders.

Dysdiadochokinesia is only partly due to akinesia. In the majority of cases it could be favorably influenced and was occasionally abolished. This result agrees with the observations during the operation that a considerable improvement of rapid alternating movements was achieved after the coagulation due to the relief of rigidity.

Although rigidity was usually abolished, *micrographia* was only slightly improved in two cases (Fig. 135c) and remained unchanged in the others (Figs. 135a, b; 136). This inadequate therapeutic effect led us to conclude that micrographia is due largely to akinesia.

In half of the cases the *activities of daily life* were improved. This agrees with results observed in cases that did not come to autopsy. In later clinical series, this percentage unequivocally increased, in correlation to the size reduction of stereotaxic lesion. In some patients the difficulty encountered in dressing and using cutlery was abolished. In all these cases, however, rigidity was 70 to 100% improved, which probably constitutes an essential part of the positive therapeutic result.

Inhibition of starting (HASSLER, 1953b), which is particularly characteristic for the Parkinson syndrome, was abolished in 2 out of 5 cases, although only on the contralateral side. This inhibition is an essential part of the akathisia when getting up from the sitting position. Improvement of rigidity is also a decisive factor in the abolition of inhibition of starting (Fig. 137a, b). Abolition of inhibition of starting was achieved in both cases by inactivation of the oral ventral nuclei. Two of the three negative cases were pallidotomies and one was a thalamotomy.

The *disorder of gait* with the typical small steps and loss of synergic movement was improved in half the cases and abolished in one. Return of the synergic movements parallels improvement of disorder of gait. In the case in which the disorder of gait was abolished, the synergic movements returned completely. Loss of synergic movements can be due to akinesia as well as to rigidity. In many cases in which the synergic movements returned after the V.o had been inactivated, we had the impression that abolition of rigidity was the important factor in the improvement of these symptoms (Fig. 137a, b). The bending forward of the trunk with increased kyphosis was also favorably influenced or almost abolished in half the cases. When this disorder of posture is accompanied by pains, the pains usually disappear even earlier than the abnormal posture. However, there are cases in whom the pains disappear even though the abnormal posture persists.

It is remarkable that the complex symptom of pulsions (retropulsions, propulsions, lateropulsions), even when severe, was often favorably influenced and often abolished. In this series propulsions occurred in the majority of cases. They were abolished by a lesion in the V.o.a (Fig. 34b) and in one case by a lesion in the pallidum (see Fig. 119). In one patient, however, inactivation of the pallidum internum did not abolish the pulsion while a previous coagulation of the contralateral V.o.p and V.o.a had the desired effect. This inactivation of the V.o at the same time abolished retropulsion, propulsion, and lateropulsion. The pulsions were the hypokinetic signs that were favorably influenced most often. In two cases in which the pulsions were not abolished by the coagulation, the effect of the operation on rigidity was 40 and

Table 8. Change of akinetic and hypokinetic symptoms in relation to inactivated structures

Case No.	Target	Akinesia	Postural disorders	Expressive movements	Aphonia, monotony of speech	Reduction of breathing movements	Gait disturbance, walking with small steps	Dysdiadochokinesia, loss of associated movements
1	V.o. lt.		trunk improved, pains abolished ([a]/∅)	amimia			walking with small steps ameliorated	
2 1. Op	Pa lt.	improved ([a])	pains abolished	hypomimia ([a])				rt. improved ([a])
2. Op	V.o. lt. Pa.i lt.			hypomimia				lt. unchanged ([a])
3. Op	V.o. rt.			hypomimia				
3	V.o. rt.		unchanged ([c])				unchanged ([c])	
4	V.o. lt.		unchanged ([c])		unchanged ([b])		unchanged ([a])	unchanged ([b])
5 1. Op	Pa.i rt.	[a]	improved ([a]/∅)				unimproved	improved ([b]/[a])
2. Op	V.o. lt.							
7 1. Op	Pa/V.o.a lt.		unimproved ([b])		unimproved ([b])		improved ([b]/[a])	unimproved ([b])
2. Op	Pa.i lt.						unimproved ([b])	
3. Op	Pa.i rt.							
12	V.o. rt.	improved ([a])	posture improved, pains abolished	amimia unimproved ([a])	unimproved ([a])		improved ([a]/∅)	
13	V.o. lt.		unimproved ([a])	unimproved ([a])	worsened ([a]/[b])		unimproved ([a])	rt. abolished ([b]/∅)
16	V.o. rt.	improved ([a])	unchanged ([a])				unchanged ([b])	abolished ([a]/∅)
17	Pa.i lt.	unimproved ([a])	unchanged ([b])	unchanged ([b])	unchanged ([b])		unchanged ([a])	unchanged ([a])

[a] slight symptom
[b] marked symptom
[c] strong symptom
/∅ abolished post operatively

100%, respectively. From the foregoing, it can at least be concluded that the therapeutic influence on the pulsions is not based solely on the abolition of rigidity.

Summarizing the changes in the hypokinetic signs in our cases that came to autopsy, we found that out of ten cases, there was good improvement in two, slight improvement in four, and the symptoms continued unchanged in four. The two best results and three of the four slight improvements were achieved by inactivating the V.o.a (sparing the V.o.p). Of the four cases with hypokinetic signs that were not improved, one had a thalamic coagulation and three, lesions of the pallidum.

Table 8 (continued)

Inhibition in starting	Pulsions			Disorders of writing and micrography	Hypokinesia in everyday activity: dressing, buttoning, shaving, using, cutlery	Postoperative assessment	Case No.
	prop.	retrop.	laterop.				
abolished ([a]/∅)	abolished ([b]/∅)				using cutlery not considerably improved ([b]/[a])	considerably improved	1
				rt. unimproved ([a]) unchanged ([b])	 dressing unchanged ([b])	unchanged unchanged unchanged	2 1. Op 2. Op 3. Op
				unchanged ([c])	dressing unchanged ([a])	short postop. course	3
					using cutlery, dressing unchanged	first postop. day slight improvement of akinisia	4
	abolished ([c]/∅)				dressing, using cutlery improved ([b]/[a])	direct improvement of akinesia and hypokinetic symptoms	5 1. Op 2. Op
	improved ([b]/[a]) unimproved ([a])	 unimproved ([a])	improved ([b]/[a]) unimproved ([a])		improved ([b]/[a]) unchanged	direct improvement of hypokinetic symptoms unchanged	7 1. Op 2. Op 3. Op
	unimproved ([b])			unimproved	dressing, using cutlery improved ([a]/∅)	marked direct improvement of hypokinetic symptoms	12
unimproved ([a])					dressing unimproved ([b]), using cutlery markedly improved ([b]/∅)	direct improvement of hypokinetic symptoms	13
abolished ([a]/∅)				slightly improved ([a])	dressing slightly improved ([a])	slight direct improvement	16
unchanged ([b])			unchanged ([a])	unchanged ([c])	unchanged ([c])	unchanged	17

6.5.3. Relevance of Extrapyramidal Systems to Akinesia and Hypokinetic Signs. Reviewing the effect of diencephalic coagulation on hypokinetic signs as checked post-mortem, shows that it is possible to achieve a therapeutic result. The extent of the influence and the number of cases are, however, much smaller than with rigidity and tremor. There is a clear difference between inactivation of the V.o and of the pallidum. Inactivating the pallidum has much less of a therapeutic effect on hypokinetic signs, in fact, as a rule no influence at all. It follows that abolition of rigidity cannot be the cause of improvement in the hypokinetic signs, because the effect on rigidity is equally good following inactivation of the pallidum and of thalamic V.o.a. Operations on the thalamus evidently can improve to a moderate degree the true akinetic components, which are independent of rigidity.

The influence on propulsion, retropulsion, and lateropulsion is remarkable. Although these movements constitute complex hypokinetic phenomena, an unilateral operation is able to make this symptom disappear in half the cases. Out of the four cases with lesions in the pallidum in which autopsies were performed, one shows improvement, although not total disappearance. In thalamic coagulations, those cases in which the part of the internal capsule adjacent to the V.o.p and the V.im.e are involved showed no result. This indicates that the additional damage of the corticospinal system militates against an improvement of propulsion, retropulsion, and lateropulsion. According to our view, given in the introduction, these symptoms are caused by the disappearance of a motor mechanism inhibiting automatisms of locomotion. This inhibiting mechanism may act in an efferent direction via a corticospinal system. In the cases for which we have autopsies, the improvement of gait is associated with a return of synergic movements and an improvement of pos-

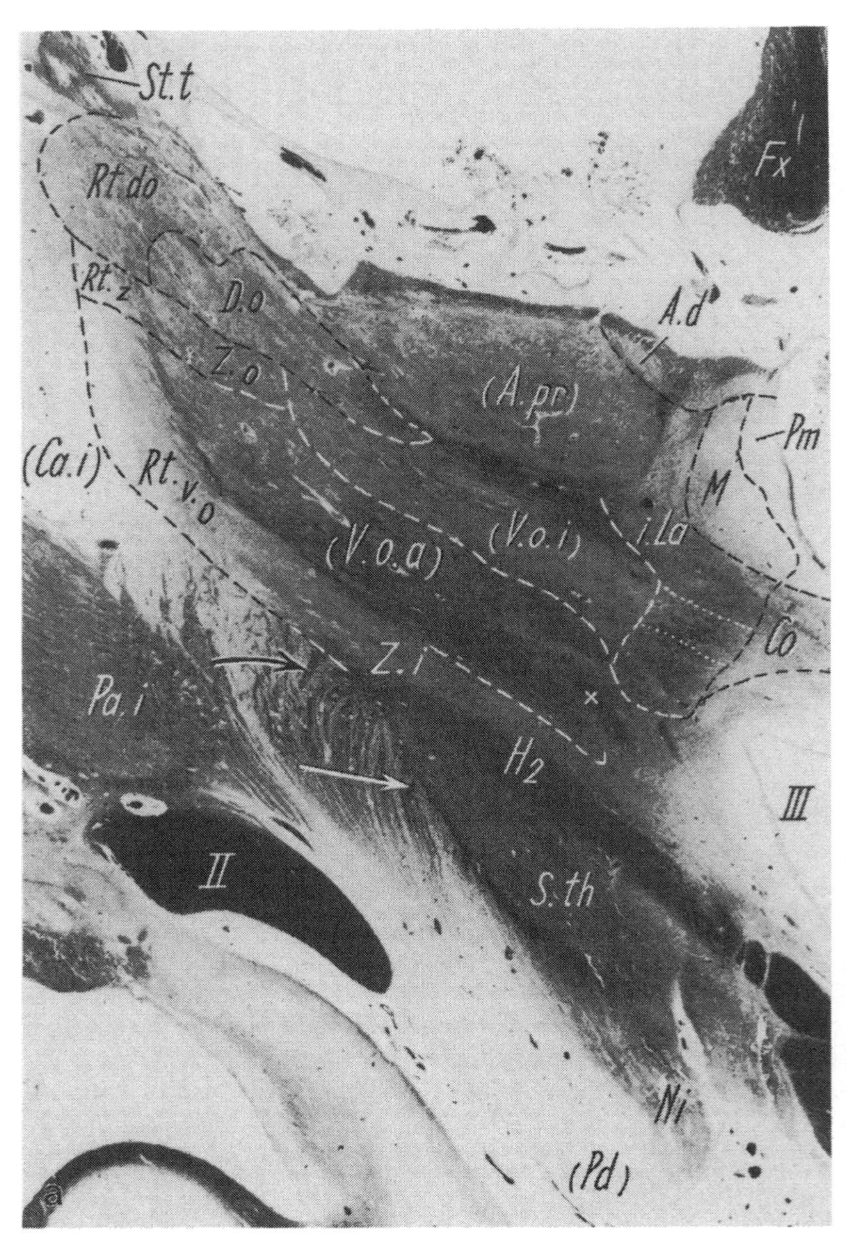

Fig. 128. (a) Cross section of left thalamus in case of complete destruction of internal capsule and cerebral peduncle (*Pd*) as well as consequent retrograde degeneration of all cortex-dependent thalamic nuclei, in this section mainly anterior principal (*A.pr*), V.o.i, and V.o.a nuclei. The degeneraction of the internal capsule fibers discover the preserved pallidothalamic fibers (*arrows*). Only gray matter of intralaminar nucleus (*i.La*) and its lower parts and commissural (*Co*) nuclei are preserved.
× = degenerated mamillothalamic bundle. Fiber stain. ×5

ture. There is a much greater chance of improving these hypokinetic signs from the thalamic target points or from their afferent fibers than there is by inactivating the pallidum.

Improvement in the activities of daily life is frequently due to a reduction of tremor and rigidity of the limbs. It is, therefore, not possible to assess quantitatively the role that pure akinesia plays in this improvement. The same applies to dysdiadochokinesia, which is a particularly multifactorial symptom. Coagulations in the thalamus are least effective against micrographia and the Parkinsonian disorder of writing (Figs. 135, 136), probably because the coordination of movements in writing is complicated and because many of those movements have been learned relatively late in life.

In this autopsied series of cases, inactivation of neither the pallidum nor thalamic V.o.a had any therapeutic influence on expressive facial movements, which is in contrast to the effects thus far discussed. In this respect one must consider that inactivation of the V.o.a and V.o.p temporarily produces a lessening of mimic innervation, even when reduction of expressive facial movements of Parkinson's disease is not already present. The effect on the facial movements caused by the oper-

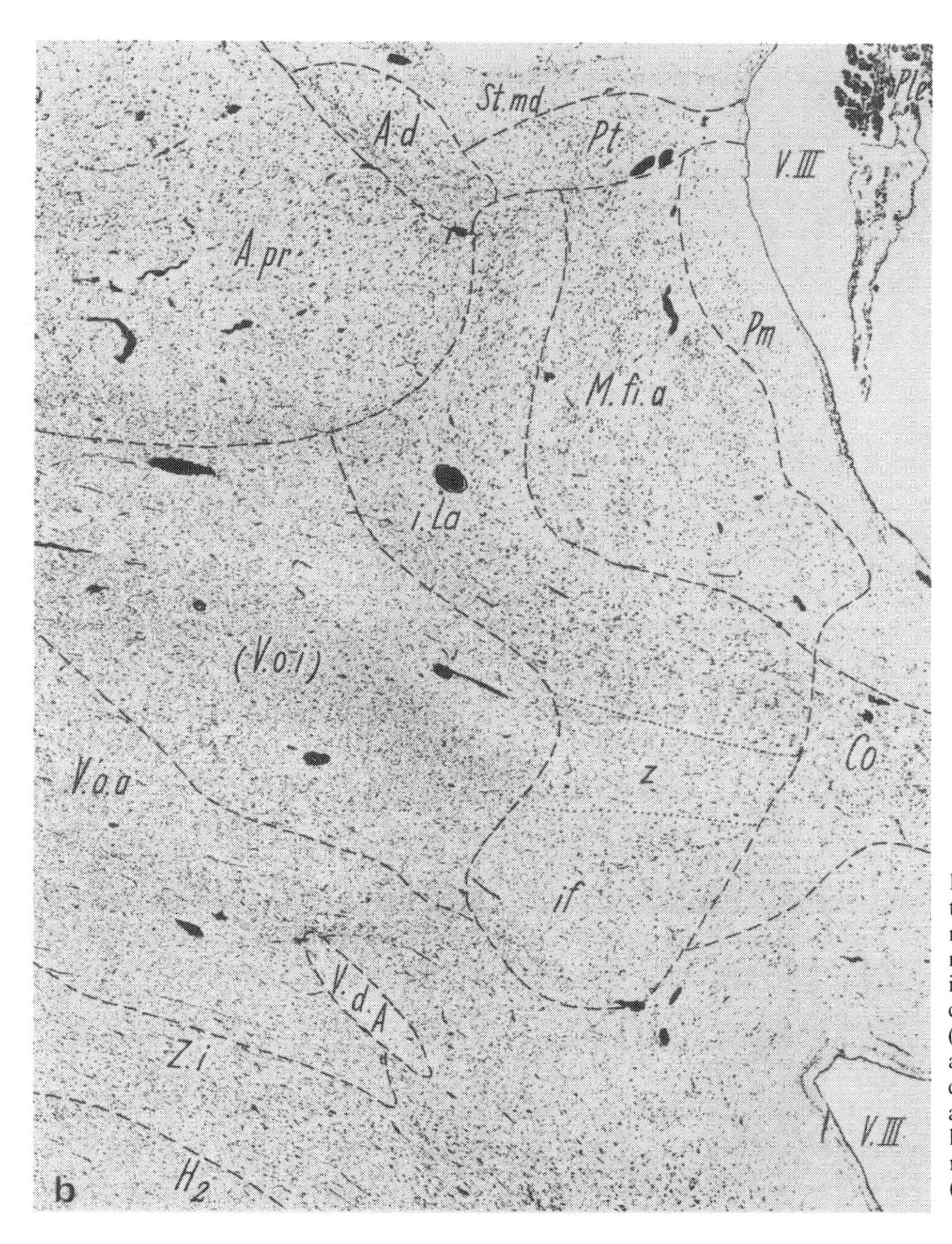

Fig. 128. (b) Adjacent section, in Nissl stain, shows marked preservation of the majority of the nerve cells in intralaminar (*i.La*) nucleus and its lower parts (*z*; *if*). Lower parts of i.La and commissural (*Co*) nuclei have lost some but not all of their nerve cells as have specific projection nuclei (*A.pr*), (*V.o.a*), (*V.o.i*). ×14 (HASSLER, 1970)

ation is manifested usually in the expressive movements; the postoperative reduction of expressive movements is due to a lesion in the pallidothalamic system, while preoperative amimia of Parkinson patients is due to nerve cell loss in the medial cell groups of the substantia nigra. So we have two different systems, the inactivation of which can result in reduction of expressive movements. Not infrequently, one sees later an increase of the expressive facial movements exceeding the preoperative condition, as observed in our large case material, published by MUNDINGER and RIECHERT (1966). In Parkinson's syndrome the voluntary movements of the facial muscles are also slowed down; their extent is reduced and, once they have appeared, they persist for a long time. Inactivation in the oral ventral nuclei, particularly in the V.o.a, and of the internal part of the pallidum is the best treatment for hyperkinesia of the facial muscles, which can be very distressing to athetotic patients. For pathophysiologic explanation see pp. 206–207

In the present material, the disorders of speech in Parkinson's disease were not improved. In one case, the disorder even became worse. To some extent the absence of an effect on the Parkinson disorder of speech and its components—monotony, festination of speech, aphonia, and dysarthria—is due to the small number of cases, since from the larger clinical series (MUNDINGER, 1963a, 1968; MUNDINGER and RIECHERT, 1966), it was shown that the Parkinsonian disorder of speech can occasionally be improved. Of the four components named previously, this is true particularly for aphonia and festination of speech, and to a lesser extent, for dysarthria (which depends on rigidity). These improvements are, however, much less constant than the improvements of rigidity and tremor of the limbs.

The larger clinical series shows in a certain percentage of cases, particularly after the second operation, an impairment of articulation and phonation. Clearly, the Parkinsonian hypokinetic signs in the bulbar area are less strongly influenced by the operation than those of limbs and trunk.

Parkinsonian akinesia as well as rigidity is usually considered to be the immediate consequence of the loss of cells in the substantia nigra, since akinesia occurs in cases of paralysis agitans in which other brain structures, with the exception of the basal nucleus and the locus caeruleus are unchanged. On the other hand, it can be stated that the most severe forms of akinesia occur with postencephalitic Parkinsonism. Here most cases, because of the acute inflammatory changes, show cell loss in other diencephalic and paramedian mesencephalic structures in addition to the Nigra. One of the most important differences between postencephalitic Parkinsonism and paralysis agitans is the particularly severe akinesia of all the muscles supplied from the medulla oblongata. This symptom is due to the loss of cells in the medial groups of the substantia nigra. In paralysis agitans, these groups are usually preserved (HASSLER, 1938). Parkinsonian akinesia was regarded as a sign of direct deficit. It occurs without disinhibition of other neuronal systems and is explained simply by lack of impulses going to the peripheral motor elements. For this reason, originally we did not expect to achieve a real improvement of the akinesia with stereotaxic inactivation.

However, our investigations of hypokinetic signs in cases in which the lesion was demonstrated at autopsy, have shown that apart from the factor of rigidity, hypokinesia can be improved to a limited degree by inactivation of the oral ventral nuclei. On the other hand, inactivations of the pallidum are practically without effect on pure akinesia. It is just this difference in localization that supports the conclusion that it is possible, at least partially, to improve pure akinetic signs.

When trying to find the pathophysiologic mechanism of these therapeutic effects on the hypokinetic signs, one must start from the anatomical fact that the oral ventral nuclei project to precentral and premotor cortex. This leads to a mechanism that is comparable to the one described previously for rigidity and tremor. As regards the hypokinetic signs, the corticobulbar and corticospinal systems normally have an inhibiting effect on the motor neurons in the spinal cord and in the brainstem. This inhibiting effect is antagonistic to the activating influence by the undamaged substantia nigra. If one accepts this interpretation, one must find another pathophysiologic mechanism for akinesia, in order to explain the fact that the therapeutic effect on akinesia is definitely less than that on rigidity and tremor. SPIEGEL *et al.* (1962, 1963) made another therapeutic attempt to influence akinesia by coagulation of the caudatum. The speed of movement was improved in three out of four cases. We do not know of any further publications about this. BECHTEREVA (1972) reported improvements also of the akinesia in patients with Parkinsonism after lesions of the centrum medianum. On the other hand, MARKHAM *et al.* (1966) insist that inactivation of the centrum medianum in cases of Parkinsonism does not have a favorable influence on akinesia.

In our opinion corticospinal activation of the peripheral motor systems is much less for normal automatic movements such as associated movements than it is for muscle tone and synchronization of the tonic innervation of muscles in the waking state. As seen in case 12, the influence on propulsion and other hypokinetic signs is much more favorable than in other coagulations of the V.o.p because of the additional interruption of corticospinal fibers in the internal capsule between the inner part of the pallidum and the V.o.p. In this case the corticospinal influence would be impaired by the additional lesion. This pathophysiologic interpretation is not yet completely satisfactory because of the small number of cases confirmed by autopsy. It still needs experimental confirmation.

6.6. Correlation of the Vegetative Symptoms with Inactivated Structures

Etiologically, the proportion of postencephalitic Parkinson patients has decreased considerably in the last twenty years while that of cases suffering from genuine paralysis agitans has increased. Because postencephalitic Parkinsonism is characterized by far more vegetative symptoms, the vegetative Parkinsonian symptoms are no longer as prominent as they were 40 to 50 years ago. We were, therefore, not as interested in evaluating the dependence of vegetative symptoms on the structures inactivated by the operation. Also, it is difficult to measure vegetative symptoms. We were more concerned with recording changes of vegetative functions by stimulation during the stereotaxic operation (see MUNDINGER and SCHOLLER, 1956a, b), using in part clinical physiologic methods (SCHMIDT, 1963).

Out of the five cases in this material in which vegetative symptoms were mentioned as obvious in the case history, an improvement of the vegetative symptoms occurred in only two. In one case (5_r), coagulation of the pallidum improved the sweating and the greasiness of the face. In the other case (2_ℓ), extensive coagulation in the V.o.a, in the L.po, and in the anterior thalamic pole abolished the disorder of circulation with unpleasant reddening of the skin and particularly the excessive sweating.

In two cases, vegetative symptoms increased after the operation, in one salivation (4_ℓ), and in the other the abnormal peripheral circulation (17_ℓ). In case 4, the medial rostral part of the peduncle was extensively coagulated because of a deviation of the electrode point by 3.5 mm in a ventral direction (see Fig. 52 and p. 74). Perhaps the increase of salivation, which had not been observed in other cases with coagulation of thalamus and pallidum, can be referred to the unusual coagulation of cerebral peduncle. In support it can be mentioned that in 18% of almost 500 cases, an increase of salivation was the only vegetative sign following coagulations in the zona incerta (MUNDINGER, 1964, 1968), although other vegetative signs were diminished when this target was reached. In one case of Parkinsonism with extensive coagulation of the pallidum internum, the peripheral disorder of circulation became worse after the operation. In this case, too, the rostral peduncle was involved in the coagulation.

As far as the small number of cases allows, it follows that coagulation in the V.o.a, V.o.p, or pallidum without further involvement of the rostral peduncle is followed by an improvement of the vegetative signs including salivation. The same result was observed in the large clinical series (evaluated by MUNDINGER and RIECHERT, 1961, 1963, 1966). However, quantitative evaluation is still lacking in clinical series. The additional damage of the rostral peduncle seems to have a particularly unfavorable influence on salivation, while in these cases the other vegetative symptoms show improvement rather than deterioration.

Bilateral mydriasis is another vegetative symptom that occurs as a result of stimulating the oral ventral nuclei and the pallidum. After subthalamic coagulation, a transient central Horner syndrome appears on the ipsilateral side, as described by KIM and UMBACH (1972). SUGITA *et al.* (1972a, b) recently found this when they investigated the results of stimulation when pushing the electrode forward, step by step. The effect was shown to be limited to the fibers entering the V.o.a. We found that stimulation of the pallidothalamic system, particularly at rates of 25–50 cps, frequently caused bilateral pupillo-dilation, reddening and sweating of the face which, a few seconds after the end of stimulation, changed to remarkable pallor. Following the inactivation of these structures, vasomotor disorders were observed only rarely and in the cases examined post-mortem, only once. In many cases (without post mortem) a strong bi-

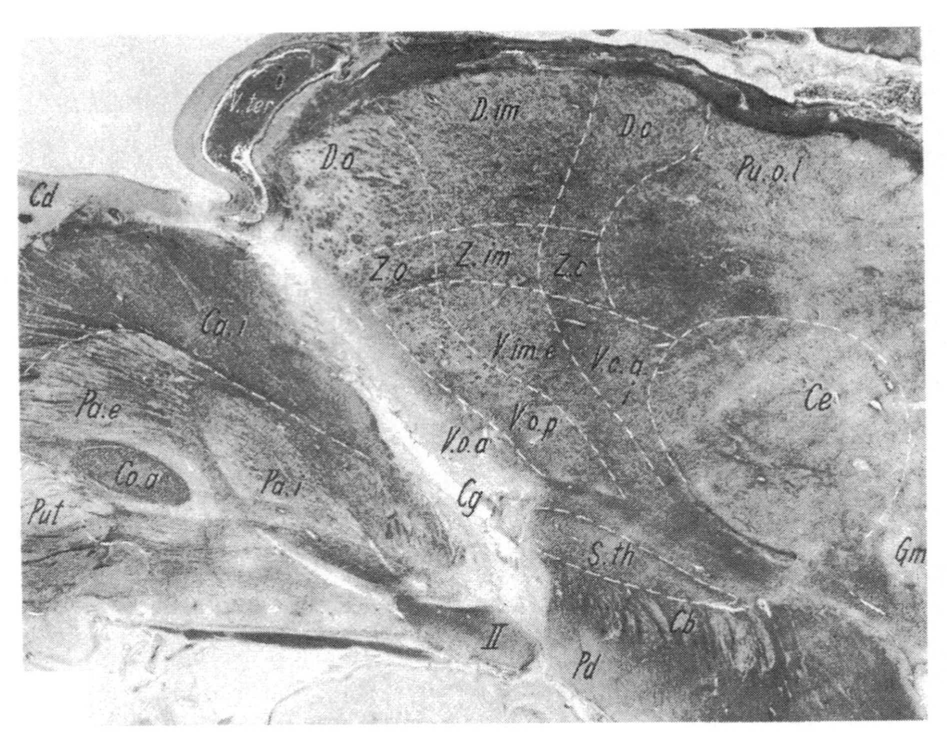

Fig. 129. Sagittal section of basal ganglia of thalamus in case No. 16. Section shows cylindrical coagulation (*C.g*) in front of thalamus from vena terminalis down to optic tract (*II*). Thalamic nucleus V.o.a and rostral part of dorsal (*D.o*) and zentrolateral (*Z.o*) nuclei are almost destroyed and demyelinated. The dark structure above centre médian (*Ce*) nucleus and posterior to *Z.c* and *D.c is* the lateral part of intralaminar nuclei. Transition zone of cerebral peduncle (*Pd*) to internal capsule (*Ca.i*) and the dorsocaudal part of *II* are destroyed by string electrode. Both segments of pallidum are strongly demyelinated, severely shrunken, and deprived of cells. Fiber stain. ×5

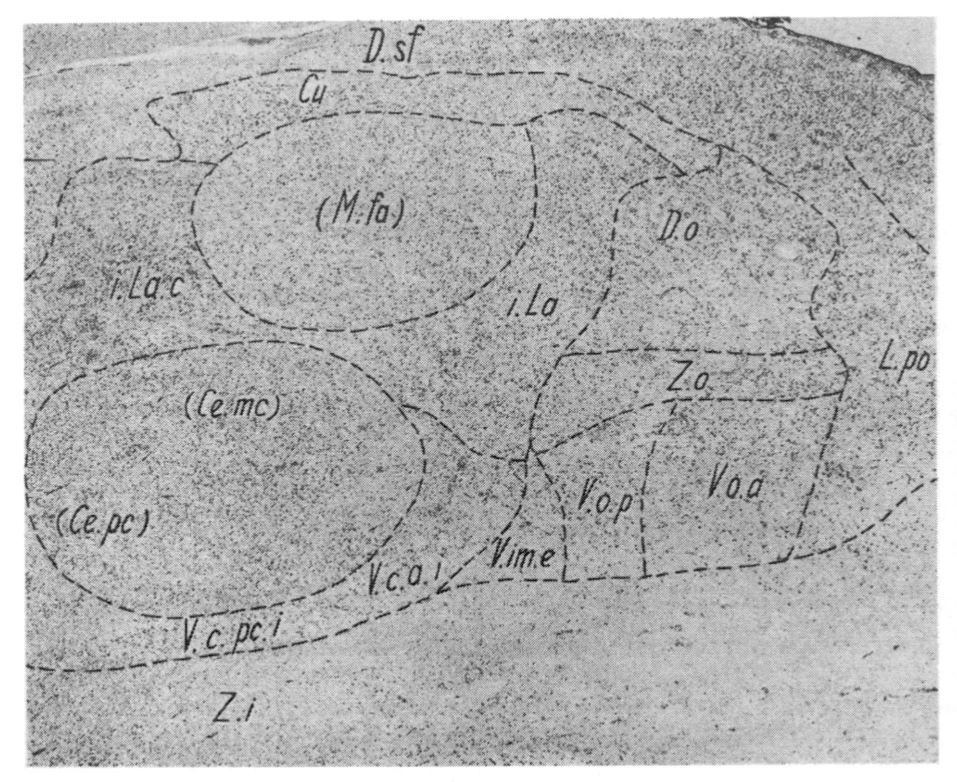

Fig. 130. Sagittal section through medial half of thalamus in case No. 16. Caudal and medial to coagulation needle track, most of the thalamic nuclei hit in this section have undergone severe nerve cell degeneration replaced by glia proliferation. Parts of medial nucleus (*M.fa*) have severe nerve cell loss and degeneration with gliosis, due to interruption of their prefrontopetal fibers. Two parts of centre médian nucleus (*Ce.mc* and *Ce.pc*) and trigeminal part of anterior somatosensory nucleus (*V.c.a.i*) show almost the same cell loss. Severe gliosis and nerve cell loss in basal part of V.o.a due to interruption of pallido-thalamic fibers by electrode track can be seen. Intralaminar nuclei (*i.La* and *i.La.c*) in front and caudal to medial nuclei are less shrunken; they show, however, strong glia proliferation and nerve cell degeneration due to the damage of their pallidar connections. All nuclear degeneration is due to interruption of their efferent or afferent pathways by coagulation (Fig. 129) in front of the thalamus. Nissl stain. ×9.2

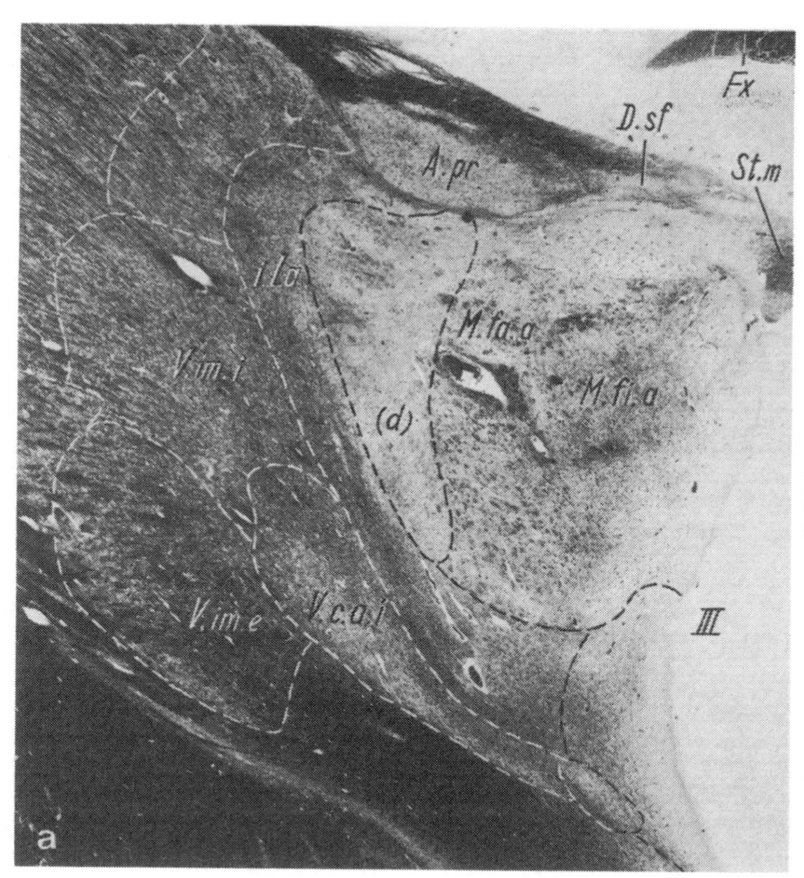

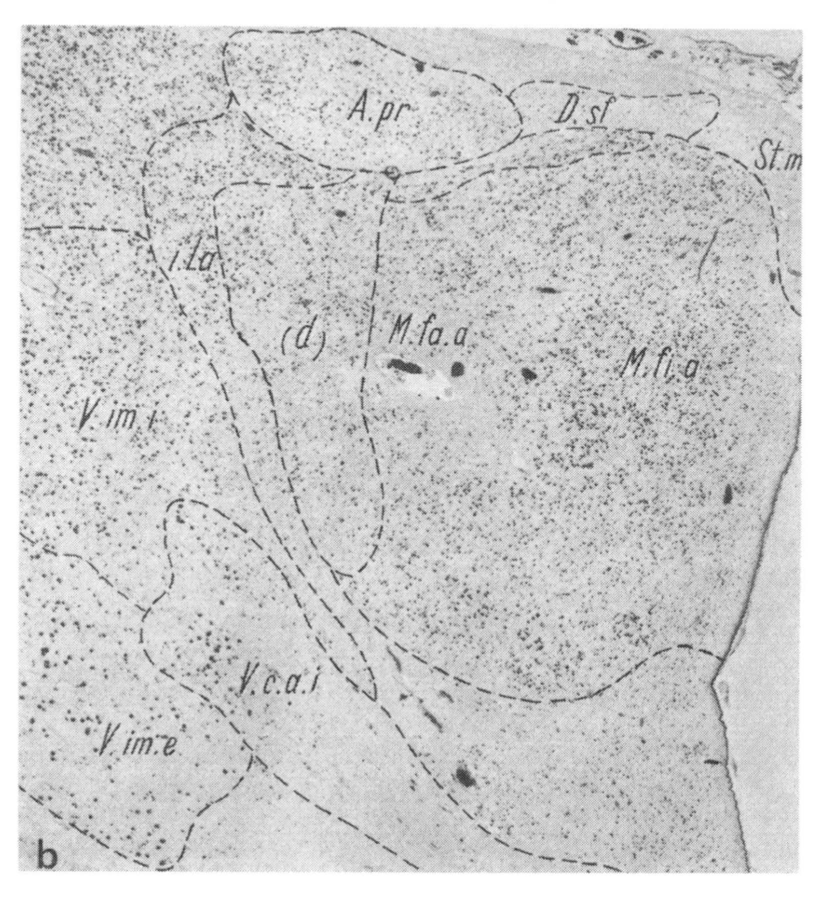

Fig. 131. (a) Cross section of thalamus of case No. 2 at level of V.im.e shows demyelination (*d*) and shrinkage of lateral part of nucleus medialis fasciculosus (*n.fa.a*) bordering on lamella medialis (*i.La*). Dorsal parts of medial nucleus at this level are also degenerated. Degeneration is due to interruption of thalamo-prefrontal fibers at level of anterior part of internal capsule (see Fig. 132a, b).

(b) Adjacent section in Nissl stain shows the nerve cell loss (*d*) in lateral M.fa.a. Also the intralaminar nucleus (*i.La*) is reduced in size and has lost a large proportion of its nerve cells by retrograde degeneration after pallidum destruction (Fig. 132a, b). ×10

lateral mydriasis appears for 2 to 3 weeks after coagulation of V.o.a and V.o.p mostly accompanied by contralateral mimic facial paresis. Occasionally patients complained for a long time after the operation about coldness of the contralateral limbs. In a case of Raynaud's disease, coagulation of the pallidum resulted in a 10-year amelioration (NARABAYASHI and NAGAO, 1972).

6.7. Correlation of Side Effects with the Damaged Structures

Despite the great accuracy with which the target can be found with the method described here, the coagulated lesion cannot be completely restricted to the target structure since the shape of the coagulated area does not coincide with that of the anatomically aimed-at structure. Consequently, smaller lesions of neighboring structures lead less frequently to clinical complications (see Tables 7, 9). Smaller lesions of the internal capsule have already been discussed extensively (p. 200). The following is a discussion of the side effects observed clinically, after partial lesions of internal capsule.

Most clinical complications are motor rather than sensory in nature, as one might expect from previous ideas regarding the thalamus. In none of these seventeen cases examined post-mortem, in which the thalamus had been coagulated, were disorders of sensation observed postoperatively. This is in agreement with our clinical experience. In

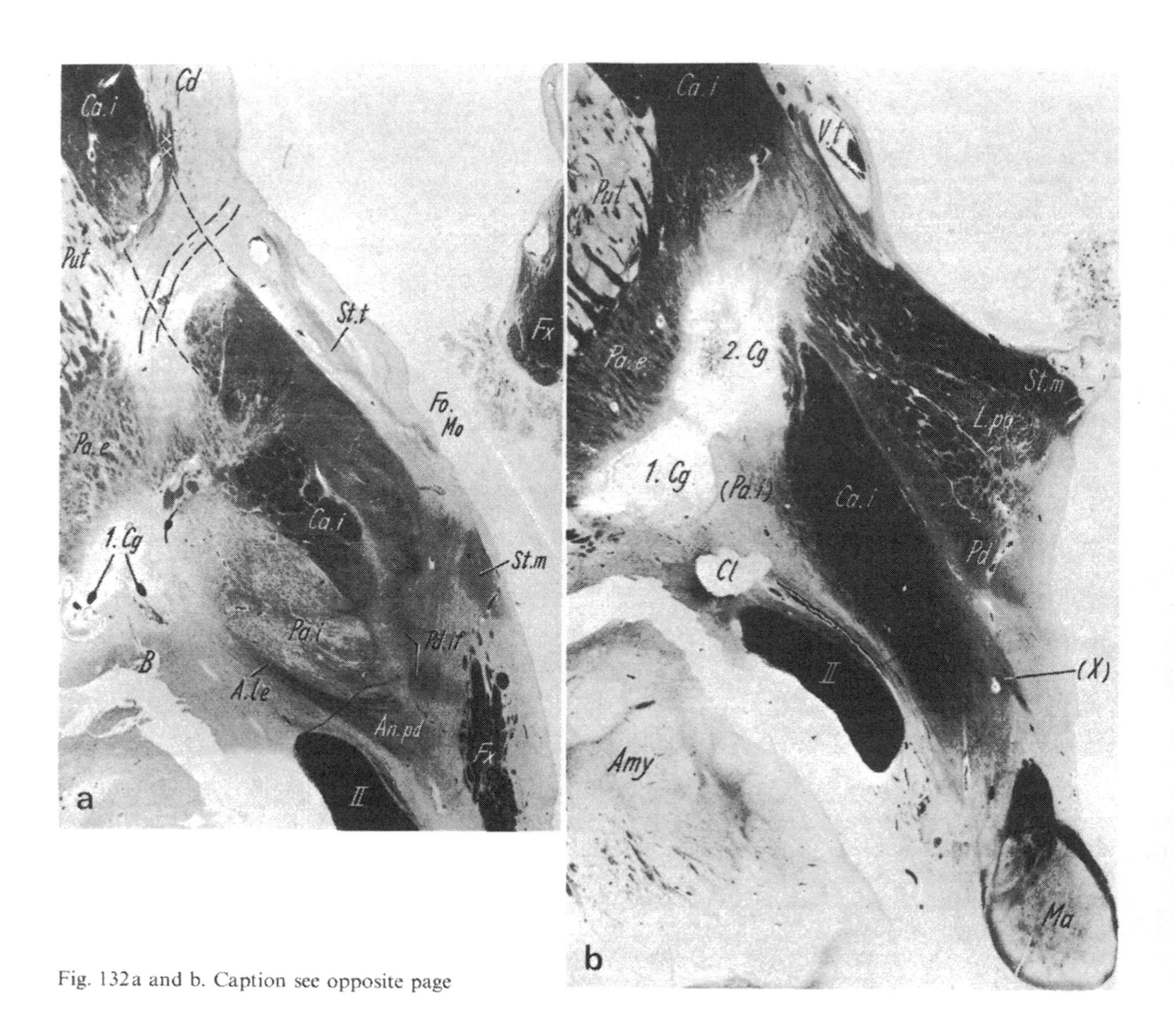

Fig. 132a and b. Caption see opposite page

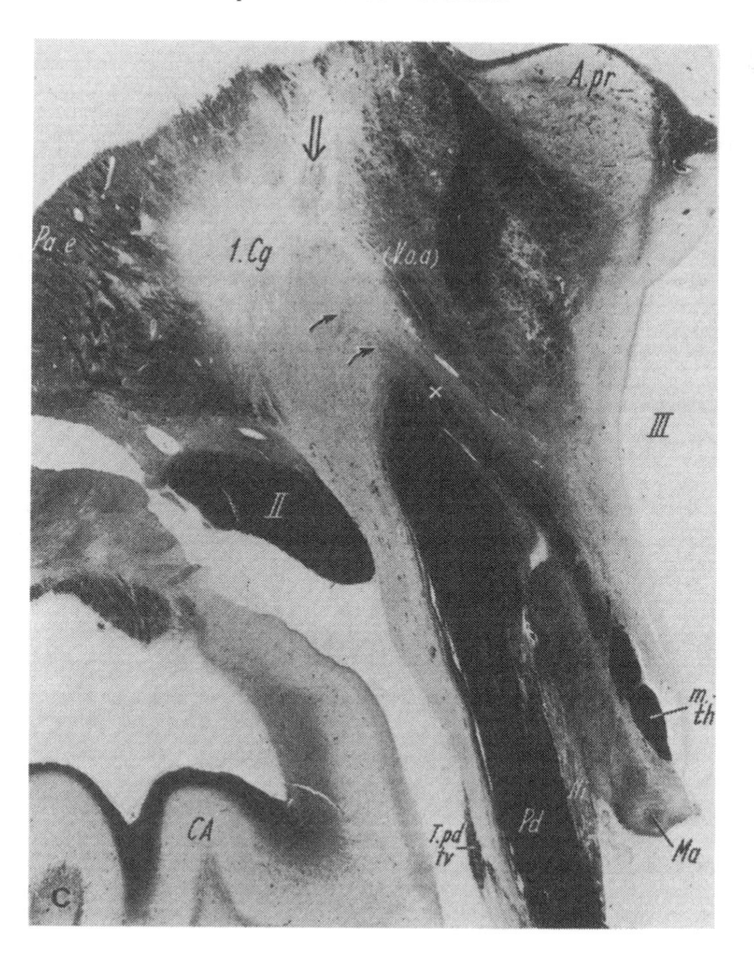

Fig. 132. (a) Cross section through interventricular foramen of case No. 2. Internal capsule is interrupted by coagulation focus above internal segment of pallidum. External segment of pallidum is also destroyed ventrally, whereas internal segment of pallidum is partially spared, and lamella pallidi interna is degenerated. Most medioventral part of capsula interna (*Ca.i*) is not destroyed. Inferior thalamic peduncle (*Pd.if*), situated on medial border on internal capsule is partially demyelinated. × 5.
(b) Section almost 2 mm further posterior than section in Figure 132a. Anterior pole of thalamus is almost completely preserved, but internal capsule is completely interrupted above outer pallidar segment and lamella pallidi interna. Internal segment (*Pa.i*) has lost almost all fiber bundles as result of 1st coagulation (*1.Cg*) and the clip (*Cl*) which was removed. 2nd coagulation (*2.Cg*) interrupted internal capsule above pallidum with shrinkage. × 5

Fig. 132. (c) In this section, 2 mm further caudally, most of internal segment of pallidum and of lamella pallidi interna is destroyed by 1.Cg. Internal capsule is completely destroyed above degenerated parts of pallidum. Focus extends to reticulate nucleus surrounding lateropolar nucleus of thalamus. Between arrows, a few pallidothalamic fibers are spared, which enter reduced H_2 bundle (×). Some degenerated fiber bundles, stained in a shade of grey, of internal capsule enter cerebral peduncle.
T.pd tv: tractus peduncularis transversus (nondegenerated). × 5

more than 2000 thalamic operations, sensory disorders were observed in less than 0.5%. In view of the variability of the extent of the lesion, the slight additional involvement of the sensory thalamic nuclei, which lie more toward the caudal end, offers an explanation. The extensive experience with thalamic lesions contradicts previous views that the thalamus is only a center for sensation. The same is shown by our numerous findings during stimulation in the oral ventral nuclei of the thalamus.

6.7.1. Reduction of Expressive Facial Movements. The most frequent side effect is a facial paresis that affects only the expressive movements (Fig. 53). Out of our post-mortem cases, case 13_r developed a severe weakness of the facial nerve affecting expressive movements. Apart from the puncture track through the rostral limb or the genu of the internal capsule, the autopsy excluded a lesion of the internal capsule (Fig. 85). Puncture tracks in the anterior limb of the internal capsule have proved to be functionally insignificant in one-sided operations since they damage only slightly the prefrontothalamic connections so that no sign of frontal lobe syndrome appears. In other cases with expressive facial weakness (13_ℓ, 15), 20 and 15%, respectively, of contralateral internal capsule were damaged lateral to the V.o.p, and in case 15 an additional 30% of the internal capsule lateral to the V.o.a was damaged (Fig. 90a). Such expressive weakness of the face exists also without damage to the internal capsule (Fig. 85, case 13_r). This weakness can be indicative of a lesion in the thalamic oral ventral (*V.o.i*) nuclei (Fig. 44), which have been inactivated in these cases, including cases 13_ℓ and 15. Evidence, which was not confirmed by autopsy, from more than 2000 coagulations in the V.o.a and V.o.p shows almost regularly a reduced contralateral innervation that lasts for only a few days (as after left-sided coagulation in Fig. 63). It is unlikely that the internal capsule is regularly involved in the coagulation when the lesions have been set with our method

Fig. 133. (a) Cross section through anterior thalamus in case No. 2, ca. 3.2 mm in front of Figure 133 b. Destroyed part of internal capsule has moved medially and downward into transition zone to cerebral peduncle (*K*). Medial needle track (*tr*) is located in V.o.a nucleus. Pallidum (*Pa*) is almost preserved at this level. ×6.
(b) Cross section through left thalamus of same case almost 8 mm further caudally. Foci in pallidum have disappeared and internal capsule is completely preserved. At this level, fiber degeneration of internal capsule appears as white strip limited to cerebral peduncle below subthalamic nucleus. Subthalamic nucleus is involved by this lesion and has an hourglass shape. Partial degeneration of subthalamic nucleus (*s.th*) continues dorsalward as cystic lesion caused by a side electrode in base of V.o.p, involving also zona incerta above subthalamic nucleus. String electrode lesion is bean-shaped and located in base of V.o.p. Above lesion, demyelination by needle track is seen almost perpendicularly to coagulation in zona incerta and base of V.o.p. Retrograde degeneration of intralaminar nuclei (*i.La*). *Cb*: comb system of the foot. ×5

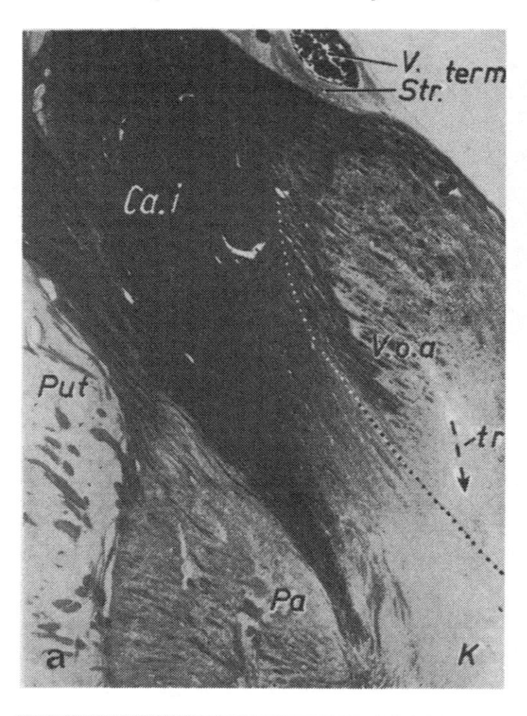

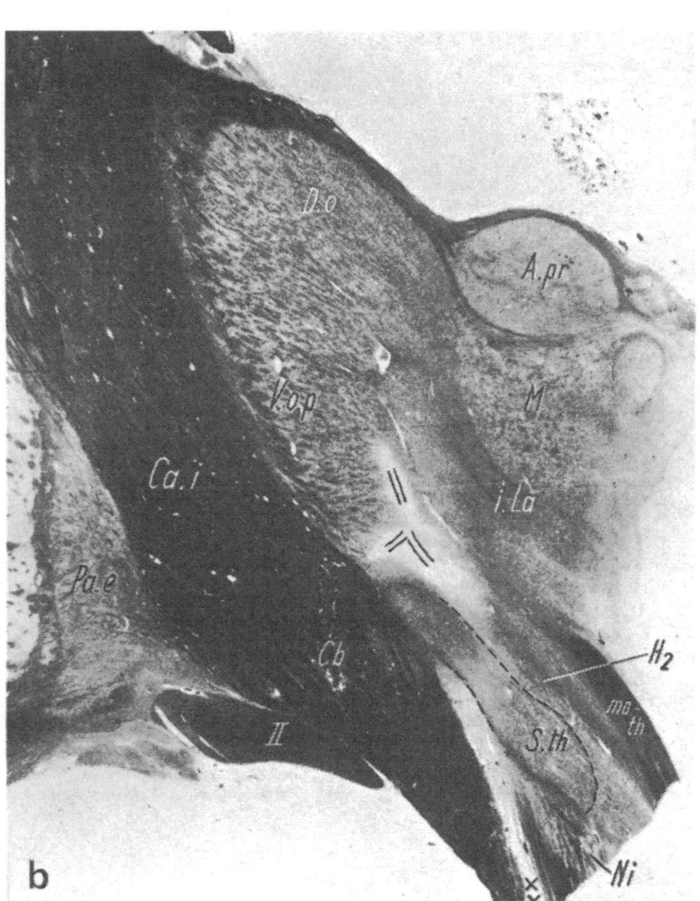

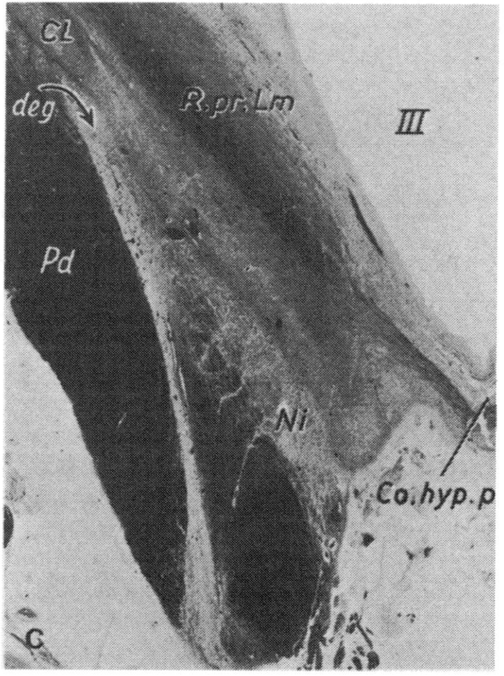

Fig. 133. (c) Cross section through thalamus almost 3 mm further caudally. Most parts of cerebral peduncle (*Pd*) are not destroyed. Degeneration of most rostral part of internal capsule has extended to cerebral peduncle, in which a 1-mm wide strip from subthalamic nucleus (*deg*) to base of peduncle is spared. This strip corresponds to weakening of right facial innervation after 3rd operation, which lasted 2 years until death. ×6

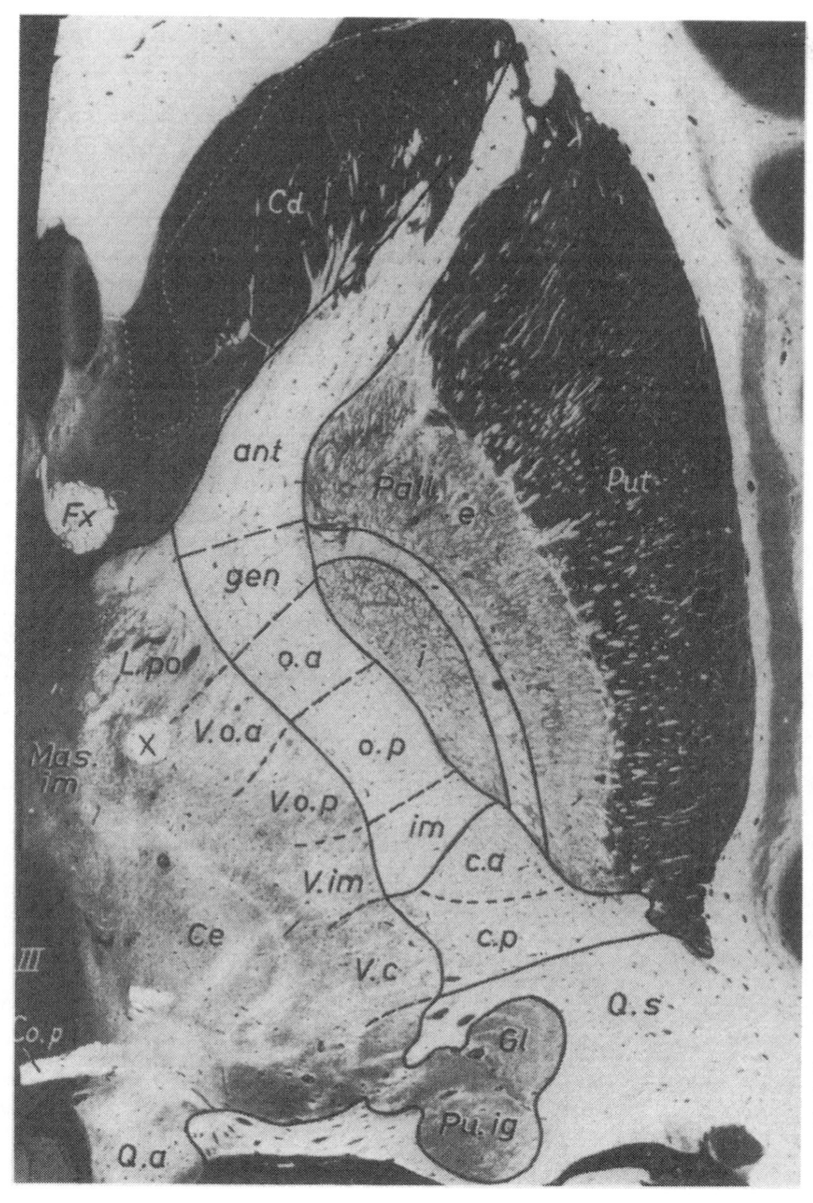

Fig. 134. Horizontal section of basal ganglia and thalamus in man shown as negative in order to render prominent the portions of internal capsule. Different portions of internal capsule are defined according to adjacent ventral thalamic nuclei. From rostral to caudal, largest anterior part is anterior limb of internal capsule containing almost exclusively prefrontothalamic and thalamoprefrontal connections. Second portion (*gen*) is pars genualis in frontal level of lateropolar nucleus of thalamus. Functionally it corresponds to cortical oculomotor pathway. Behind frontal level of mamillothalamic tract (×) follows next portion adjacent to V.o.a nucleus. This part of internal capsule is used by descending premotor extrapyramidal pathways. Involvement of o.a-part of internal capsule by coagulation results in mimical facial paresis and also motor neglect. Next portion of internal capsule (*o.p*) is adjacent to V.o.p nucleus; lesions in this part result in voluntary facial paresis and, if posterior part is also involved, in neglect, dyssynergia of contralateral side, or even in paresis of arm, mainly in distal parts. Next portion of internal capsule adjacent to V.im nucleus is passed by pyramidal fibers to lumbar and sacral segments of spinal cord. Coagulation here results in paresis, mostly of foot, and in gait disturbances. Portions of internal capsule at level of VLP or V.c nucleus are no longer related to motor function but to descending and ascending somatosensory pathways and functions to fields 3b, 1, 2 of postcentral gyrus. Next portion of internal capsule, denoted Q.S, is so-called carrefour sensitif, in which are joined somatosensory, acoustic, and visual pathways from thalamus to cortex and their feedback fibers. (After HASSLER *et al.*, 1967)

of aiming. The present post-mortem material as well as clinical experience proves the claim that LEYDEN made in 1864, that disorders of expressive movements originate in the thalamus.

6.7.2. Voluntary Weakness of the Facial Nerve was seen in 8 out of 23 operations (cases 2_ℓ, 2_r, 5_r, 7_r, 9_r, 11_r, 15_r, 17_ℓ) checked post-mortem (Table 9). In three cases (7_r, 9_r, 11_r), the facial palsy formed part of a hemiparesis. In all eight cases, 10–50% of the internal capsule lateral to the V.o.a (see also Figs. 132c, 90a) was damaged in addition to the coagulation of the V.o or the pallidum. In contrast to the findings with weakness of the expressive movements, interruption of discrete fibers in the internal capsule is always found when weakness of voluntary movements is present. This part of the internal capsule is not exclusively passed by the central neuron of the facial nerve. In case 2_ℓ, additional damage to the central neuron lateral to the L.po caused a descending degeneration in the medial peduncle (Figs. 43b, 133b, c). Functionally, a persistent right-sided facial weakness corresponds to this. The present cases show that the central neuron for voluntary innervation of the facial muscles runs through the internal capsule in the neighborhood of the V.o.a. In case 17, the paresis of the facial muscles for voluntary movements corresponded to a circumscribed interruption of the internal capsule caudal to the apex of the pallidum internum near the transition to the peduncle. Previously published results of stimulation and lesion of the internal capsule (HASSLER *et al.*, 1967) make it probable that the central pathway for the facial nerve passes in front of the central neuron for the arm through the part of the internal capsule that is lateral to the V.o.a (Fig. 134).

6.7.3. Reduction of the Postural Tone of the Arm. A lowering of the stretched-out contralateral arm, as in a slight proximal paresis, can occasionally be seen during coagulation of Pall. i. This can persist for some time after the operation without involving

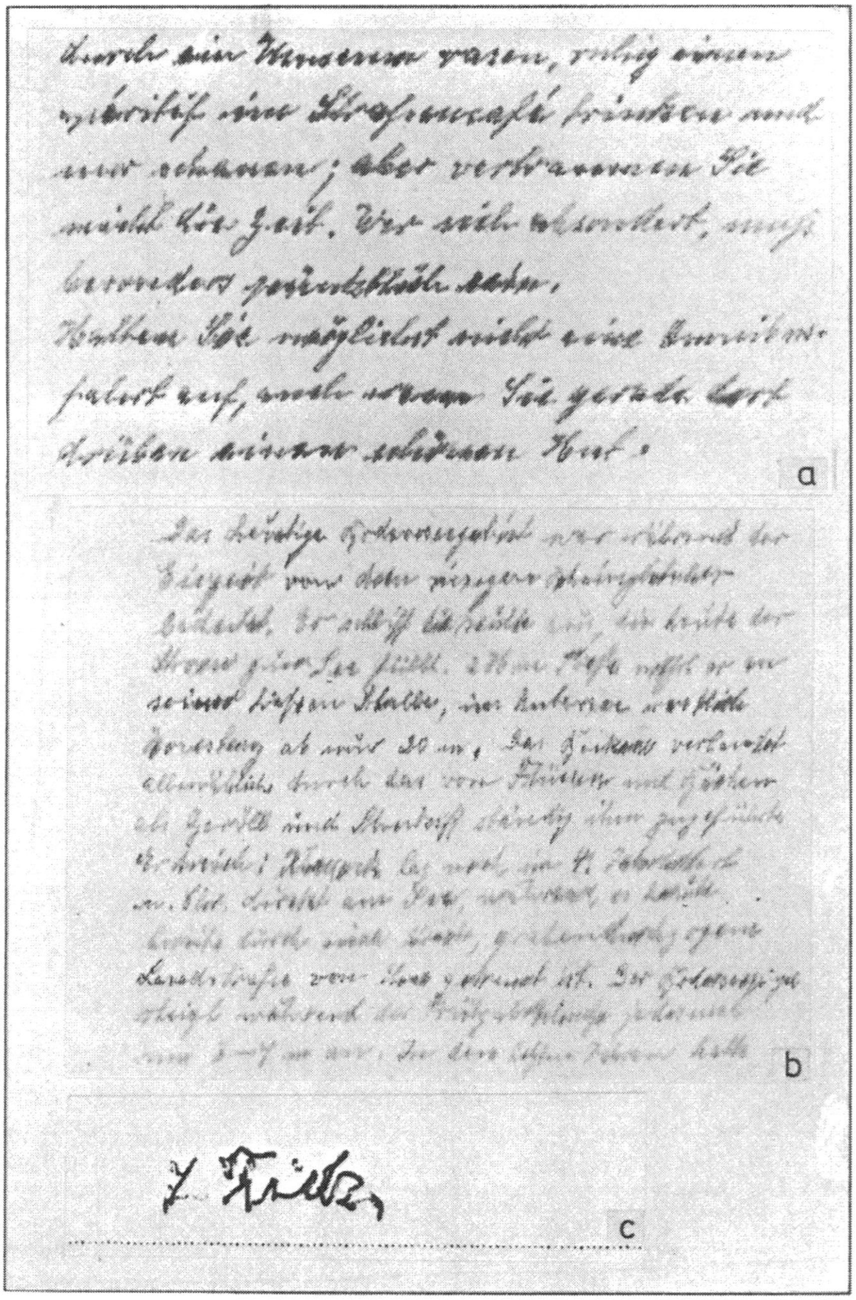

Fig. 135. (a) Writing specimen of Parkinson patient No. 2 before 1st operation. Writing is disturbed by tremor but is not micrographic. Straight lines are jagged but not too small.
(b) Four days after coagulation of left pallidum, writing of patient No. 2 is less deformed. Writing pressure changes frequently, and size of letters is reduced. After operation, however, straight lines do not show temporary disturbance caused by shaking tremor. Lack of uniformity and irregularity of writing pressure are due to acute interruption of cerebellar afferents to motor cortex by coagulation of V.o.p, which can temporarily lead to reduced coordination of fine movements.
(c) Writing of same patient is no longer micrographic 3 years after 1st and 1 year after 2nd operation in left pallidum and V.o.a because of tremor relapse. Pejoration of writing is due partly to progressive tendency of Parkinsonism, but mainly it is related to interruption of parts of internal capsule by extended lesion of basal ganglia (see also Fig. 132c, 133a)

the distal parts. Such a proximal paresis occurred five times (5_r, 7_r, 7_ℓ, 14, 16) in the present material examined post-mortem. None of these cases showed a lesion of the caudal part of the internal capsule lateral to the sensory ventral nuclei, through which the rapidly conducting pyramidal fibers run. In these five operations, there was partial damage either to the rostral part of the internal capsule lateral to the V.o.a or to its continuation into the rostral peduncle (Fig. 56a, b). These parts of the internal capsule cannot be held responsible for a central paresis of the arm that favors the distal parts, but they can be blamed for the reduction of the postural tone of the arm, which resembles neglect (Fig. 64). Without exception, the five cases with "proximal paresis" of the arm showed at autopsy an extensive contralateral lesion of the pallidum internum, pallidum externum, and lamella pallidi interna, whereby 60 to 100% of the pallidum tissue was destroyed. By the complete interruption of H_2 and H_1 bundles in case 16, the pallidum was also functionally eliminated. A reduction of the postural innervation of the body by suppressing the tonic stretch reflexes is effective in the treatment of rigidity because the proximal innervation is part of the postural innervation of the body that is due to the tonic stretch reflexes. This agrees with the complete abolishment of rigidity of the arm in the five cases. After V.o.a or V.o.p coagulation this proximal paresis is seldom seen.

Although only five cases were examined, the conclusion seems justified that an isolated lowering of the arm without weakness of clenching of the fist is not a sign of damage to the pyramidal tract in the internal capsule. During coagulations in the oral ventral nuclei and in the Forel bundles H_1 and H_2, one can also observe such a reduction of proximal innervation with a lowering of the arm during the operation, without the internal capsule being involved. Here, too, reduction of the tonic stretch reflexes, which are responsible for the proximal postural innervation, can be regarded as the pathophysiologic mechanism.

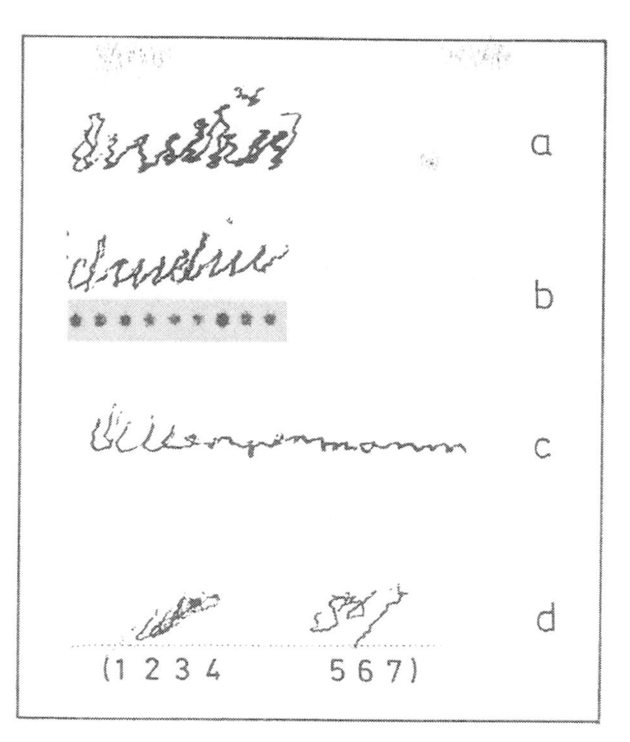

Fig. 136. (a) Writing specimen of patient No. 17 affected by paralysis agitans shows jagged letters before operation in right pallidum.
(b) Half a year after 1st coagulation in right pallidum, disfigurement of right-handed (i.e., ipsilateral) writing by tremor is markedly less because of ipsilateral less than contralateral improvement of tremor by operation.
(c) Signature of same Parkinson patient, $4^1/_2$ years after 1st operation, is less disturbed by tremor, but markedly micrographic.
(d) Writing of figures 1–4 by the same patient $4^1/_2$ years after 1st operation is totally unrecognizable because it is run together, whereas writing of figures 5–7 is less disturbed, although shaky, so that it is recognizable. Sample to show dependence of writing on momentary emotional situation and disinhibition of rapid shaking.

6.7.4. Deviation During Walking. The tendency to deviate to the side, which is influenced by the operation when walking and standing during the first few days after the operation, is a phenomenon similar to that of contralateral reduction of innervation. Lopsided walking and standing, which can be due to reduction in tone, also gradually disappears after days or a few weeks. Such a limpness or hypotonia of the muscles on the side influenced by the operation can be due to a disorder of the vestibular or cerebellar regulation of body movements and posture. By recording nystagmus during stimulation and coagulation of the thalamus as well as during caloric stimulations, FISCH and SIEGFRIED (1965) concluded that central vestibular disorders are decisively responsible for the lateropulsion. Like MAMO *et al.* (1965), we believe that this is multifactorial and is caused by hypotonia of the muscles, cerebellar disorder of coordination, vestibular disorders of posture, and the postoperative neglect of the contralateral half of the body. We do not want to underestimate the thalamic

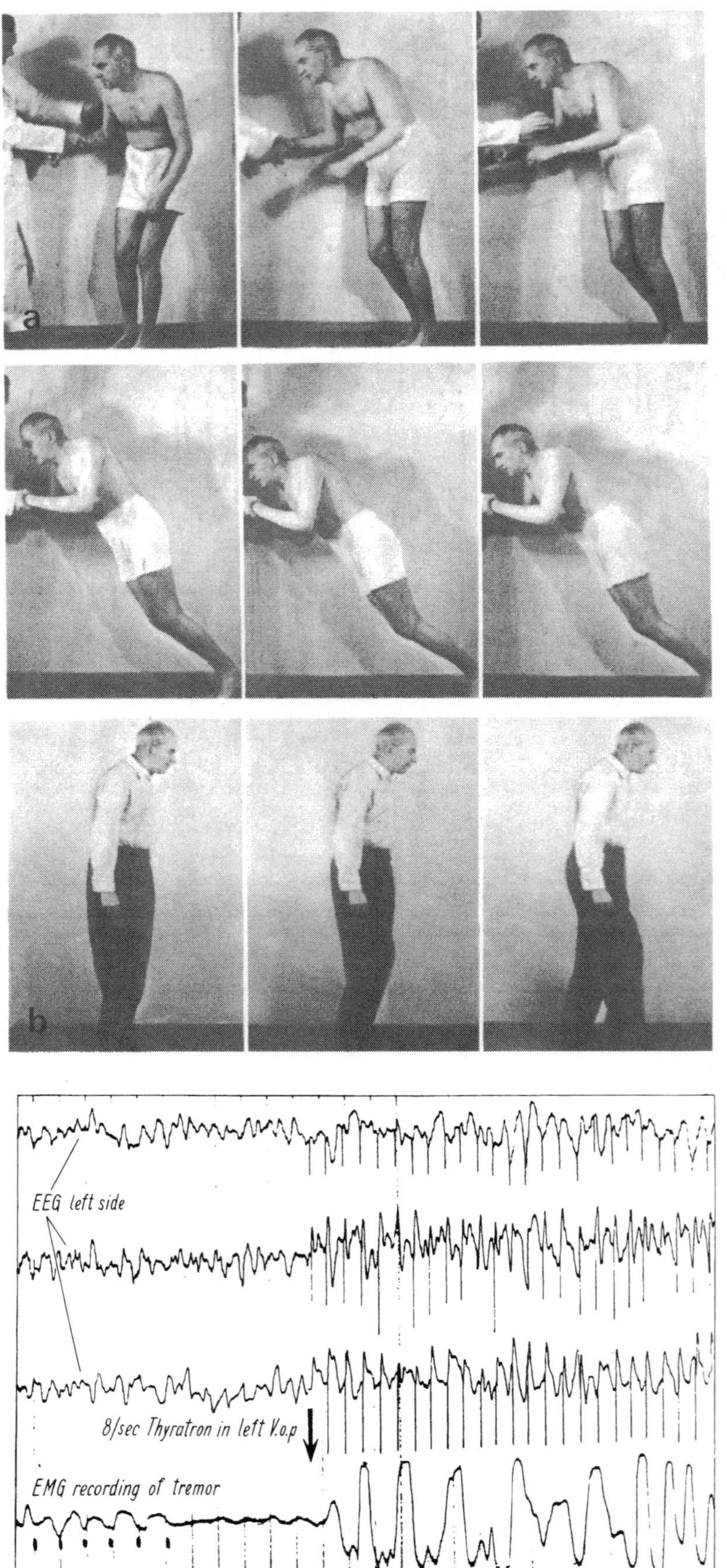

Fig. 137a–c. Patient with severe bilateral idiopathic Parkinsonism.

(a) Before operation, patient shows typical bent-forward posture of Parkinsonism. When asked to walk forward, patient is unable to lift his feet from floor, but he shifts his body forward so that he is in danger of falling and must be caught.

(a2) Extremely strong inhibition of starting movements and propulsion without movement of legs.

(b) Same patient after operation in left V.o.a. Patient is able to start normally and to lift right leg forward. (After RIECHERT, 1962)

Fig. 138. Recording of EEG and EMG of tremor of same patient during operation. Before stimulation, 5 cps tremor is present. Immediately after start of 8 cps stimulation in V.o.p, tremor registration shows large amplitude, which is expression of increase of tremor by V.o.p stimulation. At the same time in EEG, not only stimulus artifacts but also larger waves can be seen following each stimulus. They have frequency less than that of α rhythm. (After RIECHERT, 1962)

parts of the vestibular system, which were discussed previously in connection with the effect of stimulation on eye movements. The proximal lowering of the arm and the paralysis of expressive movements of the face should not be regarded as true complications in view of their temporary nature.

6.7.5. Hemiparesis. Unequivocal hemiparesis or paresis of individual segments of limbs occurred in five cases (7_r, 9_{r2}, 11, 12_ℓ, 17). In case 11, a massive hemorrhage occurring in the parieto-occipital lobe between encephalography and the stereotaxic operation caused a moderate hemiparesis that could be explained by damage to the shifted internal capsule next to the V.o.a (Fig. 119). In case 17, a smaller lesion extended into the caudal part of the internal capsule on the medial side of the caudal pole of the internal pallidum (Fig. 98). Clinically a paresis of the right foot with extensor response corresponded to this lesion of the internal capsule. In case 12_ℓ the internal capsules anterior and middle part, including the area of the adjacent ventral nuclei, V.o.a, V.o.p, and V.im.e, was mostly interrupted by a hemorrhage in the thalamus (Figs. 78a, 127a, c). This caused a contralateral hemiparesis with extensor response lasting till death. In this case considerable tremor remained on the hemiparetic side, although one could have expected the abolition of tremor because of the coagulation of the V.o.p and dentato-thalamic fibers.

In the massive hemorrhage of case 9_{r2}, the contralateral hemiparesis is adequately explained by damage to the pyramidal tract because of a neighboring hemorrhage in the internal capsule next to the V.o.p and pressure on more posterior parts of the internal capsule.

6.7.6. Dysphagia occurred twice as a complication (5_ℓ, 16). In case 5_ℓ it was explained by the direct coagulation of the medial cerebral peduncle (Fig. 123a). The coagulated area lay in the frontal plane, which corresponded to the midpoint of the length of the thalamus. In case 16, the peduncle was affected in the same frontal plane but to an even greater extent (Figs. 95, 129). The fact that in both cases the localization of the lesion was the same, is remarkable. It leads to the conclusion that a unilateral left- or right-sided lesion of the peduncle in its rostromedial part is adequate for the production of permanent dysphagia. The supranuclear corticobulbar pathway for swallowing runs close behind the genu of the internal capsule entering the rostromedial part of the peduncle (Fig. 134).

Case 16 was the only one in which postoperative zygomatico-maxillary spasms "around the mouth" were observed and complained of. This unusual postoperative hyperkinesia otherwise has been observed only in postencephalitic cases. The perioral hyperkinesia was probably due to the 20% lesion of the rostral and medial nucleus subthalamicus (corpus Luysii) found in this case. The fact that a somatotopic arrangement has been described in this nucleus (WHITTIER, 1947) makes it even more likely.

6.7.7. Aphonia was not a rare complication in our earlier series of operations. Among the cases examined post-mortem, it occurred only in case 4. The anatomical peculiarity of this case was a lesion comprising up to 50% of the medial and rostral cerebral peduncle (Fig. 52) without a simultaneous lesion in the internal capsule. Although the lesion was unilateral, the sign persisted during the seven postoperative days until the sudden massive pulmonary embolism. It is remarkable that similar to dysphagia, a unilateral lesion in the peduncle is sufficient to cause aphonia.

6.7.8. Dysarthria occurred in three cases after coagulation. In two (5_ℓ, 6), there was a lesion of the rostro-medial cerebral peduncle of about 30%. One (5_ℓ) had a bilateral inactivation with additional damage to the internal capsule next to the rostral thalamus (L.po) (Figs. 55a, 56b). In this case, therefore, there existed a bilateral lesion of the descending corticobulbar systems, which is regarded as the basis of pseudobulbar dysarthria. Because of previous coagulation of the left thalamus, case 7_r also showed bilateral damage to 40% of the internal capsule lateral to the V.o.a and V.o.p. The lesion involved the nucleus reticulatus, which lies close to the internal capsule.

Autopsies of cases following stereotaxic operations are much more suitable for the correlation of circumscribed lesions of different parts of the internal capsule (Fig. 134) with clinical signs than were earlier investigations of infantile cerebral paresis with pseudobulbar signs. Interrupting the rostro-medial fiber bundles of the cerebral peduncle on the left side only, can be the cause of aphonia (Fig. 52). Reduced respiratory excursion because of rigidity and akinesia of the respiratory muscles frequently causes aphonia in patients with long-standing Parkinson's syndrome. In case 4, however, the aphonia occurred only after the cerebral peduncle had been damaged during the operation. Such clear linking of interruption of rostral corticobulbar fibers with aphonia as in this case (HASSLER *et al.*, 1965, 1967) has not been described before in the literature except JAKOB'S (1923) case on p. 321 with a clear picture of the lesions.

In contrast, lesions of the corticobulbar system in both hemispheres were present in the three cases with postoperative dysarthria. The corticobulbar fibers were damaged in all three cases lateral to V.o.a and in two cases also lateral to L.po (Figs. 55a, 56a, 58a, b, 63, 64). This bundle of fibers is the same as that damaged in the peduncle (cases 5_ℓ and 6_r). It seems therefore that corticobulbar fibers passing through the internal capsule in the frontal plane of the rostral V.o.a are essential for the production of dysarthria. These three cases confirm and make more precise the opinion that the cause of pseudobulbar symptoms is based on cerebral paresis. [See also: Discussion on Speech Disorders, Conf. Neurol., **34**, 173–175 (1972).]

6.7.9. Postoperative Binocular Deviation to the side operated upon occurred in two cases (7_ℓ, 17). In case 7_ℓ 100% of the left pallidum was destroyed and in case 17, 90% of the internal pallidum and 50% of the lamella pallidi interna. In case 7_ℓ the binocular deviation was associated with severe clouding of consciousness. In case 17, phases of disturbed consciousness caused by recurrent pulmonary embolisms were present and were even mistaken for epileptic fits. These binocular deviations toward the side of the lesion after inactivating the pallidum (Fig. 98) can easily be explained pathophysiologically, since evidence from stereotaxic operations and animal experiments (MONTANELLI and HASSLER, 1964) has shown that stimulating the pallidum causes a turning of the eyes and head to the other side. The loss of one pallidum leads to a functional excessive influence of the other, causing a turning of the head and eyes to the side of the nonfunctioning pallidum. However, it must be noted that not every almost-complete inactivation of the pallidum causes such a conjugate deviation to the contralateral side. It may be that such a deviation was not clinically observed in some cases of hemorrhage into the pallidum with coma; on the other hand, it is known that the unilateral loss of a structure causing adversive movements is usually quickly compensated for by many other adversive structures.

6.7.10. Ballistic Hyperkinesia occurred only once (5_ℓ) and only temporarily among the cases with autopsy. There was a lesion affecting 15% of the rostral part of the subthalamic nucleus (Fig. 123a). The development of hemiballismus through damage of the subthalamic nucleus seems to be proved by WHITTIER'S (1947) summarization of the literature. It is therefore probable that in this case also the hemiballismus was due to the damage of the rostral part of the subthalamic nucleus, although only 15–20% of the total volume was affected. In an additional four cases (2_r, 13_r, 15, 16), damage of up to 20% of the volume of the subthalamic nucleus (Figs. 44, 90a, 95) occurred without hemiballismus having been noted clinically (Table 9). In cases 13_r, 15, and 16 the bundle H_2 (Figs. 85, 129) was completely interrupted. According to BALTHASAR (1930) the appearance of hemiballismus is inhibited if the bundle H_2 is interrupted simultaneously with the subthalamic nucleus.

As hemiballismus is not a rare complication, its origin after stereotaxic operations has been extensively discussed during the last few years, based on clinical and post-mortem observations (OBRADOR and DIERSSEN, 1959: 3%; RAND and MARKHAM, 1960: 9%; SCHACHTER *et al.*, 1960: 0.2–2.5%; SCHAERER, 1962: 5.5%; MUNDINGER and RIECHERT, 1963: 0.3%; GROS *et al.*, 1964: 4.6%; GILLINGHAM *et al.*, 1964: 0.3–3.6%; HUGHES, 1965: 3%; HOUDART *et al.*, 1965: 4%; SELBY, 1967: 2.4–3.1%; HOPF *et al.*, 1968; one autopsied case described). BRION *et al.* (1965: 1.4%) described twelve cases of postoperative hemiballismus in which the localization of the lesion responsible for the ballismus was either the subthalamic nucleus itself or the connections between it and the pallidum; the authors reported two postmortems as supporting evidence. The doubts (DIERSSON *et al.*, 1961) in this relationship are unfounded because the authors have not histologically checked the subthalamic nucleus.

It is noteworthy that of our seventeen cases only one had a transient contralateral ballistic disorder of movement based on a lesion of the subthalamic nucleus. The four other cases with lesions in this location (up to 20%) did not have ballistic disorders of movement. From these data, it cannot be concluded that ballismus is not due to loss of the subthalamic nucleus. As CARPENTER *et al.* (1950) showed in monkey experiments, more than 20% of the subthalamic nucleus must be destroyed in order to make "choreoid" hyperkinesia appear. In four of our cases with autopsy, it seems that

Table 9. Motor and psychic complications in relation to inactivated structures

Case No.	Partial paralysis of extremities	Facial paresis (mimic)	Facial paresis (voluntary)	Dysarthria	Aphonia	Dysphagia	Psycho-organic syndrome	Disturbance of consciousness	Case No.
2 lt.			Ca.i (L.po) 30%				bilat. destruction rost. thal., retrog. deg. in M left	bilat. destruction pallido-thalamic systems	2 lt.
2 rt.			Ca.i (V.o.a) 30%						2 rt.
4 lt.					Pd.m 50%				4 lt.
5 rt.	shoulder Ca.i (L.po) 10% Ca.i (V.o.a) 20%						experience of a double: vascular with pallidar lesions	insufficience in combination	5 rt.
5 lt.				bilateral Pd.m 30% Ca.i(L.po) 75%		Pd.m 30%	Ca.i (L.po) 75% bilateral Pa.i 50%		5 lt.
6 rt.				bilateral Pd.m 30%				episodic negativism: vigil coma Ca.i (L.po) 50%; Ca.i (V.o.a) 100%; La.m. 80%	6 rt.
7 rt.	shoulder Ca.i (V.o.a) 40% +Py		Ca.i (V.o.a) 40%, Py	Ca.i (V.o.a) 40%	(Pa 100%!)				7 rt.
7 lt.	shoulder Ca.i (V.o.a) 70%						Ca.i (L.po) 70%	amentia: bilateral Pa 100%	7 lt.
9 rt.$_2$	hemipar. Ca.i (V.o.a) 50%		Pa 100%				Ca.i (L.po) 50%	Ca.i (L.po) 50%	9 rt.$_2$
10 lt.								hemorrhage in the frontal lobe	10 lt.
11 rt.	hemipar. Babinski +: parieto-occipital hemorrhage		Ca.i (V.o.a) 10%					La.m. 100%, M 50% Ce. 50% (H_2 100%)	11 rt.
13 lt.		Ca.i (V.o.p) 20%; V.o.a 50%; V.o.p 80%							13 lt.
13 rt.		V.o.p 70%; H_1 100%						Infarction in both rostromedial thalami, disintegration of massa intermedia	13 rt.
14 lt.	shoulder Ca.i (V.o.a) 50%; Ca.i (L.po) 50%; Ca.i (V.o.p.) 20%; Pa 75%; V.o.a 40%							Pt. 100%, Cd. 100%, Pa. 100%	14 lt.
15 rt.		Ca.i (V.o.a) 30%; V.o.p 60% (100%)	Ca.i (V.o.p) 15%;					subdural hematoma, H_2 100%	15 rt.
16 rt.	slight shoulder Pd. rostr. 50%					Pd. rostr. 60%	loss of initiative, H_1, H_2 100%	pallido-thalamic fibers 100%	16 rt.
17 lt.	Babinski + Ca.i Py 5%		Ca.i Py 5%						17 lt.

the manifestation of the hemiballismus was inhibited by the additional interruption of Forel's field H_2 or the fasciculus lenticularis. Our other clinical experience agrees with this. Postoperative hemiballismus was abolished by coagulating the field H_2, or the pallidum internum, or by extensive coagulation in the V.o.a and L.po in three cases; improvement was obtained in another case without motor loss occurring. Our unequivocal clinical results are in contrast to the conclusions of BRION *et al.* (1965).

In the stereotaxic treatment of these cases of postoperative hemiballismus, we did not have a single death, so we have no post-mortem check.

According to our experience clinically and with post-mortems, we believe that the following conditions are necessary for the development of ballismus: damage to more than 15% of one subthalamic nucleus without simultaneous damage of the pallidum or the efferent pallidothalamic fibers in H_2 and its thalamic terminal nucleus V.o.a. It is of practical importance for the stereotaxic treatment that postoperative ballismus, which does not disappear spontaneously, can be abolished by further extensive inactivation of the foregoing pallidothalamic systems. The older anatomical interpretation of hemiballismus (VON SANTHA, 1928, 1932; BALTHASAR, 1930; WHITTIER, 1947) is, therefore, confirmed by our experience with stereotaxic operations and is further used for treatment.

6.7.11. Myoclonic Hyperkinesia. Myoclonic hyperkinesia has been seen only once postoperatively in the autopsied cases ($7_{\ell 2}$) apart from the ballistic postoperative hyperkinesia. The special feature of the anatomical lesion was the destruction of 75% of the caudate nucleus by a hemorrhage that originated in the substantia perforata anterior and broke through the pallidum and the internal capsule (Fig. 62). The hemorrhage did not provide us with an anatomical substrate for myoclonus because in this case large parts of other nuclei in the basal ganglia had been destroyed. HUGHES (1965) reported two comparable observations without post-mortem.

Although in two other of our post-mortem cases ($14, 9_{r2}$) damage to the caudate nucleus was 100% and 40%, respectively, due to large hemorrhages in the basal ganglia, an additional hyperkinesia could not be observed after the operation because of the patients' unconsciousness.

In three cases (5_ℓ, 9_{r2}, 17) 20 or 30% of the basal nucleus had been damaged either by coagulation or hemorrhage. No special signs resulted that could be ascribed to this lesion. This is of importance because the basal nucleus regularly shows cellular changes and loss of cells in genuine paralysis agitans and in many cases of postencephalitic Parkinsonism.

6.8. Correlation of Postoperative Psychological Disorders with the Damaged Structures

6.8.1 Disorders of Initiative. In two cases (2_{lr}, 16) a frontal lobe syndrome with loss of initiative, severe mental impairment, and incontinence occurred after the operation. In case 2 the basis of this was an interruption of the thalamo-prefrontal pathways due to extensive scar formation in the left internal capsule at the genu and caudally to it. This led to severe shrinking of the left medial thalamic nucleus because of retrograde atrophy in its rostral parts (Fig. 131a, b). If one disregards the fact that the patient left the hospital 10 days after the operation on his own accord, no obvious mental changes were caused by the severe left-sided damage of the rostral part of the thalamus (Table 9). A psycho-organic condition without impaired consciousness but with loss of self-criticism and with incontinence occurred only two years later, after the coagulation in the right rostral thalamus when, at the same time, the base of the oral ventral nuclei and the zona incerta below them, including 20% of the subthalamic nucleus, were coagulated. It might be thought that such a psychosyndrome could be explained by the bilateral inactivation of the medial thalamic nuclei. However, the psychological signs cannot be explained by this fact alone because there was no symmetrical lesion on the right side. But on the right side the thalamic connections and part of the descending system of the pallidum had been completely interrupted in the fasciculus thalamicus (H_1) and in its terminal nucleus. In addition, the mamillo-thalamic bundle had been damaged on both sides by about 20%. This additional damage of the pallidothalamic and limbic systems in the right thalamus had obviously caused the mental signs in view of the already existing severe degeneration in the left thalamo-prefrontal systems. The severe disorder of memory in particular had its morphologic equivalent in the bilateral damage to the mamillo-thalamic bundle. This bundle contains the efferent fibers from the mamillary body to the anterior nucleus of the thalamus. The psychosyndrome was caused by the bilateral functional loss of the medial thalamic nuclei. Later it was hidden by the increasing impairment of consciousness caused by repeated pulmonary embolisms.

In case 16 there were such severe mental abnormalities present before the stereotaxic operation that the brother described the inner life of the patient as a "mental heap of broken china." The coagulation in the right rostral thalamus had destroyed not only the oral ventral nuclei but also 40% of the nucleus lateropolaris through which pass some of the fiber connections from the medial thalamic nucleus to the prefrontal cortex (Figs. 130, 129). This explained the persistent reduction of initiative and self-criticism, so that the patient postoperatively became mentally so deteriorated that she lost her profession and required constant psychiatric hospitalization.

6.8.2. Experience of the Double (Doppelgänger) occurred in case 5_r after the first stereotaxic operation with inactivation of the pallidum internum during a postoperative confusional state that lasted for 6 days. In addition to the pallidum internum, only parts of the internal capsule next to the L.po (10%) and lateral to the V.o.a (20%) had been damaged (Fig. 56a, b). The experience of the double was not related to the localization of the lesions; it seems more likely that the responsibility lay with the inadequate central blood supply of the patient, who was 61 years old. This view is supported by the fact that the patient was confused for 3 nights after preoperative encephalography. In patients suffering from Parkinson's disease, visual hallucinations occurred after stereotaxic operations of the thalamus without clouding of consciousness for several months after the operation (Hartmann-von Monakow, 1959). The experience of the double can be regarded as belonging to the visual hallucinations. As the pallidum systems are also psychomotor systems, their inactivation can prepare the soil for the development of disorders of perception. In any case, there is a general tendency in diencephalic processes to develop visual hallucinations even if the visual systems are not involved (peduncular hallucination of Lhermitte and Trelles, 1932). It must be considered that, even without operation, a transient psycho-organic syndrome occurred after pneumoencephalography in 4.8% of 2,000 Parkinson patients in our series (Potthoff *et al.*, 1972; Mundinger and Riechert, 1961).

6.8.3. Negativism with Mutism or Vigil Coma was the unusual syndrome in case 6. In this case the Parkinson was combined with a tumor (spongioblastoma) that obstructed the third ventricle almost completely. This led to a hydrocephalic dilatation of the anterior parts of the third ventricle and of the lateral ventricles. In addition, the left basal ganglia contained multiple lesions due to three alcohol injections that had been previously performed elsewhere. In this case the negativistic syndrome had been linked to a hemorrhage in the right rostral thalamus. This as well as the lamella medialis and the pallidum internum had been almost completely destroyed by the hemorrhage. Negativism in the special form of resisting (*Gegenhalten*) is regarded by Kleist (1927) as being related to the thalamus. In this case negativism and mutism could be regarded as being due to the bilateral extensive lesion of the diencephalon in view of the large size of the thalamus damage. The lesion had almost completely destroyed the lamella medialis and had caused atrophy of the hypothalamic nuclei.

6.8.4. Confusional Syndrome. A severe confusional syndrome of 6 weeks' duration appeared in a case of idiopathic Parkinsonism (7_l) after the third operation. The following structures had been inactivated by the two previous operations: the left oral ventral nuclei to a large extent, together with the thalamic fasciculus, H_1 (Fig. 61), the right pallidum externum and internum, with those parts of the internal capsule that lie above them (Fig. 64). Neither disorders of consciousness nor delirium appeared after these two operations. After the third operation a hemorrhage destroyed the left pallidum internum and externum as well (Fig. 62). Only then did the bilateral inactivation of the pallidum create the neurophysiological basis for a severe disorder of consciousness, which included mistaking of the surroundings and deterioration of logical thinking. Since this confusional state lasted 45 days until the patient's death without his becoming comatose, it cannot be regarded as having been caused by pressure of the hemorrhage in the area of the basal ganglia. It seems rather that the bilateral, almost complete inactivation of the pallidum was the cerebral substrate of this confusional syndrome. This case is not unique since in a case of double athetosis the bilateral coagulation of the pallidum without hemorrhage after an initial coma (Fig. 139) produced lasting disorder of consciousness with very severe akinetic and mutistic syndrome continued for weeks (HASSLER, 1964, 1967, 1977).

6.8.5. Impairment of Consciousness from somnolence to coma occurred in 7 other cases (5, 9, 10, 12_l, 13, 14, 15) (Table 9). These impairments of consciousness were caused in two cases by hemorrhages outside the basal ganglia; in case 15 a subdural hematoma occurred, and in case 10, a hemorrhage into the frontal lobe (Fig. 71). In two other cases (9, 14) the coma developed rapidly on the operating table itself. A hemorrhage had occurred on the right side into the basal ganglia (Fig. 68b), but this led to a contralateral displacement of the both thalami. In both cases, the pallidum had been aimed at. In case 9 the hemorrhage destroyed the entire pallidum and the anterior parts of the internal capsule, 70% of the putamen, and 40% of the caudatum: this occurred in addition to the pressure of the right subdural hematoma. In case 14 destruction of three-fourths of the pallidum and the putamen and of 100% of the head of the caudatum had occurred (Fig. 88a, b). In both cases, tiredness on the operating table was the first sign of the developing hemorrhage into the basal ganglia. The size of the hemorrhage is not the only factor that determines the depth of the disorder of consciousness, for the large hemorrhage in case 11 in the parietal lobe did not cause loss of consciousness (Fig. 73a). From the two cases one can derive only a rough localization of the cerebral structures in the basal ganglia that

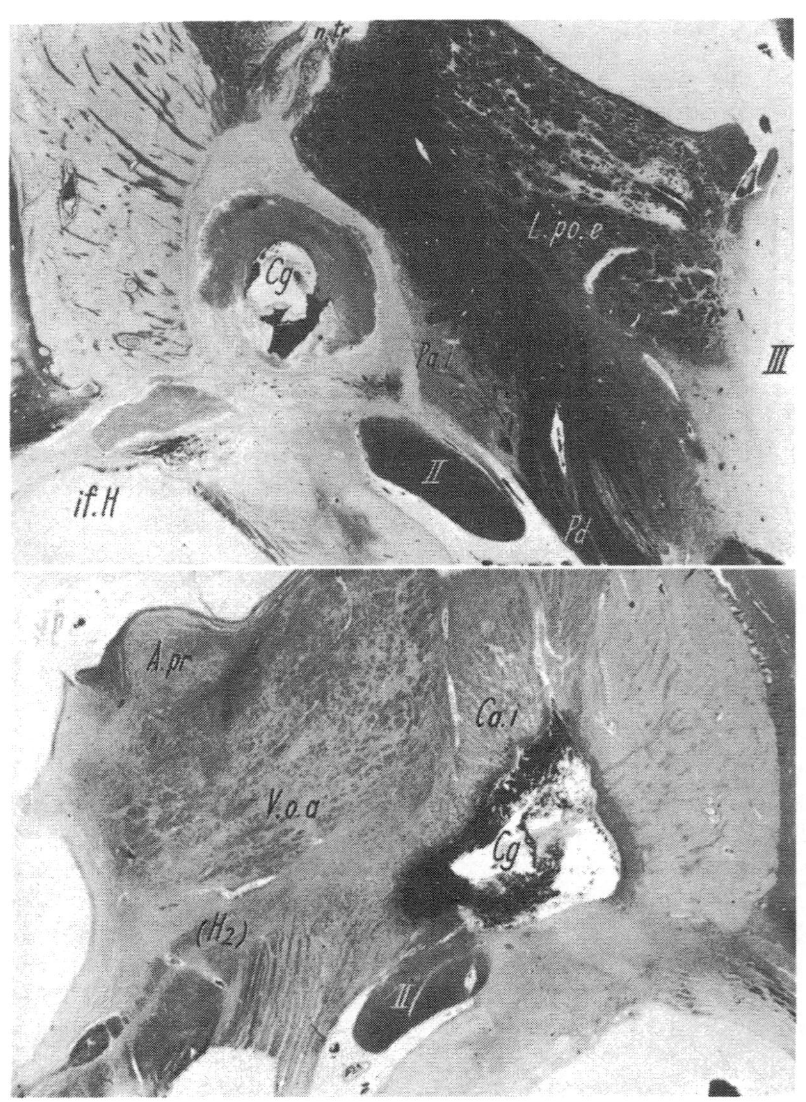

Fig. 139. Cross section through basal ganglia in case of double athetosis with bilateral coagulation in pallidum. On left side fresh coagulation destroyed lateral $^2/_3$ of pallidum internum and entire pallidum externum. On the right side the coagulation performed one year before had destroyed almost all of both segments of the pallidum (*Cg*); in consequence of this degeneration and disappearance of fields H and H_1 in V.o.a. Bilateral lesion of pallidum resulted in comatose state resembling vigil coma. Fiber stain. ×4.5

are of decisive importance for consciousness; it is not possible to correlate this disorder with individual neuronal systems.

Case 5 has already been discussed because of the experience of the double after a right-sided coagulation of the pallidum (Fig. 55a). A confusional state of varying intensity had developed after the first (right) stereotaxic operation, but particularly after the second, left-sided intervention, this state lasted until death, 21 days after the operation. Two causes can be mentioned for the confusion after the first coagulation: (1) the vascular insufficiency that had been apparent even before the first operation became manifest in the form of noctural confusion after the encephalography and (2) the extensive coagulation of the right pallidum. Regarding the severe impairment of consciousness with confusion after the second operation and lasting until death, one has to attach less significance to the inadequate cerebral circulation but primarily to the bilateral inactivation of the pallidum internum, which on the left side was more extensive, involving 50% of the latter and as well the interruption of the internal capsule lateral to L.po and below V.o.a. Because in this case the destruction of the pallidum on both sides was not complete (Fig. 56c, a), the loss of consciousness changed in its intensity and never was complete; the patient could walk about in the ward sometimes and had lucid intervals.

In case 12_{ℓ} an increasing drowsiness started on the operating table; gradually a total lack of responsiveness developed and continued with slight fluctuations until death 14 days after the operation. The anatomical correlation of this mental change was a hemorrhage in the left anterior thalamus, destroying also 100% of the lamella medialis (Fig. 78a, b).

In case 13_r an impairment of consciousness started on day 2 after the operation; it gradually became worse during the following days. Fourteen days later the patient reacted when spoken to, but remained mute and completely lacking in spontaneity. The EEG showed at that time steep delta waves in the right frontotemporal region. Apart from inactivation of the right pallidothalamic system, one found in this case unusual areas of softening in the rostromedial thalamus on both sides (Fig. 82a, b). These lesions destroyed the massa intermedia, present in older x-ray photos (Fig. 79), and, rostrally as well as dorsally to it, the commissural and paramedian as well as medial intralaminar parts of both thalami (Fig. 83a, b). The lesions were caused by the puncture track having damaged the anterior choroid artery in the right lateral ventricle (Fig. 81). As a consequence, the choroid artery which did not fill in the angiogram produced a bilateral softening of massa intermedia and of adjacent structures. However, the blood supply carried by this artery did not stop completely until the second day after the operation, which in turn caused the rapidly deteriorating impairment of consciousness.

The medial parts of the intralaminar nucleus (*i.La*) and the commissural nucleus (*Co*) were destroyed in cases 12 and 13. These nuclei (Figs. 78a, b; 82a, b; 83a, b), identical to the nucleus centralis medialis and the paracentral nucleus, belong to that part of the thalamus that projects to the external pallidum of the basal ganglia rather than to the cerebral cortex. The anatomical basis of the clouding of consciousness in these cases was obviously a permanent asymmetrical loss of the rostral segments of the unspecific system projecting to the cerebral cortex. This case can be compared to that reported by Schaltenbrand (1949), of serous meningitis with unilateral softening in the same area and to the case of Duus (1976) with an almost symmetrical butterfly-shaped lesion of the intralaminar thalamic nuclei. The relation to the lesion obtained through puncture of the anterior choroid artery was obvious in case 13.

V. Findings Regarding the Functional Anatomy of Individual Diencephalic Systems

Before the use of stereotaxic operations, all attempts to obtain functional localization in the basal ganglia consisted of transferring or extrapolating findings made in animal experiments to man. Since the first performance of a stereotaxic operation, stimulation and inactivation of discrete neuronal systems of the basal ganglia in man have been performed daily. These have revealed a wealth of information regarding the functional significance of diencephalic neuron systems in man. The information to be gained from such observations is, however, limited, because only those structures that are expected to yield a therapeutic effect are stimulated or coagulated. The knowledge regarding the functional localization in the human diencephalon has been obtained from post-mortem analyses of series of brain sections. In this chapter, the results are summarized independently of their therapeutic significance.

Not all the results concerning functional localization that have been gained from the correlation of clinical and post-mortem data are mentioned here. They have been described in detail in the preceding chapters. In this section, individual results, which thus far could not have been obtained from animal experiments or from post-mortem material are discussed. We have not attempted to summarize the functional structuring and the relationship of the extrapyramidal system. This has been tried in a survey of striatal control of the impulse generation for locomotion, for intentional actions and of integrating and perceptive activity (Hassler, 1978).

The first topic concerns the connections of the extrapyramidal nuclei and fiber tracts together with the results of their stimulation and inactivation. Then we come to the functional results regarding other fiber structures belonging to the diencephalon, e.g., the internal capsule, and we attempt to describe the functional localization in those structures of the diencephalon that have been sometimes unintentionally affected.

1. Lesions in Forel's Bundle H_1 (Fasciculus Pallido-thalamicus)

With the exception of experiments on monkeys by Carpenter and Strominger (1967) and DeLong (1973), the connections of the pallidum have been only insufficiently investigated. In nonprimate animals, the two parts of the pallidum are very small compared to the conditions in man. The connections of the pallidum to the oral ventral nuclei of the thalamus were first described only in man (C. and O. Vogt, 1907; Riese, 1925) and specified to the anterior oral ventral nucleus (V.o.a) without overlap with V.o.p in man (Hassler, 1949).

These structures belong to the most frequent targets of stereotaxic interventions. It is therefore possible to demonstrate the link from the inner segment of the pallidum crossing a narrowly circumscribed segment of the internal capsule to bundle H_2 (fasciculus lenticularis), with a turn in the H field of Forel, then passing through the bundle H_1, and ending in the V.o.a [Figs. 64 (case 7); 75a–d; 128a].

That the fiber bundles H_1 originate in the pallidum internum is also demonstrated in Fig. 119 of case 11. Because of the coagulation in the dorsal half of the pallidum internum (and externum) the field H in the base of the thalamus and the nucleus V.o.a are devoid of the pallido-thalamic fiber bundles, already 7 days after the coagulation.

The afferent ventral fibers from the pallidum branch into the lateral part of the subthalamic nucleus. These fiber connections exist also in the cat, as shown by Montanelli and Hassler (1964) with the Marchi method. It has also been demonstrated in the meantime by Carpenter and Strominger (1967), based on a large experimental series in monkeys. Thus far, little attention has been paid to the fact that these data are definite proof of the existence of a pathway for excitation from the pallidum to the rostrobasal parts of the VL which we prefer to call V.o.a for differentiating it from V.o.p. The presence of such a pathway has in the past been disputed. The stimulation of the H_1 or the V.o.a results in an increased tough and not springy muscle tension and summation of contractions of the contralateral muscle. At the same time this stimulation turns the head, the body and the binocular eye apparatus and directs the attention to the contralateral side. It is an adversive system in the motor and psychomotor sense. The effect of its unilateral destruction is an inability to turn to the contralateral side. Because the undestroyed pallidar system is still active in the waking state the eyes and head become turned and stimuli are perceived in the sensory field only on the side of the destruction. A unilateral lesion of H_1 or V.o.a also produces hypotonia of the contralateral muscles, mainly of such muscles which maintain the position of the proximal joints. After unilateral therapeutic inactivation of pallido-thalamic fibers the loss of attention and the reduced muscle tone remain but are no longer a handicap.

The functional role of the H_1 bundle in the human is the regulation of the tone of the contralateral muscles and the regulation of turning movements of the eyes, the head, the forebody and the whole body as well as directing attention to the contralateral side.

2. Change of Speed of Movement and Laughing Caused by Localized Stimulation

During stimulation of the V.o.a/p with frequencies of 25–50 cps, active movements of the limbs and articulatory movements are accelerated (Fig. 103b, c). Sometimes this acceleration increases so rapidly (Fig. 75a) that the movements of speech or limbs are arrested in a tetanic contraction, and speech stops after the voice has been raised (case 12_r). If the voltage remains near threshold value during the stimulation, the active alternating movements continue at a more rapid rate (case 1; Fig. 34b). When counting is going on, parts of the series are often left out (Figs. 52, 55a). Not all patients can themselves explain the acceleration. It is significant that the movements intended to be performed before stimulation are continued; only their speed is altered. Guiot *et al.* (1961) made similar observations. The pathophysiological interpretation of this phenomenon is that the cortical motor areas send out impulses for voluntary movements at a higher rate because their specific afferent excitation from the V.o.p is accelerated by the artificial stimulation of the V.o.p; consequently, the motor areas respond as if they had changed to a higher gear.

Slowing of alternating active movements can be induced from the V.o.p or pallidum internum by 8 cps stimuli but this phenomenon is observed much more rarely than a blockade (Figs. 90a, 103b, 119). Arrest of counting backwards was also elicited by stimulation of V.o.a/p in cases 9, 13, 15, 16 (Fig. 129) or pallidum internum in cases 8, 11, and 14 or in 5 of cerebral peduncle (Fig. 56a). This effect of the stimulation is mediated by certain areas of the motor cortex.

Although it could be expected from the result of stimulation, the functional loss of the V.o.p does not result in slowing of active movements. After really complete inactivation of one V.o.p a few cases develop a transient ataxia that is particularly obvious when pointing movements are made.

Coagulations in the pallidum internum or in the oral ventral nuclei have only a slight effect on hypokinetic disorders (see Chap. IV, Sect. 6.5.2.) apart from their improvement by the abolition of rigidity. An improvement of pure akinesia occurs to a limited degree, e.g., in the pulsions, in the gait, in the synergic movements, and activities of everyday life. In the post-mortem material, however, such an improvement was shown to occur only after inactivation of the V.o.p and the V.o.a (Figs. 34b, c, 35), but not after inactivation of the pallidum. This result is supported by the finding that improvement of rigidity or its abolition can be achieved in equal, sometimes complete, measure from the pallidum and the oral ventral nuclei, while the improvement of pure akinesia occurs only after inactivation in the thalamic V.o.a.

Production of laughing. Although Nothnagel (1873, 1889) regarded the thalamus as a center for affective movements, it is only by stimulation of the thalamus in patients subjected to stereotaxis that laughing was caused that was limited to the duration of the stimulation. This laughing (or in many cases, only a smile) frequently starts with a curling of the contralateral angle of the mouth, resulting in the impression of a one-sided smile.[1] A kind of cheerfulness or a recording of a funny experience occurs sometimes but not regularly as

[1] See also the discussion of reduction of affective or expressive movements, pp. 217–219.

a corresponding subjective equivalent to this laughing. The nucleus centralis lateralis adjacent to the internal part of the oral ventral nuclei (Fig. 129), V.o.i or Pallidum internum, are the areas in which stimulation most frequently evokes laughing in clearly conscious cooperative patients (Figs. 108, 109). The same laughing can be elicited by stimulation of the same part of the intralaminar nuclei from the medial side during psychosurgical interventions.

3. Loquaciousness

The *urge to make sounds* combined with turning of the head and eyes to the other side and the jerking up of the contralateral arm is the typical combination of movements resulting from stimulating the supplementary area as first described by PENFIELD and WELCH (1951). The same combination of movements (Figs. 109b, 110) and uttering of sound are occasionally evoked by stimulation in front of and medial to the V.o.a (Fig. 84), i.e., particularly in the medial parts of the anterior pole of the thalamus (L.po=VA) (Fig. 61). Our post-mortem findings showed that the area concerned is probably limited to the mediocaudal part of the nucleus lateropolaris L.po.mc (Fig. 61b). It seems to project to the supplementary motor area. The functional meaning of this area seems to be fast attraction of attention by a contralateral object and an individual rage reaction to it.

4. Functional Organization of the Internal Capsule Corresponding to the Neighboring Ventral Nuclei of the Thalamus

For the clinician the concept internal capsule is linked to the diencephalic part of the pyramidal tract. The previously held views regarding the internal capsule based mainly on clinical material, were as follows: the central pathway for the conjugate eye movements goes through the genu; behind it lie the upper central neurons for the nuclei of the cranial motor nerves, and further caudally are the corticospinal fibers of the pyramidal tract. A horizontal cut (Fig. 134) shows the extent of the internal capsule, which is larger than the pallidum.

The part in front of the genu consists mainly of fiber connections of the prefrontal lobe, amongst others, to the substantia nigra and the rostral pontine grey matter (HASSLER, 1948a; FREEMAN and WATTS, 1950; MEYER and BECK, 1954). Case 2_l with interruption of the anterior limb of the internal capsule (Fig. 132a, b) showed a retrograde degeneration of the corresponding medial nucleus (Fig. 131a, b). Extensions of the lesion set by coagulation have not rarely exceeded the aimed-at thalamic nucleus in a lateral direction and have damaged circumscribed parts of the internal capsule. Based on our present material it is therefore possible to come to some conclusions regarding the functional significance of individual segments of the internal capsule.

Consequently we have, in the following, divided the internal capsule into segments that are defined by the medially adjacent ventral thalamic nucleus (Fig. 134; HASSLER *et al.*, 1967a, b). In a rostrocaudal sequence we want to deal with the segments of the human internal capsule. Thereby the special horizontal level used must be considered. In our case the Fig. 134 cuts the horizontal level on the upper border of the Co.a-Co.p line.

The anterior limb of the internal capsule in front of the genu contains all ascending and descending fiber bundles from the orbital and prefrontal cortical fields, which make the connections to the medial nucleus of the thalamus, the substantia nigra and rostral pontine grey matter. In the segment of the genu that lies lateral to the L.po one finds descending corticobulbar fibers; their interruption is followed by aphonia (Fig. 43a, b). Stimulation of these parts of the internal capsule causes a feeling of oppression, with reddening of the face and apnea in inspiration (M.C. SMITH, 1967). Premotor extrapyramidal and supranuclear pathways for movements of articulation run through the parts of the internal capsule lying lateral to the V.o.a; interruption of the pathways was followed in three cases by the appearance of dysarthria mimical paresis and motor neglect. No hemiparesis is found in lesions of the internal capsule with such a localization. The segment of the internal capsule lying laterally to the V.o.p contains corticobulbar fibers for the muscles of the face. Lesions in this part of the internal capsule (Fig. 132c, 133b, c) led in three cases to a long-lasting central voluntary paresis of the facial nerve and in dyssynergia or even paresis of the contralateral fingers. In addition, pallidothalamic fibers cross ventrally through this segment of the internal capsule (see Fig. 75d). Inactivation of this segment was therefore used by GUIOT (1958; GUIOT *et al.*, 1958) and GILLINGHAM *et al.* (1960) for the treatment of the Parkinson syndrome. The part of the internal capsule lateral to the intermediate ventral nucleus (V.im) perhaps also the segment c.a contains the part of the pyramidal tract with thick fibers, in Figure 134. Stimulation of this segment at low frequency caused jerks synchronous with the stimulus and always affected the same group of muscles. A small extension of a lesion into the caudal pole of the pallidum was followed by an exclusively contralateral paresis of the foot with extensor response. Other lesions of these parts: im and c.a caused contralateral hemiparesis with extensor respons lasting about 2 to 8 weeks. Segment c.p. of the internal capsule lateral to the V.c.p (caudal VPL) contains primarily ascending somatosensory fibers and descending fibers from the sensory cortex going to the nuclei of the posterior columns or to the main sensory nucleus of the fifth cranial nerve. Our material contains no lesions in the segments of the internal capsule that lie lateral to the pulvinar. This segment consists of the fibers of the optic and acoustic radiation of the corticopetal fibers of the pulvinar and the descending fibers from the temporo-occipital region to the thalamus and to the superior colliculi.

In the present group of cases of Parkinson's disease with involvement of the internal capsule, it can be shown from the correlation between anatomical and clinical findings that the internal capsule is neither anatomically nor functionally a uniform structure. The analysis of the functional significance of all segments of the internal capsule considerably exceeds the conclusions described here.

5. Dependence of the Improvement of Tremor on Additional Coagulation of the Nucleus Lateropolaris and Neighboring Parts of the Internal Capsule

The nucleus lateropolaris of the human thalamus corresponds partly to the VA of the electrophysiologic literature. It is, in terms of functions and its afferent fibers, a complex nucleus. Similarly, its efferent projection is partly corticopetal, partly subcortical. When evaluating the therapeutic influence on the tremor, one finds that the improvement due to inactivating the dentato-thalamic system is again impaired by inactivating the nucleus lateropolaris (see p. 194). We found this an unexpected result. To explain this, one can refer to a pathway from the caudatum through the rostral pallidum to the nucleus lateropolaris. In support of this explanation is the knowledge, gained from cases of essential tremor, that patchy losses of substance in the caudatum and putamen can cause a postural tremor (HASSLER, 1939). The efferent system of the caudatum passes, after at least one neuronal relay, through the nucleus lateropolaris; damage to the system at this stage could lead to an increase of existing tremor or to a reduction in a previously achieved improvement of tremor.

6. Contribution to the Problem of Efferent Connections of the Substantia Nigra

It appeared that the old controversy over whether the efferent pathway of the substantia nigra descends to the caudal brainstem or ascends to the basal ganglia had been decided. The findings with fluorescence microscopy regarding the dopaminergic and noradrenergic fibers showed (BERTLER, 1959a; CARLSSON, 1959; DAHLSTRÖM and FUXE, 1964) that the cells of the substantia nigra lead to the putamen and caudatum. For a long time, however, these connections could not be demonstrated with many methods of degeneration or with the stains of NAUTA (1957) or FINK and HEIMER (1967), until they were shown by LLAMAS (1969) and MOORE (1970/1971) and SZABO (1971). The most detailed study by USUNOFF et al. (1976) demonstrated in cat the projection from the rostral nigra cell groups to the head of caudate; from the medial nigra group to the fundus striati and from the lateral cell groups to the putamen, whereas from the caudolateral groups also descending nigra projections arise. The transmitter of all these nigra projections is dopamine (DA) in rat and cat (BERTLER and ROSENGREN, 1959a; CARLSSON, 1959) and baboons (HASSLER, 1974b). Using a dorsal approach to the substantia nigra for the coagulating electrode, COLE *et al.* (1964) found degenerated ascending fibers that enter the posteriomedial part of the thalamic nucleus VL in cats and monkeys (AFIFI and KAELBER, 1965; CARPENTER and STROMINGER, 1967; FAULL and CARMAN, 1968; CARPENTER and

PETER, 1972; USUNOFF and DIMOV, 1972). These connections do not show a catecholamine reaction with the fluorescence method. They have been demonstrated to arise in the substantia nigra pars reticulata (RINVIK *et al.*, 1976). In our material destruction of the V.o.p and the V.o.a (corresponding to basal VL (see Fig. 46 right) has never led to retrograde degeneration of nigra cells (Fig. 47a, b). Such a degeneration of the cells in the substantia nigra should have occurred if the connection of the substantia nigra to the V.o.p existed. As Figures 47a, b, 34b, c, and 36a show, this is not the case. A pure lesion of the V.o.p (Fig. 35) is not followed by an additional degeneration of substantia nigra cells. If the internal capsule or the striatum is damaged, atrophy of the cells in the substantia nigra occurs (Fig. 47a, b). The only condition is that the stereotaxic lesions do interrupt parts of the internal capsule or the efferent pathway from the putamen and caudatum to the substantia nigra, which passes the pallidum. A severe atrophy of cells with excessive growth of glia and shrinking of the area occurs in nigra cell groups (Fig. 77c) of both the two-way connections with the striatum and the connection from the cortex, are interrupted (Fig. 75a, d). Atrophy of the cells of these groups as well as the gliosis and the shrinking of the area is less, if only either the corticonigral pathway or the two-way path between the striatum and substantia nigra are interrupted. From these findings in human serial sections in the cases without stereotaxic lesions, one of us (HASSLER, 1950c) concluded that each group of cells in the substantia nigra receives a pathway from the cortex as well as from the striatum but that the efferent pathway from the substantia nigra descends in the brainstem. This conclusion was made before the efferent pathway from the substantia nigra to the striatum became known by pharmacologic and neurochemical (EHRINGER and HORNYKIEWICZ, 1960; POIRIER and SOURKES, 1965; HORNYKIEWICZ, 1966; HASSLER and BAK, 1969; BEDARD *et al.*, 1969; KIM *et al.*, 1971; BERNHEIMER *et al.*, 1973; HASSLER, 1974b), by histochemical (ANDEN *et al.*, 1964; DAHLSTRÖM and FUXE, 1964; HÖKFELT and UNGERSTEDT, 1969); by ultra-autoradiographic (PARIZEK *et al.*, 1971) and hodologic methods (LLAMAS, 1969; LLAMAS and REINOSO-SUAREZ, 1969; MOORE *et al.*, 1970/1971; SZABO, 1971; CARPENTER and PETER, 1972; and many others). In contrast to our previous discussions HASSLER, 1950c, 1966c), the multiplicity of evidence confirms the ascending nigrostriatal pathway. However, the findings in this material support the old view that some efferent pathways of the substantia nigra (mainly of its posterolateral cell groups) ascend in a dorsomedial direction to the commissure between the superior colliculi and from there descend to the contralateral tegmentum. This is supported by the experimental data of RINVIK *et al.* (1976) and HOPKINS and NIESSEN (1976), which suggest at least a descending nigrocollicular and nigroreticular pathway. Lastly many electrophysiologic findings (ALBE-FESSARD *et al.*, 1960/64; FRIGYESI and MACHEK, 1971; HASSLER and WAGNER, 1972c; 1975; and many others) strongly suggest directly descending nigroreticular or nigrospinal fibers. Only the existence of efferent nigrocortical fibers seems not to be sufficiently supported by newer hodologic methods (MIRTO, 1896; JAKOB, 1923; LLAMAS, 1969; MOLINA, 1969) in contrast to the corticonigral fibers (FOIX and NICOLESCO, 1925; HASSLER, 1937; RINVIK, 1966). Therefore the nigrostriatal fibers, the nigrocollicular fibers, and the crossed descending nigroreticular fibers seem to be established as efferent projections from the nigra compacta, whereas the existence of nigrospinal and nigrocortical fibers and of the fibers from the nigra compacta to the socalled thalamic VL nucleus is less well supported by newer evidence.

7. Impairment of Consciousness Due to Bilateral Lesions in Nonspecific Nuclei of the Thalamus or in the Pallidothalamic System

Clouding of consciousness during a stereotaxic operation, developing quickly out of tiredness, is often due to a hemorrhage spreading into the basal ganglia. If the hemorrhage lies outside the nonspecific thalamic structures, it leads to a clouding of consciousness only if the contralateral basal ganglia are subjected to the pressure of the expanding hemorrhage. In case 13, bilateral lesions in the intralaminar thalamic nuclei (Figs. 82, 83a, b, 127b) were the anatomically relevant findings for a rapidly developing clouding of consciousness with confusion. Of particular importance was the destruction of the rostral segments of the intralaminar nucleus nonspecifically projecting to the cerebral cortex. A softening in the rostromedial thalamus below the ependyma and in the massa intermedia (Fig. 83a, b) developed due to a lesion caused by a puncture of the asymmetrical anterior choroid artery in the lateral ventricle (Fig. 81). As a result, the massa intermedia, which had been clearly visible in the x-ray before the operation, was missing at the autopsy (Figs. 79, 82a, b, 83a, b).

The post-mortem material has also produced the following results regarding the regulation of consciousness from part of the basal ganglia: bilateral, almost complete, inactivation of the pallidum leads to severe impairment of consciousness and, in some cases, to coma or akinetic mutism (Figs. 55a, 62, 64). However, the psychopathologic picture can differ: case 7 showed a severe confusional syndrome lasting 6 weeks until death although the relatives and nurses could be partly recognized. A similar confusion was also seen in case 5, in which an experience of the double occurred after the first inactivation of the right pallidum (Fig. 55a), probably under the influence of a severe cerebral vascular insufficiency. In two cases severe impairment of consciousness with a coma vigil, i.e., a coma combined with a sleeping–waking rhythm, occurred from which the patient could still be roused to a slight degree, though remaining confused and slipping quickly back into coma. Such a coma vigil occurred following a second operation for Parkinsonism, succeeded by hemorrhage into the medial thalamus. In both these cases (12, 13) the pallidothalamic system was almost completely inactivated following a previous operation (Figs. 75d, 78a, 82a, b, 85, 127a), as it was in case 2 (Fig. 43a; 45b).

The pallidothalamic system and the afferent fibers from the unspecific thalamic nuclei play an essential part in the regulation of consciousness, and are not only motor systems, as had been assumed earlier. In some of the cases for which we have no autopsy, stimulating the pallidum or the unspecific nuclei in the thalamus led to waking the patient even during general anesthesia (HASSLER, 1957a). The whole problem was extensively explored and discussed in HASSLER (1977b).

VI. Conclusions

The present analyses of cases of the Parkinson syndrome that have been treated with our methods (Hassler and Riechert, 1954) and technique of stereotaxis (Riechert and Wolff, 1951a; Riechert and Mundinger, 1956) and subsequently checked by autopsy have led to the following conclusions:

1. Determination of the Target Point in the Individual Patient

Most methods of stereotaxis try to reach a target point that is based on the computation of the average coordinates for all patients with the same symptoms. We (Hassler and Riechert, 1954) employ a more accurate method, determining the target point in the individual patient by comparing each of the three coordinates in the model brain with each of those of the individual patient's brain; this method is an extreme numerical approximation. Whether the individual target point has been reached, has in the past been checked mainly by finding radiologically the location of the electrode point. The present post-mortem examination of such cases has confirmed that the method is sufficiently reliable in reaching the target point of the individual brain. The point actually reached in the operation was compared with the intended target in the nucleus of the brain of the individual patient. The median deviation from the ideal anatomical target in the post-mortem material is: in the vertical coordinate, 1.10 ± 0.92 mm; in the frontal coordinate, 1.71 ± 1.14 mm; and in the sagittal coordinate, 1.23 ± 1.48 mm (see Fig. 101). These figures include the deviations due to the technique of individual calculation as well as those caused by the apparatus and by radiology, as determined by Mundinger *et al.* (1970) from 1837 stereotaxic operations.

The accuracy of this method of individual localization has been confirmed by autopsies. In this method three intracerebral distances are measured on coordinates that are at right angles to each other and therefore form a Cartesian spatial system of coordinates. The smallest difference between individuals is found in the vertical coordinate, which is measured from the middle of our baseline to the lower edge of the lateral ventricle (see Fig. 143b). The sagittal coordinate, which forms the baseline, goes from the posterior lower edge of the interventricular foramen to the anterior edge of the posterior commissure. It has been shown that this line can be reliably determined in a radiologic picture (Fig. 143b) and in the anatomical preparation (Fig. 143a). As was also stressed by Andrews and Watkins (1969), this is preferable for the practical determination of the target to the anterior commissure–posterior commissure (C.a-C.p) axis. For the frontal coordinate one of two lines is taken: either the width of the ventricle from the midline to the inner border of the caudatum in the plane of the posterior commissure or the width of the hemispheres from the midline of the brain to the internal table at the height where the squama temporalis reaches its greatest width. It usually lies at the upper edge of the squama temporalis and corresponds in the brain to the first temporal gyrus. The comparative radiological and anatomical evaluation shows statistically a larger deviation of the target point for the frontal than for the other coordinates. The most reliable way of determining the frontal coordinate is to use the width of the hemispheres but to subtract 1.5 mm from each hemisphere for the dura and the subarachnoid space.

Statistically significant differences have been found when the width of the third ventricle has been compared in the x-ray and at the post-mortem. These differences are to a considerable extent due to the fact that the surrounding diencephalic tissue has absorbed the C.S.F. from the third ventricle. It is known that some hours after death no more C.S.F. can be found in the ventricles. As a relating distance for the frontal coordinate the width of the third ventricle is therefore useless, particularly as neuropathologic changes in the diencephalon enlarge the third ventricle. Despite this, one is justified in measuring the frontal coordinate from the wall of the third ventricle and not from the midline of the brain, except for structures in the midbrain or adjacent to midbrain levels.

The accuracy of aiming is considerably influenced by the fact that the distances between the structures seen in the encephalogram depend on the time at which the x-ray is taken. This applies particularly to the distances in the frontal plane. The most suitable x-rays for comparison with the structures in the model brain are pictures taken less than an hour after the filling with air has begun. BRONISCH (1951, 1952) first stated that the size of the cavities in the brain continue to increase for more than 24 h, and speculated about the reasons: The dilatation of the ventricles persists for a short time even when the ventricles are again filled with C.S.F. after two or three days. This explains deviations of the point of the electrode in the frontal plane from the calculated ideal targets in patients with a dilated ventricular system in whom the operation has been carried out very soon after encephalography. The interval between encephalography and operation should therefore be somewhat extended in patients in whom the ventricles are dilated by atrophic processes.

Vertical deviations from the target point are in some cases due to the posterior commissure being as much as 4 mm too low relative to the neighboring thalamic structures. As an approximate measurement for a low position of the posterior commissure the height of the thalamus above the posterior commissure can be used. In such cases the baseline of the thalamus in its posterior part must be displaced upwards. This can not at present be done by relying on measured figures. As a rule, however, the base-line crosses the massa intermedia at the border to its lower third. It can therefore be used for reconstructing a higher point at the end of the baseline in the posterior commissure.

2. High-Frequency Coagulations

There is agreement in the literature that coagulation with high-frequency current is one of the safest methods for inactivating circumscribed parts of the brain. Alternating high-frequency current of pure frequency proved to be the best (WYSS, 1945; HUNSPERGER and WYSS, 1953). The inactivation of tissue depends on the production of heat in the cerebral tissue; one therefore measures the temperature of the tissue around the electrode used for coagulation with a thermocouple at the point of the electrode [see Figs. 114, 115b (3. electrode), 116].

Expanding massive hemorrhages occurred (Fig. 88a), only in the first few years when a spark-gap diathermy apparatus was used. With this instrument, sudden sharp rises of temperature in the wider surroundings led to severe damage of blood vessels. When performing coagulation with a current of sinusoidal pure frequency while measuring the temperature, it is sufficient to employ a temperature of 70° C; at this temperature, there is only a slight influence on the surrounding blood vessels. This method avoids massive hemorrhages almost completely. In order to reach a second target point or to enlarge the lesion in a specific direction, one uses a string electrode (Fig. 115), which can be moved out at a right angle up to 10 mm in all directions with a radius of curvature that can always be reproduced. The current to be employed should be the same as that found in the preceding coagulation when current and voltage had been determined with a temperature kept at 70° C. If the string electrode is the only one employed, then one uses empirical values. Time is the decisive variable in determining the size of the lesion when using high-frequency coagulation. Up to a duration of 30 sec, the size of the lesion increases linearly in relation to the time (Fig. 117).

For instance, when using a 4 × 2 mm thermoprobe and a coagulation of 20 sec duration, the lesion will form a cylinder about 7 mm in diameter and about 7.5 mm long. When employing the string electrode the diameter at right angles to the axis of the string is 3.4 mm (Fig. 116). These values apply to gray matter, particularly to the thalamus. If the lesion invades a fibrous structure such as the internal capsule or the peduncle, the lesion will becoma larger in that direction (Fig. 38a). A drop in the current during high-frequency coagulation led in two cases, on whom a post-mortem was performed, to an undesirable extension of the lesion, in one case with circumscribed hemorrhages. If the current drops, it is necessary to switch it off immediately in order to avoid undesirable extensions of the lesion through the development of excessive temperatures, which would damage the blood vessels. It must be regarded as a danger signal if a particularly high energy is needed to achieve the usual temperature of 70° C for coagulation. It can be explained by the cooling effect of the blood (37° C) in the vessels in the neighborhood of the electrode.

3. Check by Stimulation

Stimulations at the thalamic target points or their afferent fibers can ascertain the functionally correct localization. In the cases in which post-mortems have been carried out, the stimulation was performed with bipolar concentric electrodes; the distance between the poles was 5 mm and the frequency of stimulation, 1–50 cps. Increase of the amplitude of the spontaneous tremor or change of its frequency to that of the stimulation is specific for stimulating the V.o.p (Fig. 126) or its dentato-thalamic afferent fibers, as has been ascertained by anatomical examination. Acceleration of active alternating movements or of spontaneous speech (Fig. 103 b, c) are characteristic effects of stimulating the dentato-thalamic system. Stimulating either the V.o.a or its afferent fibers from the pallidum with higher frequencies causes a tonic contraction of the contralateral limbs and face with mydriasis, opening of the eyes, and turning of the gaze to the opposite side. If a smile on the other side of the face starts or laughing occurs during stimulation, this indicates that the electrode is located medial or lateral from the lamella medialis (Fig. 129). Turning of the head during the stimulation indicates that the electrode is located in the anterior and medial edge of the V.o.a in the direction of the L.po (Fig. 61 b).

Twitches in the facial muscles, the tongue, and the contralateral arm in rhythm with the stimulation when the intensity of the stimulation is low are suspicious of a spread of the stimulation into the neighboring lateral corticospinal system of the internal capsule, mainly lateral V.o.p and V.im.e (see Fig. 134). Higher-frequency stimulation in the internal capsule results in tetanic contractions of the contralateral limbs and facial muscles. If such stimulation effects occur the following coagulation bears the danger of a severe paresis. The disruption or interruption during such stimulation by tonic contraction in the face or tongue, can be regarded as a sign of a localization in the anterior part of the internal capsule next to the V.o.a. A coagulation at this level of internal capsule can result in a contralateral mimic weakness.

4. Further Checks on the Accuracy with which the Target Structure is Reached

Numerous attempts have been made to obtain a check of localization by electroencephalography, but they did not produce many positive results. For this reason many surgical teams have abandoned it. In the present material the most certain effect of stimulation of the V.o.a or the pallidum is the phase reversal between the high frontal and temporal bipolar leads (Fig. 65) as described by GANGLBERGER (1961 a, b, 1962 a; HESS, 1961). A scalp electrode can be influenced by potential fluctuations of the cortex that are up to 5 cm away; for this reason it is not possible to obtain an accurate check of the localization of a stimulating electrode by a scalp EEG. In Figures 66 and 125 c one observes a recruiting response of precentral evoked potentials when stimulating at a rate of 4 or 8 cps in the area of the V.o.a.

In order to obtain an indication of the localization of the electrodes in the thalamus, ALBE-FESSARD *et al.* (1962, 1966) and GUIOT *et al.* (1962 a, c) first and later JASPER and BERTRAND (1964, 1966) and many others used microelectrodes with a relatively low resistance of about 0.1–1 MegOhm.

The borderline between the intermediate and oral ventral nuclei can be more accurately determined than expected by moving the electrode from a caudal position in a rostral direction. By combining this microphysiologic method with stimulation of the skin, one can obtain information (GUIOT *et al.* (1962) about the localization of a point of the electrode in different somatotopic parts of the somatosensory ventral nuclei (V.c.p.e, V.c.a.e, V.c.p.i corresponding to VLP or VPM).

5. Special Indications for Therapeutic Results

5.1. Rigidity of Parkinsonism

Rigidity is relieved by coagulations in the pallidum internum (Figs. 64, 98, 119) or in the V.o.a (Figs. 45 c, 85, 95) or is almost abolished if this coagulations inactivate this neuronal system almost completely. The same effect on rigidity can be obtained by interrupting the fiber connections in the subthalamus between the pallidum internum and the V.o.a (Fig. 75d). It is important to place the inactivating lesion in the V.o.a near the base (Fig. 95). In the V.o.a, as in the caudal ventral nuclei, there exists a localization for head, arm, and leg from the medial toward the lateral region with a slant toward medioventral. A comparable somatotopic localization is present in the pallidum internum from its rostral toward its caudal end (Fig. 29). Such restricted lesions in the pallidum and the V.o.a can achieve complete abolition of rigidity in arm or leg (Fig. 121) without an additional lesion in the internal capsule. We do not regard a lesion of the internal capsule (Fig. 75d) as desirable, although it can increase the tonus reduction on the operating table; this lesion carries with it the danger of the later development of increase of springy muscle tone and of secondary undesired effects.

A contralateral reduction of expressive movements of the face (Fig. 53) appearing on the operating table is one of the unintended side effects of coagulation in the V.o.a or pallidum, often combined with an ipsilateral temporary increase of mimic movements. This sign is related to pallidothalamic inactivation but it need not be regarded as a serious complication since it always disappears within days or a few weeks.

The rare postoperative voluntary facial hemiparesis were often found at autopsy to be due to a lesion in the level of the internal capsule adjacent to V.o.a (Fig. 43a, b). A lowering of the contralateral arm during coagulation is not in every case a sign of a lesion of the internal capsule, but can occur upon extensive destruction of the pallidum or of the V.o.a while the internal capsule is spared. In the latter case the lowering is due to the abolition of the abnormally increased tonic stretch reflexes. Lateropulsions toward the side of the operation are based on the same abnormal mechanism. In order to avoid an impairment of active mobility to which the tonic stretch reflexes make an integral contribution, one should not attempt to replace rigidity by a marked hypotonus. Extension of the lesion to the border or into the internal capsule should, therefore, be avoided (see Figs. 34b, 35).

In most cases of Parkinsonism it is preferable for the abolition of rigidity to coagulate the V.o.a and its afferent fibers (H_1) in the subthalamus rather than to coagulate the pallidum or the part of the internal capsule where the pallidothalamic fibers cross. The reason is that a better effect on the tremor can be obtained from the neighboring V.o.p and that the hemorrhages from the basal blood vessels (Fig. 64) and their branches can be avoided. More serious than a weakness of expressive movements of the face is the postoperative reduction of spontaneous movements on the contralateral side that occurs occasionally after coagulation in the V.o.a or the pallidum even if rigidity has been abolished and no paralysis has occurred. In cases without mental change due to age, this postoperative neglect is not important because the spontaneous initiative with physiotherapeutic treatment is sufficient to overcome it. In older patients the gain of active mobility is impaired in the beginning by the postoperative neglect. It is then necessary to treat the patient for a long time, if possible as an inpatient, with physiotherapy, and to improve the cerebral blood supply.

After coagulation of the V.o.a the vegetative disorders are important, particularly the lowering of blood pressure, which in the beginning is a fairly regular event. Such patients need vigorous medical inpatient treatment and improvement of circulation by giving infusions of human serum or artificial macromolecular solutions. The postoperative lowering of the blood pressure causes the danger of an undesirable increase of the size of lesion through inadequate circulation in the small, partly obstructed blood vessels around the coagulation. It seems unjustified to make a coagulation of V.o.a only for the purpose of reducing hypertension (SCHMIDT, 1966) because this is only a side-effect of the reduction of muscle tonus.

5.2. Akinesia and Hypokinesia

Akinesia and hypokinesia that are not due to rigidity of the muscles cannot be improved by coagulation of the pallidum but to some extent by coagulating the V.o (see Table 8, p. 208, 209). After coagulation of the pallidum, akinesia on the contralateral side is occasionally at first increased; this increased postcoagulation akinesia can only be distinguished with difficulty from the primary akinesia due to nigra cell loss and from the motor reduction due to rigidity.

Aphonia and disorders of articulation can be severe after the second operation; this can be an increase of a Parkinsonian disorder of speech, or it can have developed independently. For a pathophysiologic explanation, it should be remembered that, when one coagulates the rostral part of the thalamus, the coagulating electrode pierces the internal capsule near the genu. This can cause degeneration of fibers. In order to avoid aphonia and disorders of articulation, the patients are instructed to speak in a nonautomatic way during the coagulation. For instance, they should do serial subtractions, recite poems, etc. This enables the surgeon to watch at the same time any impairment of mental concentration or attention during the coagulation. This disorder becomes evident on the operating table by tiredness and by disorders of concentration or spontaneous speech. Thus, the pallido-thalamic systems (Fig. 119) play a part in governing not only movements but also psychomotor events. The worst disorder can occur with complete bilateral inactivation of the pallidum, which can lead to akinetic mutism (see Figs. 55a, 139).

A rapidly increasing impairment of consciousness progressing to coma after the operation is an unequivocal indication of a hemorrhage in the basal ganglia (Figs. 58a, 127a, b). Such a coma can be related to bilateral damage of the pallidum (Figs. 139, 55a) or of the unspecific projection system of the thalamus (Figs. 82, 83). No success is achieved even by an immediate attempt at surgical removal of such a hemorrhage. Blood coming out of the cannula used to introduce the electrode can be the first sign of a hemorrhage. It can be treated by reintroducing the electrode and applying a high-frequency current. In the few cases so treated, we had the impression that further hemorrhage was stopped.

5.3. Tremor at Rest

As the cases with autopsy reported here show, tremor at rest, including tremor that does not belong to the Parkinson syndrome, can best be abolished by coagulation at the base of the V.o.p and by inactivating in the subthalamus the dentato-thalamic fibers (Fig. 46) before they enter the V.o.p, where they terminate. A coagulation of the nucleus reticulatus next to the V.o.p can apparently be equally effective (see ROSENSCHON and WECHSLER, 1964), but damage to the internal capsule must be avoided. The improvement caused by complete inactivation of the pallidum never exceeds two-thirds. The V.im was (Fig. 35) either not involved at all or only up to 5% in all cases of tremor that were completely or almost completely improved (Fig. 122). Thus the simultaneous inactivation of the so-called tremorogenous zone in the base of V.im has no part in abolition of tremor according to our post-mortem material. This must be emphasized in contradiction to NARABAYASHI (1972, 1977) whose opinion is not supported by post-mortem examinations.

When coagulating the dentato-thalamic fibers in the zona incerta, occasionally the bundle H_2 or the edge of the subthalamic nucleus (corpus Luysii) is also coagulated (Figs. 35, 44). This leads a few days or weeks after abolition of the tremor to the appearance of nonrhythmic rapid small twitches of the fingers or the hand (myoclonic bursts; HASSLER, 1977a). The rare cases of ballistic hyperkinesia following operations for Parkinsonism are caused by a larger lesion of the subthalamic nucleus (corpus Luysii) destroying more than 20% of its volume as seen also by HOPF *et al.* (1968).

In cases with extensive inactivation of the V.o.p, one can find a considerable slowing down of the relearning of coordinated movements and synergy of muscles. This is evident in many everyday activities, e.g., writing or buttoning, without the existence of paralysis. According to the autopsy findings, the absence of synergy is more frequent when the pallidum has been extensively inactivated (Figs. 43a, 98). These disorders can create difficulty on the dominant side. The best means of avoiding their appearance is probably to restrict the lesion to the size that is essential for abolition of tremor. In order to recognize slight disorders of synergic movements, it proved useful to make the patients either stretch their arms out or to perform alternating movements during the coagulation.

6. Pathophysiology of Parkinson Syndrome

6.1. Tentative Interpretations of Parkinson Symptoms:

a) Disorders of Nigrostriatal Circuit Function and

b) Antagonism of Descending Nigral and Pyramidal Impulses in Anterior Horn Apparatus

All types of Parkinson syndrome are due to a slow, progressive destruction of nigra neurons. The dopaminergic nigra neurons contain dopa-decarboxylase by which they form dopamine (DA) from L-dopa. The neurons do not contain dopamine-β-hydroxylase, so the DA is not further transformed into noradrenaline (NA) in the nigra. The nigra neurons take up tritium-labeled NA (^{3}H-NA) injected into the brain ventricles in the same amount as tritium-labeled DA (^{3}H-DA) in the perikarya and dendrites. Both ^{3}H-NA and ^{3}H-DA are stored in nigra neurons (Fig. 13a, b) (PARIZEK *et al.*, 1971). Tritium-labeled serotonin (^{3}H-5HT), however, injected into the brain ventricles is taken up in another pattern and stored mainly in boutons containing dense-core vesicles (Fig. 13c, d) that emit a yellow fluorescence indicative of serotonin.

The majority of nigra neurons conduct the transmitter DA upward to the striatum; there DA acts by axo-spinous boutons, type I (Figs. 16a, 17b) in the interneuronal apparatus and by axo-dendritic or axo-somatic boutons en passant on the large projecting neurons (Fig. 16b). The therapeutic action of DA can be exerted in Parkinson patients only if some of the DA-ergic neurons are preserved. The DA-ergic boutons converge within the striatal spiny interneurons with glutamatergic type III boutons (Figs. 16c, 17c) originating in different cortical areas and with the probably cholinergic type IV boutons originating from the center median nucleus of the thalamus (HASSLER, 1975; CHUNG *et al.*, 1977). The great convergence of terminal boutons on the spiny striatal interneurons (Fig. 4) is the morphologic basis for the regulatory, predominantly inhibitory influence exerted by the cholinergic synapses of type IX on the large "aspiny" striatal neurons (HASSLER *et al.*, 1977).

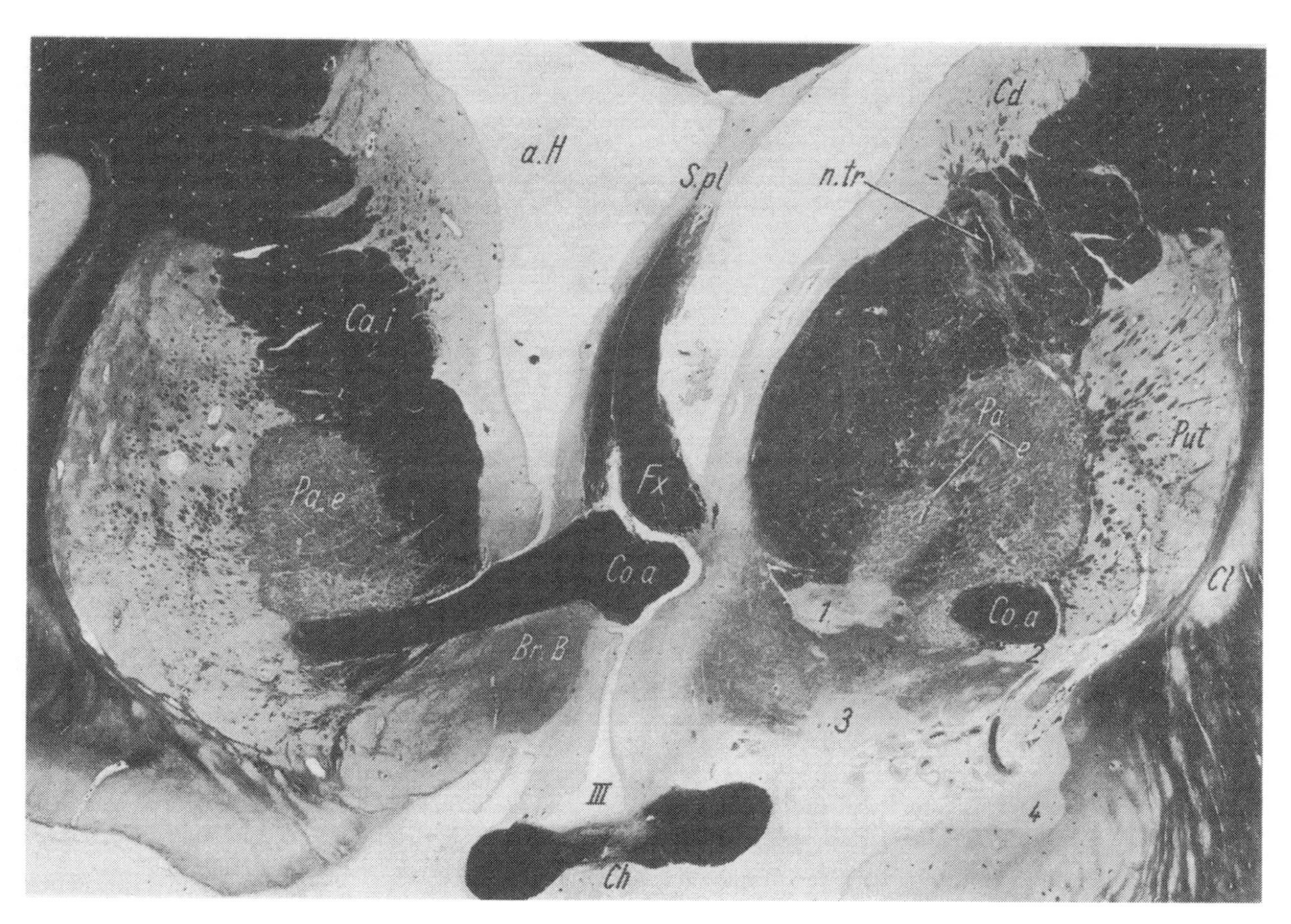

Fig. 140. Section through basal ganglia of case of combined status marmoratus and polysclerotic Parkinsonism. Right putamen is much more reduced in size by status marmoratus than left. Conspicuous foci of demyelination in substantia innominata (*2, 3*), in limen insulae (*4*), and in pallidum (*1*). In right capsula interna is electrode track (*n.tr*), through which action myoclonus has been relieved. *a.H.* anterior horn of lateral ventricle; *Br.B* bundle of Broca; *Ch* chiasma; *Cl* claustrum; *Co.a* anterior commissure; *Fx* fornix; *Pa.e* external segment of pallidum; *Pa.i* internal segment of pallidum; *S.pl* septum pellucidum; *Cd* caudatum; *Put* putamen; *III* third ventricle. Fiber stain. ×2.5 (HASSLER *et al.*, 1975)

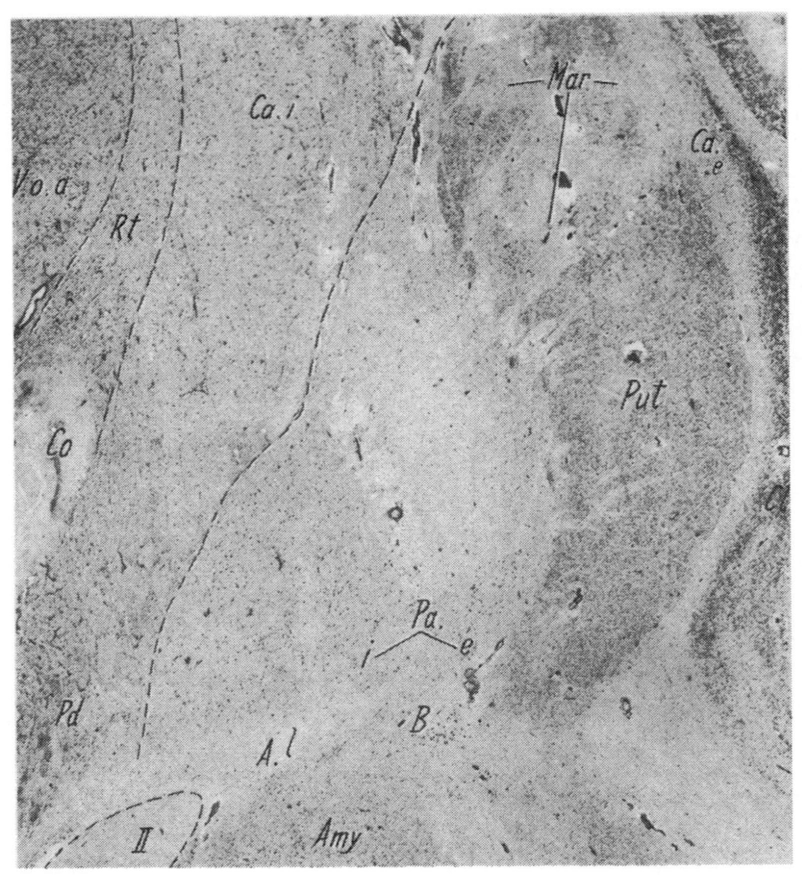

Fig. 141. Cross section through anterior thalamus of same case as in Figure 140. Dorsal putamen (*Put*) is interspersed by irregular glia scars due to status marmoratus (*Mar*). External segment of pallidum (*Pa.e*) shows extensive nerve cell loss without gliosis, possibly secondary effect of status marmoratus. Coagulation focus (*Co*) is located in reticulate nucleus (*Rt*) around V.o.a nucleus of thalamus, whereby action myoclonus was abolished.
A.l ansa lenticularis; *Amy* amygdaloid nucleus; *B* nucleus basalis; *Ca.e* external capsule; *Cl* claustrum; *II* optic tract. Nissl stain. ×5.6 (HASSLER *et al.*, 1975)

In addition to the spiny interneuronal apparatus, the three mentioned afferent pathways to the striatum regulate the large GABA-ergic projecting neurons directly by synapses. After nigra cell loss in Parkinson syndrome, the DA-ergic input to the striatal interneurons is strongly reduced (HORNYKIEWICZ, 1966) so that the striatal output does not inhibit the tonic pallidar action. This results in the appearance of Parkinson symptoms such as rigidity and akinesia. An unusual case of combined unilateral severe loss of striatal tissue due to status marmoratus and bilateral nigra cell loss due to polysclerotic demyelinating foci (Figs. 140, 141, 142) has clearly demonstrated that the manifestation of Parkinson symptoms requires an almost undamaged striatum with an undiminished population of interneurons. Otherwise, instead of Parkinson rigidity, akinesia, and tremor at rest, an intention-myoclonus occurs (HASSLER *et al.*, 1975). If the striatal parenchyma is not reduced, the nigra cell loss causes a severe increase in muscle tone and the appearance of tremor-like or myoclonic movements. The striatum on its side exerts its regulating, mainly inhibitory, influence on three efferent structures: (1) the substantia nigra with the feedback loop to the striatum; (2) the pallidum externum with the feedback loop through the subthalamic nucleus; and (3) the pallidum internum with the thalamo-cortico-thalamic feedback loop (Figs. 2, 6, 124a–c).

For the pathophysiologic explanation, however, it is also necessary to deal with the descending as well as with the ascending fibers of the substantia nigra. For many decades normal anatomical findings (WEISSCHEDEL, 1937), myelogenetic findings (HASSLER, 1950, 1966), and experimental data (HASSLER and WAGNER, 1975) yielded evidence that mainly the caudolateral cell groups of the nigra have descending projections. A recent silver impregnation study (USUNOFF *et al.*, 1976) and horseradish peroxidase investigations on the nigra projections have confirmed this idea of descending efferent nigra fibers, though whether they relay in the reticular substance, optic tectum or spinal cord has not been established. An interpretation and explanation of tremor at rest or of some forms of akinesia appear impossible without assuming the recently well-substantiated existence of such a descending pathway (RINVIK *et al.*, 1976; HOPKINS and NIESSEN, 1976).

6.2. Tremor

The normal substantia nigra has a desynchronizing action on the anterior horn motor apparatus (Fig. 124a), which consists of the moto- and fusimotor neurons intermingled with a great number

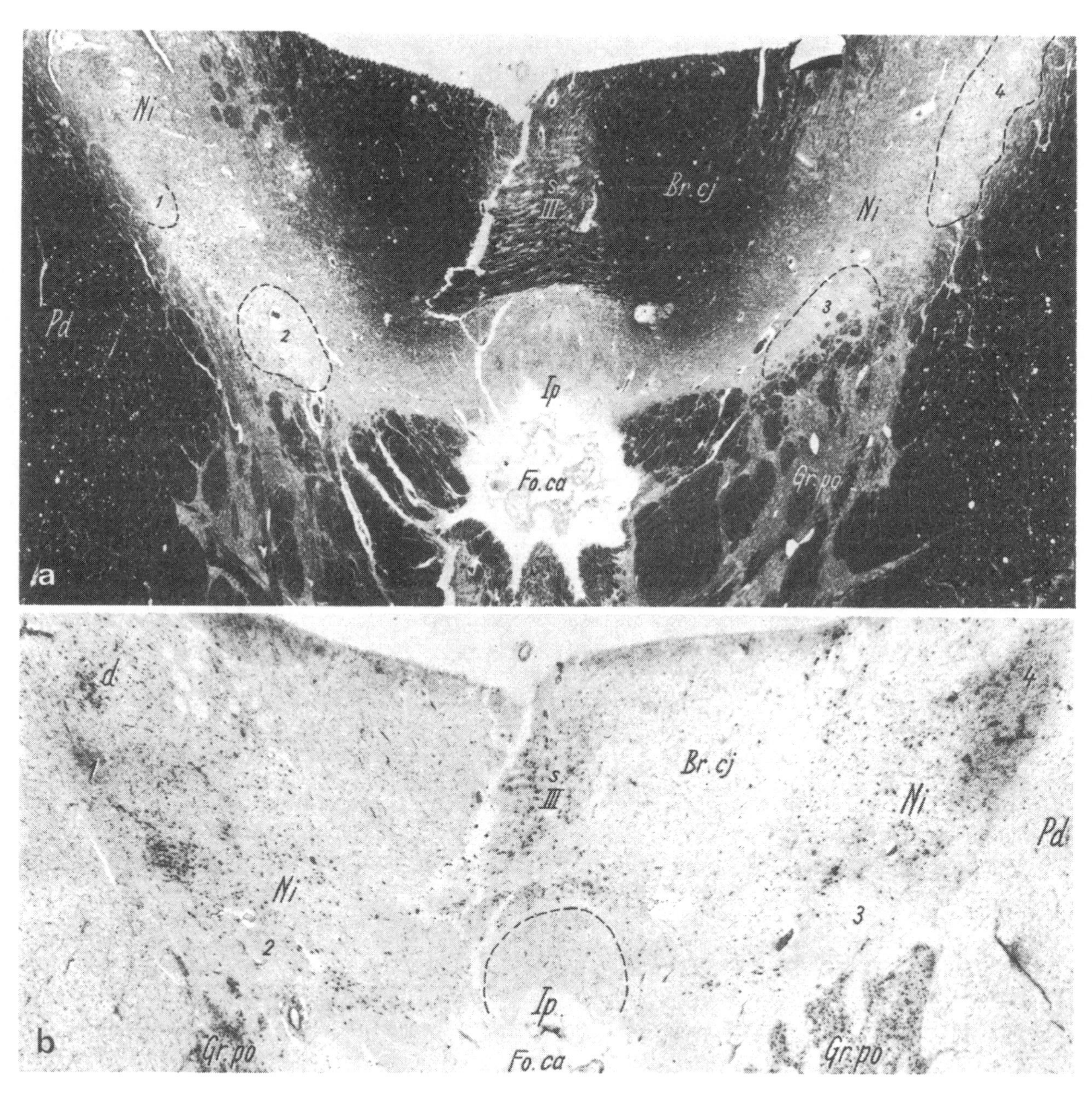

Fig. 142. (a) Cross section through midbrain in case of combined status marmoratus and polysclerotic Parkinsonism. Both substantia nigra (*Ni*) contain poorly defined demyelination foci (*1,2,3,4*) of different ages. Fiber stain.
(b) Adjacent section in Nissl stain shows dense glia proliferation in lateral cell groups (*1,4*) with almost complete nerve cells on right side, whereas in older demyelination foci (*2,3*) nerve cells have disappeared without gliosis. *Br.cj* brachia conjunctiva; *Fo.ca* Foramen caecum pontis; *Gr.po* pontine gray matter; *Ip* interpeduncular nucleus; *Pd* cerebral peduncle; *s.III* nucleus linearis subocu-lomotorius (HASSLER *et al.*, 1975)

of interneurons. In the resting state a steady weak flow of pyramidal impulses induces the motoneurons to feed to the motor units a randomly distributed series of impulses that result in a normal weak muscle tone at rest, without jerky muscular contraction. During an intended movement, however, the pyramidal impulses produce an alert moderate muscle tone by a synchronization of motoneuron activity with a preferential stronger activation of the groups of motoneurons involved in the aimed movement. In Parkinson cases the loss of nigra nerve cells, mainly in the posterolateral group, which project toward the spinal cord, causes the desynchronizing nigra action to decrease and the synchronizing influences of the pyramidal tract to prevail (Fig. 124b). In the alert state of Parkinson patients, this steady flow of pyramidal impulses down to the motor apparatus is the basis for tremor at rest, as soon as the desynchronizing descending action of the substantia nigra is inactivated. Then the alternating synchronized pyramidal impulses to the agonistic and antagonistic muscles predominate, resulting in the tremor at rest with a rhythm of 4 to 6 cps. Since the pyramidal impulses at rest cease during sleep, the tremor at rest also ceases during sleep (Fig. 124a–c).

Such pyramidal impulses at rest originate in the primary motor area 4γ and are there induced by specific afferents that conduct impulses from specific ventral nuclei of the thalamus, especially the posterior part of the oral ventral nuclei (V.o.p) to area 4γ. Because the V.o.p nucleus (the posterior basal part of the VL nucleus, which is the same) is the terminal nucleus of dentato-thalamic fibers originating from the contralateral parvocellular dentate nucleus (HASSLER, 1950), the primary motor area 4γ is the effector of this main output nucleus of the cerebellum in primates (Figs. 1, 2). By stimulating the V.o.p of Parkinson patients with low frequency electrical impulses (below 6 cps), it is possible to synchronize and increase the amplitudes of tremor at rest and, with stimuli of 8 or more cps, to accelerate and increase the tremor until an interference of the tremor and stimulation rhythm occurs.

To abolish the Parkinson tremor, it is sufficient to inactivate in the V.o.p the specific thalamic afferents to the primary motor cortex (HASSLER, 1953b; HASSLER and RIECHERT, 1954a), and it is not necessary to excise area 4γ in the central fissure (BUCY and CHASE, 1939) or to interrupt the pyramidal tract in the spinal cord (PUTNAM, 1940). These two operations, which result in the actual destruction of pyramidal tract neurons, are less effective for tremor alleviation than is the deafferentation. The latter operation does not cause motor paresis or impairment. Also post-mortem studies (HASSLER *et al.*, 1970) show that to achieve tremor relief, the most effective target for coagulation is the specific afferents projecting to area 4γ either in the V.o.p neurons or in their specific afferents, the dentato-thalamic neurons. This intervention results in a strong reduction of the synchronizing pyramidal action, so that a new balance (see Fig. 124c) between it and the lost desynchronizing nigra action is established after the operation.

A 5- to 7-cps pacemaker rhythm has been recorded in Parkinson– (ALBE FESSARD *et al.*, 1962; JASPER and BERTRAND, 1964) and in other patients from the base of the V.im nucleus of the thalamus, from the V.o.p part of the VL, and from the "LP" nucleus, by which parts of the dorsal or pulvinar nuclei are labeled (ALBE FESSARD *et al.*, 1963), even if no tremor at rest is present. This pacemaker rhythm has also been recorded in monkey experiments with experimental postural tremor (GYBELS, 1963) or without tremor (LAMARRE *et al.*, 1971). Because of the lack of its constant correlation to tremor at rest, this rhythm cannot be the pacemaker for tremor in the contralateral side of the body. Moreover the tremor rhythms in the contralateral extremities are not synchronous but differ in frequency in the arm and even in the hand, leg, and face (JUNG, 1941). The meaning of this rhythm in the base of the thalamic V. im is nothing else but the reflection of the impulse rhythm of the anterior horn interneuronal apparatus by ascending pathways under the condition of the Parkinson syndrome.

The γ-loop, consisting of the fusimotor neuron, the innervated muscle spindle, and its afferent fiber IA to the motoneuron, has a stabilizing and amplitude-diminishing action on tremor at rest. If the γ-loop is blocked by injecting Novocain into the muscles responsible for the shaking (WALSHE *et al.*, 1924), *increased* amplitudes of tremor at rest appear. The same is more clearly shown by dorsal root sections of one arm in a postencephalitic Parkinson patient, performed by POLLOCK and DAVIS (1930), which immediately after the operation had resulted in increased tremor amplitudes. If the reflex loop in Parkinson patients is interrupted, as in chronic polyneuritis (clinical observation of HASSLER) with loss of monosynaptic reflexes, the tremor at rest is extremely strong and fluttering, like "wing-beating" and the rigidity in the parts with impaired reflectory sensibility abolished. From these observations the following conclusion is drawn: the pacemaker for the rhythms of tremor of different parts of the body is situated in the anterior horn interneuronal apparatus, which is subjected to the inhibitory influence of the γ-loop and the central antagonistic (desynchronizing–synchronizing) influences of the substantia nigra and the pyramidal tract with its afferents from the V.o.p nucleus (see Figs. 2, 124a–c).

A very important question is the enhancement of tremor amplitudes and appearance of tremor caused by the additional lesions of the putamen and caudate nucleus due to pericapillary loss of nerve cells which the French neurologists have called, for a long time, état précriblé. The pathoanatomic cause of the essential tremor without akinesia and rigidity (hereditary or senile essential tremor HASSLER, 1939) is the état précriblé of the putamen and less that of the caudate nucleus. If such an état précriblé is combined with nigra cell loss, the Parkinsontremor is especially strong and has large amplitudes.

6.3. Rigidity

The loss of the DA-forming nigra cells mainly in the caudolateral cell groups affects the anterior horn interneuronal apparatus so that the DA-ergic inhibition of the tonic stretch reflexes (Fig. 121) is abolished and the increased stretch reflexes cause rigidity. At the same time, nigra nerve cell loss abolishes the facility of the fast γ-neurons for starting and accelerating movements, which is mediated by the nigra. The latter results in a slowing down of new movements and "start-inhibition." Both act to increase rigidity. The reflex reinforcement of the Jendrassik manoeuvre is lost (see Fig. 103a); the fast alternation of to and fro movements are abolished (dysdiadochokinesia). The activity of the preserved slow and tonic motoneurons prevails under the facilitating cholinergic action of the corticospinal or pyramidal tract neurons. The degree of muscle rigidity can be reduced by anticholinergic drugs, whether they are the natural ones such as atropine, scopolamine, belladonna, or the synthetic ones such as trihexyphenidyl or benzatropine. By raising the availability of dopamine for the undestroyed nigra neurons, the activity of these remaining neurons can be raised to an almost normal level. Thus, rigidity can be abolished mainly by the combination of L-dopa and anticholinergic drugs (in contrast to the insufficient action against Parkinson tremor, because the V.o.p transmitter is not known).

To achieve long-term relief from rigidity, it is necessary to abolish the facilitating action of pyramidal tract neurons on the anterior horn interneuronal apparatus by inactivation of these neurons, or better, of the afferents to area 6a α of the motor cortex originating in the nucleus V.o.a of the thalamus (Fig. 121). The same result can be obtained from the preceding neurons in the pallidum internum or the pallidothalamic bundle H_1 by coagulation (Figs. 45, 119). The activity of this pallidum internum, whose transmitter is probably acetylcholine (ACh), is controlled by the putamen output neurons (Figs. 6, 4).

It is also possible to explain the pathophysiology and the operative relief of rigidity in terms of the ascending mechanism from the substantia nigra to the putamen (Fig. 121). The loss of nigra nerve cells, the common pathoanatomical basis of all kinds of Parkinsonism, causes a severe loss of nigrostriatal dopaminergic input to the interneuronal apparatus of the putamen and a loss of nigrostriatal input by type II en passant synapses to the "aspiny" output neurons. Thereby, the physiologic control of the striatal output to the pallidum internum is severely disturbed and reduced. Consequently the efferent influence of the pallidum internum on the V.o.a is strongly disinhibited, i.e., reinforced. This can be suppressed by a stereotaxic coagulation of the V.o.a, so that the facilitating influence to the area 6a α efferents is abolished. The nigrostriatal loop can also explain the restitution between the facilitating and inhibitory influences on the tonic stretch reflex in the anterior horn by V.o.a coagulation (Fig. 121).

The neuronal circuits with their transmitters of the mechanism of rigidity are much better known than those of tremor, and therefore the medical treatment of rigidity is much more effective. The disturbed ratio of DA/ACh in the anterior horn gray matter can be balanced by L-dopa and the exaggeration of striopallidar output can be reduced by anticholinergic drugs.

6.4. Akinesia and Festination

Parkinson akinesia consisting of loss of automatic, involuntary supporting movements, of associated movements, of expressive and reactive movements, results from the loss of nigra nerve cells. However, this is not the only existing form of akinesia. In the Parkinson syndrome a special form of akinesia also occurs. This is the inability to interrupt ongoing automatic movements that accelerate themselves, known as festination of gait, of speaking, of writing, and retro-, antero-, lateropulsions, etc. The slowing down of movements, the typical retarded actions with starting inhibition are other symptoms of psychomotor akinesia. The pathomechanism descends mainly from the nigra to the motor interneuronal apparatus. Therefore, most of the hypokinetic symptoms cannot be restored by a stereotaxic operation in V.o.a or V.o.p, although the associated movements involved in walking can reappear in some cases. Small akinetic symptoms can be improved after V.o.a—not after pallidar—coagulation. This is probably due to the abolishment of rigidity in which small automatisms are "smothered". Thus, it seems that the true hypokinetic symptoms are due to a conduction from the nigra cells to the anterior horn motor apparatus. They can be influenced by replacing L-dopa for the remaining nondestroyed descending nigra neurons and otherwise only by removing rigid muscle tensions, which disturb the easy and smooth performance of the automatic movements.

If in Parkinson therapy an overdose of L-dopa is given against akinesia or rigidity, toxic symptoms in the form of myoclonic, choreatic, ballistic, and dystonic movements occur. Because of the complexity of these movements, their neuronal apparatus seems to be the interneuronal set of the striatum. There is one toxic symptom of L-dopa overdosage that has not been described except by one of us (Hassler and Wagner, 1975). It is hyperactivity of tendon reflexes, which is very evident on the nonoperated side in Parkinson cases in which on the other side a stereotaxic coagulation of the V.o.a and the V.o.p has been performed. For explanation: as previously described, DA enhances the activity of fast γ-neurons, so that the sensitivity of the muscle spindle is increased (Fig. 121) with respect to the mechanical stretch exerted by the reflex-hammer (Fig. 103).

6.5. Vegetative Symptoms

The pathophysiology of the vegetative Parkinson symptoms, mainly hyperhidrosis, sialorrhea, and "grease face," is not clear. Even 38 years ago (as early as 1938), Hassler tried to correlate them with the almost constant nerve cell loss and nerve cell alterations found in locus caeruleus and in dorsal nucleus of vagus in Parkinson brains. In the meantime it has become obvious that the locus caeruleus produces great amounts of noradrenaline (NA) and transports it upward to most parts of the cerebrum. Probably there is also a transportation of NA downward to the bulb.

6.6. Bradyphrenia

The most controversial point in Parkinson pathophysiology is the psychic symptomatology—slowness of thinking, lack of initiative, difficulty of decision-making, tendency to compulsive behavior. It can perhaps be correlated to the glassy degeneration of the nerve cells of the basal nucleus, including the Broca band, which belongs to the constant neuropathologic pattern of Parkinson syndromes. The large nerve cells seem to be the pacemakers of the Ammons horn, according to Stumpf (1965). One might speculate that psychic Parkinson symptoms could also be correlated with functional disturbances in the interneuronal apparatus of the striatum (see Cools and van den Bercken, 1977; Hassler, 1977b; 1978).

6.7. Pathomorphologic Differential Diagnosis of the Different Etiologic Forms of Parkinsonism

The nerve cell losses in different forms of Parkinsonism are not evenly distributed through the discrete nerve cell groups of substantia nigra. Therefore an attempt at a quantitative approximation had to be made. The distribution of the nerve cell losses in the individual nigra cell groups of the idiopathic (and hereditary) cases of Parkinsonism follows a distinct pattern of sequence, whereas in other etiologies there is no constant pattern. This characteristic pattern of involvement of groups established by Hassler (1938) on 2 hereditary and 10 cases of idiopathic Paralysis agitans was confirmed in this monograph by 8 additional Paralysis agitans cases: *1* (Fig. 36a, b), *2* (Figs. 47a, b; 133b, c), *3* (Fig. 150c), *5* (Fig. 55b, c), *7, 10, 12* (Figs. 76a, b; 77a, b, c) and *13* (Figs. 84; 85).

In all 6 cases of postencephalitic Parkinsonism (cases 4, 8, 14, 15, 16 [Fig. 96a, b] and 17) of this monograph as in 11 postencephalitic cases published by HASSLER (1938), however, the involvement of the Nigra nerve cell groups does not follow a distinct pattern and is easily to distinguished from the paralysis agitans pattern mainly because of the nerve cell losses in the medial cell groups of the posterior main division. Also in case 6, where the Parkinson syndrome was due to a hamartoma expanding the third ventricle, a pattern (Figs. 60a, b, c) completely different from that of idiopathic and hereditary Parkinson syndromes appeared, as in one former case with a midbrain tuberculoma (HASSLER, 1953c).

The cytopathologic differential diagnosis of idiopathic Paralysis agitans is based on the presence of so called LEWY's (layered inclusion) bodies in remaining Nigra cells (HASSLER, 1938; 1955b); they have been found only in 4 of the 8 new histologically checked paralysis agitans cases. They are not present in any of the new postencephalitic cases nor in the tumor case 6. Cytopathologic characteristics of postencephalitic Parkinsonism are the neurofibrillary tangles as described by FENYES (1933) and HALLERVORDEN (1935). This cytopathologic feature, as well as the appearance of LEWY's bodies in Nigra cells is, however, extremely rare and therefore not sufficient for morphological diagnosis. The distinct different pathoarchitectonic pattern of nigra nerve cell loss, however, allows in almost all cases such a differential diagnosis. This must be stated, although almost all other authors who described Nigra cell lesions in Parkinson cases, compilated by H. JACOB (1978), did not refer to this patho-architectonic pattern of paralysis agitans cases, detected by HASSLER.

Arteriosclerotic Parkinsonism. At later stages the idiopathic Parkinson syndrome is often combined with arteriosclerosis of many brain vessels and functionally with a reduced cerebral blood circulation in the Kety-Schmidt or radioisotope tests (K. SCHMIDT, 1966). This does not justify to call these Parkinson-syndromes "arteriosclerotic" because most of these cases have the typical Paralysis agitans pattern of Nigra nerve cell loss and occasional inclusion bodies. The arteriosclerosis of some cerebral vessels is only accidental. Only if within the substantia nigra a focal vascular destruction is present, as in two "senile" cases of HASSLER's (1938) it is justified to speak of arteriosclerotic Parkinsonism.

The described two forms of nigra cell degeneration have been interpreted as precocious senile involutions of Nigra cells (HALLERVORDEN, 1935; HASSLER, 1938; 1955b; SPATZ, 1938; C. and O. VOGT, 1942). They must be neurochemically different types of presenile involution, of which the neurofibrillary tangles can now experimentally imitated by Aluminium salts (KLATZO *et al.*, 1965; WISNIEWSKI *et al.*, 1970). Especially evident is that the basal nucleus or nucleus substantiae innominatae which in normal ageing undergoes an early change of cytologic structure (HASSLER, 1938) in the form of glossy cell alteration (LEWY, 1923) with some nerve cell disintegration; this is enhanced and appears earlier in cases of paralysis agitans, which is also a genuine precocious senile involution in contrast to most of the nonspecific cell degenerations due to vascular disorders in many other cerebral structures, cortical and subcortical.

6.8. Are the Described Cases Representative?

That the discussion of side effects and complications fills a considerable space in this monograph is unavoidable because these investigations and correlations are based only on cases coming to autopsy. These represent almost always a negative selection and can be regarded only to a limited degree as examples of the stereotaxic treatment of the Parkinson syndrome. The anatomical variations and the exact anatomical localization of the locus of stimulation and of the coagulated lesions and the neighboring structures involved were investigated in serial section. It was thus possible to examine the correlations between clinical results, physiologic observations, and the discrete responsible neuronal systems. The reliability of using stereotaxic interventions in the pallido-thalamo(V.o.a)-cortical system as treatment of rigidity and in the dentato-thalamo(V.o.p)-cortical system as treatment of tremor at rest was thereby confirmed. At the same time, this monograph has provided many theoretical insights into the functional significance of many central neuronal systems that could not be found in animal experiments. The present investigation can represent the therapeutic justification for and localization of the findings in cases which have not come to autopsy. This material consists of 5323 stereotaxic operations that we have performed for extrapyramidal disorders of movement, particularly for the Parkinson syndrome.

6.9. The Future of Therapy in Parkinsonism

Although all the cases described above were operated on before the era of treatment with L-dopa, the latter seems to us to supplement rather than substitute for the stereotaxic treatment of Parkinsonism. L-dopa therapy alone or combined with decarboxylase inhibitor is almost generally considered as less effective against tremor at rest than against rigidity and akinesia. Consequently the tremor, not effectively controlled by L-dopa treatment, was the most frequent indication for stereotaxic Parkinson therapy in the last ten years. Over the years L-dopa treatment of Parkinson symptoms has proved to be progressively less effective after periods of 3 up to 7 years (BARBEAU, 1976; DIAMOND *et al.*, 1976; SELBY, 1976; FISCHER *et al.*, 1978) even for akinesia and tremor at rest. The side effects of long-term L-dopa treatment are: lack of tolerance with inner restlessness in spite of enhanced brachykinesia and developing organic psychosyndromes and paranoia. L-dopa acts on the remaining intact ascending and descending dopaminergic nigra neurons. However, Parkinson syndromes can be favourably influenced also via the antagonistic cholinergic pallido-thalamo-cortical and pallido-reticulo-spinal neuronal pathways.

Since in contrast to the dopaminergic nigra system, these systems are not primarily damaged in Parkinsonism, a Parkinson therapy acting on these antagonistic system does not become ineffective after a period of several years. The treatment may consist of the administration of anticholinergic drugs or of a circumscribed stereotaxic inactivation of the pallido-thalamo(V.o.a)-cortical systems. The surgical therapy has the longest lasting therapeutic effect; we have only exceptionally seen a relapse of rigidity and tremor at rest on the operated side even in patients we have operated on 25 years ago. In view of the long-lasting reliefs of rigidity following a stereotaxic coagulation of V.o.a and of pallido-thalamic fibers (H_1) one may speculate that the operation had a prophylactic effect in that it prevented the appearance of Parkinson symptoms that might first have set in later years. The stereotaxic operation has also the advantage that, in addition to the inactivation of the–probably cholinergic–pallido-thalamo-cortical pathway in V.o.a for the relief of rigidity, also the dentato-thalamo-cortical pathway in V.o.p can be inactivated for the relief of tremor at rest. The transmitter of this latter neuronal chain is not acetylcholine. It may therefore be predicted that in the long run the stereotaxic treatment of Parkinsonism will be supplementary to treatment with L-dopa but not replaced by it, at least until the other transmitter involved in tremor at rest and its receptor becomes known.

Atlas of the Basal Ganglia in Parkinsonism

By R. Hassler

The stereotaxic treatment of the Parkinson syndrome is based on a functional anatomical localization in the structures of the diencephalon and particularly of the thalamus. It is necessary to illustrate a series of cerebral sections from this area in order to determine known as well as new target points, as has been done in all three planes of the sections by one of us (HASSLER, 1959b, 1977c).

In the past, in order to determine the individual target points in stereotaxic operations on patients suffering from Parkinson's disease, brains or pictures from an atlas of brains of young adults who died suddenly but not of a brain disease, were used as models. Such a series of sections does not enable one to determine with a high degree of accuracy the changes of the diencephalic system of coordinates that occur in the cerebral disease of Parkinson. As a result, a wrong target point could be determined. For this reason we have chosen for the series of frontal sections the hemisphere of a Parkinson patient who was not operated on. The pictures must extend beyond the area of the selected target structure in order to enable the reader to orient himself as well as possible to the structures surrounding the coagulation and the unintended extensions of the coagulated area. In the illustrated series the cell and fiber structures have been demonstrated by staining of adjacent (20 μm thick) section. It is inevitable to consider the fiber structures during stereotaxic interventions as targets to be inactivated and their avoidance as functionally important neuronal structures, more important than the nerve cell nuclei since these structures in the thalamus can be distinguished only to a very limited degree in a fresh cerebral section, or even when it is considerably enlarged in the cytoarchitectonic picture.

1. Architectonic Differentiation and Methods of Staining

We therefore show frontal serial sections of a Parkinson brain at intervals of 2 mm, with staining of myelinated fibers according to HEIDENHAIN-WÖLCKE at a magnification of 5. Abbreviations of the names of the individual nuclei of the thalamus and fiber bundles are indicated directly on the picture. To enable the reader to use this series for measurements, a grid has been drawn onto the pictures. The distance between the lines is 2.0 mm; with the magnification of 5:1, the distance equals 10 mm. In addition, the grid has been subdivided on the edge of the pictures at distances of 2 mm (in the enlargement, the subdivision equals 10 mm). The reader can use this series for the choice of a target point by comparing the coordinates in the brain of the individual patient with those of this atlas series to avoid the problematic "average" target.

A precise differentiation of the individual thalamic nuclei is possible only in a cytoarchitectonic preparation. We have, therefore, added these in the middle of the thalamus at a distance of 4 mm from anterior to posterior in order to achieve such a differentiation. In order to make cytoarchitectonic differences visible, the magnification has to be twice as powerful (10). The frontal sections are supplemented by three pairs of sagittal sections with Nissl and fiber stains at distances from the edge of the third ventricle of 9.5, 6, and 14 mm.

The boundaries of all nuclear and fiber structures are drawn in, as is done in the previously illustrated cases with post mortems, in order to enable the reader to see clearly the exact extension of the aimed-at structure, to avoid unintended lesions of neighboring structures, and to determine whether the structure has been completely destroyed. It must be borne in mind that the boundaries of these structures have been determined after detailed microscopic study.

2. Arrangement of the Planes of Section

The frontal plane of the section lies at a right angle to the "baseline of the thalamus," which we introduced (1954). The latter line runs in the median plane from the anterior edge of the posterior commissure to the ventral border of the interventricular (Monro) foramen, which corresponds anatomically to the caudal edge of the fornix where it enters the hypothalamus. The length of this baseline in this Parkinson brain is 24 mm.

In the serial sections, the length of the thalamus as determined in the encephalogram, forms the basis for the calculation of shrinkage. The distances between the sections are given corresponding to the original length of the thalamus. In the myelin fiber pictures, a distance of 2 mm has been selected between individual sections of the series. At a thickness of 20 μm, this corresponds to every

100th or to every 87th 23 µm. In this series the baseline of the thalamus has been entered in order to make it possible to determine the vertical coordinates; + indicates the mm above, − indicates the mm below the baseline. We have illustrated neighboring cellular sections in those planes where areas meet which, in extrapyramidal motor diseases, are of pathophysiologic significance. These illustrations have double magnification, × 10, so they show, for technical reasons, smaller areas.

Most atlases, even of the diencephalon, are based on planes at right angles to the ventral surfaces of the anterior and posterior commissure. They are, therefore, at an angle of about 8.5° to the planes of section we have chosen. The plane that ANDREWS and WATKINS (1969) used in their atlas is at right angles to the line connecting the upper edge of the interventricular foramen with the posterior commissure, and does not quite coincide with ours. It will be explained later why we preferred the line connecting the lower edge of the interventricular foramen to the most anterior point of posterior commissure as our baseline of the thalamus.

The frontal series illustrated here was taken from a 65-year-old patient suffering from Parkinson's disease who died from a lung embolism following a stereotaxic operation in the nondemonstrated diencephalon. We chose this series of frontal sections deliberately because the relationships in the area of the basal ganglia correspond more to those of patients suffering from Parkinson's disease who are candidates for operation than the usually selected series of so-called normal brains of younger individuals killed, as a rule, in accidents.

We have not investigated the variability of the individual thalamic nuclei and fiber structures because it seems unnecessary when using the method described below of determining the individual target point by comparing each patient's brain with one particular model brain. Our method, therefore, differs from that of ANDREWS and WATKINS (1969) and others, who in their atlas calculated the average boundaries of individual nuclei based on 20 normal brains.

3. Introduction to the Determination of Individual Target Points

The following steps are taken in practice when determining the target point for the stereotaxic operation, according to HASSLER and RIECHERT (1954a, 1955): see Fig. 143a, b.

1. Determination of the target points in the serial sections; this is best done in a series from a patient who suffered from Parkinson's disease, as is the case in the accompanying atlas series;

2. Measuring the distance of the target point
a) from the wall of the third ventricle (frontal coordinate)
b) from the baseline of the thalamus (vertical coordinate) and,
c) measuring the distance of the section containing the target point from the interventricular foramen (s. Fig. 143b) or from the posterior commissure (sagittal coordinate);

3. Calculating the real values of baseline, height of the thalamus (s. Fig. 143a), and width of the ventricles or the hemispheres in the brain of the patient by reducing the distortions of the x-ray in the lateral and anterior-posterior (a.p.) encephalogram;

4. Calculating the proportion factors between the brain in the atlas and that of the patient for each of the three coordinates;

5. Calculating the individual coordinates of the target point in the patient by multiplying by the factors obtained in step 4;

6. Enlarging the three coordinates of the individual target point by the radiologic distortions and transfering them to the radiologic pictures of the patient on the day of the operation by using bony reference points;

7. Determining the three coordinates of the individual target point relative to the zero points of the basal ring as it is fixed on the day of the operation;

8. Locating the target point on the phantom basal ring and fixing the stimulating electrode onto this target point with the help of the aiming bow;

9. Transferring the aiming bow with the angles fixed on it to the basal ring fixed to the patient's head.

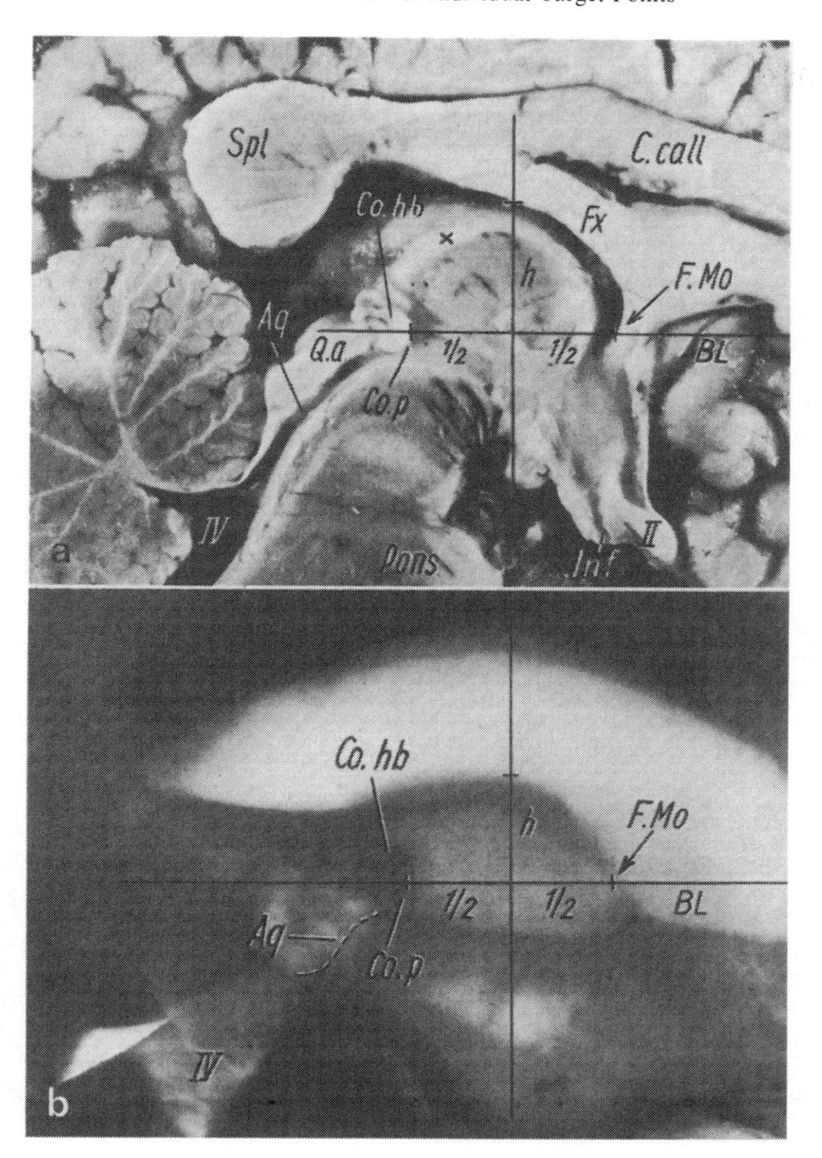

Fig. 143. (a). Coordinate system of diencephalon for stereotaxic operation from medial aspect of hemisphere. Baseline of thalamus starts in Monro foramen (interventricular foramen, *F.Mo*) and extends to most rostral point of posterior commissure (*Co.p*). Aquaeductus cerebri (*Aq*) starts below Co.p. Aq is continued by the IVth ventricle. Height of thalamus is measured from middle of baseline (from *F.Mo* and *Co.p*) to superior border of thalamus.
(b). Comparison of this anatomic preparation with lateral x-ray of brain in which ventricles are filled with air. Baseline is drawn in between inferior border of interventricular foramen and most anterior point of shadow of posterior commissure. Height of thalamus in middle of baseline is marked (*h*).
BL baseline (of thalamus); *Co.hb* commissura habenularum; *C.call* corpus callosum; *Fx* fornix; *Inf* infundibulum; *Q.* superior colliculus; *Spl* splenium of corpus callosum; *II* optic chiasma; × stria medullaris thalami

4. Description of the Series of Frontal Sections

The following is a description of frontal serial brain sections with fiber stain in a case of Parkinson's syndrome. The sections were taken at distances of 2.0 mm at a right angle to the baseline from the lower edge of the interventricular foramen to the anterior edge of the posterior commissure with a magnification of 5× (Fig. 144).

The basal length of the thalamus from the posterior lower edge of the interventricular foramen to the posterior commissure is 24.0 mm, the height of the thalamus at the middle of this distance measures 150 mm.

Fig. 145. *Frontal section* in the plane of the interventricular (Monro) foramen (*Fo.Mo*) at a right angle to the baseline: The starting point of the baseline is the caudobasal point of the fornix before it disappears in the hypothalamus. In this plane both parts of the pallidum (*Pa.e* and *Pa.i*) are seen. Below them is the rostral part of the ansa lenticularis (*A.le*). Parallel to the ansa lenticularis, forms an accumulation of fibers above the optical tract (*II*) in the rostral continuation of the ansa peduncularis (*A.pd*), root of the inferior thalamic peduncle (*Pd.if*). Above the genu of the internal capsule (*Ca.i*), which is seen here, lies the stria terminalis (*Str.t*) with the terminal vein (*Ve.t*) *B* nucleus basalis. ×5.

Fig. 144. Medial aspect of unoperated hemisphere of Parkinson patient, showing baseline of thalamus and perpendicular to it, levels of sections showing distance from interventricular foramen. Hatched line is commissural line between anterior (*C.a*) and posterior commissure (*C.p*). Arrow in brainstem shows Meynert's axis. *Pi* pineal gland; *Pu* pulvinar *Spl* splenium of corpus callosum; *Fx* Fornix. Scale 1:1

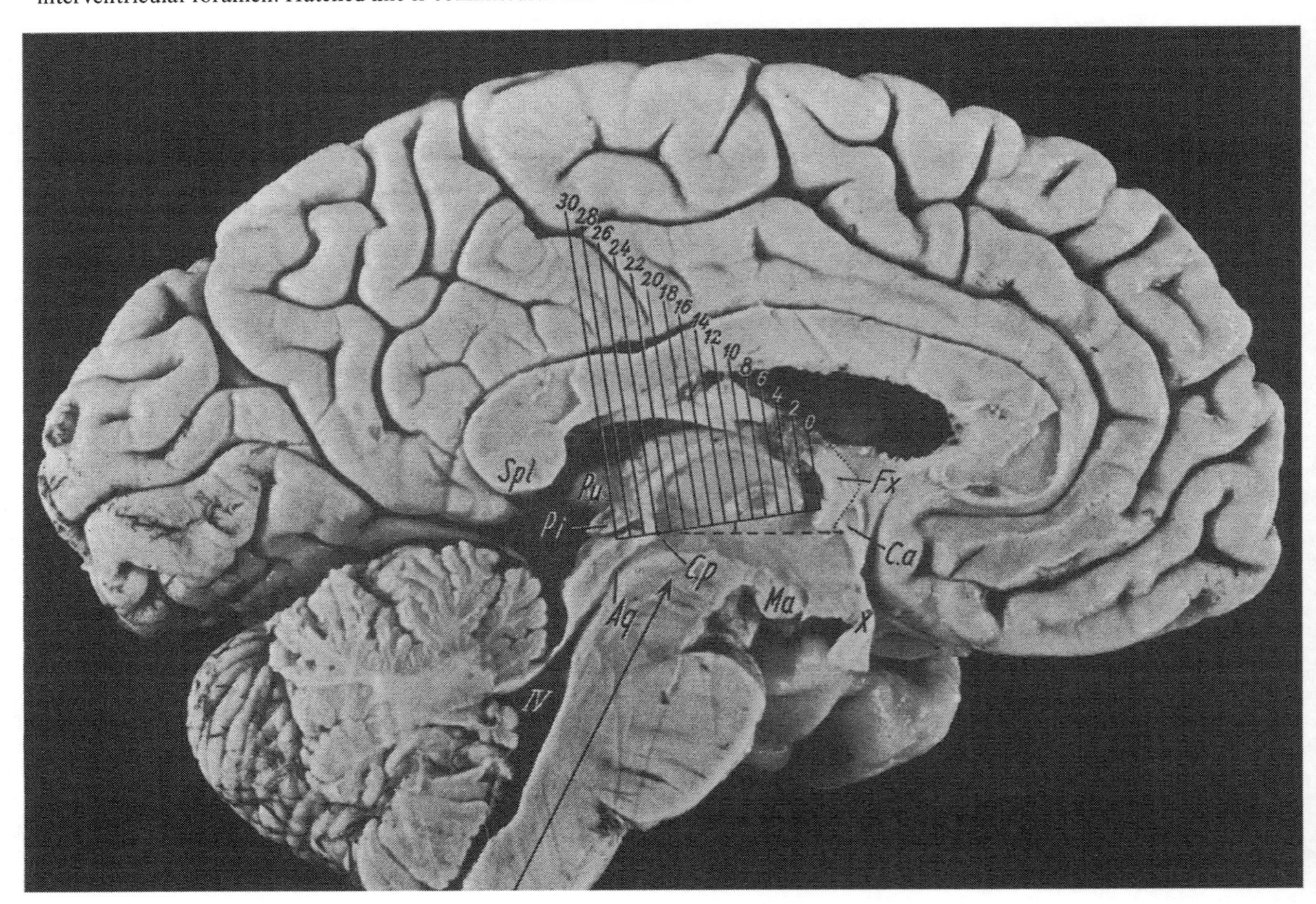

Fig. 145. For description see opposite page

Fig. 146. For description see p. 260

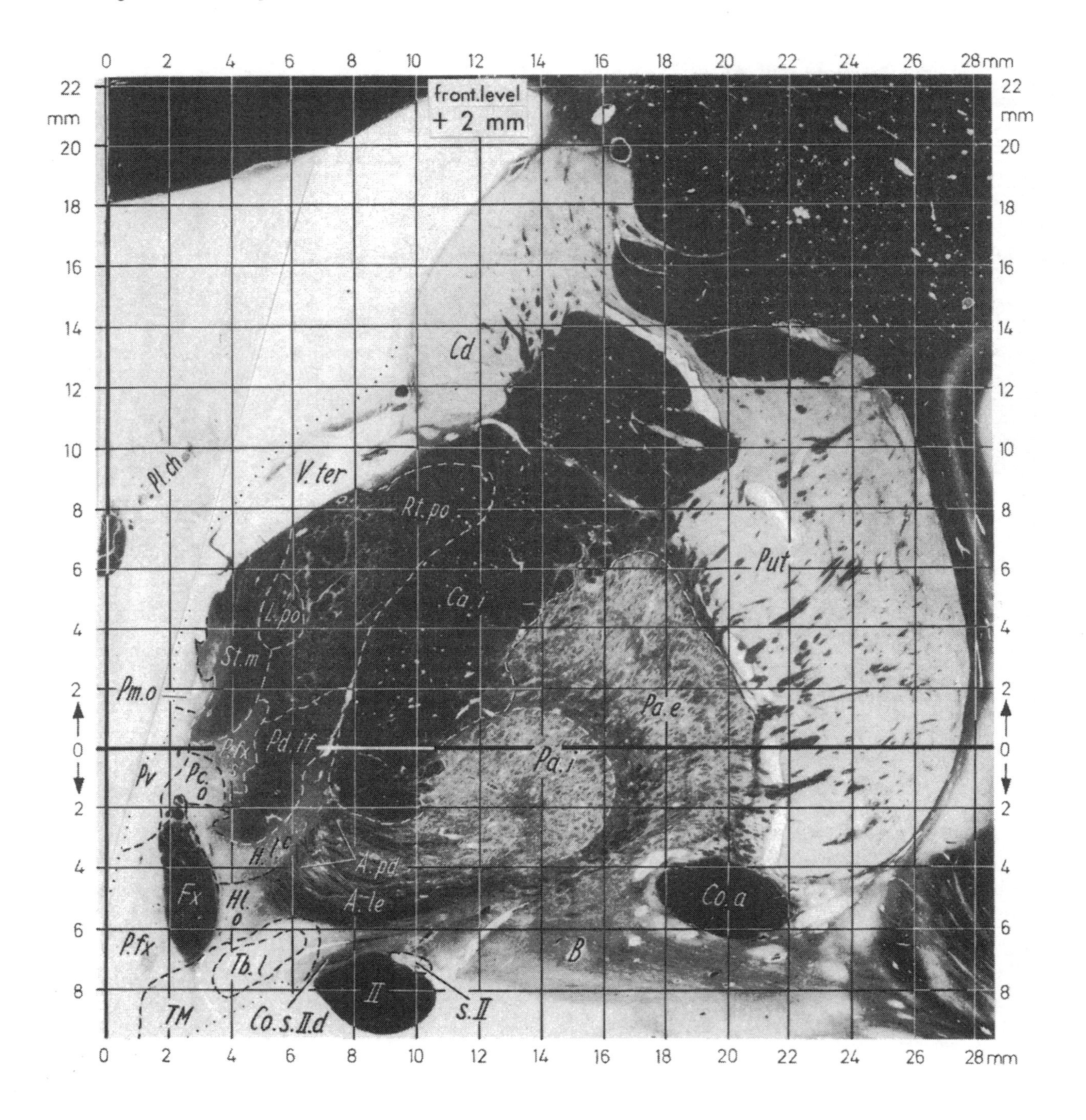

Fig. 147. For description see p. 260

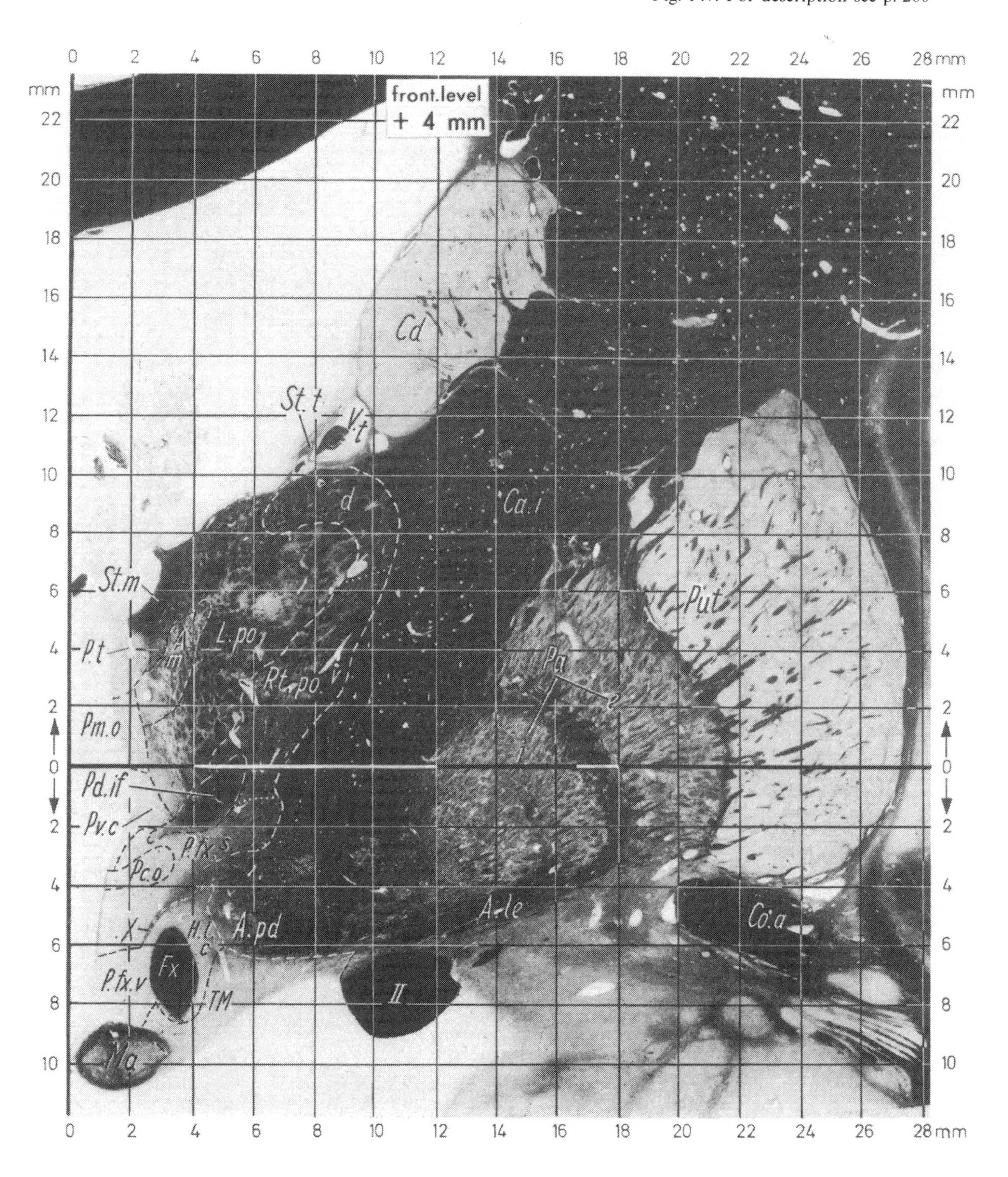

Fig. 148. (a) For description see p. 260

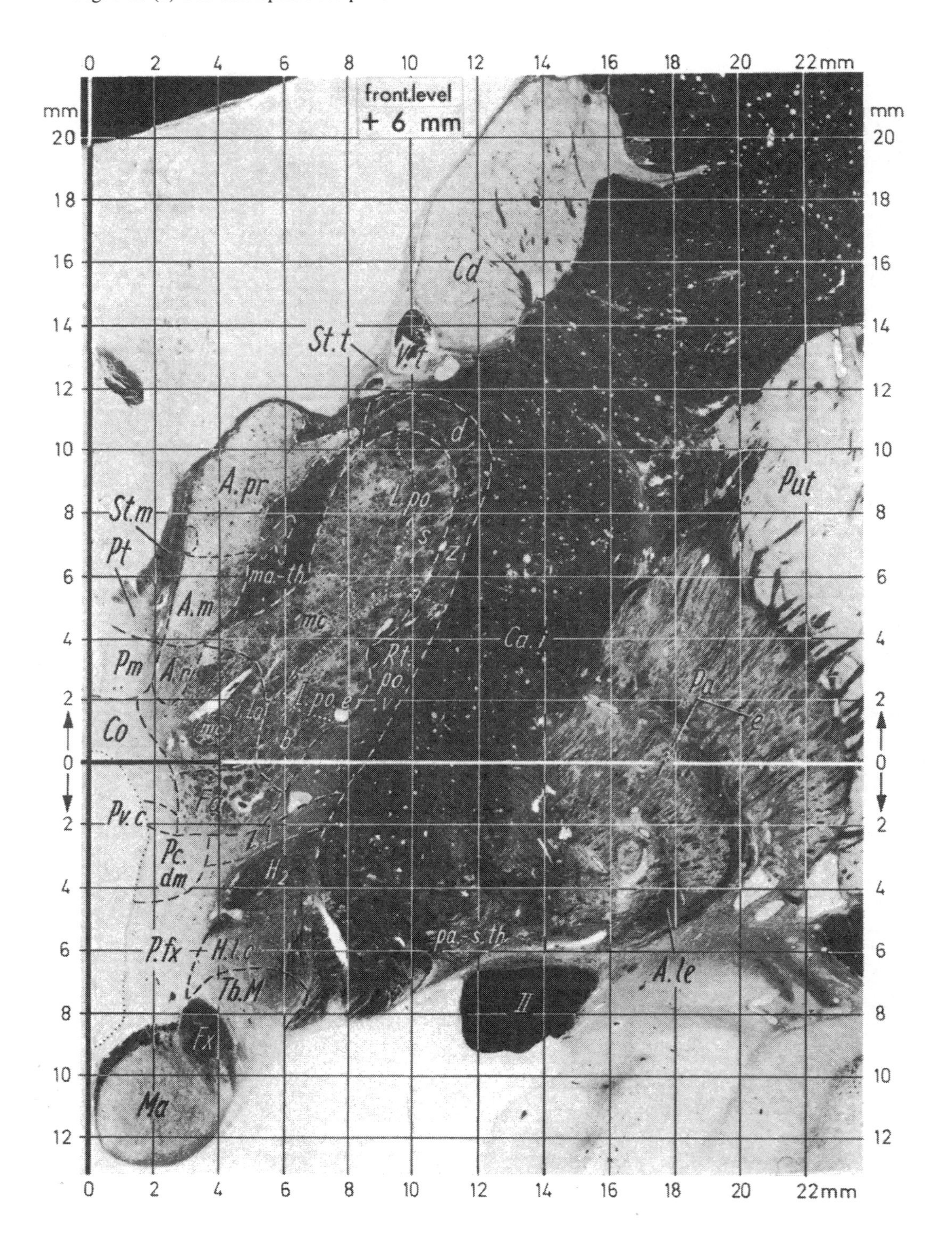

Fig. 148. (b) For description see p. 260

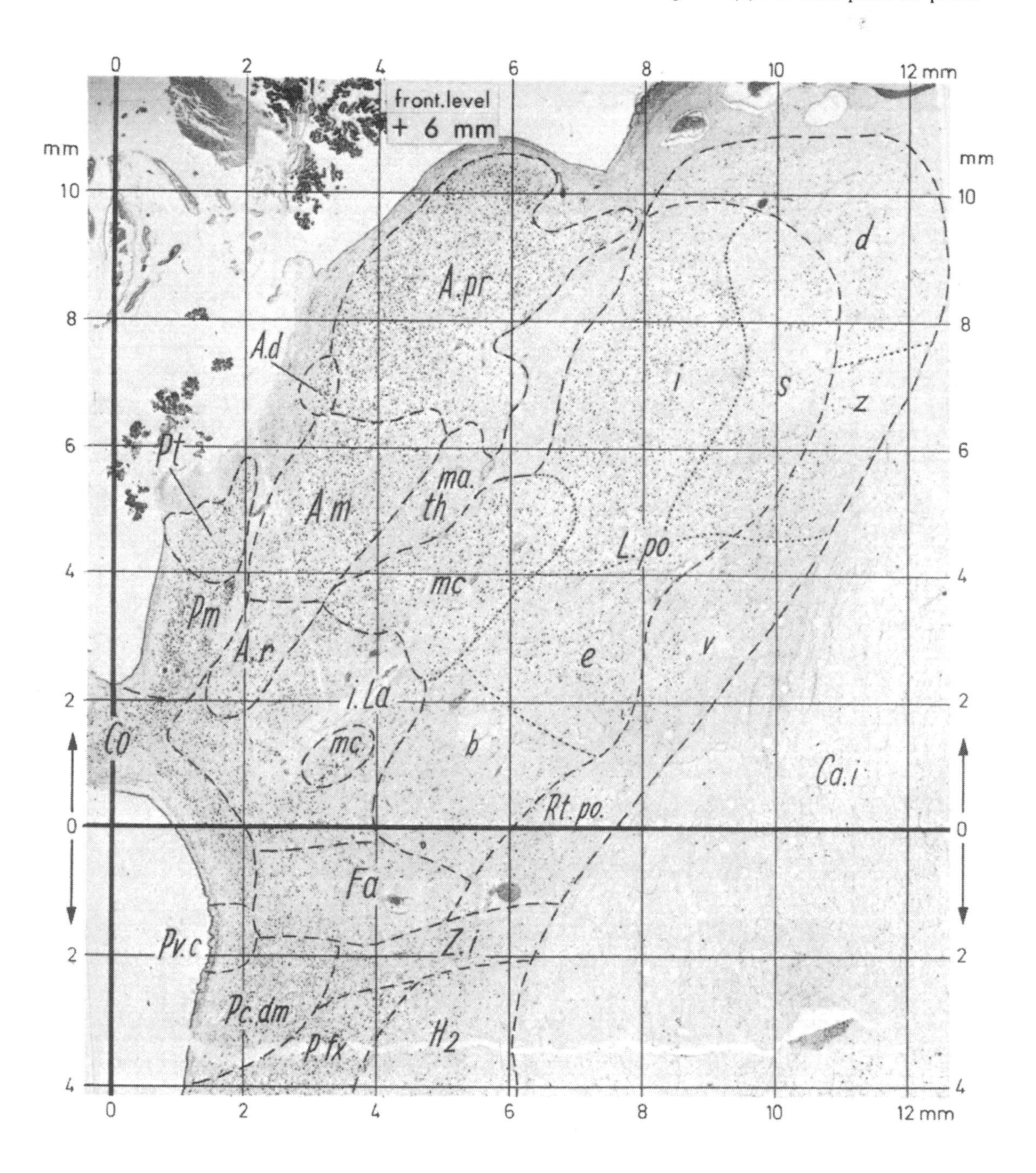

Fig. 146. *Frontal section*, 2 mm behind Fig. 145. The interventricular foramen is now filled by the choroid plexus (*Pl.ch*). The anterior pole of the thalamus stands on the inferior peduncle (*Pd.if*) and extends laterally. The thalamus is formed at this level mainly by the polar reticulate nucleus (*Rt.po*). Between Rt.po and the stria medullaris (*St.m*), the rostral first section of the lateropolar nucleus (*L.po*) is cut. The wall of the third ventricle is occupied by the oral paramedian nucleus (*Pm.o*). It is replaced ventrally by the paraventricular nucleus (*Pv*) and by the parafornical nucleus (*P.fx.s*) of the hypothalamus. The fornix and the inferior peduncle of the thalamus are surrounded by many hypothalamic nuclei. The tuberomamillar-complex (*TM*) surrounds the nucleus tuberis lateralis (*Tb.l*). × 5.

Fig. 147. *Frontal section*, 4.0 mm behind the interventricular foramen. The genu of the internal capsule (*Ca.i*) is pushed laterally away from the third ventricle by the anterior thalamic pole. The fibers of the inferior thalamic peduncle (pedunculus inferior = *Pd.if*) lie in the basis of the thalamus and fan out in a dorsomedial direction. In the basis of the internal part of the pallidum (*Pa.i*) the ansa lenticularis is more dense. Some fibers coming from the ansa have crossed the lower edge of the internal capsule and run as a bundle (fasciculus pallido-hypothalamicus = *X*) across the fornix (*Fx*) into the hypothalamus. The optic tract (*II*) has moved more laterally and is only $^1/_2$ mm distant from the ansa lenticularis (*A.le*). × 5.

Fig. 148a. *Frontal section*, 6.0 mm behind the interventricular foramen, goes through the anterior edge of the massa intermedia. The main anterior nucleus (*A.pr*) occupies the mediodorsal part in the thalamus and bulges as the tuberculum anterius into the lateral ventricle. The bundles on its lateral surface belong to the tractus mamillothalamicus (*ma.th.*). Ventrally the A.pr is continued by nucleus anteromedialis (*A.m*) and anteroreuniens (*A.r*), which includes the anteroinferior further ventral by nucleus. Ventral to the stria medullaris thalami (*St.m*) lies the thalamic gray substance, consisting of 3 nuclei: parataenialis (*Pt*), paramedianus (*Pm*), and commissuralis (*Co*), which occupies the massa intermedia. The lateral and basal part of the thalamus is occupied by the nucleus lateropolaris (*L.po*) and the nuclei belonging to it. The thalamus is separated from the internal capsule by nucleus reticulatus polaris (*Rt.po*). The bottom of the internal capsule is crossed by the rostral pallido-subthalamic fibers (*pa.-s.th*) of the ansa lenticularis (*A.le*). They curve up in Forel's bundle H_2 and melt away in the parafornical nucleus (*P.fx*) above the mamillary body (*Ma*). × 5

Fig. 148b. *Frontal section*, 6.0 mm behind the interventricular foramen, cellular stain: In the massa intermedia and ventrally to lies the nucleus commissuralis (*Co*). Above this, but covered by the ependyma, are located the nuclei paramedianus (*Pm*) and parataenialis (*Pt*) below the stria medullaris thalami (*St.m*). The principal anterior nucleus (*A.pr*) which includes the first cut of nucleus anterodorsalis (*A.d*) continues dorsally to the nucleus anteromedialis (*A.m*) and anteroreuniens (*A.r*) which includes the anteroinferior. The medial thalamus is here occupied by nuclei containing many fibers: fasciculosus (*Fa*) and intralaminaris (*i.La*). In the lateropolar nuclear territory, one finds below the tractus mamillothalamicus (*ma.-th*), the nuclei (1) lateropolaris magnocellularis (*L.po.mc*) with a heterotopic island in the intralaminar nucleus (*i.La*), (2) the lateropolaris basalis (*L.po.b*) containing few cells, and (3) the lateropolares externus, internus, and superior (*L.po.e*, *L.po.i*, and *L.po.s*). Against the internal capsule (*Ca.i*), they are surrounded by the nucleus reticulatus polaris (*Rt.po*), with 3 subnuclei (*d, z, v*). × 10.

Fig. 149. *Frontal section*, 8.0 mm behind the interventricular foramen. The tractus mamillothalamicus (*ma.th*) has separated from the nucleus anterior (*A.pr*). On its lateral side the nucleus ventrooralis anterior (*V.o.a*) extends with pallidothalamic fibers ending in it. Medially to the mamillothalamic tract (*ma.th*) lie the anterior poles of the intralaminar thalamic nuclei (*i.La*), which fuse with the nucleus of the massa intermedia (*Co* commissuralis). In its most ventral part lies the caudal most extension of the nucleus lateropolaris magnocellularis (here: *mc*). Below the thalamus are the zona incerta (*Z.i*) and the fasciculus lenticularis (H_2) in their major extensions. The link with the capsule of the mamillary body is only apparent and not real. The pallido-subthalamic (*pa.-s.th*) fibers, which cross the rostral border of the cerebral peduncle and terminate in the subthalamic nucleus (*S.th*), contrast with the much lighter strionigral (*st.-ni*) fibers, which are located ventral to them and terminate in the substantia nigra (*Ni*). × 5.

Fig. 149. For description see opposite page

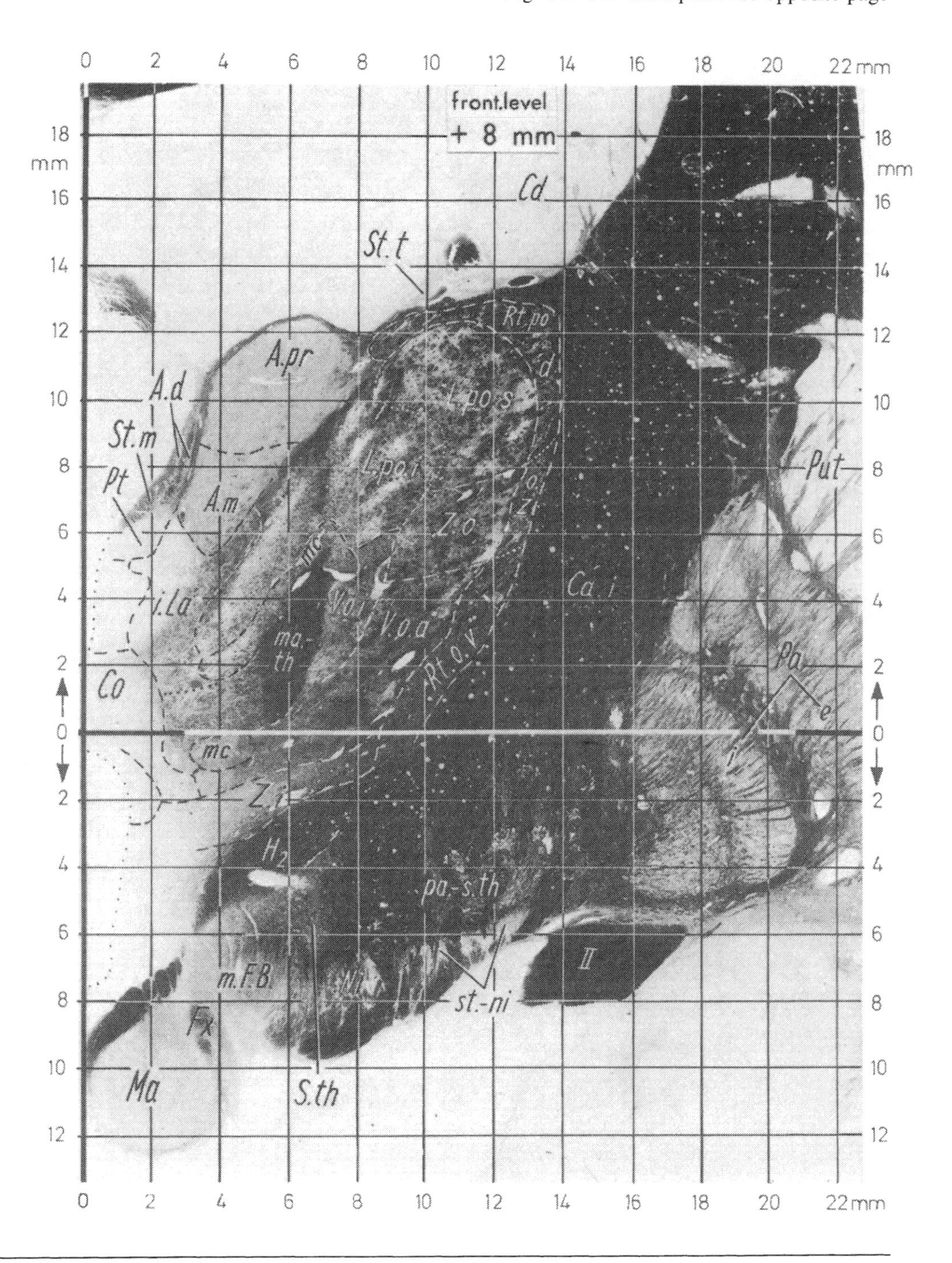

Fig. 150. (a) For description see p. 264

Fig. 150. (b) For description see p. 264

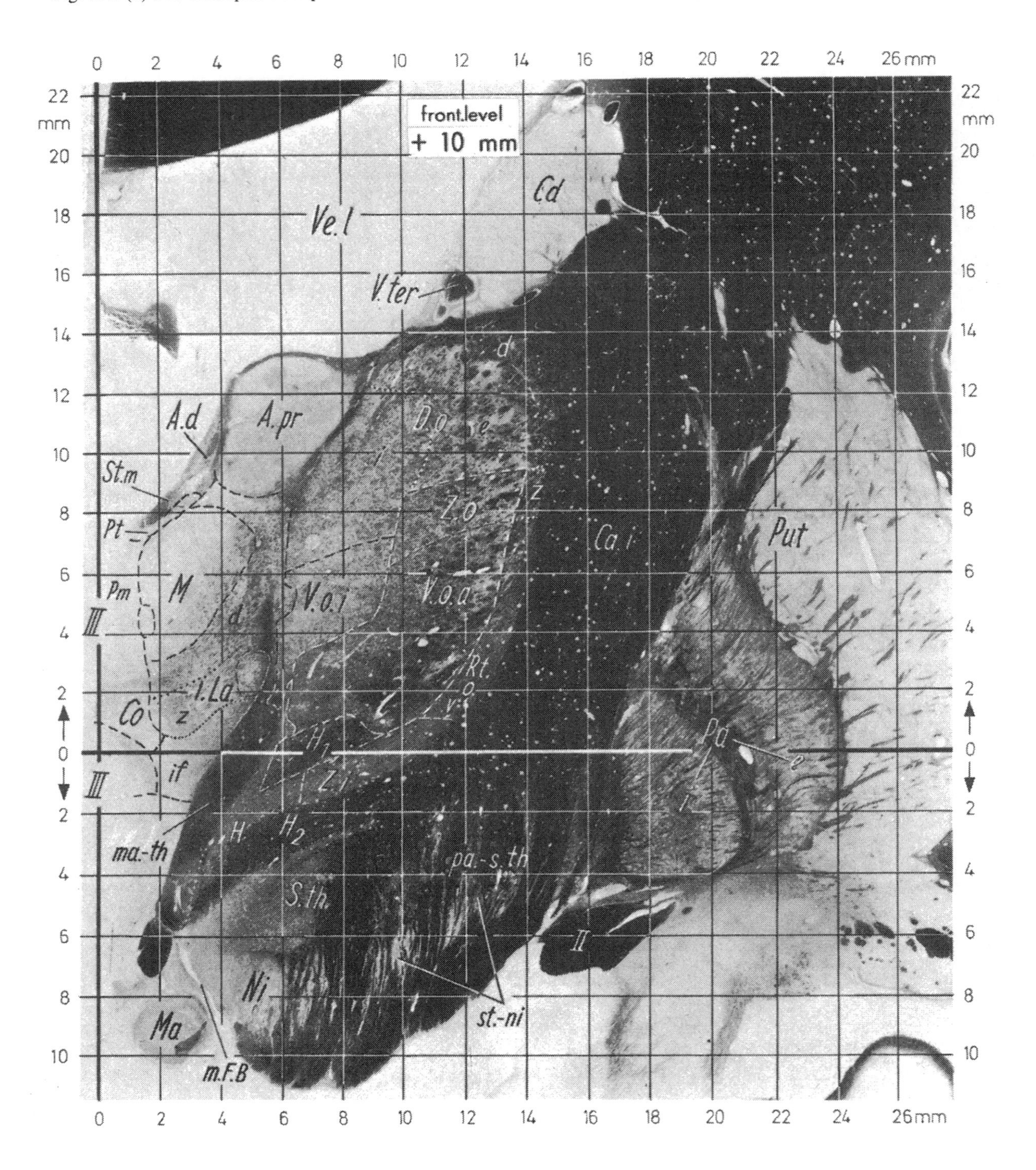

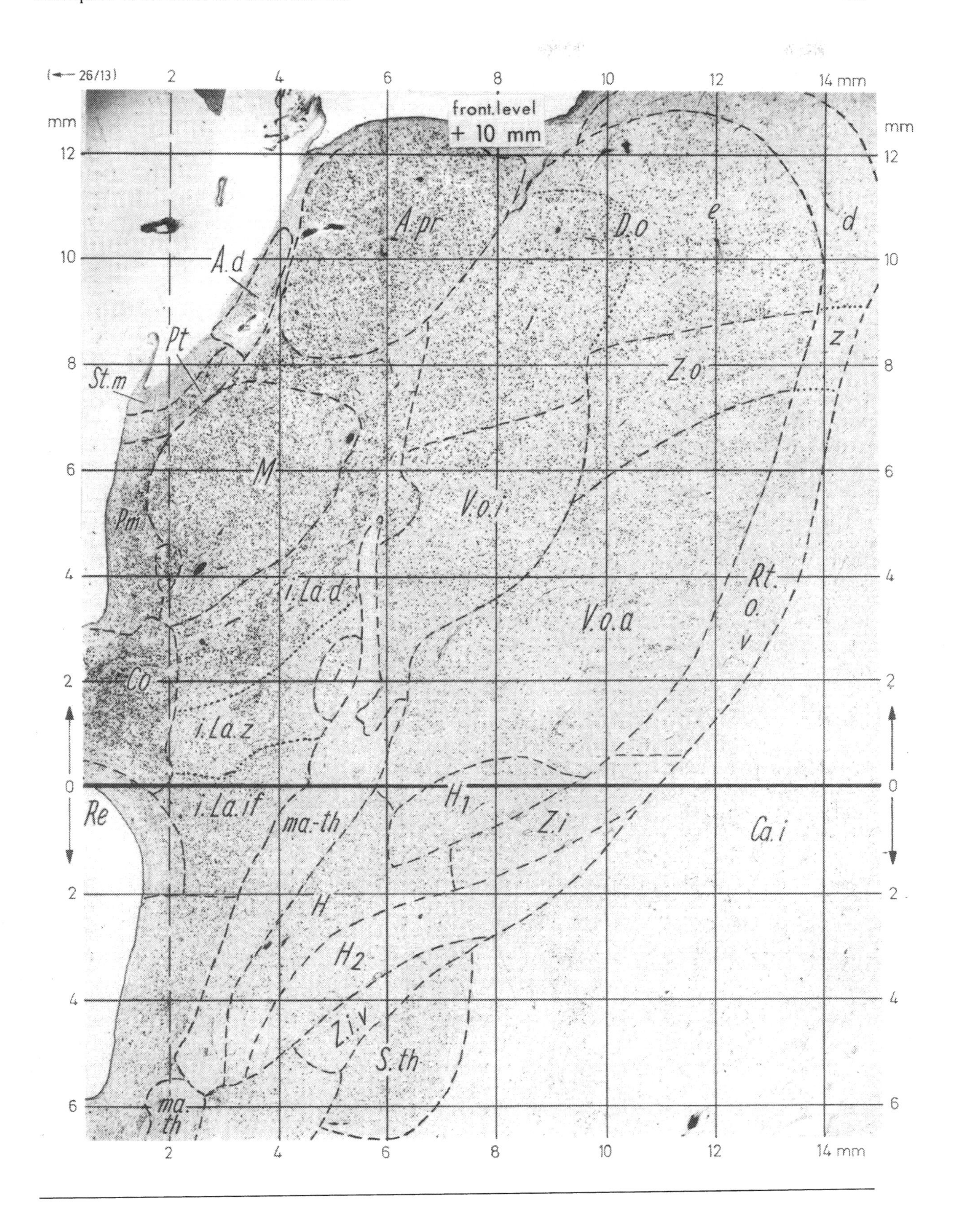
(← 26/13)
front. level
+ 10 mm
mm
14 mm
A.d
A.pr
D.o
e
d
Pt
St.m
Z.o
z
M
Pm
V.o.i
Rt.
o.
v
i.La.d
V.o.a
Co
i.La.z
Re
i.La.if
ma-th
H1
Z.i
Ca.i
H
H2
Zi.v
S.th
ma.
th

Fig. 150 a. *Frontal section*, 10.0 mm behind the interventricular foramen. Maximal extension of the nucleus ventro-oralis anterior (*V.o.a*) containing fiber bundles from H_1. The zona incerta (*Z.i*) extends between bundles H_2 and H_1. Forel's field H lies close to the lateral side of the mamillothalamic tract (*ma.th*). The rostral pole of the subthalamic nucleus (*S.th*) extends ventrally to the H_2. Further ventrally the rostral pole of the substantia nigra (*Ni*) lies laterally to the mamillary body (*Ma*). Note the dark pallido-subthalamic (*pa.-s.th*) and the light strio-nigral (*st.-ni*) fibers. Medially to the substantia nigra is the so-called medial forebrain bundle (*mFB*) constituting a connection of thin fibers between septum and midbrain. The intralaminar thalamic nuclei (*i.La*) with the subnuclei dorsalis (*d*), entralis (*z*) and inferior (*if*) and the nucleus commissuralis (*Co*) have their greatest extension lateral to and in the massa intermedia. ×5

Fig. 150 b. *Frontal section*, 10.0 mm, *in Nissl stain* shows how the intralaminar nuclei (*i.La*) are framed around the frontal pole of the medial thalamic nucleus (*M*). Medially to this lies the dense nucleus commissuralis (*Co*) in the massa intermedia. Scattered clusters of medium-sized nerve cells fill the greater part of the V.o.a while more mediodorsally the V.o.i is characterized by more densely grouped, larger nerve cells. Above V.o.i and V.o.a the oral dorsal nuclei D.o.i and D.o.e have replaced the superior lateropolar nucleus. Forel's tegmental field H contains many nerve cells. ×10.

Fig. 151. *Frontal section*, 12.0 mm behind the interventricular foramen. The section passes through field H of Forel, in which many fibers of the bundle H_2 bend around to form the dense bundle H_1 and terminate in the V.o.a. Close to the medial side of field H is a section of the mamillothalamic tract (*ma.th*) and again, medial to it the most rostral fibers of the fasciculus longitudinalis posterior terminate in the nucleus praestitialis. Ventrally to this are seen the fibers of the commissura hypothalamica posterior of Forel (*Co.hyp.p*), which joins the two subthalamic nuclei (*S.th*). Below the subthalamic nucleus the substantia nigra (*Ni*) is slightly larger than in the previous section. Dorsally the nucleus dorso-oralis internus (*Do.i*) follows the nucleus ventro-oralis internus (*V.o.i*). Over the V.o.a is a narrow oral zentrolateral nucleus (*Z.o*) and above this the nucleus dorso-oralis externus (*D.o.e*). Both parts of the pallidum (*Pa*) have moved further laterally and have become smaller. The numerous aberrant fiber bundles in the subthalamic nucleus (*S.th*) are unusual. The white arrows indicate the path of the pallido-subthalamic (*pa.-s.th*) fibers in the comb system of the peduncle, and ventral to them the lighter strio-nigral fibers (*st.-ni*), which run to the substantia nigra (*Ni*). Medially to the H-field runs the fiber bundle of the mamillo-thalamic tract (*ma.th*). Ventrally ma.th is surrounded by the prestitial nucleus (*Pr.st*). ×5

Fig. 151. For description see opposite page

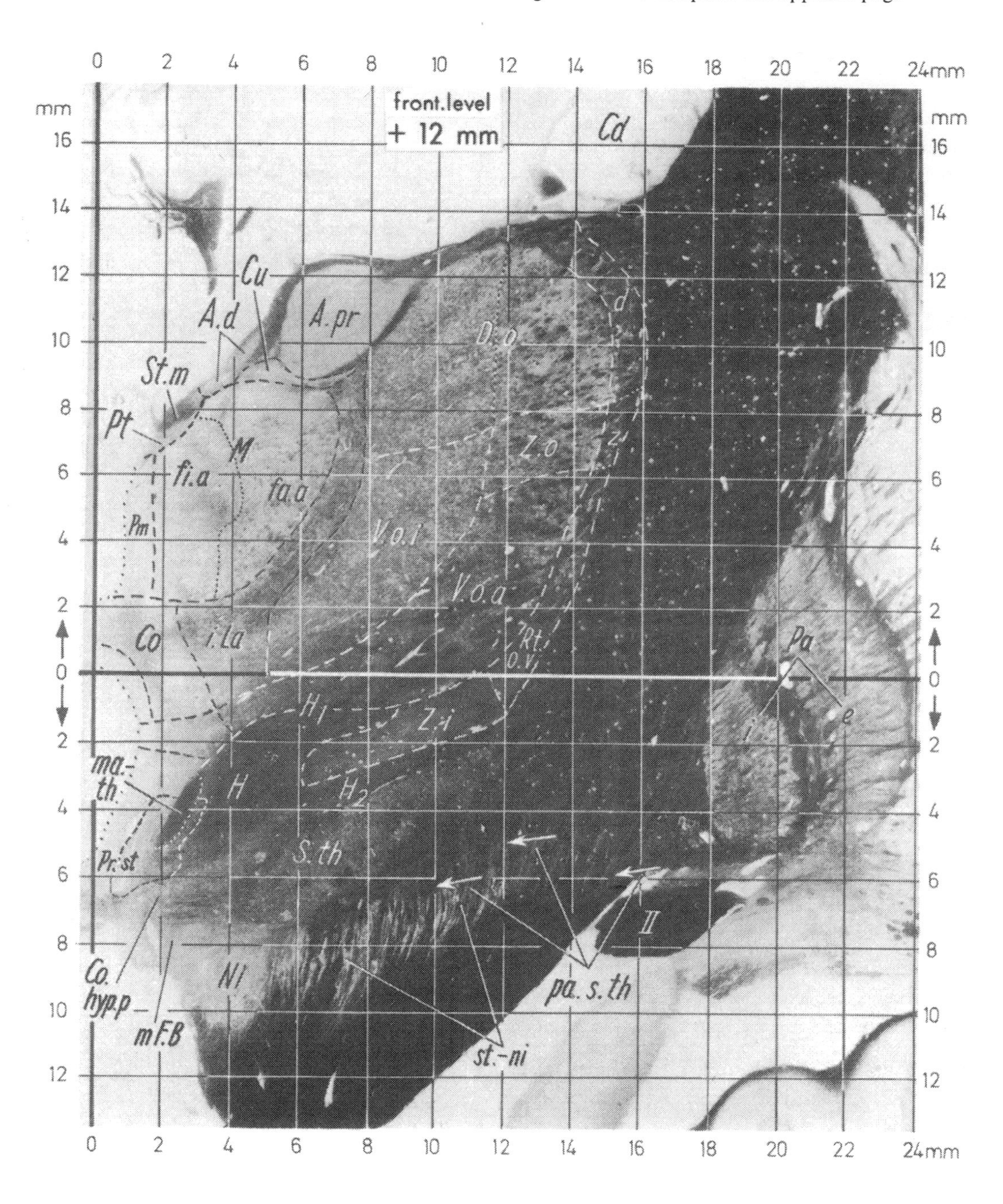

Fig. 152. (a) For description see p. 268

Fig. 152. (b) For description see p. 268 ▶

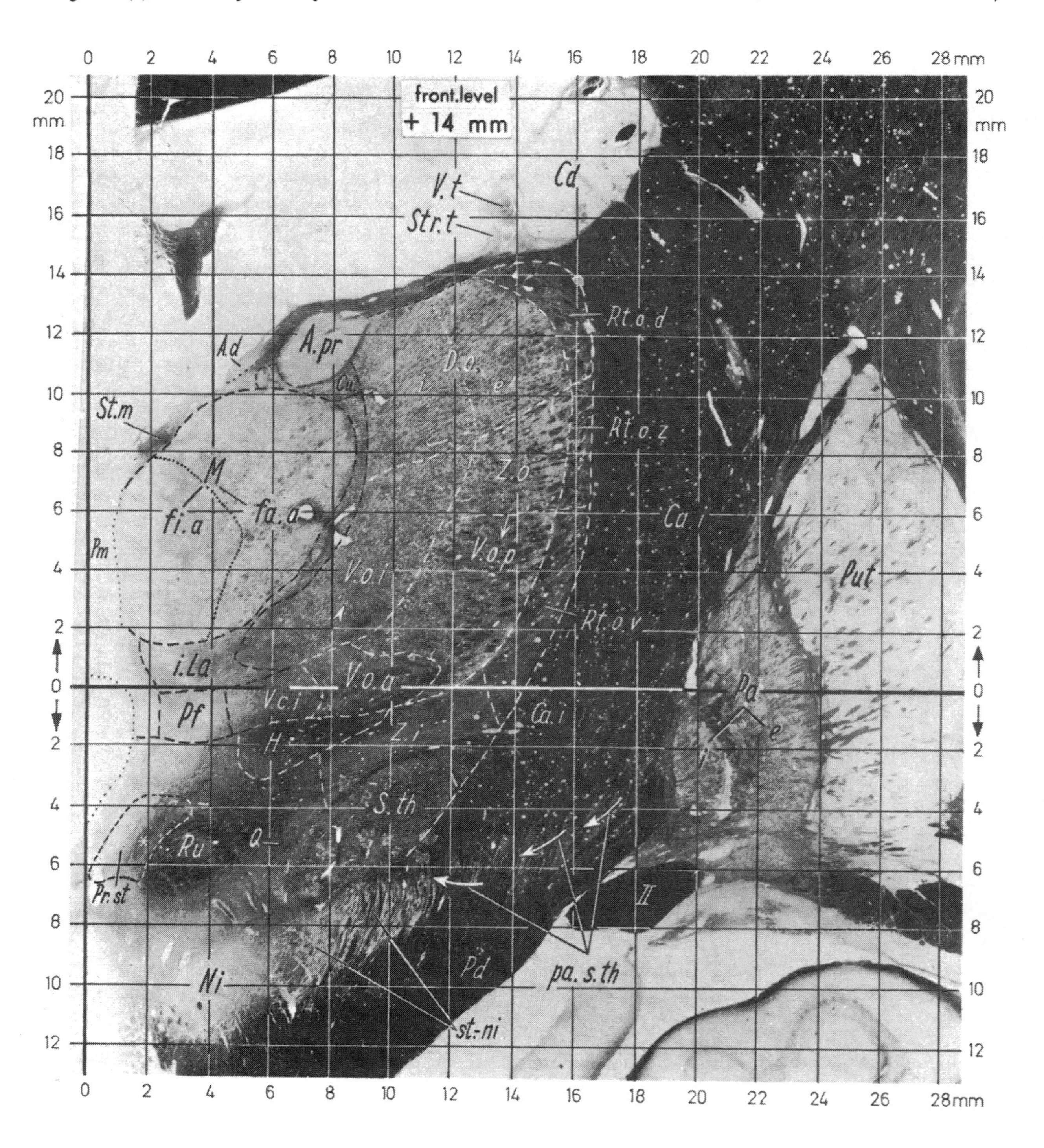

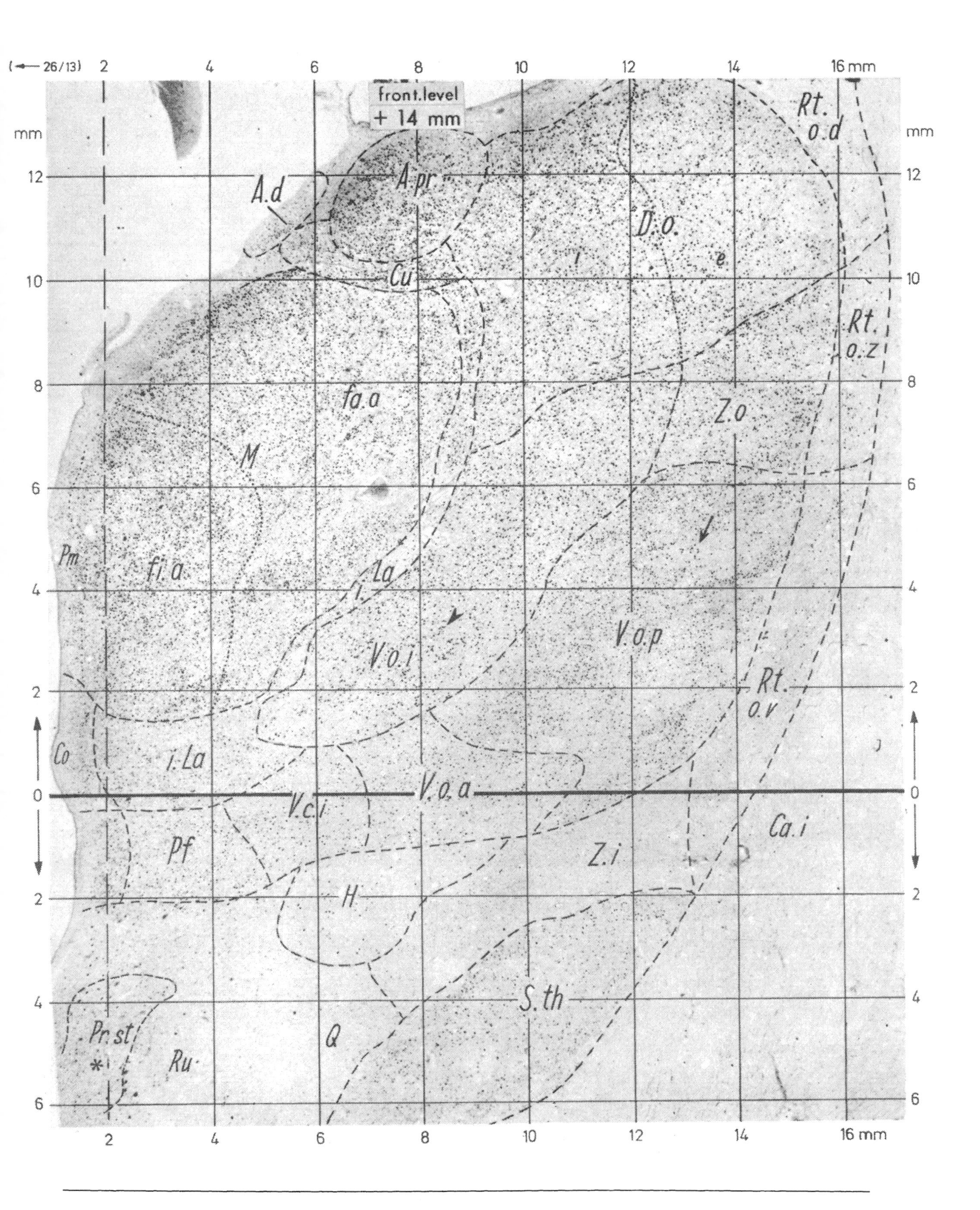
(← 26/13)
front.level
+ 14 mm
mm
16 mm
Rt.
o.d
A.d
A.pr
D.o.
e
Cu
Rt.
o.z
fa.a
Z.o
M
Pm
fi.a
La
V.o.p
V.o.i
Rt.
o.v
Co
i.La
V.o.a
V.c.i
Ca.i
Pf
Z.i
H
S.th
Pr.st
Q
Ru
*

Fig. 152a. *Frontal section*, fiber stain, 14.0 mm behind the interventricular foramen. Most of the V.o.a is replaced by the nucleus ventro-oralis posterior (*V.o.p*), which is characterized by the fewer fiber bundles, particularly next to the lamella lateralis. Medially in the base is found the hindmost part of field H and above H appears the rostral pole and fiber capsule of V.c.i. In the midbrain one meets the anterior pole of the red nucleus (*Ru*) bordered medially by the nucleus praestitialis around the Tractus retroflexus (*). Here the subthalamic nucleus (*S.th*) has its largest extension; the pallidum internum (*Pa.i*) is much reduced and the lamella pallidi interna is wider. Within the area of V.o.i extensions of the nucleus ventrointermedius internus (*V.im.i*) with much larger nerve cells (*arrow head*). The white arrow in V.o.p marks the place of a few larger nerve cells of V.im.e. Above the remaining of the H-fields the rostral pole and fiber capsule of the nucleus ventrocaudalis internus (*V.c.i*). The lamella medialis (*i.La*) is narrower while dorsally the medial nucleus (*M.fi.a*+*M.fa.a*) has become larger. ×5.

Fig. 152b. *Frontal section*, adjacent section cellular stain: On the lateral side of the narrow lamella medialis (*i.La*) lie (*arrow head*) some larger cells of the nucleus ventrointermedius internus (*V.im.i*) within the V.o.i, which is clearly distinguished from the V.o.p by its evenly distributed, medium-sized nerve cells. In the V.o.p fairly dorsally (*arrow*), the rostral point of the Ventrointermedius externus (*V.im.e*) with a few larger cells has been cut. The nucleus reticulatus next to the V.o.p (*Rt.o.v*) contains very few cells but is rather wide in the fibril stain. Medially over the fiber field H of Forel are the rest of the nucleus ventro-oralis anterior (*V.o.a*) and the rostral pole and fiber capsule of the nucleus ventrocaudalis internus (*V.c.i*). Medially to i.La the commissural nucleus (*Co*); medially to the Pf the subnucleus parafascicularis medialis (*Pf.m*) is not labelled. Medially to the rostral pole of the red nucleus (*Ru*) the prestitial nucleus (*Pr.st*) is situated. *: Tractus retroflexus. ×10.

Fig. 153. *Frontal section*, fiber stain, 16 mm behind the interventricular foramen. In the midbrain, the large red nucleus (*Ru*) stands out. Its medial capsule is enlarged by the Tractus retroflexus (*). Ventrally, the fiber bundles of the oculomotor nerve (*nIII*) enter the fossa interpeduncularis. Here one finds the largest extension of the anterior main section of the substantia nigra (*Ni*). Medial to the Ru a branch of the bundle of Fasciculus longitudinalis medialis terminates in the prestitial nucleus (*Pr.st*) which is responsible for raising movements. Between the red nucleus and the subthalamic nucleus (*S.th*), which has already become smaller, lies the Q-bundle, which probably contains dark pallidomesencephalic fibers. At the base of the V.o.p one sees between two white arrows the entry of dentato-thalamic fibers (*dt-t*). The V.o.p is dorsally clearly differentiated from the intermediate zentrolateral nucleus (*Z.im*) and from the intermediate dorsal nucleus (*D.im*). The lateral part of the V.o.p is already replaced by the nucleus ventrointermedius externus (*V.im.e*). The V.im.i containing many fibers has replaced the V.o.i next to the narrow dense lamella medialis (*i.La*). Below, one sees the rostral beginning of the nucleus arcuatus (*V.c.i*). This contains the representation of the fifth cranial nerve in the thalamus = nucleus ventrocaudalis internus (*V.c.i*); in the fibril stain it is very light. In the medial nucleus of the thalamus the posterior fibrous part (*M.fi.p*) is clearly separated from the lateral M.fa.p containing a few fiber bundles. The wide lamella pallidi interna (*L.pa.i*) lies here behind the caudal pole of the internal part of the pallidum. Over the lamella medialis (*i.La*) the caudal pole of the anterior main nucleus (*A.pr*) is extended. This is replaced medially as well as caudally by the nucleus dorsalis superficialis (*D.sf*). Ventrally to the remains of the intralaminar nuclei (*i.La*) is the nucleus parafascicularis (*Pf*). ×5.

Fig. 153. For description see opposite page

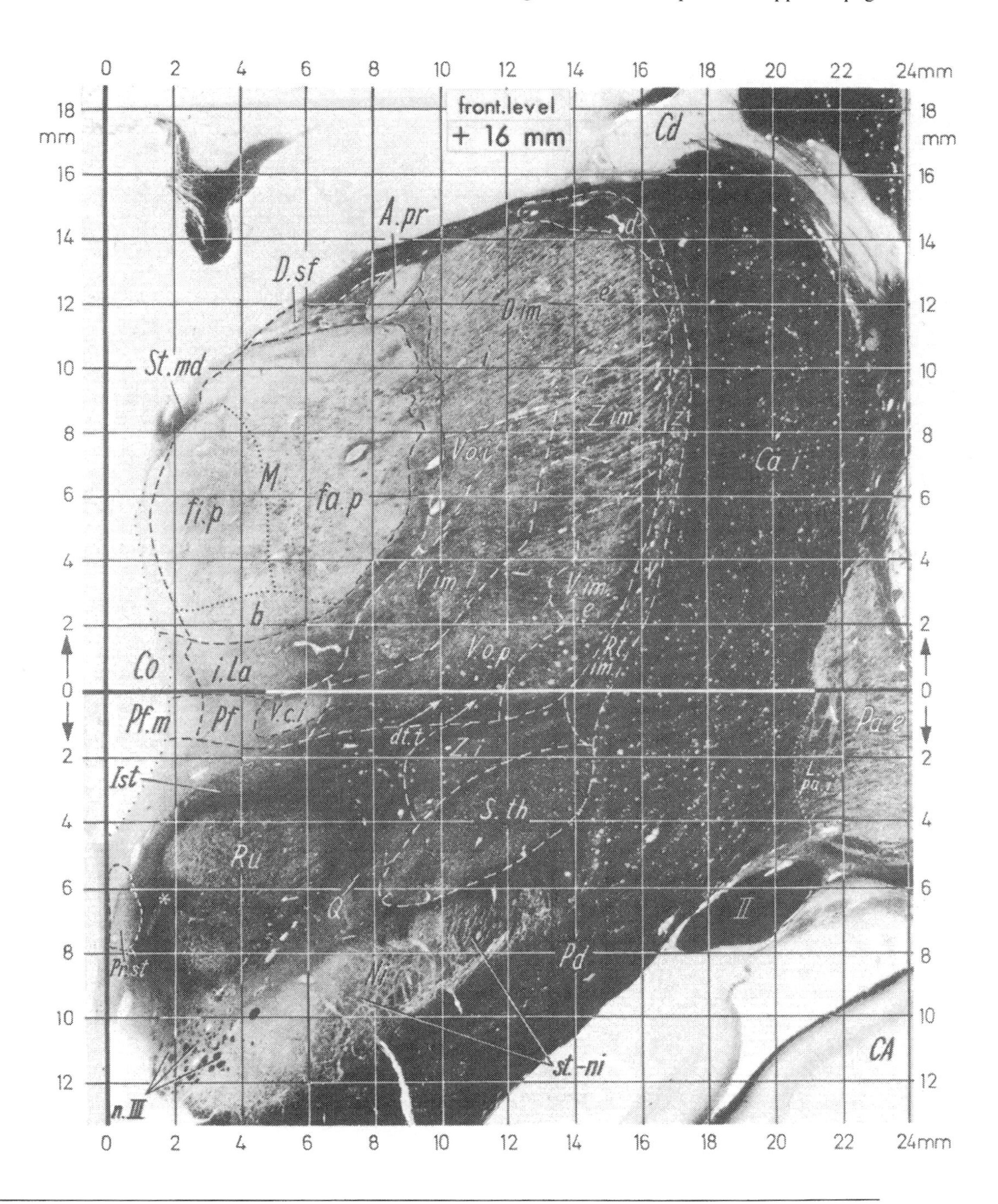

Fig. 154. (a) For description see p. 272

Fig. 154. (b) For description see p. 272 ▶

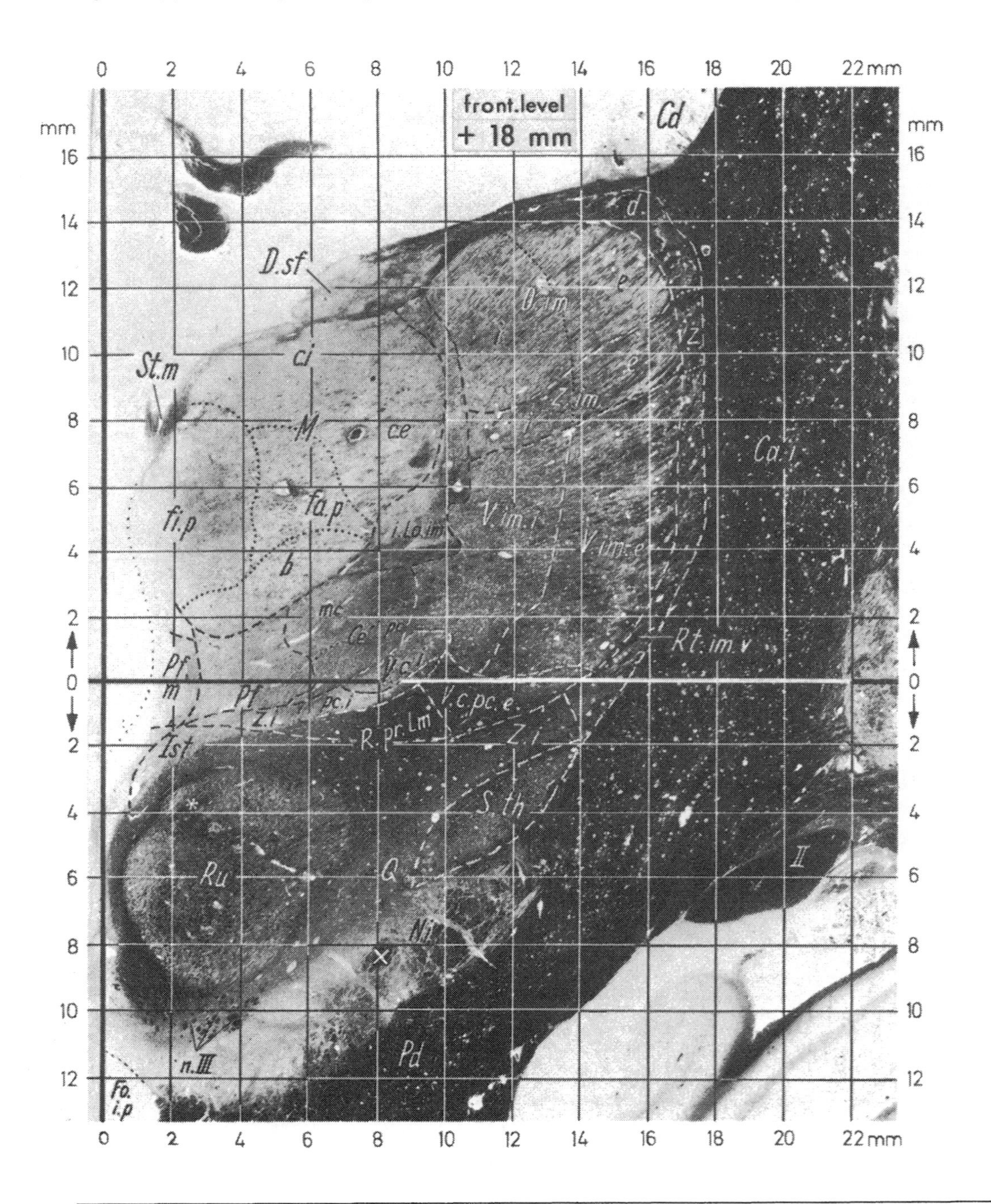

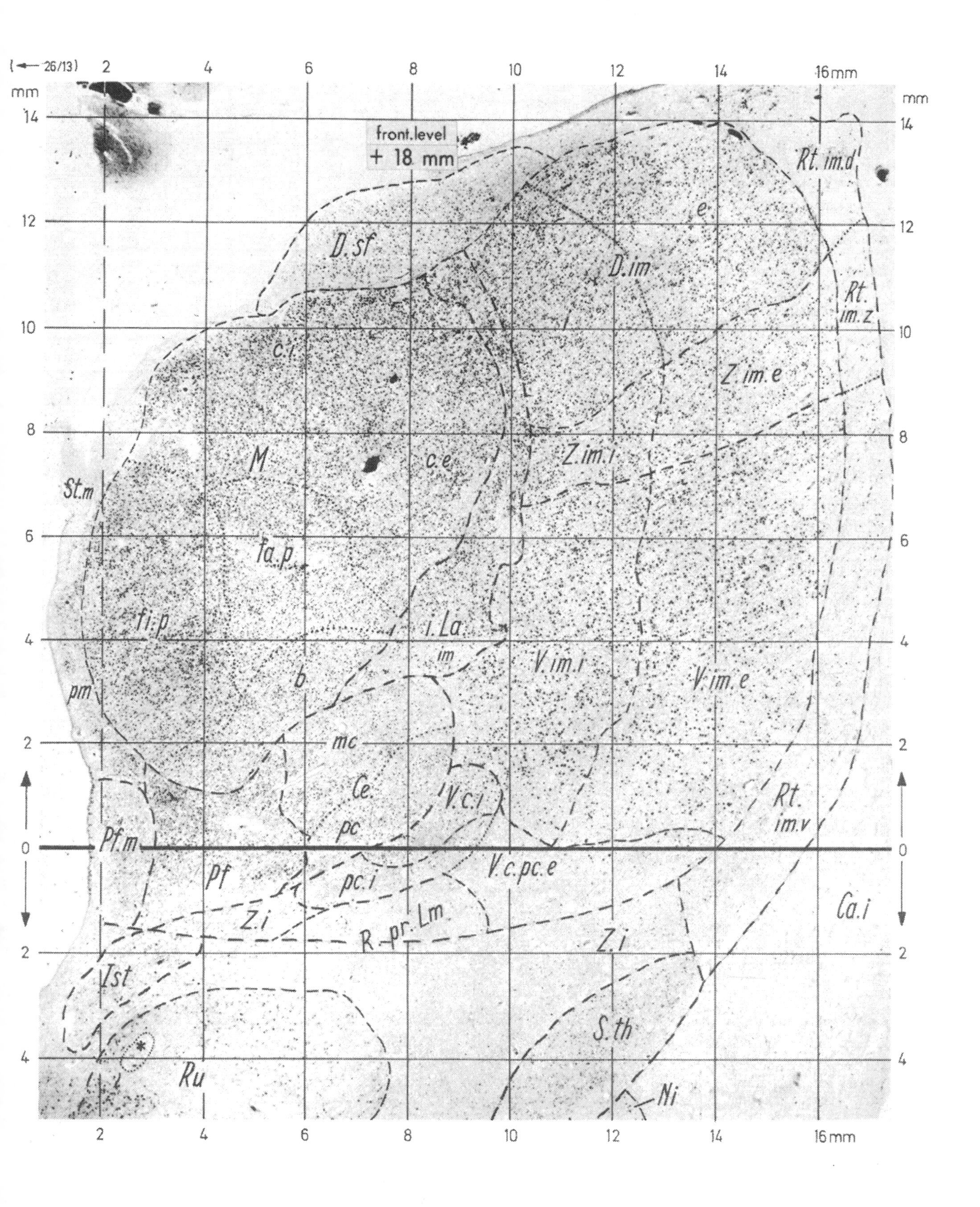

(← 26/13)
front. level
+ 18 mm
D.sf
D.im
e
Rt. im.d
Rt. im.z
c.i
Z. im.e
M
c.e
Z. im.i
St.m
fa.p
fi.p
i.La
im
V.im.i
V.im.e
b
pm
mc
Ce
V.c.i
Rt. im.v
Pf.m
pc
Pf
pc.i
V.c.pc.e
Z.i
R. pr. Lm
Ca.i
Ist
S.th
Ru
Ni

Fig. 154a. *Frontal section*, fiber stain, 18.0 mm behind the interventricular foramen. The V.o.p has been completely replaced by the nucleus ventrointermedius externus (*V.im.e*). The latter is characterized by containing particularly numerous fibers. The radiatio praelemniscalis (*R.pr.Lm*) which carries the vestibulo-thalamic, the dentato-thalamic, rubro-thalamic and nigro-thalamic fibers increases between the red nucleus and the base of the V.im.e. Below V.im.e appears the nucleus ventrocaudalis parvocellularis externus (*V.c.pc.e*), in which spino-thalamic terminate. The dorsal nuclei of the thalamus have also altered. The intermediate zentrolateral (*Z.im*) and the intermediate dorsal nuclei (*D.im*) are found here. In place of the main anterior nucleus one finds the nucleus dorsalis superficialis (*D.sf*). The nucleus *centre médian* (*Ce*) is developed and distinguished from the nucleus parafascicularis (*Pf*) by its denser fiber content. Above it, there is a further increase of size and differentiation of the medial territory [*M.fi.p, M.fa.p, M.b, M.c.i, M.c.e* bordering on the intermediate part of intralaminar nucleus (*i.La.im*)]. The zona incerta (*Z.i*) above the much diminished subthalamic nucleus (*S.th*) is linked to the capsule and shell of the nucleus ruber (putamen nuclei rubri).*: Tractus retroflexus. In the substantia nigra (*Ni*), the main posterior segment already has a different fiber structure and different distribution of cells, × aberrant fiber bundle in Ni. In the dorsomedial capsule of Ru and in the central gray above it extends the interstitial nucleus (*Ist*). ×5.

Fig. 154b. *Frontal section*, adjacent section in Nissl stain: The magnocellular part of the centre médian nucleus (*Ce.mc*) and its parvocellular part (*Ce.pc*) are clearly distinct from the dense, dark parafascicularis (*Pf*). The ventral area contains the nuclei ventrointermedii externus and internus (*V.im.e; V.im.i*). Below them the nucleus ventrocaudalis parvocellularis externus (*V.c.pc.e*) appears. Above this lie a layer of intermediate zentrolateral nuclei (*Z.im.e; Z.im.i*) and the large nucleus dorsointermedius (*D.im*). The dorsalis superficialis (*D.sf*) is clearly distinguished by the sparse number of nerve cells. Between the ependyma of the third ventricle and the M.fi.p lies the paramedian (*Pm*) nucleus. ×10.

Fig. 155. *Frontal section*, myelin stain, 20 mm behind the interventricular foramen. Above the radiatio praelemniscalis (*R.p.L*) lies the well-formed nucleus ventrocaudalis internus (*V.c.i*), which surrounds the centre médian nucleus (*Ce*) from the ventrolateral side. Its lowest part contains small cells – ventrocaudalis parvocellularis internus (p.ci) and externus (*V.c.pc.e*). It is distinguished from the V.im.e by its fewer fibers. Dorsally to the substantia nigra (*Ni*) the subthalamic nucleus ends and its place is occupied by the caudal zona incerta (*Z.i.c*). The white arrow indicates the fasciculi tegmenti Foreli, which consist among others of vestibulothalamic fibers. *: Fasciculus retroflexus in dorsomedial fiber capsule of red nucleus (*Ru*). ×5.

Fig. 155. For description see opposite page

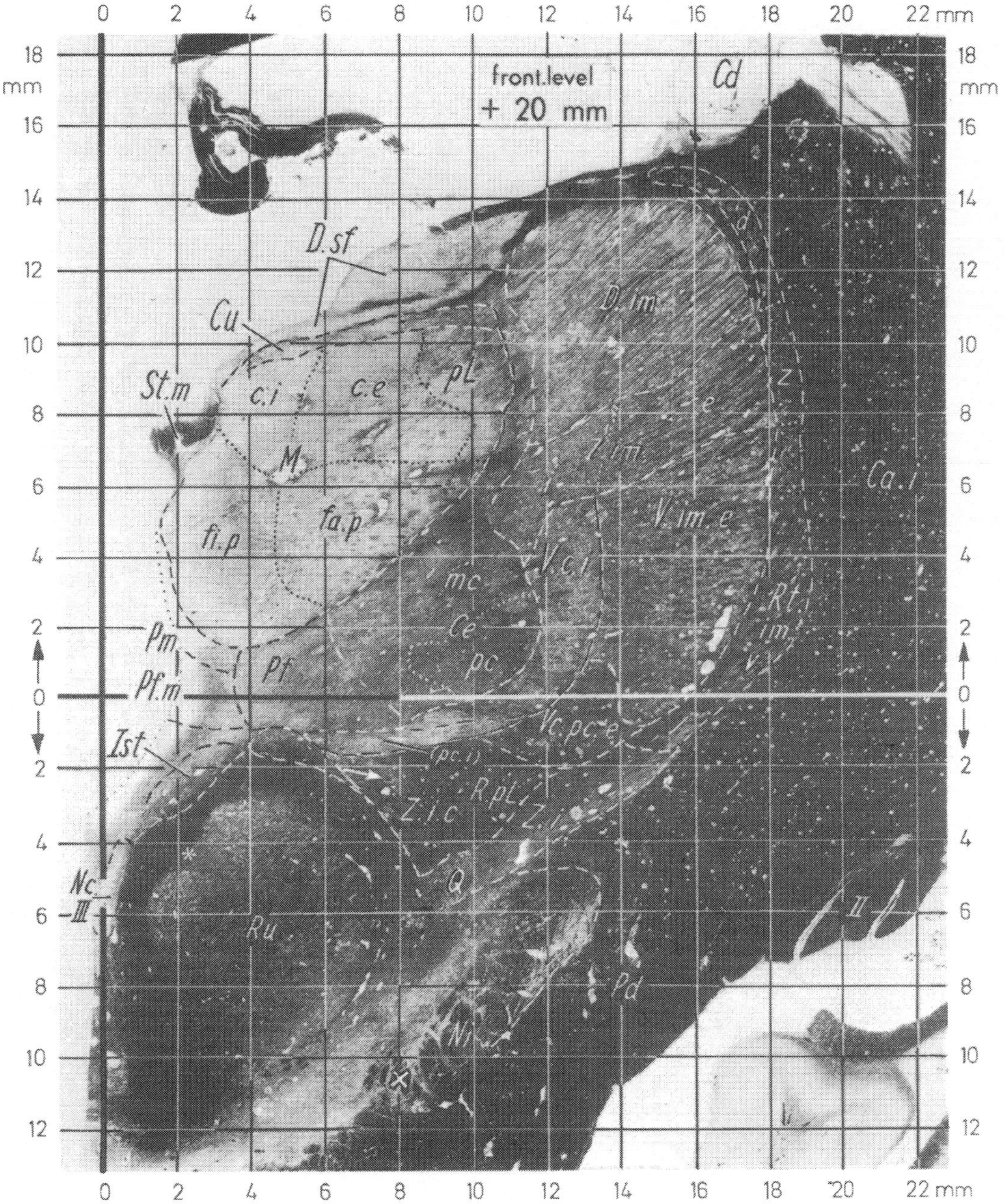

Fig. 156. (a) For description see p. 276

Fig. 156. (b) For description see p. 276 ▶

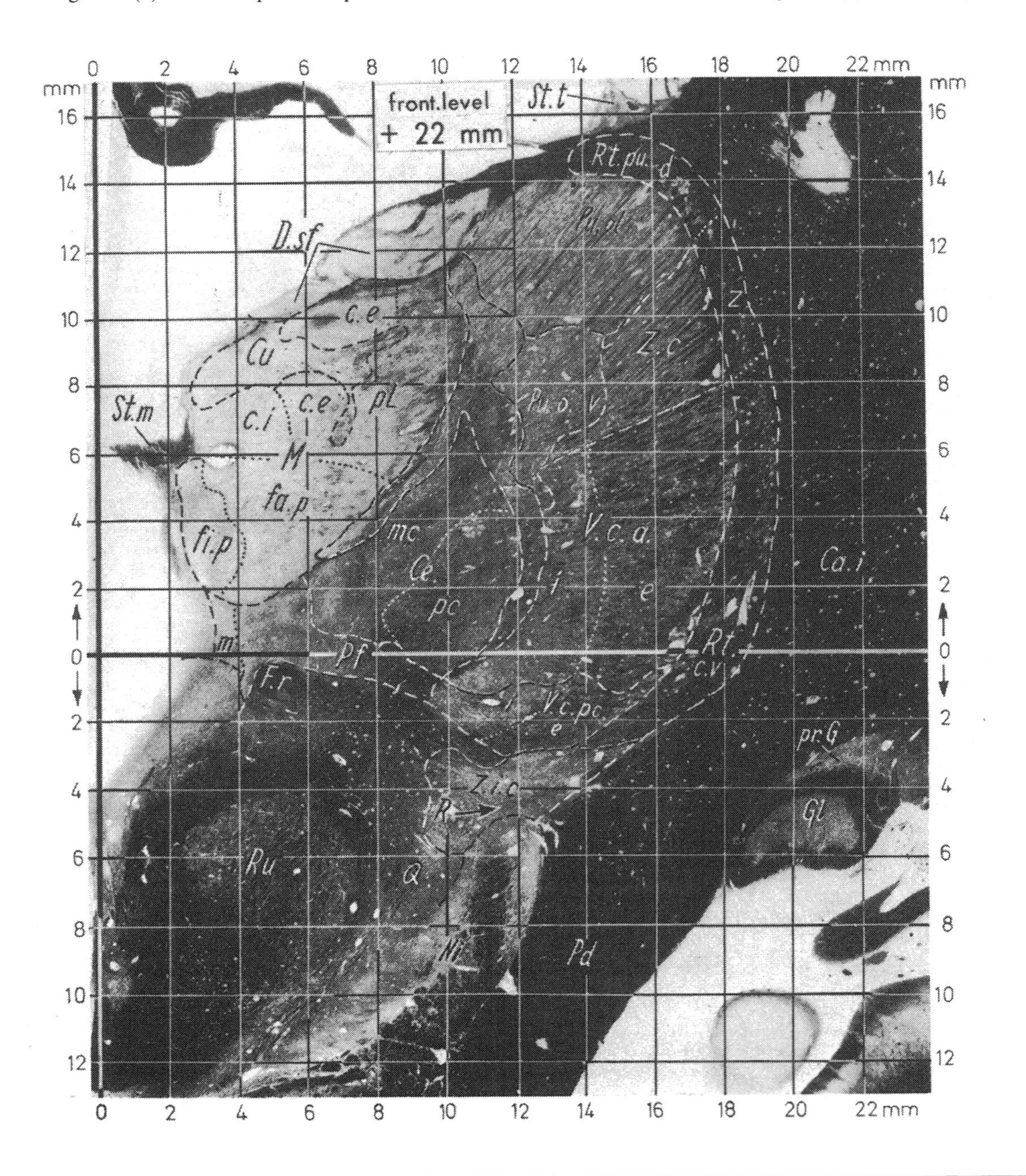

front. level
+ 22 mm

2 4 6 8 10 12 14 16 18 mm

Rt. pu. d

D. sf

Pu. ol

Rt. c. z

M. c. e

Cu

i. La. c

Z. c

pL

Pu. o. v

St. m

c. e

c. i

M.

fa. p

fi. p

V. c. a.

Ce.

La. im

mc.

e

pc

Rt. c. v

Pf

m

F. r

i

V. c. pc.

e

Ist

Z. i. c

Ru

Ni

2 4 6 8 10 12 14 16 18 mm

Fig. 156a. *Frontal section*, fiber stain, 22.0 mm behind the interventricular foramen. Laterally to the further enlarged central nucleus (*Ce*), one finds only sensory ventral nuclei: *V.c.a.i* (=VPM) and the nucleus ventrocaudalis anterior externus (*V.c.a.e*=VPL). The caudal zona incerta (*Z.i.c*) containing few fibrils delimited from the midbrain by the bundle R of SANO. Above the already diminished red nucleus, which receives cerebellar fibers from the decussatio brachiorum conjunctivorum (*Dc.br.cj*), appears a section of the fasciculus retroflexus (*F.r*=habenulointerpeduncularis (Meynert)), which gave the name to the nucleus parafascicularis (*Pf*). The upper part of medial territory (*M*) is occupied by the nuclei medialis internus (*c.i*), externus (*c.e*) and paralamellaris (*pL*) and infiltrated by strauds of the nucleus cucullaris (*Cu*). The lower part of M is occupied by M.fi.p and M.fa.p. ×5.

Fig. 156b. *Frontal section*, adjacent section in Nissl stain: The centre médian nucleus with its magnocellular (*Ce.mc*) and parvocellular (*Ce.pc*) component nuclei has its greatest extension here. Further laterally, the somatosensory anterior ventrocaudal nuclei are clearly divided into an inner nucleus (*V.c.a.i*) for the face and an external (*V.c.e*) for the extremities. Below each of them are the external and internal parvocellular ventral nuclei (*V.c.pc.e* and *V.c.pc.i*), in which the major part of the trigeminothalamic or the spinothalamic fibers terminate. The nucleus cucullaris, a dorsal part of the intralaminar nuclei (*i.la*) scatter and are found as dense cellular islands in the caudal parts of the medial nuclei (*M.c.e*). ×10.

Fig. 157. *Frontal section*, myelin stain, 24.0 mm behind the interventricular foramen, passes through the anterior edge of the posterior commissure (*Co.p*). The stria medullaris thalami (*St.m*) has descended to the ganglion habenulae (*Hb*). Lateral to this is the caudal edge of the medial nuclear region (*M.c.i*). Above this lies the nucleus pulvinaris superficialis (*Pu.sf*). Between the diminished central nucleus (*Ce*) and the V.c.p.i a pulvinar nucleus, the Pu.o.v has pushed in from the dorsal side. Bundles of thick fibers of the lemniscus medialis (*Lm*) enter into the V.c.p.e. The cerebral peduncle (*Pd*) is separated from the internal capsule (*Ca.i*) and from the lateral geniculate body (*G.l*) by the nucleus peripeduncularis (*p.Pd*) and the nucleus praegeniculatus (*pr.Gl*). In the midbrain one sees the broadening of the lemniscus medialis (*Lm*) medially to the substantia nigra (*Ni*), the crossed brachium conjunctivum (*Br.cj*) and the fascicules centralis tegmenti (*F.c.tg*). It runs besides the interstitial nucleus (*Ist*) and beneath that periaqueductal reticular substance (*Rt.aq*) and Fasciculi Foreli (*F.Fo*) which contain the vestibulothalamic fibers. Pulvinar nuclei (*Pu.o.l; Pu.o.v*) now lie above the V.c.p.e. ×5.

Fig. 157. For description see opposite page

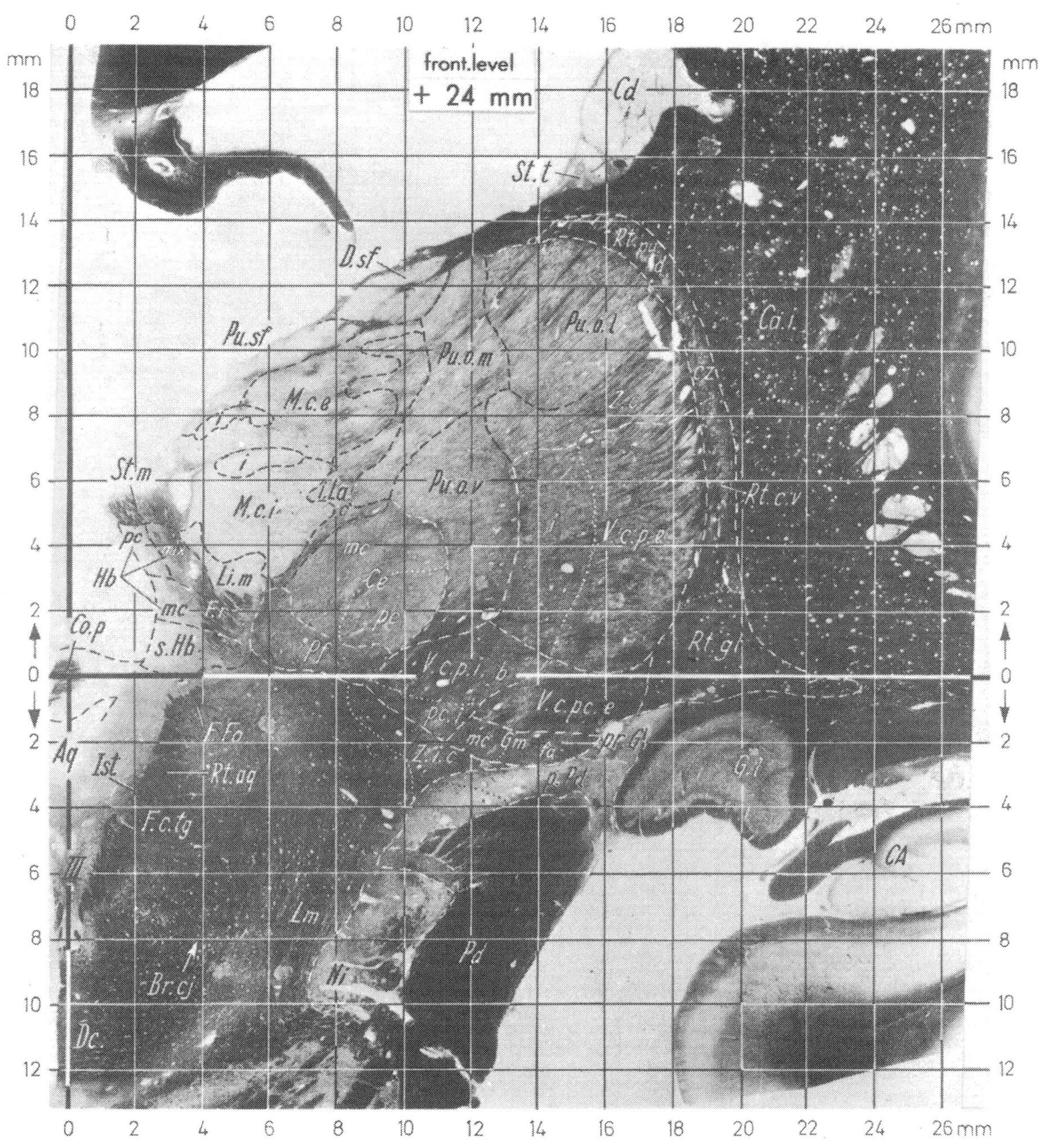

Fig. 158. (a) For description see p. 280

Fig. 158. (b) For description see p. 280

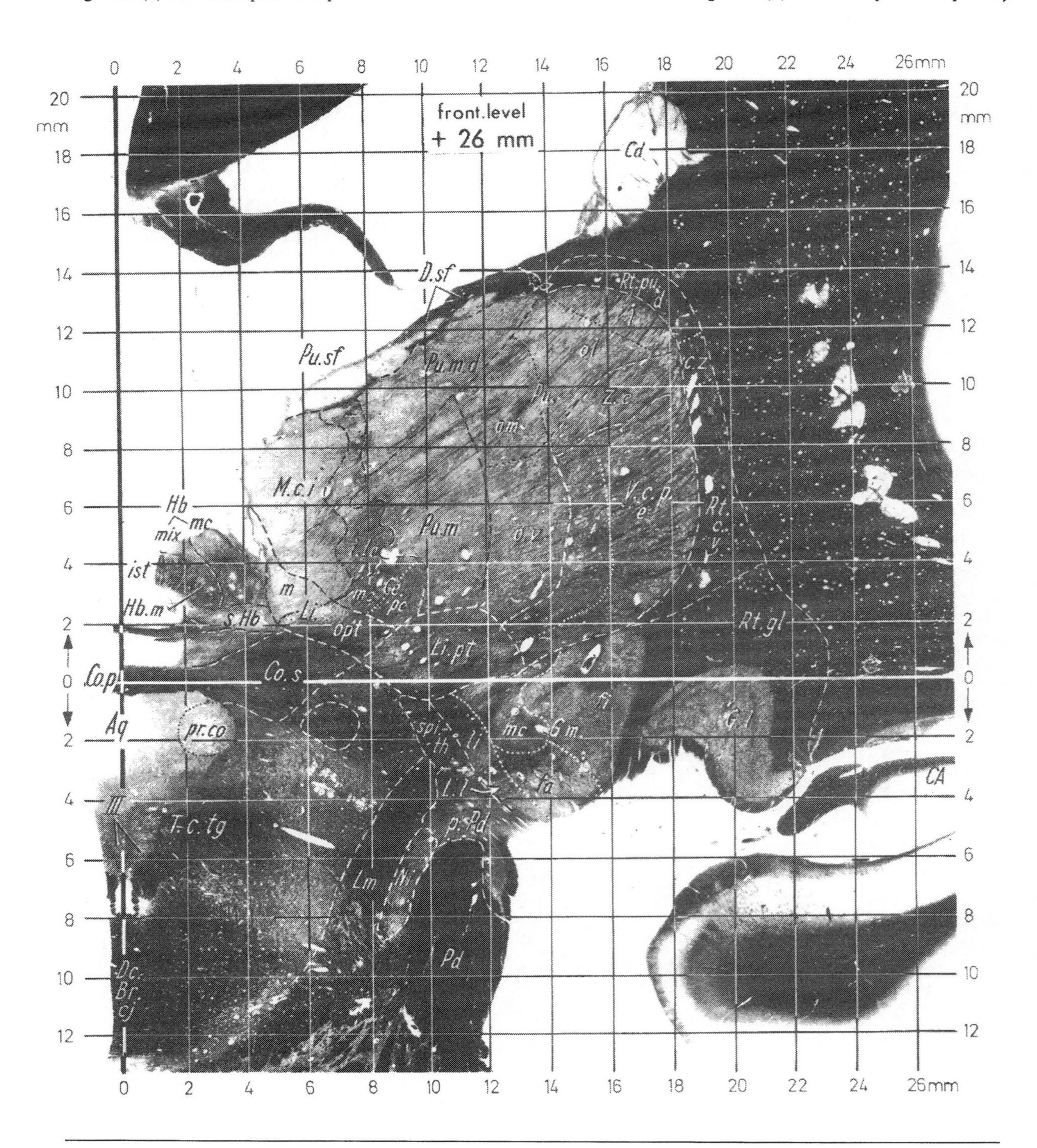

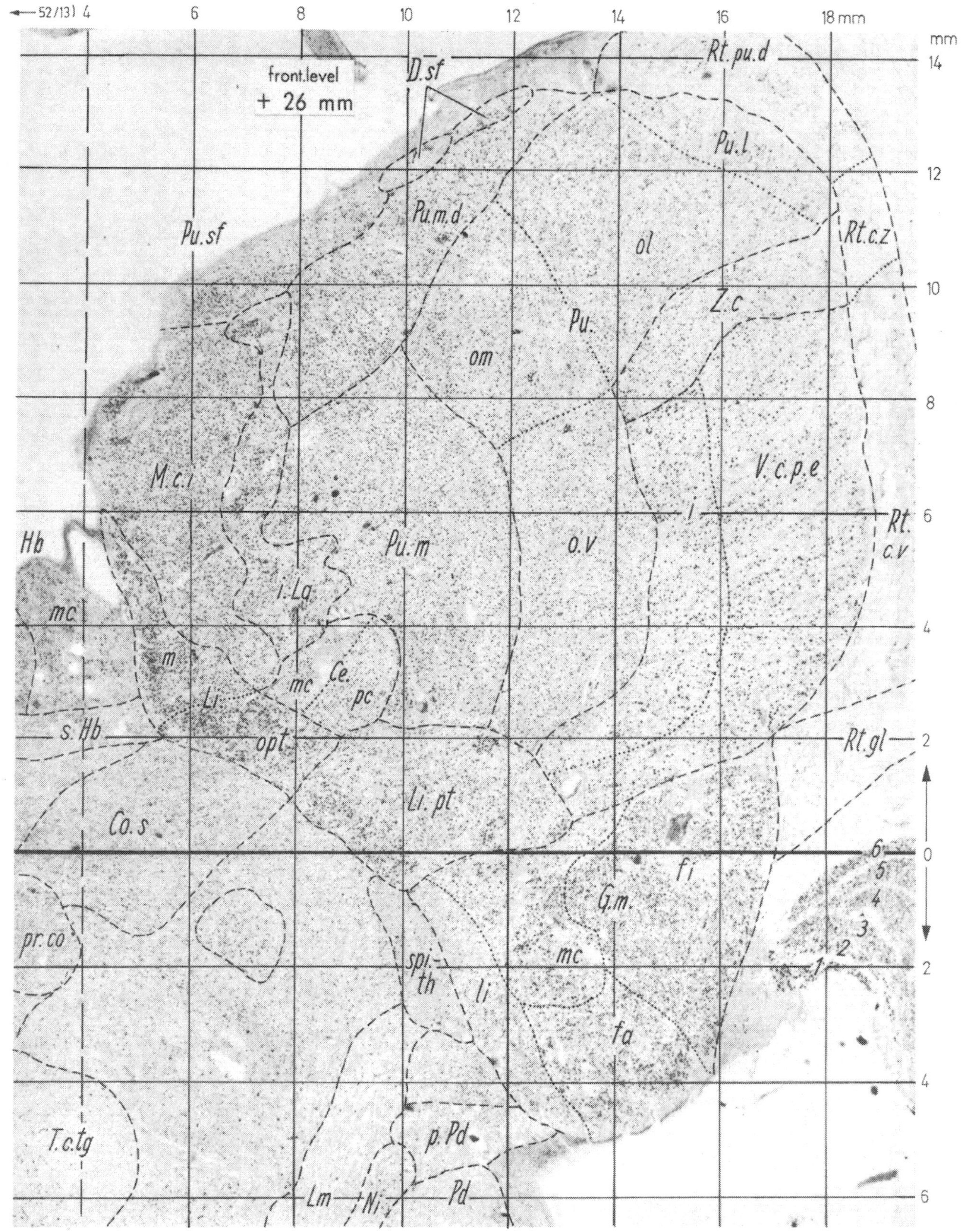

← 52/13) 4 6 8 10 12 14 16 18 mm

mm

front.level
+ 26 mm

D.sf

Rt.pu.d

14

Pu.l

12

Pu.m.d

Pu.sf

ol

Rt.c.z

10

Z.c

Pu.

om

8

M.c.i

V.c.p.e

i

Rt. c.v

6

Hb

Pu.m

o.v

i.La

mc

4

m

Ce.

mc

pc

Li

s.Hb

opt

Rt.gl

2

Li.pt

Co.s

0

6 5 4 3 2 1

fi

G.m.

pr.co

spi.- th

li

mc

2

fa

4

T.c.tg

p.Pd

Lm

Ni

Pd

6

4 6 8 10 12 14 16 18 mm

Fig. 158a. *Frontal Section*, myelin stain, 26.0 mm behind the interventricular foramen. The section passes through the posterior commissure (*Co.p*) and the commissura habenularum (*Co.hb*). The Co.hb seems to originate from the nucleus habenularis medialis (*Hb.m*) and the nucleus subhabenularis (*s.Hb*). Above it lies the caudal pole of the ganglion habenulae (*Hb*). The nucleus limitans (*Li*) extends from this in a basolateral direction toward the medial geniculate body (*G.m*) immediately at the boundary between midbrain tegmentum and the thalamus. The medial (*G.m*) and the lateral geniculate bodies (*G.l*) lie close together at the base of the brain. The G.m is composed of the nuclei limitans (*li*), magno cellular (*mc*) fasciculosus (*fa*) and fibrosus (*fi*). In the thalamus, lying only latterally, are the hindmost parts of the sensory nuclear areas = *V.c.p.e* or VPL. Medially one encounters the caudal pole of the medial nuclear area (*M.c*). In the tegmentum of the midbrain over the crossing of the brachia conjunctiva lie the descending fasciculus centralis tegmenti (*F.c.tg*) and the ascending fasciculi Foreli (*F.Fo*). Laterally in the central gray matter lies the precommissural nucleus (*pr.co*) and above it the rostralmost part of the superior colliculus (*Co.s*). Here is seen the largest extension of the lemniscus medialis (*L.m*) from the dorsal border of the pons to the nucleus limitans portae (*Li.pt*). The red nucleus is replaced by the fiber mass of brachium conjunctivum on both sides of the Decussatio brachiorum conjunctivorum (*Dc.Br.cj*). ×5

Fig. 158b. *Frontal section*, cellular stain, 26.0 mm behind the interventricular foramen. The caudal pole of the centromedian nucleus is surrounded by the nucleus limitans (*Li.opt* and *Li.pt*) as well as by the intralaminar nucleus (*i.La*) and the medial pulvinar nucleus (*Pu.m*). The section includes segments of the medial geniculate body (*G.m*), with its parts: magnocellular (*mc*), fibrous (*fi*), fascicular (*fa*), and limitans (*li*). The oral nuclei of pulvinar (*Pu.o.v, Pu.o.m, Pu.o.l*) extend into the upper part of the thalamus. The Nucleus precommisuralis (*pr.co*) lies in the lateral arc of the periaqueductal gray matter medio-ventral to the cut through the rostral pole of the superior colliculus (*Co.s*). In the lateral geniculate body only the 6 layers are numbered. ×10.

Fig. 159. *Frontal section*, myelin stain, 28 mm behind the interventricular foramen. Over the midbrain in the region of the superior colliculi are the pulvinar and the corpora geniculata. Between the medial (*Pu.m.d, Pu.m.z*) and lateral pulvinar nuclei (*Pu.l.s, Pu.l.if*) the oral pulvinar nuclei are situated (*Pu.o.v, Pu.o.m*). The nucleus pulvinaris intergeniculatus (*Pu.ig*) extends between the geniculatum mediale (*G.m*) and laterale (*G.l*). At the boundary between the colliculus superior (*Q.s*) and the pulvinar, the nucleus limitans (*Li*) continues. The border between midbrain tegmentum and thalamus is occupied by the limitans nucleus with three subnuclei: *m*, *opt* and *pt*. ×5.

Fig. 159. For description see opposite page

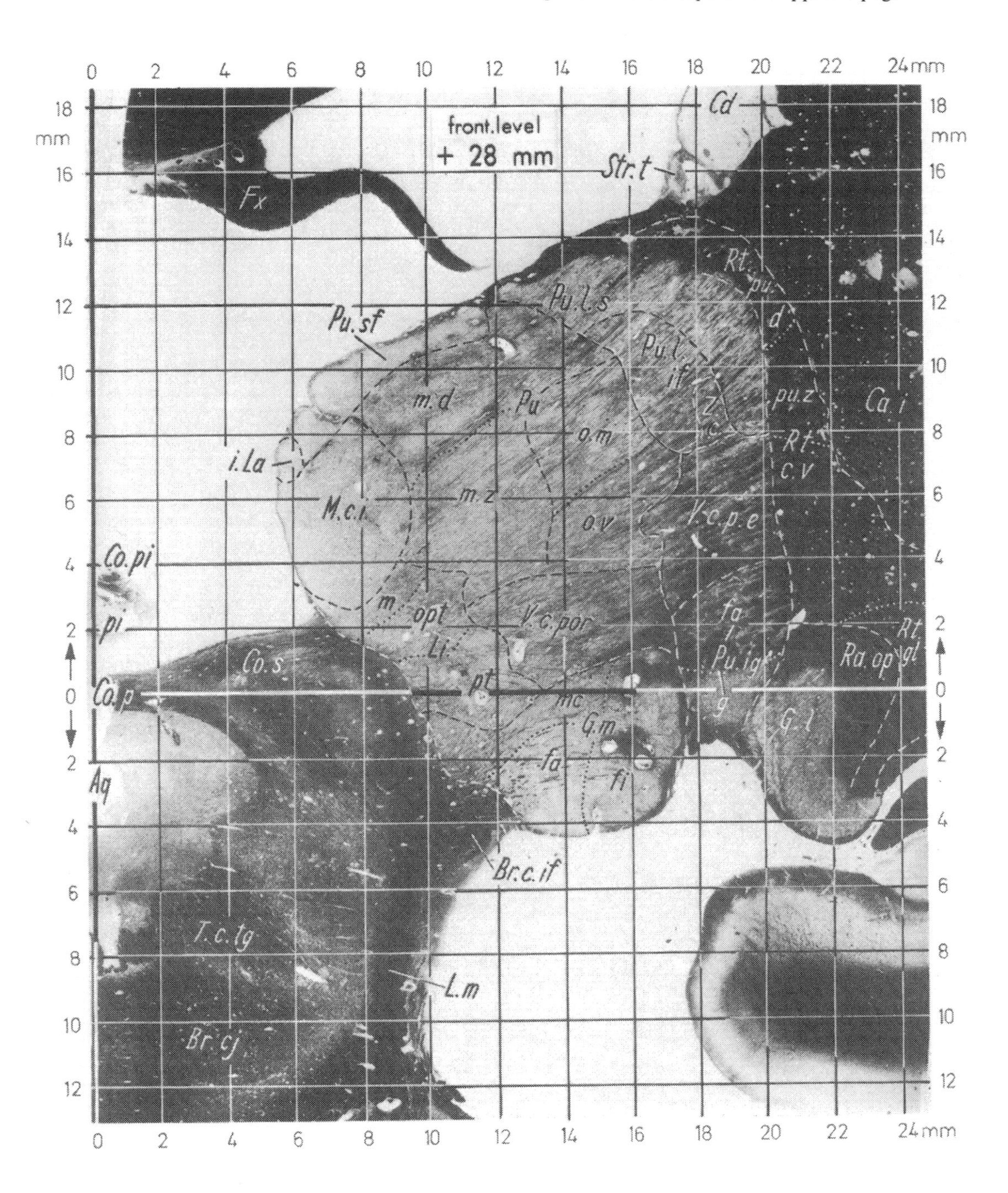

Fig. 160. *Frontal section*, myelin stain, 30.0 mm behind the interventricular foramen. The midbrain is linked to the thalamus by the brachium colliculi superioris (*Br.co.s*). The caudal poles of the medial (*G.m*) and the lateral (*G.l*) geniculate bodies are below the pulvinar. The nucleus pulvinaris intergeniculatus (*Pu.ig*) shows its greatest extension here. In the base of the pulvinar one sees longitudinally cut bundles of fibers of the radiatio corticomesencephalica lateral from Pu.sb and above Br.co.s. Between Pu-nuclei and Ca.i the three parts of reticulatus pulvinaris (*Rt.pu*). ×5.

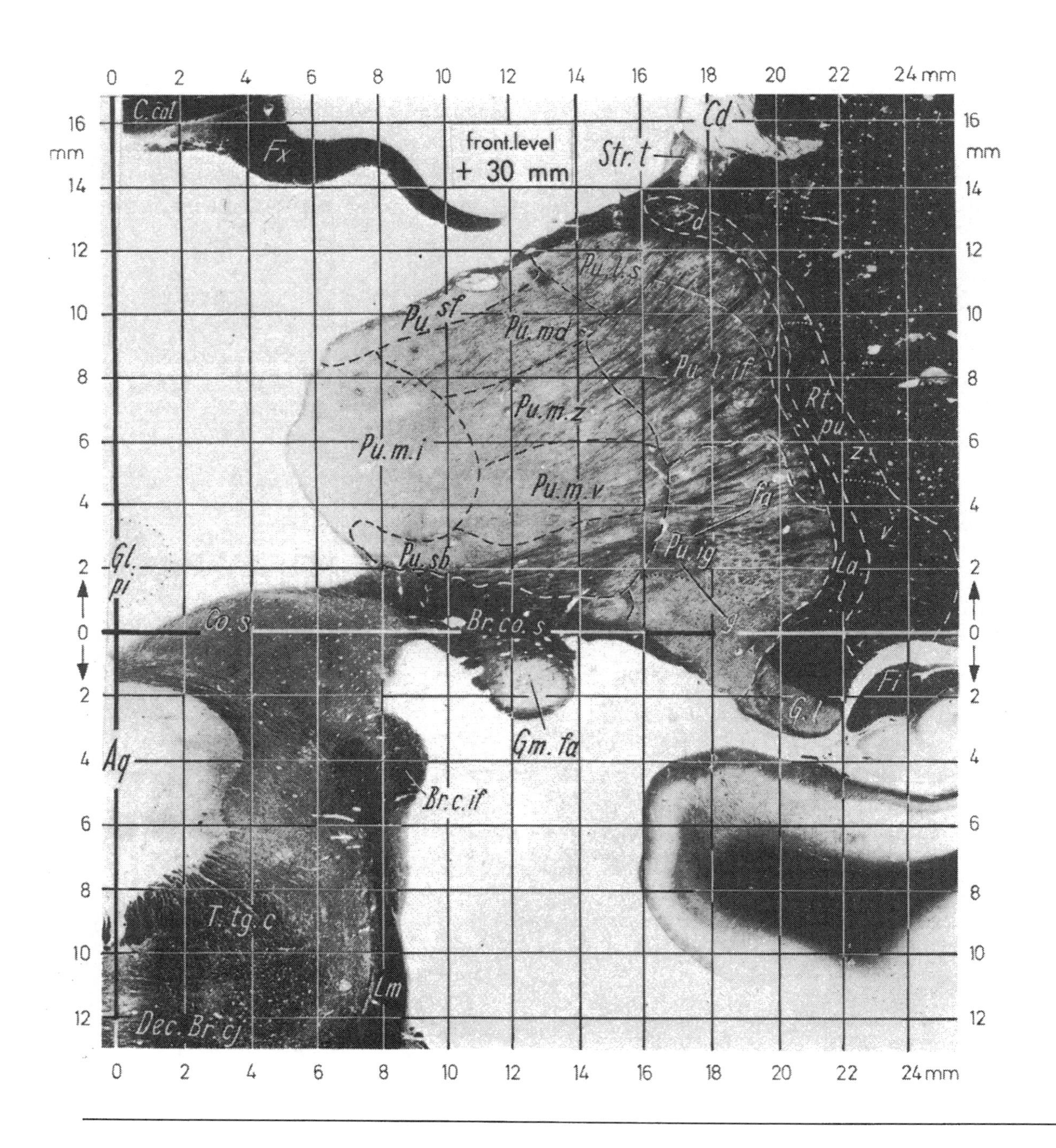

5. Description of Serial Sagittal Sections

For knowledge concerning the correct approach of the electrode to the target of stimulation as well as for coagulation, the sagittal sections are both instructive and informative. That is the reason that three levels of sagittal sections are shown: 6 mm, 9,5 mm, and 13 mm from the border of the third ventricle. Each level is shown in both fiber and cell (Nissl) stains.

Fig. 161 a. Sagittal section, 6 mm from the border of the third ventricle at right angles to the midline. The level of the C.a-C.p line is indicated by a thick line o—o. The main part of the thalamus contains the different nuclei of the medial territory of the thalamus. Shown here are the nuclei medialis fasciculosus anterior (*M.fa.a*), medialis fasciculosus posterior (*M.fa.p*), and medialis caudalis externus (*M.c.e*) and the nucleus cucullaris (*Cu*). Dorsally they are surrounded by the dorsal superficial nucleus (*D.sf*) and the anterior principal nucleus (*A.pr*); rostrally, however, the medial territory is enclosed by the intralaminar nucleus (*i.La*) and the parafascicular nucleus (*Pf*) below. In front of the intralaminar nucleus the nuclei ventro-oralis anterior and posterior (*V.o.a* and *V.o.p*) occupy the basal part of the thalamus; the H_1 fiber bundles of Forel end exclusively in V.o.a. The extension of the zona incerta (*Z.i*) from the inferior thalamic peduncle (*Pd.if*) to the parafascicular nucleus (*Pf*) is interrupted by the *H*-field of Forel. Note the sagittal course of the fibers running below the zona incerta, caudally through H_2 and curving upward in the H field to the V.o.a nucleus. The dorsal part of the rostral thalamus is occupied by the nuclei zentrolateralis oralis (*Z.o*), dorsalis oralis internus (*D.o.i*), and lateropolaris, mainly its superior part (*L.po.s*). In the hypothalamus the tuber nuclei (*Tu*) and supraoptic commissure formed by minor bundles are evident. In front of the hypothalamus the Broca band (*Br*) appears in this section. ×5.

Fig. 161 b is the adjacent section in Nissl stain (6 mm from the wall of the third ventricle). It shows most of the rostral pole of the thalamus and its surroundings. The intralaminar nucleus (*i.La*) extends in front of the nucleus medialis fasciculosus (*M.fa*). The V.o.p nucleus is surrounded rostrally by the V.o.a nucleus with the afferent bundles of H_1. The ventral oral nuclei are covered by the nucleus zentrolateralis oralis (*Z.o*). The dorsal oral nucleus (*D.o.i*) appears above this nucleus, and the rostral pole of the thalamus is occupied by the lateropolar nucleus (*L.po.s*), mainly by the superior and external parts (*e*). This picture clearly demonstrates that the course of the zona incerta (*Z.i*) is interrupted from rostral to caudal by the H-field. The caudal zona incerta (*Z.i.c*) surrounds the parafascicular nucleus (*Pf*) from below. The rostral pole of the thalamus is surrounded by a layer of the reticular nucleus, mainly its polar part (*Rt.po.v*). *d*=*Rt.po.d* ×10.

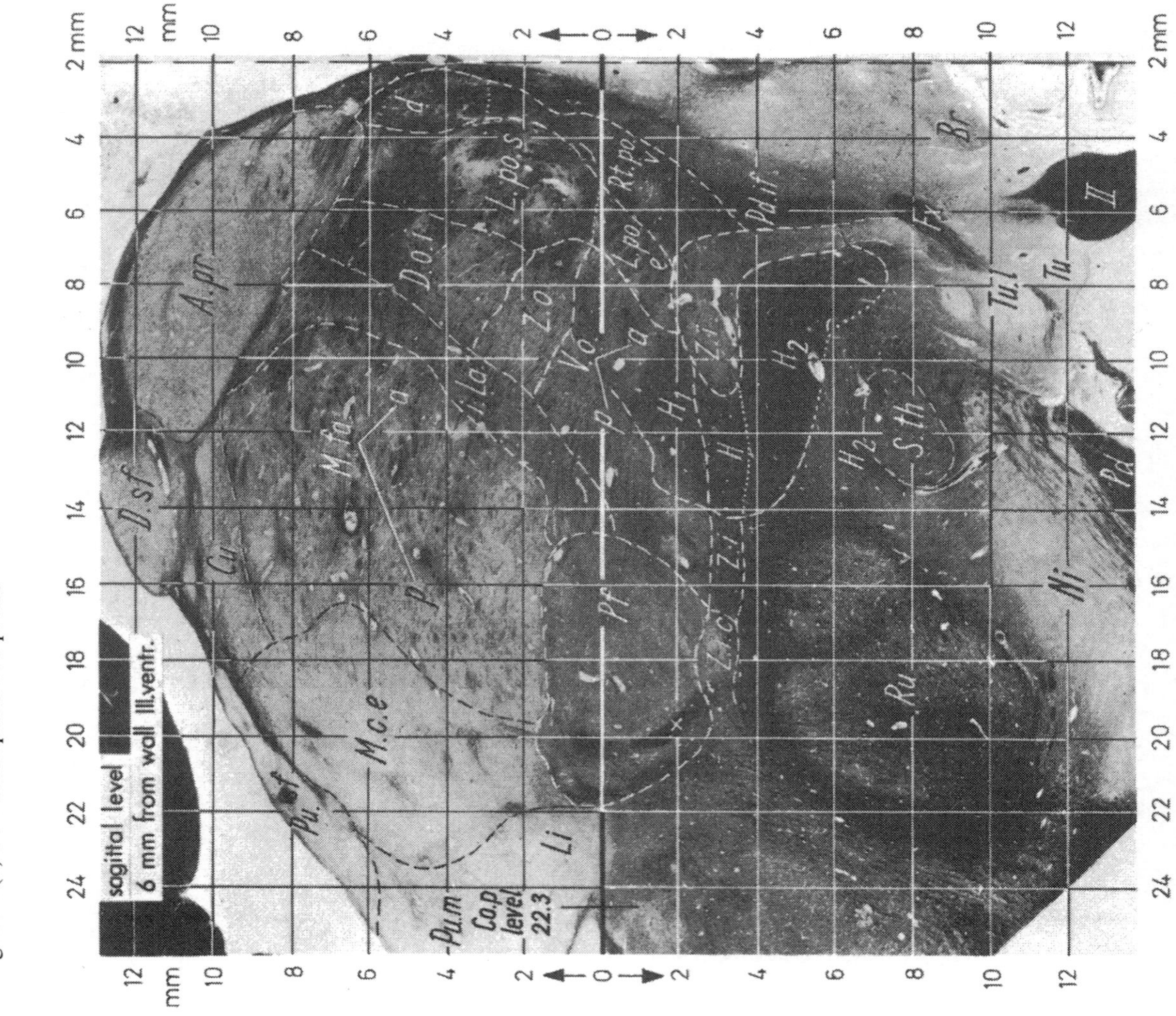

Fig. 161. (a) For description see p. 283

Fig. 161. (b) For description see p. 283

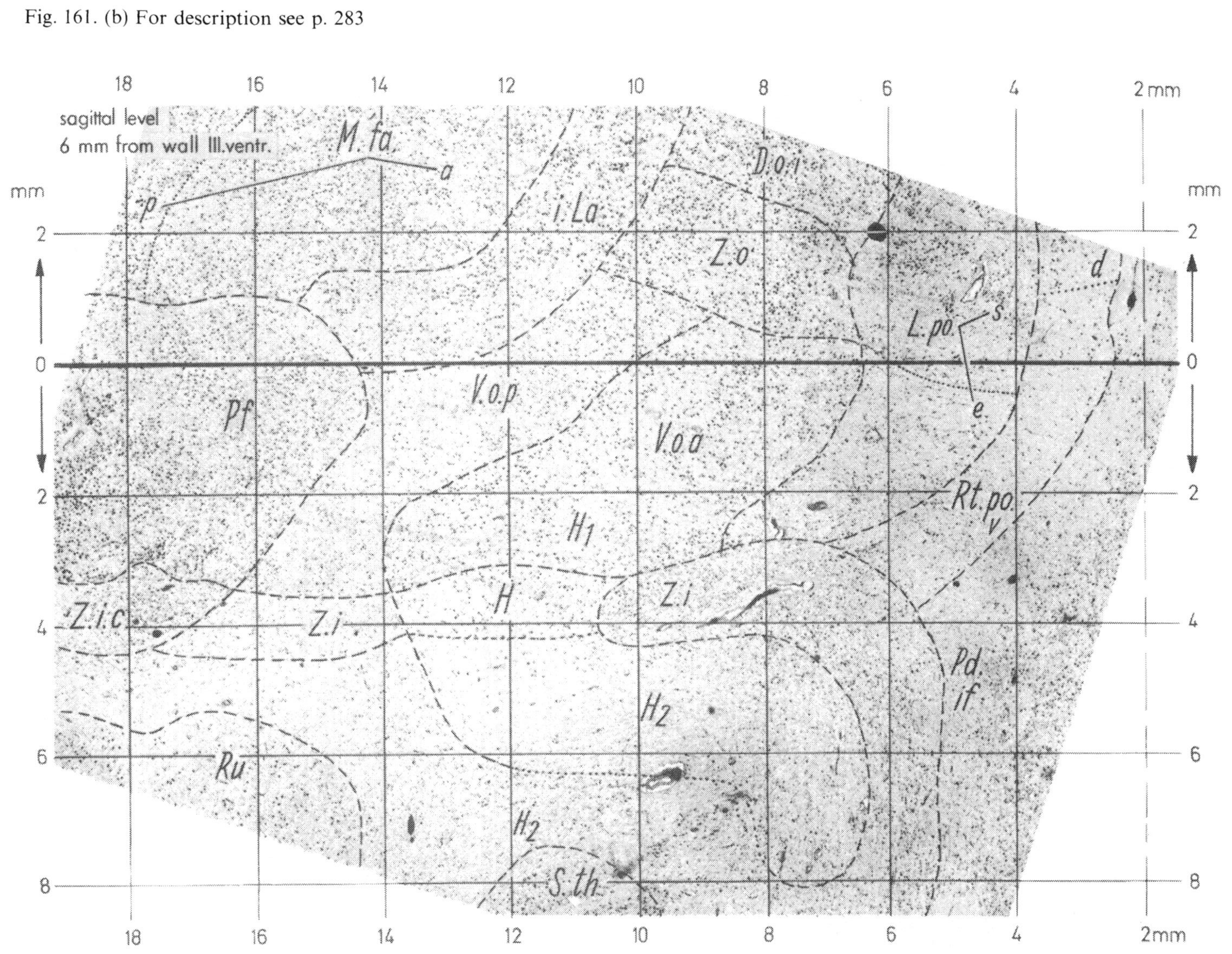

Fig. 162a. Sagittal section, 9.5 mm from the wall of the third ventricle, in fiber stain. This sagittal section traverses all the ventral nuclei of the thalamus and behind them the centre médian nucleus. The very thin layer of the nucleus ventrocaudalis anterior (*V.c.a.i*) is located in front of both parts of the centre médian nucleus (*Ce.mc* and *Ce.pc*). Frontally follow the intermediate ventral nucleus (*V.im.e*), and the V.o.p and V.o.a nuclei. Dorsally these are covered by three zentrolateralis nuclei: caudal (*Z.c*), intermedius (*Z.im*), and oral (*Z.o*). The latter nuclei are covered by the nuclei dorsalis oralis (*D.o*) and by dorsalis intermedius (*D.im*). The frontal pole of the thalamus is occupied by the nucleus lateropolaris (*L.po*) and surrounded by the polar reticulate (*Rt.po.z*) thalamic nucleus. The nucleus intralaminaris caudalis (*i.La.c*) appears large because it is cut tangentially. Caudally to it are two oral pulvinar nuclei, pulvinaris oralis lateralis (*Pu.o.l*) and pulvinaris oralis ventralis (*Pu.o.v*). ×5.

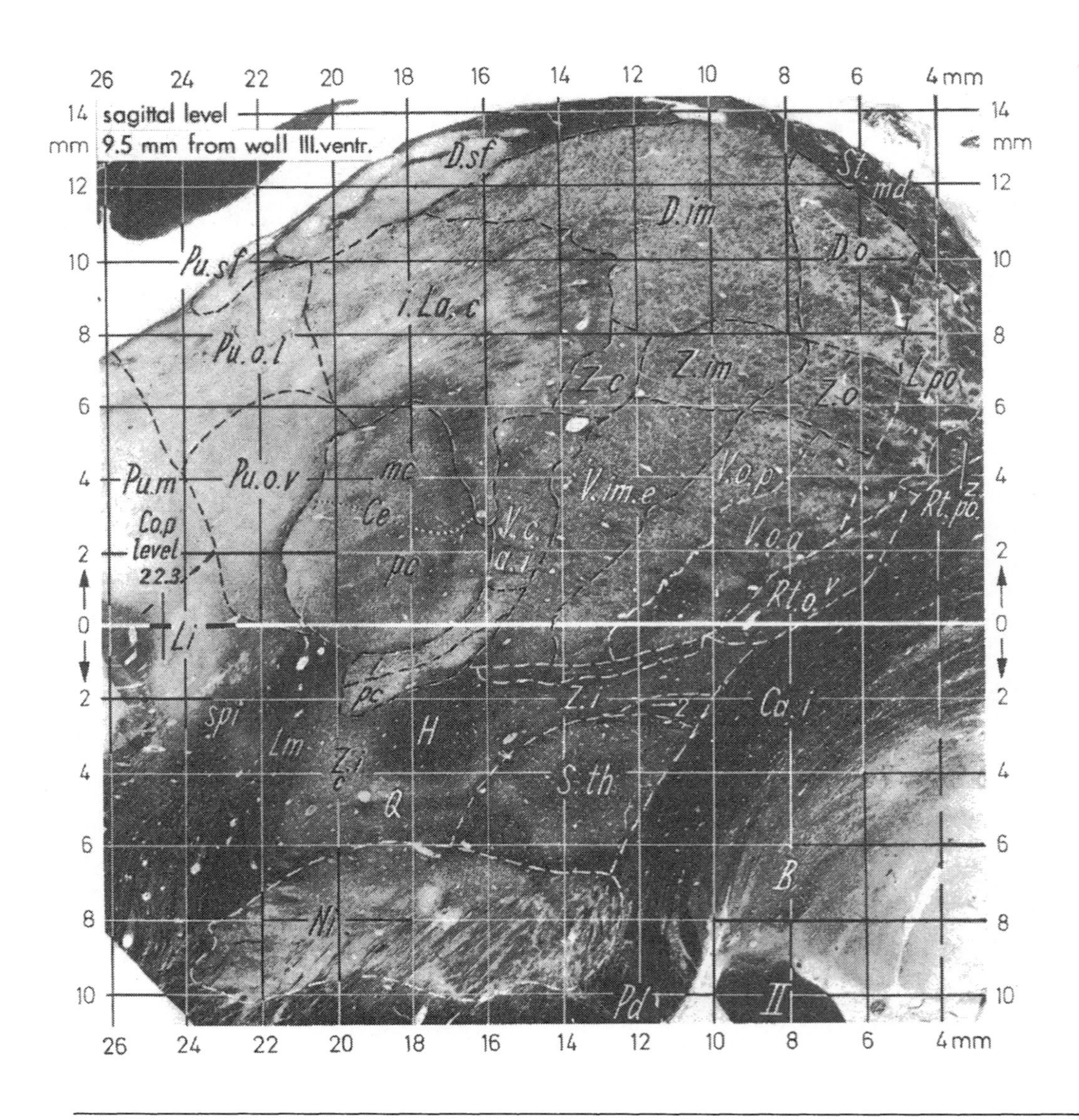

Fig. 162b. The neighboring section 9.5 mm from the wall of third ventricle in Nissl stain, at a higher magnification, shows almost the same borderlines as the fiber picture. The cellular arrangement is visible: the largest nerve cells in the nucleus ventrointermedius externus (*V.im.e*). The dorsal (*D.im* and *D.o*) and the zentrolateral nuclei (*Z.c*, *Z.im*, *Z.o*) can be seen in their typical position above all ventral nuclei. Notice the cytoarchitectonic difference between V.o.p and V.o.a. × 10.

Fig. 163a. This section, 14 mm lateral to the wall of the third ventricle, includes many regions of the thalamus, especially the ventral and dorsal nuclei, the pulvinar nuclei, and the medial geniculate body (*Gm*). The place of the nucleus pulvinaris oralis medialis (*Pu.o.m*) is occupied here by the pulvinaris oralis lateralis (*Pu.o.l*). The pulvinar oralis ventralis (*Pu.o.v*) appears below it. In front of these nuclei appear nuclei dorsalis intermedius externus (*D.im.e*) and dorsalis oralis externus (*D.o.e*). Below them three zentrolateral nuclei are located: zentrolateralis caudalis (*Z.c*), zentrolateralis intermedius (*Z.im*), and zentrolateralis oralis (*Z.o*), which is replaced rostrally by the reticulate nucleus (*Rt.o.z*). The row of ventral nuclei starts with the ventrocaudalis posterior (*V.c.p*), characterized by thick afferent fibers. Next is the ventrocaudalis anterior (*V.c.a*) (appears lighter), supplied by the cervicothalamic fibers. In front of the V.c.a is situated the ventrointermedius externus (*V.im.e*), which is one of the darkest-staining nuclei of the thalamus because of its extrathalamic afferents. In front of the V.im.e, the V.o.p appears considerably lighter; it is supplied by the dentato-thalamic fibers. A small section of the V.o.a is situated in front of the V.o.p. ×5

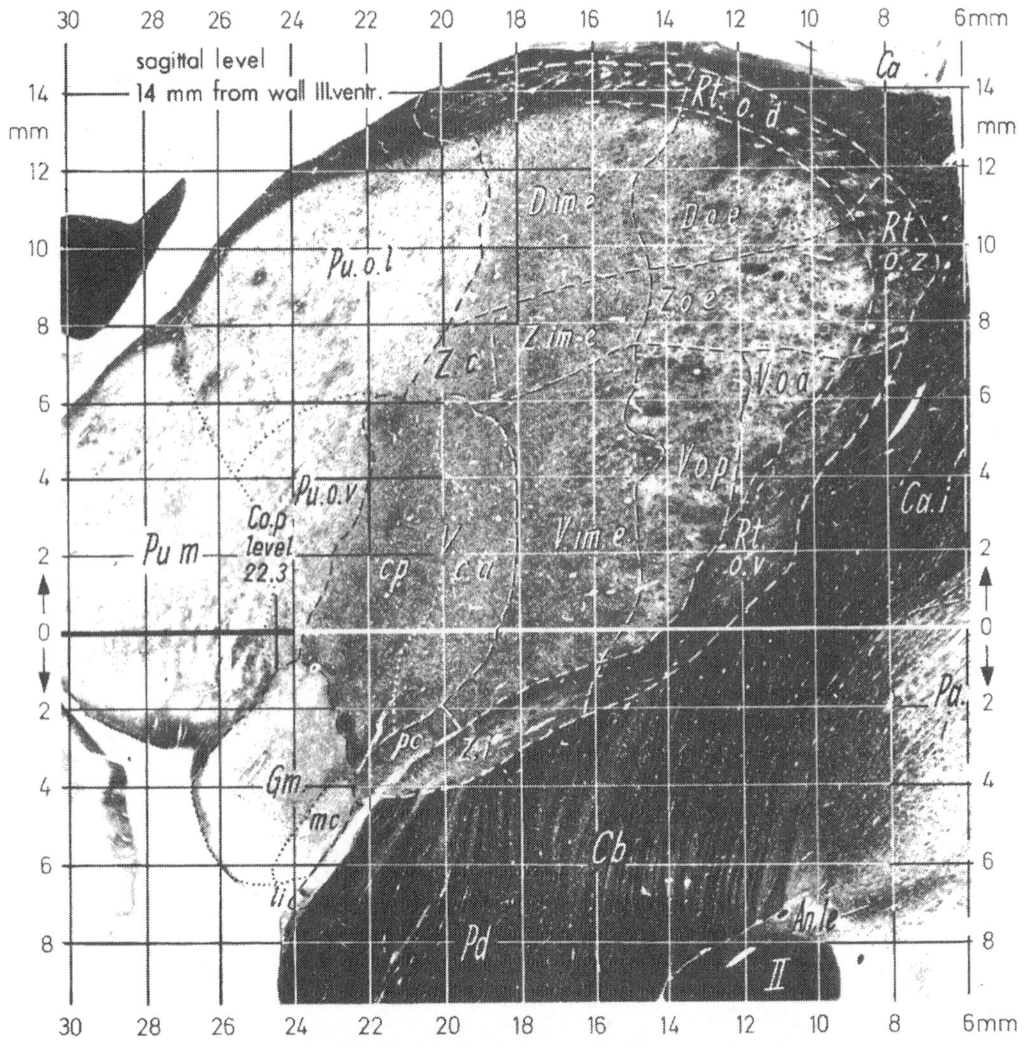

Fig. 163b. (Nissl stain), neighboring section to Fig. 163a, 14 mm lateral to the wall of the third ventricle, shows only the ventral nuclei. The nuclei ventrocaudalis posterior (*V.c.p*) and ventrocaudalis anterior (V.c.a) can be differentiated because of the difference in nerve cell size. Nucleus ventrointermedius externus (V.im.e) is prominent because of its brightness and its very large and clearly stained cell bodies. In front of it, the V.o.p nucleus begins with a sharp borderline and is characterized by a much denser population of nerve cells. The V.o.a nucleus in front of it is not so densely packed with nerve cells. The Z.o, Z.im, and Z.c appear above. ×10

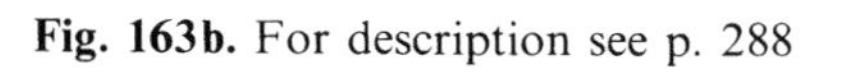

Fig. 163b. For description see p. 288

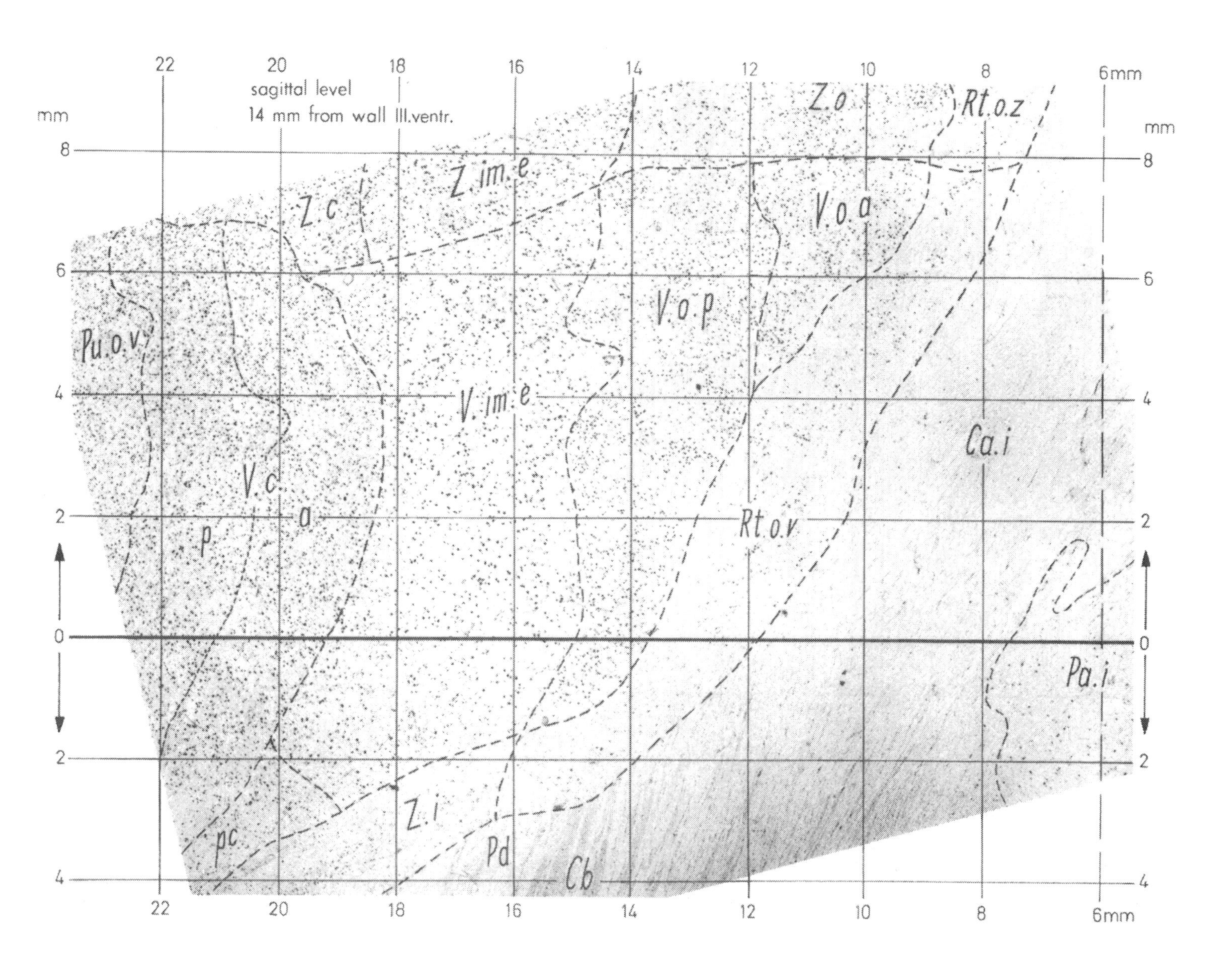

References

ACHARD, CH.: L'encéphalite léthargique. Paris: Baillière 1921.

AFIFI, A., KAELBER, W.W.: Efferent connections of the substantia nigra in the cat. Exp. Neurol. **11**, 474–482 (1965).

AKERT, K., ANDERSSON, B.: Experimenteller Beitrag zur Physiologie des Nucleus caudates. Acta physiol. scand. **22**, 281–298 (1951).

ALBE-FESSARD, D., ARFEL, G., GUIOT, G.: Activités électriques charactéristiques de quelques structures cérébrales chez l'homme. Ann. Chir. **17**, 1185–1214 (1963).

ALBE-FESSARD, D., ARFEL, G., GUIOT, G., HARDY, J., VOURC'H, G., HERTZOG, E., ALEONARD, P., DEROME, P.: Dérivations d'activités spontanées et évoquées dans les structures cérébrales profondes de l'homme. Rev. neurol. **106**, 89–105 (1962).

ALBE-FESSARD, D., GUIOT, G., LAMARRE, Y., ARFEL, G.: Activation of thalamocortical projections related to tremorogenic processes. In: The Thalamus. D.P. Purpura, M.D. Yahr (eds.), pp. 237–253. New York and London: Columbia Univ. Press 1966.

ALBE-FESSARD, D., KRAUTHAMER, G.: Inhibition of units of the non-specific afferent system by stimulation of the basal ganglia. J. Physiol. (Lond.) **175**, 54–55 (1964).

ALBE-FESSARD, D., LIEBESKIND, J.: Origine des messages somato-sensitifs activant les cellules du cortex moteur chez le singe. Exp. Brain Res. **1**, 127–146 (1966).

ALBE-FESSARD, D., OSWALDO-CRUZ, E., ROCHA-MIRANDA, C.E.: Activités évoquées dans le noyau caudé du chat en réponse à des types divers d'afférences. I. Etude macrophysiol. Electroenceph. clin. Neurophysiol. **12**, 405–420 (1960).

ALBE-FESSARD, D., ROCHA-MIRANDA, C.E., OSWALDO-CRUZ, E.: Activités évoquées dans le noyau du chat en réponse à des types divers d'afférences. II. Etude microphysiol. Electroenceph. clin. Neurophysiol. **12**, 649–661 (1960).

ALBE-FESSARD, D., ROUGEUL, A.: Activité d'origine somesthésique évoquées sur le cortex non spécifique du chat anesthèsié à chloralose: rôle du centre median du thalamus. Electroenceph. clin. Neurophysiol. **10**, 131–152 (1958).

ALVORD, E.C., Jr.: The pathology of Parkinsonism. II. An interpretation with special reference to other changes in the aging brain. Contemp. Neurol. Ser. **8**, 131–161 (1971).

ALZHEIMER, A.: Über die anatomische Grundlage der Huntingtonschen Chorea und die choreatischen Bewegungen überhaupt. Neurol. Zbl. **30**, 981 (1911).

AMADOR, L.V., BLUNDELL, J.E., WAHREN, W.: Description of coordinates of the deep structures. In: Einführung in die stereotaktischen Operationen mit einem Atlas des menschlichen Gehirns. G. Schaltenbrand, P. Bailey (eds.), pp. 20–28. Stuttgart: Thieme 1959.

ANDÉN, N.E., CARLSSON, A., DAHLSTRÖM, A., FUXE, K., HILLARP, N.S., LARSSON, K.: Demonstration and mapping out of nigroneostriatal dopamine neurons. Life Sci. **3**, 523–530 (1964).

ANDÉN, N.E., DAHLSTRÖM, A., FUXE, K., LARSSON, K.: Further evidence for the presence of nigro-neostriatal dopamine neurons in the rat. Amer. J. Anat. **116**, 329–333 (1965).

ANDERSSON, S.A., LANDGREN, S., WOLSK, D.: The thalamic relay and cortical projection of group I muscle afferents from the forelimb of the cat. J. Physiol. (Lond.) **183**, 576–591 (1966).

ANDREWS, J., WATKINS, E.S.: A Stereotaxic Atlas of the Human Thalamus and Adjacent Structures. Baltimore: Williams & Wilkins 1969.

ANSCHÜTZ, W.: Hirnoperation bei Hemiathetose. Berl. klin. Wschr. **47**, 1687 (1910).

ANTON, G.: Über die Beteiligung der großen basalen Gehirnganglien bei Bewegungsstörungen, insbesondere bei Chorea. Jber. Psychiat. **14**, 141–182 (1896).

ARONSON, L.R., PAPEZ, J.W.: The thalamic nuclei of pithecus (macacus) chesm. II. Dorsal thalamus. Arch. Neurol. (Chic.) **32**, 27–44 (1934).

BAILEY, P., DAVIS, E.W.: Effects of lesions of the periaqueductal grey matter in the cat. Proc. Soc. exp. Biol. (N.Y.) **51**, 305–6 (1943).

BAILEY, P., DAVIS, E.W.: Effects of lesions of the periaqueductal gray matter of the Macaca mulatta. J. Neuropath. exp. Neurol. **3**, 69–72 (1944).

BAK, I.J.: The ultrastructure of the substantia nigra and caudate nucleus of the mouse and the cellular localization of catecholamines. Exp. Brain Res. **3**, 40–57 (1967).

BAK, I.J., CHOI, W.B., HASSLER, R., USUNOFF, K.G., WAGNER, A.: Fine structural synaptic organization of the corpus striatum and substantia nigra in rat and cat. In: Adv. Neurol. **9**: Dopaminergic mechanisms. D.B. Calne, Th.N. Chase, A. Barbeau (eds.), pp. 25–41, 1975.

BAK, I.J., HASSLER, R.: Wirkung von Iproniazid und L-Dopa auf die dens core vestricles in der Substantia nigra bei der Maus. Naturwissenschaften **54**, 47 (1967).

BAK, I.J., HASSLER, R., KIM, J.S.: Differential monoamine depletion by oxypertine in nerve terminals. Z. Zellforsch. **101**, 448–462 (1969).

BAK, I.J., HASSLER, R., KIM, J.S.: Fine localization of dopamine and other chemical transmitter substances in basal ganglia and their possible roles in akinesia.

In: Parkinson's Disease. Rigidity, Akinesia, Behavior. J. Siegfried (ed.), Vol. 2, pp. 151–161. Bern: H. Huber 1973.

BAK, I.J., MARKHAM, C.H., COOK, M.L., STEVENS, J.G.: Ultrastructural and immunoperoxidase study of striatonigral neurons by means of retrograde axonal transport of herpes simplex virus. Brain Res. **143**, 361–368 (1978).

BALTHASAR, K.: Über das Syndrom des Corpus Luys an Hand eines anatomisch untersuchten Falles von Hemiballismus. Z. Neurol. **128**, 702–720 (1930).

BALTHASAR, K.: Gezielte Kälteschäden in der Großhirnrinde der Katze. Dtsch. Z. Nervenheilk. **176**, 173–199 (1957).

BARBEAU, A.: Six years of high-level levodopa therapy in severely akinetic parkinsonian patients. Arch. Neurol. (Chic.) **33**, 333–338 (1976).

BATES, J.A.V.: The significance of tremor phasic units in the human thalamus. In: 3rd Symp. on Parkinson's Disease, Edinburgh. F.J. Gillingham, I.M.L. Donaldson (eds.), pp. 118–124. Edinburgh and London: Livingstone 1969.

BECHTEREVA, N.P., BONDARTCHUK, A.N., GRETCHIN, V.B., ILIUKHINA, V.A., KAMBAROVA, D.K., MATVEEV, YI.K., PETUSHKOV, E.P., PODZDEEV, D.K., SMIRNOV, V.M., SHANDURINA, A.N.: Structural-functional organization of the human brain and the pathophysiology of the parkinsonian type hyperkineses. Confin. neurol. (Basel) **34**, 14–17 (1972).

BECHTEREW, W. v.: Die Funktionen der Nervenzentra. Vol. I–III. Jena: Fischer 1909–1911.

BECK, E.: The origin, course and termination of the prefronto-pontine tract in the human brain. Brain **73**, 368–391 (1950).

BEDARD, P., LAROCHELLE, L., PARENT, A., POIRIER, L.J.: The nigrostriatal pathway—A correlative study based on neuroanatomical and neurochemical criteria in the cat and the monkey. Exp. Neurol. **25**, 365–377 (1969).

BEHEIM-SCHWARZBACH, D.: Lebensgeschichte der melaninhaltigen Nervenzellen des nucleus caeruleus unter normalen und pathogenen Bedingungen. J. Hirnforsch. **1**, 61–95 (1954).

BEHEIM-SCHWARZBACH, D.: Pathokline Niger-Veränderungen. J. Hirnforsch. **2**, 94–126 (1955/56).

BERNHEIMER, H., BIRKMAYER, W., HORNYKIEWICZ, O., JELLINGER, K., SEITELBERGER, F.: Brain dopamine and syndromes of Parkinson and Huntington—Clinical, morphological and neurochemical correlations. J. neurol. Sci. **20**, 415–455 (1973).

BERNHEIMER, H., EHRINGER, H., HEISTRACHER, P.: Zur Biochemie des Parkinson-Syndroms des Menschen. Einfluß der Monoaminoxydase-Hemmer-Therapie auf die Konzentration des Dopamins, Noradrenalins und 5-Hydroxytryptamins im Gehirn. Klin. Wschr. **41**, 465–469 (1963).

BERTLER, A., HILLARP, N.A., ROSENGREN, E.: "Bound" and "free" catecholamines in the brain. Acta physiol. scand. **50**, 113–118 (1960).

BERTLER, A., ROSENGREN, E.: Occurrence and distribution of chatecholamines in brain. Acta physiol. scand. **47**, 350–361 (1959a).

BERTLER, A., ROSENGREN, E.: Brain catecholamine content after sectioning the adrenergic nerves to the brain vessels. Acta physiol. scand. **47**, 362–364 (1959b).

BERTRAND, G.: A pneumotaxic technique for producing localized cerebral lesions and its use in the treatment of Parkinson's disease. J. Neurosurg. **15**, 251–264 (1958).

BERTRAND, G.: Stimulation during stereotactic operations for dyskinesias. J. Neurosurg. **24**, Suppl., 419–423 (1966).

BERTRAND, G., BLUNDELL, J., MUSELLA, R.: Electrical exploration of the internal capsule and neighbouring structures during stereotaxic procedures. J. Neurosurg. **22**, 333–343 (1965).

BERTRAND, G., HARDY, J., MOLINA-NEGRO, P., MARTINEZ, S.N.: Tremor of attitude. Confin. neurol. (Basel) **31**, 37–41 (1969).

BERTRAND, G., JASPER, H.: Microelectrode recording of unit activity in the thalamus. Confin. neurol. (Basel) **26**, 205–208 (1965).

BERTRAND, G., JASPER, H., WONG, H., MATHEWS, G.: Microelectrode recording during stereotactic surgery. Clin. Neurosurg. **16**, 328–355 (1969).

BIRKMAYER, W., HORNYKIEWICZ, O.: Der L-Dioxyphenylalanin (=-L-DOPA)-Effekt beim Parkinson-Syndrom des Menschen: Zur Pathogenese und Behandlung der Parkinson-Akinese. Arch. Psychiat. Nervenkr. **203**, 560–574 (1962)

BISHOP, G.H., CLARE, M.H., PRICE, J.: Patterns of tremor of normal and pathological conditions. J. appl. Physiol. **1**, 123–147 (1948).

BLOCQ, P., MARINESCO, G.: Sur un cas de tremblement parkionsonien hémiplégique symptomatique d'une tumeur du pédoncule cérébral. C.R.Soc.Biol. (Paris) **45**, 105 (1893).

BOSTROEM, A.: Der amyostatische Symptomenkomplex und verwandte Zustände; klinischer Teil. Verh. Ges. dtsch. Nervenärzte **11**, 92 (1922).

BOWSHER, D.: Termination of the central pain pathway in man: the conscious appreciation of pain. Brain **80**, 606–622 (1957).

BOYD, I.A.: The structure and innervation of the nuclear bag muscle fibre system and the nuclear chain muscle fibre system in mammalian muscle spindles. Phil. Trans. B. **245**, 81–136 (1962).

BRIERLEY, J.B., BECK, E.: The significance in human stereotactic brain surgery of individual variation in the diencephalon and globus pallidus. J. Neurol. Neurosurg. Psychiat. **22**, 287–298 (1959).

BRION, S., GUIOT, G., DERÔME, P., COMOY, C.: Hemiballismes post-opératoires au cours de la chirurgie stéréotaxique: à propos de 12 observations dont 2 anatomo-cliniques dans une série de 850 interventions. Rev. neurol. **112**, 410–443 (1965).

BRISSAUD, E.: Lecons sur les maladies nerveuses. Paris 1895.

BROCKHAUS, H.: Zur feineren Anatomie des Septum und des Striatum. J. Psychol. Neurol. (Lpz.) **51**, 1–56 (1942).

BRONISCH, F.W.: Über das 24-Std-Encephalogramm. Dtsch. Z. Nervenheilk. **166**, 65 (1951).

BRONISCH, F.W.: Über das 24-Std-Encephalogramm. Weitere Ergebnisse. Nervenarzt **23**, 188–190 (1952).

BROWN, C.W., HENRY, F.M.: The central nervous mechanism for emotional responses. II. A technique for destroying the deeper nuclear regions within the cerebrum with a minimal destruction of the intervening cortex. Proc. nat. Acad. Sci. (Wash.) **20**, 310–315 (1934).

Brown, C.W., Henry, F.M.: The central nervous mechanism for emotional responses. III. A combination headholder and goniometer-manipulator for controlling movements of a print electrode within the brain. J. comp. Neurol. **95**, 349–370 (1935).

Brunzema, F.R.: Elektromyographische Veränderungen nach stereotaktischen Operationen an Parkinson Patienten. Dissertation Freiburg 1960.

Bucy, P.C.: Electrical excitability and cyto-architecture of the premotor cortex in monkeys. Arch. Neurol. Psychiat. (Chic.) **30**, 1205–1225 (1933).

Bucy, P.C.: Areas 4 and 6 of the cerebral cortex and their projection systems. Arch. Neurol. Psychiat. (Chic.) **35**, 1396–1400 (1936).

Bucy, P.C.: Studies of the human neuromuscular mechanism II. Effect of ventromedial cordotomy on muscular spasticity in man. Arch. Neurol. Psychiat. (Chic.) **40**, 639–662 (1938a).

Bucy, P.C.: Athetose und Tremor. Ihr physiologischer Mechanismus und ihre Beeinflussung durch chirurgische Maßnahmen. Nervenarzt **11**, 562–568 (1938b).

Bucy, P.C.: Cortical extirpation in the treatment of involuntary movements. Res. Publ. Ass. nerv. ment. Dis. **21**, 551–595 (1940).

Bucy, P.C.: The neural mechanism of athetosis and tremor. J. Neuropath. exp. Neurol. **1**, 224–239 (1942).

Bucy, P.C.: Surgical relief of tremor at rest. Ann. Surg. **122**, 933–941 (1945).

Bucy, P.C.: Relation to abnormal involuntary movements. In: The Precentral Motor Cortex, pp. 395–408. Chicago 1949.

Bucy, P.C.: Effects of extirpation in man. In: The Precentral Motor Cortex. Bucy (eds.), pp. 353–394. Urbana Ill.: University Press 1949.

Bucy, P.C., Case, R.J.: Athetosis. II. Surgical treatment of unilateral athetosis. Arch. Neurol. Psychiat. (Chic.) **37**, 983–1020 (1937).

Bucy, P.S., Case, J.T.: Tremor: physiologic mechanism and abolition by surgical means. Arch. Neurol. Psychiat. (Chic.) **41**, 721–746 (1939).

Bürgi, S.: Reizung und Ausschaltung des Brachium conjunctivum I+II. Helv. physiol. pharmacol. Acta **1**, 359–380, 380–487 (1943).

Burckhardt, G.: Über Rindenexcisionen als Beitrag zur operativen Therapie der Psychosen. Allg. Z. Psychiat. **47**, 463–548 (1891).

von Buttlar-Brentano, K.: Das Parkinsonsyndrom im Lichte der lebensgeschichtlichen Veränderungen des Nucleus basalis. J. Hirnforsch. **2**, 55–76 (1955/56).

Carlsson, A.: The occurence, distribution and physiological role of catecholamines in the nervous system. Pharmacol. Rev. **11**, 490–493 (1959).

Carpenter, M.B., McMasters, R.E.: Lesions of the substantia nigra in the rhesus monkey. Efferent fiber degeneration and behavioral observations. Amer. J. Anat. **114**, 293–320 (1964).

Carpenter, M.B., Nakano, K., Kim, R.: Nigrothalamic projections in the monkey demonstrated by autoradiographic technics. J. comp. Neurol. **165**, 401–416 (1976).

Carpenter, M.B., Peter, P.: Nigrostriatal and nigrothalamic fibers in the rhesus monkey. J. comp. Neurol. **144**, 93–116 (1972).

Carpenter, M.B., Strominger, N.L.: Efferent fiber projections of the subthalamic nucleus in the rhesus monkey. A comparison of the efferent projections of the subthalamic nucleus substantia nigra and globus pallidus. Amer. J. Anat. **121**, 41–72 (1967).

Carpenter, M.B., Whittier, J.R.: Study methods for producing experimental lesion of the central nervous system with special reference to stereotaxique technique. J. comp. Neurol. **97**, 73–132 (1952).

Carpenter, M.B., Whittier, J.R., Mettler, F.A.: Analysis of choreoid hyperkinesia in the rhesus monkey. J. comp. Neurol. **92**, 293–331 (1950).

Chase, T.N.: Central monoamines and parkinsonian signs. In: Parkinson's Disease. Rigidity, Akinesia, Behavior. J. Siegfried (ed.), Vol. 2, pp. 141–149. Bern: H. Huber 1973.

Chow, K.L., Dement, W.C., Mitchell, S.A., Jr.: Effects of lesions of the rostral thalamus on brain waves and behavior in cats. Electroenceph. clin. Neurophysiol. **11**, 107–120 (1959).

Chung, J.W., Hassler, R., Wagner, A.: Degenerated boutons in the fundus striati (Nucleus accumbens septi) after lesion of the parafascicular nucleus in the cat. Cell Tiss. Res. **172**, 1–14 (1976).

Chung, J.W., Hassler, R., Wagner, A.: Degeneration of two of nine types of synapses in the putamen after center median coagulation in the cat. Exp. Brain Res. **28**, 345–361 (1977).

Clakrke, R.H.: Investigation of the central nervous system. Methods and instruments. Johns Hopk. Hosp. Rep. spec. Vol. 1–162 (1920).

Coe, J., Ommaya, A.K.: Evaluation of local lesions of the central nervous system produced by extreme cold. J. Neurosurg. **27**, 433–444 (1964).

Cole, M., Nauta, J.H., Mehler, W.R.: The ascending efferent projections of the substantia nigra. Trans. Amer. neurol. Ass. **89**, 74–78 (1964).

Cools, A.R., van den Bercken, J.H.L.: Cerebral organisation of behaviour and the neostriatum. In: Psychobiology of the Striatum. A.R. Cools, A.H.M. Lohmann, J.H.L. van den Bercken (eds.). Amsterdam: Elsevier 1977.

Cooper, I.S.: Chemopallidectomy; an investigate technique in geriatric parkinsonians. Science **121**, 217–218 (1955).

Cooper, I.S.: Parkinsonism. Its Medical and Surgical Therapy. Springfield, Ill.: C.C Thomas 1961.

Cooper, I.S., Bravo, G.J.: Chemopallidectomy and chemothalamectomy. J. Neurosurg. **15**, 244 (1958).

Cordeau, J.P., Gybels, J., Jasper, H., Poirier, L.J.: Microelectrode studies of unit discharges in the sensorimotor cortex. Neurol. **10**, 591–600 (1960).

Crouch, R.L.: The nuclear configuration of the thalamus of macacus rhesus. J. comp. Neurol. **59**, 451–485 (1934).

D'Abundo, E.: Experimenteller Beitrag zum Studium des Hypothalamus und der Hirnschenkel. Arb. neurol. Inst. Univ. Wien **27**, 229–234 (1925).

Dahlström, A., Fuxe, K.: Evidence for the existence of monoaminecontaining neurons in the central nervous system. I. Demonstration of monoamines in the cell bodies of brain stem neurons. Acta physiol. scand. **62**, 1–55 (1964).

DÉJERINE, J.: Anatomie des centres nerveux. Vol. 2, Paris: J. Rueff 1901.

DELONG, M.R., COYLE, J.T.: Globus pallidus lesions in the monkey produced by kainic acid: Histologic and behavioral effects. In: Basal Ganglia – Cellular and Functional Aspects. R. Hassler, J.F. Christ (eds.). Appl. Neurophysiol. **42**, 95–97 (1979).

DEMPSEY, E.W., MORISON, R.S.: The production of rhythmically recurrent cortical potentials after localized thalamic stimulation. Amer. J. Physiol. **135**, 293–300 (1942a).

DEMPSEY, E.W., MORISON, R.S.: The interaction of spontaneous and induced cortical potentials. Amer. J. Physiol. **135**, 301–308 (1942b).

DENNY-BROWN, D.: The basal ganglia and their relation to disorders of movement. Oxford Neurological Monographs. Oxford Univ. Press 1962.

DIAMOND, SH.G., MARKHAM, C.H., TRECIOKAS, L.J.: Long-term experience with L-dopa: Efficacy, progression and mortality. In: Advances in Parkinsonism. W. Brinkmayer, O. Hornykiewicz (eds.), pp. 444–455. Basel: Editiones „Roche" 1976.

DIECKMANN, G., GABRIEL, E., HASSLER, R.: Size, form and structural peculiarities of experimental brain lesions obtained by thermocontrolled radiofrequency. Confin. neurol. (Basel) **26**, 134–142 (1966).

DIECKMANN, G., HASSLER, R.: Opening and closing of eyes and signs of psychomotor excitation resulting from stimulation of the putamen in cats. Nature (Lond.) **216**, 580–581 (1967).

DIECKMANN, G., HASSLER, R.: Reizexperimente zur Funktion des Putamen der Katze. J. Hirnforsch. **10**, 187–225 (1968).

DIECKMANN, G., HASSLER, R.: Electrophysiological correlates of nonspecific cortical activation by electrical stimulation of the putamen and pallidum in cats. In: Surgical Approaches in Psychiatry (Proc. 3. Internat. Congr. Psychosurg., Cambridge 1972), Laitinen, Livingstone (eds.), Lancaster MTP 1973 pp. 257–265.

DIECKMANN, G., SASAKI, K.: Recruiting responses in the cerebral cortex produced by putamen and pallidum stimulation. Exp. Brain Res. **10**, 236–250 (1970).

DIERSSEN, G., BERGMANN, L.C., GIONO, G., COOPER, I.S.: Hemiballism following surgery for Parkinson's disease. Arch. Neurol. Psychiat. (Chic.) **5**, 627–637 (1961).

Discussion on speech disorders in Parkinsonism, psychosurgery. Confin. neurol. (Basel) **34**, 173–175 (1972).

DOMESICK, V.B., BECKSTEAD, R.M., NAUTA, W.J.H.: Some ascending and descending projections of the substantia nigra and ventral tegmental area in the rat. Neurosci. Abstr. **2**, 61 (1976).

DUUS, P.: Neurologisch-topische Diagnostik. Anatomie, Physiologie, Klinik. Stuttgart: Georg Thieme 1976.

EBIN, J.: Combined lateral and ventral pyramidotomy in treatment of paralysis agitans. Arch. Neurol. (Chic.) **62**, 27–47 (1949).

VON ECONOMO, C.: Die zentralen Bahnen des Kau- und Schluckaktes. Pflügers Arch. ges. Physiol. **91**, 629 (1902).

VON ECONOMO, C.: Die Encephalitis lethargica. Jber. Psychiat. **38**, 202–242 (1917).

VON ECONOMO, C.: Encephalitis lethargica. Wien. klin. Wschr. **30**, 581–587 (1917).

EHRINGER, H., HORNYKIEWICZ, O.: Verteilung von Noradrenalin und Dopamin (3-Hydroxytyramin) im Gehirn des Menschen und ihr Verhalten bei Erkrankungen des extrapyramidalen Systems. Klin. Wschr. **38**, 1236–1239 (1960).

FALCK, B.: Observations on the possibilities of the cellular localization of monoamines by a fluorescence method. Acta physiol. scand. **56**, suppl., **197**, 1–26 (1962).

FALCK, B., HILLARP, N.-A., THIEME, G., TORP, A.: Fluorescence of chatecholamines and related compounds condensed with formaldehyde. J. Histochem. Cytochem. **10**, 348–354 (1962).

FALCK, B., OWMAN, C.: A detailed methodological description of the fluorescence method for the cellular demonstration of biogenic monoamines. Acta Univ. Lund, Sect. II, **7**, 1–23 (1965).

FAULL, R.L.M., CARMAN, J.B.: Ascending projections of the substantia nigra in the rat. J. comp. Neurol. **132**, 73–92 (1968)

FÉNYES, I.: Alzheimer Fibrillenveränderungen im Hirnstamm einer 28jährigen Postencephalitikerin. Arch. Psychiat. Nervenkr. **96**, 700–717 (1932).

FERNANDEZ DE MOLINA, A., HUNSPERGER, R.W.: Organization of the subcortical system governing defence and flight reactions in the cat. J. Physiol. (Lond.) **160**, 200–213 (1962).

FINK, R.P., HEIMER, L.: Two methods for selective silver impregnation of degenerating axons and their synaptic endings in the central nervous system. Brain Res. **4**, 369–374 (1967).

FISCH, U., SIEGFRIED, J.: Prä- und postoperative Untersuchungen über die Vestibularfunktion beim Parkinsonismus. Schweiz. Arch. Neurol. Psychiat. **96**, 286–305 (1965).

FISCHER, P.-A., SCHNEIDER, E., JACOBI, P.: Die Langzeitbehandlung des Parkinson-Syndroms mit L-Dopa. Befunde und Probleme. In: Langzeitbehandlung des Parkinson-Syndroms. P.-A. Fischer (ed.). Stuttgart-New York: Schattauer 1978.

FOIX, C.: Les lésions antomiques de la maladie de Parkinson. Rev. neurol. **37**, 593–600 (1921).

FOIX, C., NICOLESCO, J.: Anatomie cérébrale: les noyaux gris centraux et la région mésencephalo-sousoptique. Paris: Masson 1925.

FOX, C.A., ANDRADE, A.N., HILLMAN, D.E., SCHWYN, R.C.: The spiny neurons in the primate striatum: A Golgi and electron microscopic study. J. Hirnforsch. **13**, 181–201 (1971).

FOX, C.A., ANDRADE, A.N., SCHWYN, R.C., RAFOLS, J.A.: The aspiny neurons and the glia in the primate striatum: A Golgyi and electron microscopic study. J. Hirnforsch. **13**, 341–362 (1971/72).

FOX, C.A., RAFOLS, J.A.: The striatal efferents in the globus pallidus and in the substantia nigra. In: The Basal Ganglia. M.D. Yahr (ed.), pp. 37–55. New York: Raven Press 1976.

FREEMAN, W., WATTS, J.W.: Psychosurgery. Oxford: Blackwell 1950.

FREUND, C.S., VOGT, C.: Ein neuer Fall von État marbré des Corpus striatum. J. Psychol. Neurol. (Suppl.) **18**, 489–500 (1911).
FRIGYESI, T.L.: An electrophysiogical analysis of effects of 6-hydroxydopamine induced nigrectomy on sensorimotor integration. In: Parkinson's Disease. Rigidity, Akinesia, Behavior. J. Siegfried (ed.), Vol. 2, pp. 87–100. Bern: H. Huber 1973.
FRIGYESI, T.L., MACHEK, J.: Basal-ganglia-diencephalon synaptic relations in the cat. II. Intracellular recordings from dorsal thalamic neurons during low frequency stimulation of the caudato-thalamic projection systems and the nigrothalamic pathway. Brain Res. **27**, 59–78 (1971).
FRY, W.J., MOSBURG, W.H., BARNARD, J.W., FRY, F.J.: Production of focal destructive lesions in the central nervous system with ultrasound. J. Neurosurg. **11**, 471–478 (1954).

GANGLBERGER, J.A.: Über die Beeinflussung des α-Rhythmus durch stereotaktische Operationen an den Basalganglien. Arch. Psychiat. Nervenkr. **199**, 630–642 (1959).
GANGLBERGER, J.A.: The EEG in parkinsonism and its alteration by stereotaxically produced lesions in pallidum or thalamus. Electroenceph. clin. Neurophysiol. **13**, 828 (1961a).
GANGLBERGER, J.A.: Vorübergehende Herdveränderungen im EEG nach stereotaktischen Operationen an den Basalganglien. Arch. Psychiat. Nervenkr. **201**, 528–548 (1961b).
GANGLBERGER, J.A.: The effect of stereotaxic lesions in pallidum or thalamus upon the electro-encephalogram in parkinsonian disease. Excerpta med., Int. Congr. Ser. No. 37, 71–72 (1961c).
GANGLBERGER, J.A.: Über EEG-Veränderungen nach stereotaktischer Ausschaltung subcorticaler Strukturen bei 800 Parkinson-Kranken. Arch. Psychiat. Nervenkr. **203**, 519–544 (1962a).
GANGLBERGER, J.A.: Das EEG beim Parkinson-Syndrom und seine Veränderungen nach stereotaktischen Operationen. Klin. Wschr. **40**, 551–552 (1962b).
GANGLBERGER, J.A.: Elektromyographie während stereotaktischer Eingriffe im Stammganglienbereich. Klin. Wschr. **41**, 524 (1963a).
GANGLBERGER, J.A.: Effects of circumscribed diencephalic electrical stimulation upon the EMG in man. Proc. Int. Meet. EMG Copenhagen, pp. 94–98 (1963b).
GANGLBERGER, J.A.: Wirkungen umschriebener Reizungen im menschlichen Zwischenhirn auf das EMG. In: Progress in Brain Research. W. Bargmann, J.E. Schadé (eds.), Vol. 5: Lectures on the diencephalon, pp. 33–45. Amsterdam: Elsevier 1964.
GANGLBERGER, J.A.: Stereotaktische Operationen und neuere Hirnforschung. Wien: Brüder Hollinek 1970.
GANGLBERGER, J.A., BRUNZEMA, R.: Elektromyographische Untersuchungen bei Parkinsonismus vor und nach stereotaktischen Eingriffen an den Basalganglien. Neurochirurgie (Stuttg.) **5**, 59–74 (1962).
GANGLBERGER, J.A., HAIDER, M.: Computer analysis of cortical responses to thalamic stimulation and of thalamo-cortical relationship of contingent negative variation in man. In: 3nd Symp. on Parkinson's Disease. F.J. Gillingham, I.M.L. Donaldson (eds.), pp. 138–141. Edinburgh and London: Livingstone 1969.
GANGLBERGER, J.A., PRECHT, W.: Tremorregistrierung als ein Mittel zur Objektivierung des Reizeffektes während stereotaktischer Eingriffe. Arch. Psychiat. Nervenkr. **206**, 1–16 (1964).
GILDENBERG, P.L.: Studies in stereoencephalotomy. X. Variability of subcortical lesions produced by heating electrode and Cooper's balloon cannula. Confin. neurol. (Basel). **20**, 53–65 (1960).
GILLINGHAM, F.J.: Small localized lesions of the internal capsule in the treatment of the dyskinesias. Confin. neurol. (Basel) **22**, 385–392 (1962).
GILLINGHAM, F.J., KALYNARAMAN, S., DONALDSON, A.A.: Bilateral stereotactic lesions in the management of Parkinsonism and the dyskinesias. Brit. med. J. **1964 II**, 656–659.
GILLINGHAM, F.J., WATSON, W.S., DONALDSON, A.A., NAUGHTON, J.A.L.: The surgical treatment of Parkinsonism. Brit. med. J. **1960 II**, 1395–1402.
GILMAN, S.: Cerebellar, thalamic, and cerebral control of proprioceptive responses in relation to Parkinson's disease. In: Parkinson's Disease. Rigidity, Akinesia, Behavior. J. Siegfried (ed.), Vol. 2, pp. 47–58. Bern: H. Huber 1973.
GOLDSTEIN, K.: Über anatomische Veränderungen (Atrophie der Substantia nigra) bei postencephalitischen Parkinsonismus. Z. ges. Neurol. Psychiat. **76**, 627 (1922).
GREENFIELD, J.G., BOSANQUET, F.D.: The brainstem lesions in Parkinsonism. J. Neurol. Neurosurg. Psychiat. N.S. **16**, 213–226 (1953).
GREY WALTER, W.: Theoretical properties of diffuse projection systems in relation to behaviour and consciousness. In: Brain Mechanism and Consciousness. J.F. Delafresnaye (ed.), pp. 345–373. Oxford: Blackwell 1954.
GROFOVA, I.: The identification of striatal and pallidal neurons projecting to substantia nigra. An experimental study by means of retrograde axonal transport of horseradish peroxidase. Brain Res. **91**, 286–291 (1975).
GROFOVA, I.: Types of striato-nigral neurons labeled by retrograde transport of horseradish peroxidase. Appl. Neurophysiol., in press (1978)
GROFOVA, I., RINVIK, E.: An experimental electron microscopic study on the striatonigral projection in the cat. Exp. Brain Res. **11**, 249–262 (1970).
GROFOVA, I., RINVIK, E.: Effect of reserpine on large dense-core vesicles in the boutons in the cat's substantia nigra. Neurobiology **1**, 5–16 (1971).
GROS, C., VLAHOVITCH, B., NGHIA, N.S.: La zona «ciblé» dans la chirurgie de la maladie de Parkinson. Etude d'une série de 407 opérations stéréotaxiques. Neurochirurgie **10**, 413–426 (1964).
GUILLAIN, G., MOLLARET, P.: Deux cas de myoclonies synchrones et rythmées vélo-pharyngolaryngo-oculo-diaphragmatiques. Le problème anatomique et physiologique. Rev. neurol. **2**, 545–566 (1931).
GUIOT, G.: Le traitement des syndromes parkinsoniens par la destruction du pallidum interne. Neurochirurgie (Stuttg.) **1**, 94–98 (1958).
GUIOT, G., ALBE-FESSARD, D., ARFEL, G., HARDY, J., HERTZOG, E., VOURC'H, G., DEROME, P., ALEONARD, P.: Investigations électrophysiologiques et chirurgie stéréotaxique. Rev. neurol. **107**, 84–86 (1962a).

GUIOT, G., ALBE-FESSARD, D., ARFEL, G., HERTZOG, E., VOURC'H, G., HARDY, J., DEROME, P., ALEONARD, P.: Interprétation des effets de la stimulation du thalamus de l'homme par chocs isolés. C.R. Acad. Sci. (Paris) **254**, 3581–3583 (1962b).

GUIOT, G., BRION, S.: Traitement neurochirurgical des syndromes choréo-athétosique et parkinsonien. Sém. Hôp. Paris **28**, 2095–2099 (1952).

GUIOT, G., BRION, S.: Traitement des mouvements anormaux par la coagulation pallidale. Technique et résultats. Rev. neurol. **89**, 578–580 (1953).

GUIOT, G., DEROME, P., TRIGO, J.-C.: Le tremblement d'attitude. Indication la meilleur de la chirurgie stéréotaxique. Presse méd. **75**, 2513–2518 (1967).

GUIOT, G., HARDY, J., ALBE-FESSARD, D.: Délimitation précise des structures sous-corticales et identification des noyaux thalamiques chez l'homme par électrophysiologie stéréotaxique. Neuro-chirurgica **5**, 1–18 (1962c).

GUIOT, G., HERTZOG, E., RONDOT, P., MOLINA, P.: Arrest or acceleration of speech evoked by thalamic stimulation in the course of stereotaxic procedures for Parkinsonism. Brain **84**, 363–379 (1961).

GUIOT, G., PECKER, J.: Tractotomie mésencephalique antérieure pour tremblement parkinsonien. Rev. neurol. **81**, 387–388 (1949).

GUIOT, G., ROUGERIE, J., SACHS, M., HERTZOG, E.: La stimulation capsulaire chez l'homme. Sous intérêt dans la stéréotaxie pallidale pour syndromes parkinsoniens. Rev. neurol. **98**, 222–224 (1958).

GYBELS, J.M.: On the neural mechanism of Parkinsonian tremor. Thesis Antwerpen (1963).

HAIDER, M., GANGLBERGER, J.A., GROLL-KNAPP, E.: Computer analysis of subcortical and cortical evoked potentials and of slow potential phenomena in humans. Confin. neurol. (Basel) **34**, 224–229 (1972).

HAJDU, F., HASSLER, R., BAK, I.J.: Electron microscopic study of the substancia nigra and the strio-nigral projection in the rat. Z. Zellforsch. **146**, 207–221 (1973).

HALLERVORDEN, J.: Zur Pathogenese des postencephalitischen Parkinsonismus. Klin. Wschr. **12**, 692–695 (1933).

HALLERVORDEN, J.: Anatomische Untersuchungen zur Pathogenese des postencephalitischen Parkinsonismus. Dtsch. Z. Nervenheilk. **136**, 68–76 (1935).

HALLERVORDEN, J.: Erkrankungen mit vorwiegender Lokalisation im extrapyramidalen Apparat. In: Handbuch der speziellen und pathologischen Anatomie und Histologie, Vol. 13, part A, pp. 783–933. Berlin-Göttingen-Heidelberg: Springer 1957.

HANIEK, A., MALONEY, A.F.J.: Localization of stereotaxic lesions in the treatment of Parkinsonism. A clinicopathological comparison. J. Neurosurg. **31**, 393–399 (1969).

HARTMANN-VON MONAKOW, K.: Halluzinosen nach doppelseitiger stereotaktischer Operation bei Parkinson-Kranken. Arch. Psychiat. Nervenkr. **199**, 477–486 (1959).

HARTMANN-VON MONAKOW, K.: Histological and clinical correlations in 29 Parkinson patients with stereotaxic surgery. Confin. neurol. (Basel) **34**, 210–217 (1972).

HASSLER, R.: Zur Normalanatomie der Substantia nigra. J. Psychol. Neurol. (Lpz.) **48**, 1–55 (1937).

HASSLER, R.: Zur Pathologie der Paralysis agitans und des postencephalitischen Parkinsonismus. J. Psychol. Neurol. (Lpz.) **48**, 367–476 (1938).

HASSLER, R.: Zur pathologischen Anatomie des senilen und parkinsonistischen Tremor. J. Psychol. Neurol. (Lpz.) **49**, 193–230 (1939).

HASSLER, R.: Über die Thalamus-Stirnhirnverbindungen beim Menschen. Nervenarzt **19**, 9–12 (1948a).

HASSLER, R.: Forels Haubenfaszikel als vestibuläre Empfindungsbahn mit Bemerkungen über einige andere sekundäre Bahnen des Vestibularis und Trigeminus. Arch. Psychiat. Nervenkr. **180**, 23–53 (1948b).

HASSLER, R.: Über die afferenten Bahnen und Thalamuskerne des motorischen Systems des Großhirns. Arch. Psychiat. Nervenkr. **182**, 759–818 (1949a).

HASSLER, R.: Über die Rinden- und Stammhirnanteile des menschlichen Thalamus. Psychiat. Neurol. med. Psychol. (Lpz.) **1**, 181–187 (1949b).

HASSLER, R.: Über die afferente Leitung und Steuerung des striären Systems. Nervenarzt **20**, 537–541 (1949c).

HASSLER, R.: Über Kleinhirnprojektionen zum Mittelhirn und Thalamus beim Menschen. Dtsch. Z. Nervenheilk. **163**, 629–672 (1950a).

HASSLER, R.: Die Anatomie des Thalamus. Arch. Psychiat. Nervenkr. **184**, 249–256 (1950b).

HASSLER, R.: Contribution morphologique à la physiologie des lobes frontaux. Compt. rend. I. Congr. mond. Psychiat. Paris 1950c, Vol. III, pp. 119–127.

HASSLER, R.: Quelques considérations sur la spécificité des différents circuits thalamo-frontaux. Compt. rend. I. Congr. mond. Psychiat. Paris 1950d, Vol. IV, pp. 450–454.

HASSLER, R.: Erkrankungen des Kleinhirns. In: Handbuch der inneren Medizin, 4th ed., Vol. V/3, pp. 620–675. Berlin-Göttingen-Heidelberg: Springer 1953a.

HASSLER, R.: Extrapyramidal-motorische Syndrome und Erkrankungen. In: Handbuch der inneren Medizin, 4th ed., Vol. V/3, pp. 676–904. Berlin-Göttingen-Heidelberg: Springer 1953b.

HASSLER, R.: Erkrankungen der Oblongata, der Brücke und des Mittelhirns. In: Handbuch der inneren Medizin, 4th ed., Vol. V/3, pp. 552–619. Berlin-Göttingen-Heidelberg: Springer 1953c.

HASSLER, R.: Functional anatomy of the thalamus. Acta y Trabajos. VI. Congr. lat.-amer. Neurochir., Montevideo, 1955a, pp. 754–787.

HASSLER, R.: The pathological and pathophysiological basis of tremor and parkinsonism. Proc. II. Int. Congr. Neuropathol. London, 1955b, Vol. 1, pp. 29–48.

HASSLER, R.: The influence of stimulations and coagulations in the human thalamus on the tremor at rest and its physiopathologic mechanism. Proc. 2nd Internat. Congr. Neuropathol. London 1955c, pp. 637–642.

HASSLER, R.: Die extrapyramidalen Rindensysteme und die zentrale Regelung der Motorik. Dtsch. Z. Nervenheilk. **175**, 233–258 (1956a).

HASSLER, R.: Die zentralen Apparate der Wendebewegungen, I. und II. Arch. Psychiat. Nervenkr. **194**, 456–516 (1956b).

HASSLER, R.: Weckeffekte und delirante Zustände durch elektrische Reizungen bzw. Ausschaltungen im

menschlichen Zwischenhirn (Film). I. Congr. int. Neurochir., Brüssel, Acta med. belg. **1957a**, 179–181.
HASSLER, R.: Über die Bedeutung der pallidären Systeme für Parkinson-Syndrom und Psychomotorik nach Erfahrungen bei gezielten Hirnoperationen. I. Congr. int. Neurochir., Brüssel, Acta med. belg. **1957b**, 171–178.
HASSLER, R.: Über die pathologische Anatomie der Paralysis agitans und der verschiedenen Formen des Parkinsonismus einschl. des essentiellen Tremors. I. Congr. int. Sci. neurol. Brüssel. Journée commune. **1**, 49–56 (1957c).
HASSLER, R.: Anatomy of the thalamus. In: Introduction to Stereotaxic Operations with an Atlas of the Human Brain. G. Schaltenbrand, P. Bailey (eds.), pp. 230–290. Stuttgart: Thieme 1959a.
HASSLER, R.: Stereotactic brain surgery for extrapyramidal motor disturbances. In: Einführung in die stereotaktischen Operationen. G. Schaltenbrand, P. Bailey (eds.), pp. 472–488. Stuttgart: Thieme 1959b.
HASSLER, R.: Die zentralen Systeme des Schmerzes. Acta neurochir. **8**, 353–423 (1960a).
HASSLER, R.: Thalamo-corticale Systeme der Körperhaltung und der Augenbewegungen. In: Structure and Function of the Cerebral Cortex. D.B. Tower, J.P. Schadé (eds.), pp. 124–130. Amsterdam: Elsevier 1960b.
HASSLER, R.: Motorische und sensible Effekte umschriebener Reizungen und Ausschaltungen im menschlichen Zwischenhirn. Dtsch. Z. Nervenheilk. **183**, 148–171 (1961).
HASSLER, R.: Spezifische und unspezifische Systeme des menschlichen Zwischenhirns. In: Lectures on the Diencephalon. W. Bargmann, J.P. Schadé (eds.), pp. 1–32. Amsterdam: Elsevier 1964.
HASSLER, R.: Extrapyramidal control of the speed of behavior and its change by primary age processes. In: Behavior, Aging and the Nervous System. A.T. Welford, J.E. Birren (eds.), pp. 284–306. Springfield, Ill.: C.C Thomas 1965.
HASSLER, R.: Extrapyramidal motor areas of the cat's frontal lobe: their functional and architectonic differentiation. Int. J. Neurol. (Montevideo) **5**, 301–316 (1966a).
HASSLER, R.: Thalamic regulation of muscle tone and the speed of movements. In: The Thalamus. D.P. Purpura, M.D. Yahr (eds.), pp. 418–438. New York: Columbia University Press 1966b.
HASSLER, R.: Comparative Anatomy of the central visual systems in day- and night-active primates. In: Evolution of the Forebrain. R. Hassler, H. Stephan (eds.), pp. 419–434. Stuttgart: Thieme 1966c.
HASSLER, R.: Funktionelle Neuroanatomie und Psychiatrie. In: Psychiatrie der Gegenwart. H.W. Gruhle, R. Jung, W. Mayer-Gross, M. Müller (eds.), Vol. I/1A, pp. 152–285. Berlin-Heidelberg-New York: Springer 1967.
HASSLER, R.: Myoclonies extrapyramidales traitées par coagulation stéréotaxique de la voie dentato-thalamique et leur mécanisme physiopathologique. Rev. neurol. **119**, 409–418 (1968).
HASSLER, R.: Anatomical and functional studies of the efferent mechanisms of the cerebellum. (Abstr.) 4th Symp. Fulton Soc. on Cerebellum, New York, Sept. 1969.
HASSLER, R.: Dichotomy of facial pain conduction in the diencephalon. In: Trigeminal Neuralgia (Pathogenesis and Pathophysiology). R. Hassler, A.E. Walker (eds.), pp. 123–138. Stuttgart: Thieme 1970.
HASSLER, R.: Physiopathology of rigidity. In: Parkinson's Disease. J. Siegfried (ed.), Vol. I, pp. 20–45. Bern: H. Huber 1972a.
HASSLER, R.: Hexapartition of inputs as a primary role of the thalamus. In: Corticothalamic Projections and Sensorimotor Activities. T. Frigyesi, E. Rinvik, M.D. Yahr (eds.), pp. 551–579. New York: Raven Press 1972b.
HASSLER, R.: Corrélation entre les médiateurs chimiques (dopamine sérotonine, acétylcholine et gaba), tonus musculaire et rapidité du mouvement. Rev. neurol. Paris **127**, 51–63 (1972c)
HASSLER, R.: Pathophysiologie der Bewußtlosigkeit. In: Der Notfall: Bewußtlosigkeit. H.J. Streicher, J. Rolle (eds.), pp. 1–13. Stuttgart: Thieme 1974a.
HASSLER, R.: Fiber connections within the extrapyramidal system. Confin neurol. (Basel) **36**, 237–255 (1974b)
HASSLER, R.: A cholinergic centro-thalamic input to the striopallidal systems. Abstr. 10th Int. Congr. Anat. Tokyo, p. 136, 1975a.
HASSLER, R.: The central organization of adversive movements as the main direction of locomotion on land. In: Cerebral Localization. K.J. Zülch, O.D. Creutzfeld, G.C. Galbraith (eds.), pp. 79–97. Berlin-Heidelberg-New York: Springer 1975b.
HASSLER, R.: Die neuronalen Systeme der exrapyramidalen Myoklonien und deren stereotaktische Behandlung. In: Aktuelle Neuropädiatrie; Myoklonien, Ataxie, das unruhige Kind. H. Doose (ed.), pp. 20–48. Stuttgart: Thieme 1977a.
HASSLER, R.: Basal ganglia systems regulating mental activity. Int. J. Neurol. **12**, 53–72 (1977b)
HASSLER, R.: Architectonic organization of the thalamic nuclei. In: Atlas for Stereotaxy of the Human Brain. 2nd revised and enlarged Edit. G. Schaltenbrand, W. Wahren (eds.). Stuttgart: Thieme 1977c.
HASSLER, R.: Striatal control of locomotion, intentional actions and of integrating and perceptive activity. J. neurol. Sci. **36**, 187–224 (1978).
HASSLER, R., AHN, E.T., WAGNER, A., KIM, J.S.: Experimenteller Nachweis von intrastriatalen Synapsentypen und Axon-Kollateralen durch Isolierung des Fundus striati von allen extrastriatalen Verbindungen. Anat. Anz. **143**, 413–436 (1978)
HASSLER, R., BAK, I.J.: Submikroskopische Catecholaminspeicher als Angriffspunkte der Psychopharmaka, Reserpin und Mono-Amino-Oxydase-Hemmer. Nervenarzt **37**, 493–498 (1966a)
HASSLER, R., BAK, I.J.: Effects of amine-depleting and amine-storing substances on the axon terminals of the pineal gland. Abstract. 6th Internat. Congr. for Electron Microscopy pp. 521–522. 1966b.
HASSLER, R., BAK, I.J.: Unbalanced ratios of striatal dopamine and serotonin after experimental interruption of strio-nigral connection in rat. In: 3rd Symp. on Parkinson's Disease. F.J. Gillingham, I.L.M. Donaldson (eds.), pp. 29–38. Edinburgh and London: Livingstone 1969.
HASSLER, R., BAK, I.J., KIM, J.S.: Unterschiedliche Entleerung der Speicherorte für Noradrenalin, Dopamin und

Serotonin als Wirkungsprinzip des Oxypertins. Nervenarzt **41**, 105–118 (1970).

HASSLER, R., BAK, I.J., USUNOFF, K.J., CHOI, W.B.: Synaptic organization of the descending and ascending connections between the striatum and the substantia nigra in the cat. Neuropsychopharmacology (Proc. IX. Cong. Collegium Internationale Neuropsychopharmacologicum, Paris, 7–12 July 1974). J.R. Boissier, H. Hippius, P. Pichot (eds.), pp. 397–411. Amsterdam: Excerpta Medica 1975.

HASSLER, R., CHUNG, J.W.: The discrimination of nine different types of synaptic boutons in the fundus striati (nucleus accumbens septi). Cell Tiss. Res. **168**, 489–505 (1976).

HASSLER, R., CHUNG, J.W., RINNE, U., WAGNER, A.: Selective degeneration of two out of the nine types of synapses in cat caudate nucleus after cortical lesions. Exp. Brain Res. **31**, 67–80 (1978)

HASSLER, R., CHUNG, J.W., WAGNER, A.: Electron-microscopical study of the different types of striatal synapses especially of the thalamo-striatal connections. In: Synapses. Cottrell, Usherwood (eds.), pp. 355–356. Glasgow: Blackie, a. London 1977.

HASSLER, R., CHUNG, J.W., WAGNER, A., RINNE, U.: Experimental demonstration of intrinsic synapses in cat's caudate nucleus. Neurosci. Letters **5**, 117–121 (1977).

HASSLER, R., DIECKMANN, G.: Arrest reaction, delayed inhibition and unusual gaze behavior resulting from stimulation in awake, unrestrained cats. Brain Res. **5**, 504–508 (1967).

HASSLER, R., DIECKMANN, G.: Locomotor movements in opposite directions induced by stimulation of pallidum or of putamen. J. neurol. Sci. **8**, 189–195 (1968).

HASSLER, R., HAJDU, F., BAK, I.J.: Über die 6 verschiedenen Typen von Synapsen im elektronenmikroskopischen Längsschnitt der Substantia nigra bei der Ratte. Verh. anat. Ges. (Jena) **68**, 315–318 (1974).

HASSLER, R., MUHS-CLEMENT, K.: Architektonischer Aufbau des sensomotorischen und parietalen Cortex der Katze. J. Hirnforsch. **6**, 377–420 (1964).

HASSLER, R., MUNDINGER, F., RIECHERT, T.: Correlations between clinical and autoptic findings in stereotaxic operations of Parkinsonism. Confin. neurol. (Basel) **26**, 282–290 (1965).

HASSLER, R., MUNDINGER, F., RIECHERT, T.: The functional importance of individual portions of the internal capsule on the basis of stimulation and coagulation in stereotaxic operations. Comm. III. Europ. Congr. of Neurosurgery, Madrid 1967a, pp. 23–26.

HASSLER, R., MUNDINGER, F., RIECHERT, T.: Functional differentiation of the internal capsule based on stimulation and elimination effects during stereotaxic operations. Abstr. 3rd Europ. Congr. of Neurosurgery, Madrid 1967, Internat. Congr. Series 139, Excerpta Med. **189** (1967b).

HASSLER, R., MUNDINGER, F., RIECHERT, T.: Target reliability of the stereotaxic method as compared with autopsy findings. In: 3rd Symp. on Parkinson's Disease. F.J. Gillingham, I.M.L. Donaldson (eds.), pp. 206–210. Edinburgh and London: Livingstone 1969a.

HASSLER, R., MUNDINGER, F., RIECHERT, T.: Pathophysiology of tremor at rest derived from the correlation of anatomical and clinical data. Abstr. 4th Symp. Intern. Soc. Res. in Stereoencephalotomy, New York, 1969b, pp. 2–3.

HASSLER, R., MUNDINGER, F., RIECHERT, T.: Pathophysiology of tremor at rest derived from the correlation of anatomical and clinical data. Confin. neurol. (Basel) **32**, 79–87 (1970).

HASSLER, R., RIECHERT, T.: Indikationen und Lokalisationsmethode der gezielten Hirnoperationen. Nervenarzt **25**, 441–447 (1954a).

HASSLER, R., RIECHERT, T.: Clinical effects produced by stimulation of different thalamic nuclei in humans. Electroenceph. clin. Neurophysiol. **6**, 518 (1954b).

HASSLER, R., RIECHERT, T.: A special method of stereotactic brain operation. Proc. roy. Soc. Med. **48**, 469–470 (1955).

HASSLER, R., RIECHERT, T.: Über einen Fall von doppelseitiger Fornicotomie bei sogenannter temporaler Epilepsie. Acta neurochir. **5**, 330–340 (1957).

HASSLER, R., RIECHERT, T.: Klinische und anatomische Befunde bei stereotaktischen Schmerzoperationen im Thalamus. Arch. Psychiat. Nervenkr. **200**, 93–122 (1959).

HASSLER, R., RIECHERT, T.: Wirkungen der Reizungen und Koagulationen in den Stammganglien bei stereotaktischen Hirnoperationen. Nervenarzt **32**, 97–109 (1961).

HASSLER, R., RIECHERT, T., MUNDINGER, F.: Anatomische Treffsicherheit der stereotaktischen Parkinson-Operationen auf Grund autoptischer Bestimmung. Arch. Psychiat. Nervenkr. **212**, 97–116 (1969).

HASSLER, R., RIECHERT, T., MUNDINGER, F., UMBACH, W., GANGLBERGER, J.A.: Physiological observations in stereotaxic operations in extrapyramidal motor disturbances. Brain **83**, 337–356 (1960).

HASSLER, R., SCHMIDT, K., RIECHERT, T., MUNDINGER, F.: Stereotactic treatment of action myoclonus in a case of combined status marmoratus and multiple sclerosis. A contribution to the Pathophysiology of Basal Ganglia with Multiple Lesions in both the striatum and the substantia nigra. Confin. neurol. (Basel) **37**, 329–356 (1975).

HASSLER, R., USUNOFF, K.G., WAGNER, A., BAK, I.J.: Über die doppelläufigen Verbindungen zwischen Striatum und Substantia nigra im licht- und elektronenmikroskopischen Bild bei der Katze. Anat. Anz. **137**, 357–368 (1975).

HASSLER, R., WAGNER, A.: Locomotor activity and speed of movements in relation to monoamine-acting drugs. Int. J. Neurol. (Montevideo) **10**, 80–97 (1975).

HAWRYLYSHYN, P.A., RUBIN, A.M., TASKER, R.R., ORGAN, L.W., FREDERICKSON, J.M.: Vestibulo-thalamic projections in man – a sixth primary sensory pathway. J. Neurophysiol. **41**, 394–401 (1978).

HESS, R.: The influence of stereotactic lesions in the EEG. Electroenceph. clin. Neurophysiol., Suppl., **19**, 166–171 (1961).

HESS, W.R.: Hirnreizversuche über den Mechanismus des Schlafes. Arch. Psychiat. Nervenkr. **86**, 287–292 (1928).

HESS, W.R.: Die Motorik als Organisationsproblem. Biol. Zbl. **61**, 545–572 (1941).

HESS, W.R.: Experimenteller Beitrag zur Frage der extrapyramidalen Motorik. Z. Neurol. **72**, 639 (1941).

HESS, W.R.: Charakter der im Zwischenhirn ausgelösten

Bewegungseffekte. Ein Beitrag zur extrapyramidalgesteuerten Motorik. Pflügers Arch. ges. Physiol. **244**, 767–786 (1941).

HESS, W.R.: Physiologische Aspekte der extrapyramidalen Motorik. Nervenarzt **15**, 457–466 (1942).

HESS, W.R.: Vegetative Funktionen und Zwischenhirn. Basel: Schwabe 1948.

HESS, W.R.: Das Zwischenhirn. Syndrome, Lokalisationen und Funktionen. 2nd ed. Basel: Schwabe 1954.

HESS, W.R., WEISSCHEDEL, E.: Die höheren Zentren der regulierten Körperhaltung. Helv. physiol. pharmacol. Acta **7**, 451–469 (1949).

HÖHNE, P.: Über akute Veränderungen vegetativer Funktionen beim Parkinsonsyndrom während Reizung und Ausschaltung im Stammganglienbereich. Inaug. Diss. Freiburg 1964.

HÖKFELT, T., KELLERTH, J.O., NILSSON, G., PERNOW, B.: Substance P localization in the central nervous system and in some primary sensory neurons. Science **190**, 889–890 (1975).

HÖKFELT, T., UNGERSTEDT, U.: Electron and fluorescence microscopical studies on the nucleus caudatus putamen of the rat after unilateral lesions of ascending nigro-neostriatal dopamine neurons. Acta physiol. scand. **76**, 415–426 (1969).

HOPF, A., WORINGER, E., HAMOU, I.: Postoperativer Hemiballismus. Neurochirurgica (Stuttgart) **11**, 1–8 (1968).

HOPKINS, D.A., NIESSEN, L.W.: Substantia nigra projections to the reticular formation, superior colliculus and central gray in the rat, cat and monkey. Neurosci. Letters **2**, 253–259 (1976).

HORNYKIEWICZ, O.: Dopamin (3-Hydroxytyramin) im Zentralnervensystem und seine Beziehung zum Parkinson-Syndrom des Menschen. Dtsch. med. Wschr. **87**, 1807–1810 (1962).

HORNYKIEWICZ, O.: Die topische Lokalisation und das Verhalten von Noradrenalin und Dopamin (3-Hydroxytyramin) in der Substantia nigra des normalen und Parkinsonkranken Menschen. Wien. klin. Wschr. **75**, 309–312 (1963).

HORNYKIEWICZ, O.: Metabolism of brain dopamine in human Parkinsonism – Neurochemical and clinical aspects. In: Biochemistry and Pharmacology of the Basal Ganglia. E. Costa, L.J. Côté, M.D. Yahr (eds.), pp. 171–185. New York: Raven Press 1966.

HORNYKIEWICZ, O.: Biochemical and pharmacological aspects of akinesia. In: Parkinson's Disease. Rigidity, Akinesia, Behavior. J. Siegfried (ed.), Vol. I, pp. 127–149. Bern: H. Huber 1972.

HORROBIN, D.F.: The lateral cervical nucleus of the cat: an electrophysiological study. Quart. J. exp. Physiol. **51**, 351–371 (1966).

HORSLEY, V.: The function of the so-called motor area of the brain. Brit. med. J. **2**, 125–132 (1909).

HORSLEY, V., CLARKE, R.H.: The structure and functions of the cerebellum, examined by a new method. Brain **31**, 45–124 (1908).

HOUDART, R., MAMO, H., DONDEY, M., COPHIGNON, J.: Résultats des coagulations sousthalamiques dans la maladie de Parkinson (à propos de 50 cas). Rev. neurol. **112**, 521–529 (1965).

HOUSEPIAN, E.M., POOL, J.L.: The accuracy of human stereoencephalotomy as judged by histological confirmation of roentgenographic localisation. J. nerv. ment. Dis. **130**, 520–525 (1960).

HUGHES, B.: Involuntary movements following stereotactic operations for Parkinsonism with special reference to hemichorea (ballismus). J. Neurol. Neurosurg. Psychiat. **28**, 291–303 (1965).

HUNSPERGER, R.W., WYSS, O.A.M.: Qualitative Ausschaltung von Nervengewebe durch Hochfrequenzkoagulation. Helv. physiol. pharmacol. Acta **11**, 283–304 (1953).

IBATA, Y., NOJYO, Y., MATSUURA, T., SANO, Y.: Nigroneostriatal projection – A study with Fink-Heimer impregnation, Fluorescence histochemistry and electron microscopy. Z. Zellforsch. 138–344 (1973).

IKEDA, K., IKEDA, S., YOSHIMURA, T., KATO, H., NAMBA, M.: Idiopathic Parkinsonism with Lewy-Type-Inclusions in cerebral cortex. Acta neuropath. **41**, 165–168 (1978).

JACOB, H.: Neuropathologie des Parkinson-Syndroms und die Seneszenz des Gehirns. In: Langzeitbehandlung des Parkinson-Syndroms. P.-A. Fischer (ed.), pp. 5–25. Stuttgart-New York: Schattauer 1978.

JAKOB, A.: Die extrapyramidalen Erkrankungen. Berlin: Springer 1923.

JASPER, H.H., BERTRAND, G.: Stereotaxic microelectrode studies of single thalamic cells and fibers in patients with dyskinesias. Trans. Amer. neurol. Ass. **89**, 79–82 (1964).

JASPER, H.H., BERTRAND, G.: Thalamic units involved in somatic sensation and voluntary and involuntary movements in man. In: The Thalamus. D.P. Purpura, M.D. Yahr (eds.), pp. 365–390. New York: Columbia University Press 1966.

JASPER, H.H., DROOGLEVER-FORTUYN, J.: Experimental studies on the functional anatomy of petit mal epilepsy. Res. Publ. Ass. nerv. ment. Dis. **26**, 272–298 (1946).

JUNG, R.: Physiologische Untersuchungen über den Parkinsontremor und andere Zitterformen beim Menschen. Z. ges. Neurol. Psychiat. **173**, 263–322 (1941).

JUNG, R.: Correlation of bioelectrical and autonomic phenomena with alterations of consciousness and arousal in man. In: Brain Mechanism and Consciousness. J.F. Delafresnaye (ed.), pp. 310–344. Oxford: Blackwell 1954.

JUNG, R., HASSLER, R.: The extrapyramidal motor system. In: Handbook of Physiol. Neurophysiol., J. Field, H.W. Magoun, V.E. Hall (eds.), Vol. II, pp. 863–927. Washington, D.C.: Amer. Physiol. Soc., 1960.

JUNG, R., RIECHERT, T.: Eine neue Methodik der operativen Elektrocorticographie und subcorticalen Elektrographie. Acta neurochir. (Wien) **2**, 164–186 (1952).

JUNG, R., RIECHERT, T.: EEG-Befunde bei Thalamusreizung am Menschen. Nervenarzt **26**, 35 (1955).

JUNG, R., RIECHERT, T., HEINES, K.-D.: Zur Technik und Bedeutung der operativen Elektrocorticographie und subcorticalen Hirnpotentialableitung. Nervenarzt **22**, 433–436 (1951).

JUNG, R., RIECHERT, T., MEYER-MICKELEIT, R.W.: Über intracerebrale Hirnpotentialableitungen bei hirnchirurgischen Eingriffen. Dtsch. Z. Nervenheilk. **162**, 52–58 (1950).

JURMAN, N.: Anatomische und physiologische Untersuchungen der Substantia nigra Soemmringii. Neurol. Zbl. **19**, 510 (1900).

KATAOKA, K., NAKAMURA, Y., HASSLER, R., BAK, I.J., KIM, J.S.: Activity of L-glutamic acid decarboxylase in different regions of baboon brain. Folia psychiat. neurol. jap. **29**, 361–370 (1975).

KENNARD, M.A.: Experimental analysis of the functions of the basal ganglia in monkeys and chimpanzees. J. Neurophysiol. **7**, 127–148 (1944).

KIM, J.S.: Wirkungen von 6-Hydroxydopamin auf den Acetylcholin- und γ-Aminobuttersäure-Stoffwechsel im Striatum und in der Substantia nigra in der Ratte. Brain Res. **55**, 472–475 (1973).

KIM, J.-S.: Transmitters for the afferent and efferent systems of the neostriatum and their possible interactions. Advanc. biochem. Psychopharmacol. **19**, 217–233 (1978).

KIM, J.S., BAK, I.J., HASSLER, R., OKADA, Y.: Role of γ-aminobutyric acid (GABA) in the extrapyramidal motor system. 2. Some evidence for the existence of a type of GABA-rich strionigral neurons. Exp. Brain Res. **14**, 95–104 (1971).

KIM, J.-S., HASSLER, R., HAUG, P., PAIK, K.-S.: Effect of frontal cortex ablation on striatal glutamic acid level in rat. Brain Res. **132**, 370–374 (1977).

KIM, J.S., HASSLER, R., KUROKAWA, M., BAK, I.J.: Abnormal movements and rigidity induced by harmaline in relation to striatal acetylcholine, serotonin and dopamine. Exp. Neurol. **29**, 189–200 (1970).

KIM, Y.K., UMBACH, W.: The effects of stereotaxic subthalamotomy on sympathetic tonus. Confin. neurol. (Basel) **34**, 156–160 (1972).

KJELLBERG, R.N., PRESTON, W.M.: Anwendung des Bragg-Spitzeneffektes des Protonenstrahles für intracerebrale Läsion. II. Int. Congr. Neurol. Surg., Washington D.C., 14.–20. October 1961.

KLATZO, I., WISNIEWSKI, H., STREICHER, E.: Experimental production of neurofibrillary degeneration. I. Light microscopic observations. J. Neuropath. exp. Neurol. **24**, 187–199 (1965).

KLEIST, K.: Untersuchungen zur Kenntnis der psychomotorischen Bewegungsstörungen bei Geisteskrankheiten, pp. 1–171. Leipzig 1908.

KLEIST, K.: Zur Auffassung der subkortikalen Bewegungsstörungen (Chorea, Athetose, Bewegungsausfall, Starre, Zittern). Arch. Psychiat. Nervenkr. **59**, 790 (1918).

KLEIST, K.: Gegenhalten (motorischer Negativismus), Zwangsgreifen und Thalamus opticus. Mschr. Psychiat. Neurol. **65**, 317–396 (1927).

KLEIST, K.: Gehirnpathologie. Leipzig: Johann Ambrosius Barth 1934, 1065 pp.

KRAYENBÜHL, H., AKERT, K., HARTMANN, K., YASARGIL, M.G.: Étude de la corrélation anatomo-clinique chez des malades opérés de Parkinsonisme. Neuro-chirurgie **10**, 397–412 (1964).

KRAYENBÜHL, H., YASARGIL, M.G.: Ergebnisse der stereotaktischen Operationen beim Parkinsonismus, insbesondere der doppelseitigen Eingriffe. Dtsch. Z. Nervenheilk. **182**, 530–541 (1961).

KRIENITZ, E.: Elektromyographische Untersuchungen über die eigenreflektorische Erregbarkeit der rigiden Muskulatur bei Parkinsonkranken. Diss. Freiburg/Br. 1961.

KROTZ, U.: Objektive und subjektive Wirkungen von Reizungen und Ausschaltungen des Pallidum internum und der oralen Ventralkerne des Thalamus bei 400 Parkinson-Patienten. Diss. Freiburg/Br. 1962.

LAFORA, G.R.: Contribution à la histopathologie de la paralysie agitante. Trab. Lab. Invest. Biol. Madrid **11**, 43 (1913).

LAFORA, G.R., GLUECK, B.: Beitrag zur Histopathologie der myoklonischen Epilepsie. Z. ges. Neurol. **6**, 2–14 (1911).

LAITINEN, L.A., OHNO, Y.: H reflex in Parkinsonism. A pre- and postoperative study. Confin. neurol. (Basel) **32**, 88–92 (1970).

LAMARRE, Y., FILION, M., CORDEAU, J.P.: Neuronal discharges of the ventrolateral nucleus of the thalamus during sleep and wakefulness in the cat. I. Spontaneous activity. Exp. Brain Res. **12**, 480–498 (1971).

LANCE, J.W., ANDREWS, C.J., GILLIES, J.D., BURKE, D., ASHBY, P.: The use of the tonic vibration reflex in the study of supraspinal control of tonic mechanisms in cat and man. In: Parkinson's Disease. Rigidity, Akinesia, Behavior. J. Siegfried (ed.), Vol. 2, pp. 59–64. Bern: H. Huber 1973.

LANDAU, W.M.: Spacticity and rigidity. In: Recent Advances in Neurology VI. F. Plum (ed.), pp. 1–32. Philadelphia: Davis Comp. 1969.

LANDGREN, S., SILFVENIUS, H., WOLSK, D.: Vestibular, cochlear and trigeminal projections to the cortex in the anterior suprasylvian sulcus of the cat. J. Physiol. (Lond.) **191**, 561–573 (1967).

LAPRAS, C.: Chirurgie stéréotaxique des dyskinésies. Thèse Fac. méd. Lyon. Lyon: M. et A. Audin 1960.

LAROCHELLE, L., BÉDARD, P., BOUCHER, R., POIRIER, L.J.: The rubro-olico-cerebello-rubral loop and postural tremor in the monkey. J. neurol. Sci. **11**, 53–64 (1970).

LAROCHELLE, L., BÉDARD, P., POIRIER, L.J., SOURKES, T.L.: Correlative neuroanatomical and neuropharmacological study of tremor and catatonia in the monkey. Neuropharmacology **10**, 272–288 (1971).

LEKSELL, L.: A stereotaxic apparatus for intracerebral surgery. Acta chir. scand. **99**, 229–233 (1949a).

LEKSELL, L.: The action potential and excitatory effects of the small ventral root fibres to skeletal muscle. Acta physiol. scand. **10**, Suppl. 31, 1–84 (1949 b).

LEKSELL, L.: Gezielte Hirnoperationen. In: Handbuch der Neurochirurgie. Vol. 6, pp. 178–199. Berlin-Göttingen-Heidelberg: Springer 1957.

LEKSELL, L.: Stereotaxis and Radiosurgery. Springfield, Ill: C.C Thomas 1971.

LEKSELL, L., LARSSON, B., ANDERSSON, B., REXED, B., SOURANDER, B., MAIR, W.: Lesions in the depth of the brain produced by a beam of high energy protons. Acta radiol. (Stockh.) **54**, 251–264 (1960).

LEWY, F.H.: Die Lehre vom Tonus und der Bewegung. Berlin: Springer 1923.

LEYDEN, E. v.: Fall von Paralysis agitans des rechten Armes infolge der Entwicklung eines Sarkoms im linken Thalamus. Virchows Arch. path. Anat. **29**, 202–205 (1864).

LHERMITTE, J., TRELLES, J.-O.: Physiologie et physiopathologie du corps strié et des formations sousthalamiques. Encéphale **27**, 235 (1932).

LI, C.L., CULLEN, C., JASPER, H.H.: Laminar microelectrode analysis of cortical unspecific recruiting resposes and spontaneous rhythms. J. Neurophysiol. **19**, 131–143 (1956a).

LI, C.L., CULLEN, C., JASPER, H.H.: Laminar microelectrode studies of specific somatosensory cortical potentials. J. Neurophysiol. **19**, 111–130 (1956b).

LIM, T.-H., MATSUMOTO, K., COOPER, I.S.: Electroencephalographic changes in parkinsonian patients following cryogenic lesions in the ventrolateral nucleus of the thalamus. J. int. Coll. Surg. **42**, 281–286 (1964).

LLAMAS, A.: Conexiones efferentes de la substancia negra y del area tegmental ventral de Tsai. Anales Anat. **18**, 355–392 (1969).

LLAMAS, A., REINOSO-SUAREZ, F.: Projections of the substantia nigra and ventral tegmental mesencephalic area. In: 3rd Symp. on Parkinson's Disease. F.T. Gillingham, I.M.L. Donaldson (eds.), pp. 81–87. Edinburgh and London: Livingstone 1969.

LUCKSCH, F., SPATZ, H.: Die Veränderungen des Zentralnervensystems bei Parkinsonismus in den Spätstadien der Encephalitis epidemica. Münch. med. Wschr. **70**, 1245 (1923).

LUSSANA, F.: Physio-pathologie du cervelet. Arch. ital. Biol. **7**, 145–157 (1886).

LUSSANA, F., LEMOIGNE, A.: Fisiologia dei centri nervosi encefalici. Padova: Prosperini 1871.

LUYS, J.: Recherches sur le système nerveux cerebro-spinal: sa structure, ses fonctions et ses maladies. Paris: Baillière & Fils 1865.

MACCHI, G., DALLE ORE, G., DA PIAN, R.: Reperti anatomici in soggetti operati di talamotomia stereotassica. Sist. nerv. **16**, 193–223 (1964).

MAGENDIE, F.: Leçons sur les fonctions et les maladies du système nerveux. Paris: Ebrard 1841.

MAMO, H., DONDEY, M., COPHIGNON, J., PIALOUX, P., FONTELLE, P., HOUDART, R.: Latéro-pulsion transitoire au décours de coagulations sousthalamiques et thalamiques chez des parkinsoniens. Rev. neurol. **112**, 509–520 (1965).

MARK, V.H., CHATO, J.C., EASTMAN, F.G., ARONOW, S., ERVIN, F.R.: Localized cooling in the brain. Science **134**, 1520–1521 (1961).

MARK, V.H., ERVIN, F.R., YAKOVLEV, P.J.: Stereotactic thalamotomy III. The verification of anatomical lesion sites in the human thalamus. Arch. Neurol. (Chic.) **8**, 528–538 (1963).

MARKHAM, C.H., BROWN, W.J., RAND, R.W.: Stereotaxic lesions in Parkinson's disease: Clinicopathological correlations. Arch. Neurol. (Chic.) **15**, 480–497 (1966).

MARKHAM, C.H., RAND, R.W.: Stereotactic surgery in Parkinson's disease. Arch. Neurol. (Chic.) **8**, 621–631 (1963).

MARTIN, J.P.: Hemichorea resulting from a local lesion of the brain (syndrome of the body of Luys). Brain **50**, 637–651 (1927).

MARTIN, J.P., ALCOCK, N.S.: 1. Hemichorea associated with lesions of the Corpus Luysii. Brain **57**, 504–516 (1934).

MEHLER, W.H., FEFERMAN, M.E., NAUTA, W.J.H.: Ascending axon degeneration following anterolateral cordotomy. An experimental study in the monkey. Brain **83**, 718–750 (1960).

MELLA, H.: The experimental production of basal ganglion symptomatology in Macacus rhesus. Arch. Neurol. (Chic.) **11**, 405–417 (1924).

MELZACK, R., STOTLER, W.A., LIVINGSTON, W.K.: Effect of discrete brainstem lesions in cats on perception of noxious stimulation. J. Neurophysiol. **21**, 353–367 (1958).

METTLER, F.A.: Substantia nigra and Parkinsonism. Arch. Neurol. (Chic.) **11**, 529–542 (1964).

METTLER, F.A., ADES, H.W., LIPMAN, E., CULLER, E.A.: The extrapyramidal system. Arch. Neurol. (Chic.) **41**, 984–995 (1939).

METTLER, F.A., METTLER, C.C.: The effects of striatal injury. Brain **65**, 242–255 (1942).

MEYER, A., BECK, E.: Prefrontal leucotomy and related operations. Anatomical aspects of success and failure. Edinburgh-London: Oliver and Boyd 1954.

MEYER, A., BECK, E., MCLARDY, T.: Prefrontal leucotomy: A neuroanatomical report. Brain **70**, 18–49 (1947).

MEYERS, R., FRY, W.J., FRY, F.J., DREYER, L.L., SCHULTZ, D.F., NOYES, F.R.: Early experiences with ultrasonic irradiation of the pallidofugal and nigral complexes in hyperkinetic and hypertonic disorders. J. Neurosurg. **16**, 32–54 (1959).

MINKOWSKI, M.: Über den Verlauf, die Endigung und die zentrale Repräsentation von gekreuzten und ungekreuzten Sehnervenfasern bei einigen Säugetieren und beim Menschen. Schweiz. Arch. Neurol. Psychiat. **6**, 201–252 (1920).

MINKOWSKI, M.: Etude sur les conexions anatomiques des circonvolutions rolandiques, pariétales et frontales. Schweiz. Arch. Neurol. Psychiat. **14**, 255–278 (1924).

MIRTO, D.: Contributo alla fina anatomia della substantia nigra di Soemmering e del peduncolo cerebral dell'uomo. Riv. sper. Freniat. **22**, 197–210 (1896).

MOLINA, P.: Conexiones nigro-corticales. Tesis Doctoral. Univ. Navarra 1965.

MOLINA, P.: Conexiones nigrocorticales. Un estudio en el gato con método de la degeneración retrograda. An. Anat. **18**, 285–354 (1969).

MOLINA, P., REINOSO-SUAREZ, F.: Mesencephalo-cortical connections. Abstracts 8th Internat. Cong. Anatomists, Wiesbaden 1965.

MONAKOW, C. v.: Experimentelle und pathologisch-anatomische Untersuchungen über die Beziehungen der sog. Sehsphäre zu den intracorticalen Opticuszentren und zum N. opticus. Arch. f. Psychiat. **16**, 151–199 (1885).

MONTANELLI, R.P., HASSLER, R.: Stimulation effects of the globus pallidus and nucleus entopeduncularis of the cat. Excerpta Med. Internat. Congr. Ser. **48**, Nr. 1091. XXII. Internat. Congr. Physiol. Sci. Leiden 1962.

MONTANELLI, R.P., HASSLER, R.: Motor effects elicited by stimulation of the pallido-thalamic system in the cat. In: Lectures on the Diencephalon. W. Bargmann, J.P. Schadé (eds.), pp. 56–66. Amsterdam: Elsevier 1964.

MOORE, R.Y.: The nigrostriatal pathway in the cat. Abstr. 9th Internat. Congr. Anatomists, Leningrad, 1970, p. 89.

MOORE, R.Y., BHATNAGAR, R., HELLER, A.: Anatomical and chemical studies of a nigro-neostriatal projection in the cat. Brain Res. **30**, 119–135 (1971).

MORISON, R.S., DEMSEY, E.W.: A study of thalamocortical relations. Amer. J. Physiol. **135**, 281–292 (1942).

MORISON, R.S., DEMSEY, E.W.: Mechanism of thalamocortical augmentation and repetation. Amer. J. Physiol. **138**, 297–308 (1943).

MORISON, R.S., DEMPSEY, E.W., MORISON, B.R.: On the propagation of certain cortical potentials. Amer. J. Physiol. **131**, 744–751 (1941).

MOUNTCASTLE, V.B.: The neural replication of sensory events in the somatic afferent system. In: Brain and Conscious Experience. J.C. Eccles (ed.), pp. 85–115. Berlin: Springer 1966.

MOUNTCASTLE, V.B., POWELL, T.P.S.: Central nervous mechanisms subserving position sense and kinesthesis. Bull. Johns Hopk. Hosp. **105**, 173–200 (1959).

MULLAN, S., MOSELEY, R.D., JR., HARPER, P.V., JR.: The creation of deep cerebral lesions by small beta-ray sources implanted under guidance of fluoroscopic image intensifiers (as used in the treatment of Parkinson's disease). Amer. J. Roentgenol. **82**, 613–617 (1959).

MUNDINGER, F.: Contributo alle indicationi, alla dosimetria et alla tecnica di applicazione di radioisotopi per l'irradiazione interstiziale dei tumori cerebrali. Anat. e Chir. **3**, 21–28 (1958).

MUNDINGER, F.: Die stereotaktisch-operative Behandlung des Parkinson-Syndroms. Pathophysiologie, Ergebnisse und Indikationsstellung. Med. Klin. **29**, 1181–1186 (1963a).

MUNDINGER, F.: Die interstitielle Radio-Isotopen-Bestrahlung von Hirntumoren mit vergleichenden Langzeitergebnissen zur Röntgentiefentherapie. Acta neurochir. (Wien) **9**, 89–109 (1963b).

MUNDINGER, F.: Progress in the stereotaxic treatment of extrapyramidal motor disturbances and long-term follow up results after 1500 uni- and bilateral interventions. New York, Univ. Postgrad. Medical School, New York, 15. 9. 1964; Columbia-Univ. Neurolog. Inst., New York, 16. 9. 1964.

MUNDINGER, F.: Stereotaxic interventions on the zona incerta area for treatment of extrapyramidal motor disturbances and their results. Confin. neurol. (Basel) **26**, 222–230 (1965).

MUNDINGER, F.: Klinische und operative Erfahrungen mit der Subthalamotomie bei unwillkürlichen Bewegungsstörungen. Acta 25. Conventus Neuropsychiatrici et EEG Hungarici, Bundapestini, 1966, pp. 625–633.

MUNDINGER, F.: Subthalamische Reizeffekte und klinische Resultate der stereotaktischen Hochfrequenzausschaltung in der Zona incerta-Region beim Parkinsonsyndrom. Sowjetisch-westdeutsches Symposium über stereotaktische Neurochirurgie. Moskau, 31. 5.–1. 6. 1968; Leningrad, 4. u. 5. 6. 1968.

MUNDINGER, F.: Brain tumor therapy by interstitial application of radioactive isotopes. In: The Treatment of Brain Tumors with Interstitially Applied Radioactive Isotopes. Paoletti, P. von, Yen Wang (eds.), pp. 199–265. Springfield, Ill.: C.C Thomas 1970.

MUNDINGER, F., MINTERT, F., GURKMANN, O., KUTLAR, B.: Wie genau können vorausberechnete tiefliegende Hirnstrukturen mit dem stereotaktischen Zielverfahren mit einer Sonde erreicht werden? Dtsch. med. Wschr. **95**, 2224–2233 (1970).

MUNDINGER, F., POTTHOFF, P.: Encephalographische und klinische Untersuchungen zur funktionellen Somatotopik des Pallidum internum bei stereotaktischen Pallidotomien. Arch. Psychiat. **201**, 151–164 (1960).

MUNDINGER, F., POTTHOFF, P.: Messungen im Pneumencephalogramm zur intracerebralen Korrelationstopographie bei stereotaktischen Hirnoperationen unter besonderer Berücksichtigung der stereotaktischen Pallidotomie. Acta neurochir. (Wien) **9**, 196–214 (1961).

MUNDINGER, F., RIECHERT, T.: Ergebnisse der stereotaktischen Hirnoperationen bei extrapyramidalen Bewegungsstörungen aufgrund postoperativer und Langzeit-Untersuchungen. Dtsch. Z. Nervenheilk. **182**, 542–576 (1961).

MUNDINGER, F., RIECHERT, T.: Die stereotaktischen Hirnoperationen zur Behandlung extrapyramidaler Bewegungsstörungen (Parkinsonismus und Hyperkinesen) und ihre Resultate. Fortschr. Neurol. Psychiat. **31**, 1–66, 70–120 (1963).

MUNDINGER, F., RIECHERT, T.: Indikation und Langzeitergebnisse von 1400 uni- und bilateralen stereotaktischen Eingriffen beim Parkinson-Syndrom. Wien. Z. Nervenheilk. **23**, 147–177 (1966).

MUNDINGER, F., RIECHERT, T., GABRIEL, E.: Untersuchungen zu den physikalischen und technischen Voraussetzungen einer dosierten Hochfrequenzkoagulation bei stereotaktischen Hirnoperationen. Zbl. Chir. **85**, 1051–1063 (1960).

MUNDINGER, F., SCHOLLER, K.L.: Reaktionen des hämatopoetischen Systems bei gezielten Hirnoperationen. Acta neurochir. (Wien) Suppl. **III**, 147–152 (1956a).

MUNDINGER, F., SCHOLLER, K.L.: Hämatopoetische Reaktionen bei stereotaktischen Operationen. Proc. Cong. Europ. Ges. f. Haematologie, Freiburg, pp. 252–255. Berlin-Göttingen-Heidelberg: Springer 1956b.

MUNDINGER, F., UHL, H.: Über die Genauigkeit der röntgenologischen Zielpunktbestimmung bei stereotaktischen Hirnoperationen. Fortschr. Röntgenstr. **103**, 419–431 (1965).

MUNDINGER, F., UHL, H.: Investigations into possible roentgenographic errors in stereotaxics. Confin. neurol. (Basel) **28**, 24–28 (1967).

MUNDINGER, F., ZINSSER, O.: Variationen der oralen Ventralkerne des Thalamus. Untersuchungen zur Stereotaxie extrapyramidal-motorischer Erkrankungen. Arch. Psychiat. Nervenkr. **207**, 342–359 (1965).

MURATOFF, W.: Secundäre Degeneration nach Zerstörung der motorischen Sphäre des Gehirns in Verbindung mit der Frage von der Localisation der Hirnfunktionen. Arch. Anat. Physiol., Anat. Abt., 97–116 (1893).

NAKANO, K., HASSLER, R.: The thalamo-cortical connections of the subareas of the motor cortex in the cat, demonstrated by horseradish peroxydase technique. (In preparation 1979).

NARABAYASHI, H.: Procaine oil blocking of pallidum in cases of athetose double. Psychiat. neurol. jap. **54**, 672–677 (1952).

NARABAYASHI, H.: Neurophysiological ideas on pallidotomy and ventrolateral thalamotomy for hyperkinesis. Confin. neurol. (Basel) **22**, 291–303 (1962).

NARABAYASHI, H.: Physiological analysis of ventrolateral thalamotomy for rigidity and tremor. Confin. neurol. (Basel) **26**, 264–268 (1965).

NARABAYASHI, H.: Further results of thalamic surgery (in the sub-ventrolateral area and the CM) in athetosis and spasticity. Confin. neurol. (Basel) **29**, 256 (1967).

NARABAYASHI, H.: Functional differentiation in and around the ventrolateral nucleus of the thalamus based on experience in human stereoencephalotomy. Johns Hopk. med. J. **122**, 295–300 (1968).

NARABAYASHI, H.: Importance of muscle tone in production or modification of tremorous movements. In: Parkinson's Disease. Rigidity, Akinesia, Behavior. J. Siegfried (ed.), Vol. 2, pp. 27–36. Bern: H. Huber 1972.

NARABAYASHI, H.: Abnormal movements and muscle tone – clinical and physiological observation. Abstract No. 11. World Congr. Neurol., Amsterdam, Sept. 1977.

NARABAYASHI, H., NAGAHATA, M., NAGAO, T., SHIMAZU, H.: A new classification of cerebral palsy based upon neurophysiologic considerations. Confin. neurol. (Basel) **25**, 378–392 (1965).

NARABAYASHI, H., NAGAO, T.: Ten year follow-up of a case of Raynaud's disease treated by pallidotomy. Confin. neurol. (Basel) **34**, 152–155 (1972).

NARABAYASHI, H., OKUMA, T.: Procaine oil blocking of the globus pallidus for the treatment of rigidity and tremor of parkinsonism. Proc. Jap. Acad. **29**, 134–137 (1953).

NARABAYASHI, H., OKUMA, T., SHIKIBA, S.: Procaine oil blocking of the globus pallidus. Arch. Neurol. **75**, 36–48 (1956).

NASHOLD, B.S., WILSON, W.P.: Olfactory hallucinations evoked from stimulation of human thalamus. Confin. neurol. (Basel) **32**, 298–307 (1970).

NAUTA, W.J.H.: Silver impregnation of degenerating axonas. In: New Res. Techniques of Neuroanatomy. W.F. Windle (ed.), pp. 17–26. Springfield Ill.: C.C Thomas 1957.

NISHIMOTO, A., MATSUMOTO, K.: Bilateral, three negative, cortical evoked potentials after unilateral stimulation of the thalamic VL nucleus in man. Confin. neurol. (Basel) **32**, 349–366 (1970).

NOTHNAGEL, H.: Experimentelle Untersuchungen über die Funktionen des Gehirns. Virchows Arch. path. Anat. **57**, 184–227 (1873).

NOTHNAGEL, H.: Zur Diagnose der Sehhügelerkrankungen. Z. klin. Med. **16**, 424–430 (1889).

OBRADOR, S.: A simplified neurosurgical technique for approaching and damaging the region of the globus pallidus in Parkinson's disease. J. Neurol. Neurosurg. Psychiat 10, 20–47 (1947).

OBRADOR, S., DIERSSEN, G.: Personal experience in the treatment of parkinsonian symptoms with subcortical operations. In: 1st Intern. Congr. Neurol. Sci. L. van Bogaert, J. Radermecker (eds.), Vol. 2, pp. 119–122. London: Pergamon 1959.

OLIVIER, A., PARENT, A., SIMARD, H., POIRIER, L.J.: Cholinesterasic striatopallidal and striatonigral efferents in the cat and the monkey. Brain Res. **18**, 273–282 (1970).

PAGNI, C.A., WILDI, E., ETTORRE, G., INFUSO, L., MAROSSERO, F., CABRINI, G.P.: Anatomic verification of lesions which abolished tremor and rigor in Parkinsonism. Confin. neurol. (Basel) **26**, 291–294 (1965).

PARIZEK, J., HASSLER, R., BAK, I.J.: Light and electron microscopic autoradiography of substantia nigra of rat after intraventricular administration of tritium labelled norepinephrine, dopamine, serotonin and the precursors. Z. Zellforsch. **115**, 137–148 (1971).

PARKINSON, J.: An essay on the shaking palsy. London 1817. [Reprinted: Arch. Neurol. (Chic.) **7**, 682–710 (1922).]

PAYR, E.: Unterscheidung eines motorischen Rindenfeldes bei Athetose. Münch. med. Wschr. **68**, 1570 (1921).

PELLEGRINO DE IRALDI, A., ZIEHER, L.M., DE-ROBERTIS, E.: Ultrastructure and pharmacological studies of nerve endings in the pineal organ. In: Progress in Brain Research. A. Kappers, T.P. Schadé (eds.), Vol. 10, pp. 389–422. Amsterdam: Elsevier 1965.

PENFIELD, W., WELCH, K.: The supplementary motor area of the cerebral cortex. Arch. Neurol. Psychiat. (Chic.) **66**, 289–317 (1951).

PETERSON, E.E., MAGOUN, H.W., MCCULLOCH, W.S., LINDSLEY, D.B.: Production of postural tremor. J. Neurophysiol. **12**, 371–384 (1949).

PHILLIPS, C.G., POWELL, T.P., WIESENDANGER, M.: Projections from low-threshold muscle afferents of hand and forearm to area 3a of baboons cortex. J. Physiol. (Lond.) **217**, 419–446 (1971).

POIRIER, L.J.: Experimental and histological study of midbrain dyskinesias. J. Neurophysiol. **23**, 534–545 (1960).

POIRIER, L.J.: Production expérimentale du tremblement postural. Rev. canad. Biol. **20**, 137–142 (1961).

POIRIER, L.J.: Neuroanatomical study of an experimental postural tremor in monkeys. 2nd Symp. on Parkinson's Disease. Suppl. to J. Neurosurg., Jan. 1966, Part II, 191–199 (1966).

POIRIER, L.J.: Physiopathology of akinesia. In: Parkinson's Disease. Rigidity, Akinesia, Behavior. J. Siegfried (ed.), Vol. 1, pp. 115–126. Bern: H. Huber 1972.

POIRIER, L.J., SOURKES, T.L.: Influence du Locus niger sur la concentration des catécholamines du striatum. J. Physiol. (Paris) **56**, 426–427 (1964).

POIRIER, L.J., SOURKES, T.L.: Influence of the substantia nigra on the catecholamine content of the striatum. Brain **88**, 181–192 (1965).

POIRIER, L.J., SOURKES, T.L., BOUVIER, G., GARABIN, S.: Striatal amines, experimental tremor and the effect of harmaline in the monkey. Brain **89**, 37–52 (1966).

POLLOCK, L.J., DAVIS, L.: Studies in decerebration. II. An acute decerebrate preparation. Arch. Neurol. (Chic.) **12**, 288–293 (1924).

POLLOCK, L.J., DAVIS, L.: The reflex activities of a decerebrate animal. J. comp. Neurol. **50**, 377–411 (1930a).

POLLOCK, L.J., DAVIS, L.: Muscle tone in parkinsonian states. Arch. Neurol. Psychiat. (Chic.) **23**, 303–317 (1930b).

POTTHOFF, P.C., TETTEH, J., RIECHERT, T.: Postencephalographic psycho-organic syndrome in Parkinsonism. Confin. neurol. (Basel) **34**, 285–294 (1972).

PRECHT, W.: Tremorregistrierung als ein Mittel zur Objektivierung des Reizeffektes während stereotaktischer Eingriffe beim Parkinsonsyndrom. Dissertation Freiburg, 1963.

PRUS, J.: Die Leitungsbahnen und Pathogenese der Rindenepilepsie. Wien. klin. Wschr. **11**, 857–863 (1898).

PUTNAM, T.J.: Results of treatment of athetosis by section of extrapyramidal tracts in the spinal cord. Arch. Neurol. Psychiat. (Chic.) **39**, 258–275 (1938).

PUTNAM, T.J.: Treatment of unilateral paralysis agitans by section of the lateral pyramidal tract. Arch. Neurol. Psychiat. (Chic.) **44**, 950–976 (1940).

RAND, R.W., MARKHAM, C.H.: Hyperkinetic syndromes following thalamectomy and pallidectomy. Surg. Forum **10**, 800–803 (1960).

RAND, R.W., MARKHAM, C.H.: Cryothalamectomy of Parkinson's disease. Calif. Med. **101**, 248–252 (1964).

RASMUSSEN, T.B., HARPER, P.V., YUHL, E., BERGENSTAL, D.M.: The destruction of the pituitary gland in metastatic carcinoma with Yttrium 90 pellets. Semiannual Rep. Atomic Energ. Comm. Argonne Cancer Res. Hosp. Univ. Chic. **3**, 17 (1955).

RIBSTEIN, M.: Exploration du cerveau humain par électrodes profondes. Electroenceph. clin. Neurophysiol. **16**, 1–129 (1960).

RIECHERT, T.: Fortschritte und Ausblicke der Neurochirurgie. Monatsk. ärztl. Fortb. **12**, 210–212 (1962a).

RIECHERT, T.: Das stereotaktische Operationsverfahren bei der Schmerzbekämpfung. Dtsch. med. Wschr. **87**, 1177–1179 (1962b).

RIECHERT, T.: The stereotactic technique and its application in extrapyramidal hyperkinesia. Confin. neurol. (Basel) **34**, 325–330 (1972).

RIECHERT, T.: The contribution of the stereotactic method to brain research and neurosurgical technique. Modern Aspects of Neurosurgery. Excerpta med. (Amst.) **3**, 4–12 (1973).

RIECHERT, T., GABRIEL, E.: Eine neue chirurgische Methode zur Ausschaltung von biologischem Gewebe durch induktive Erwärmung (Indukoagulation). Dtsch. med. Wschr. **92**, 513–516 (1967).

RIECHERT, T., HASSLER, R.: Die Methodik der gezielten Hirnoperationen (Film). Zbl. ges. Neurol. Psychiat. **122**, 26 (1953).

RIECHERT, T., HASSLER, R., MUNDINGER, F., BRONISCH, F., SCHMIDT, K.: Pathologic-anatomical findings and cerebral localization in stereotactic treatment of extrapyramidal motor disturbances in multiple sclerosis. Confin. neurol. (Basel) **37**, 24–40 (1975).

RIECHERT, T., KRAINICK, J.U.: Application of inductive coagulation to produce reversible nerve tissue damage. Special topics in Stereotaxis. Hippokrates (Stuttg.) 121–129 (1971).

RIECHERT, T., MUNDINGER, F.: Beschreibung und Anwendung eines Zielgerätes für stereotaktische Hirnoperationen (II. Modell). Acta neurochir., Suppl. III, 308–337 (1956).

RIECHERT, T., MUNDINGER, F.: Ein kombinierter Zielbügel mit Bohraggregat zur Vereinfachung stereotaktischer Hirnoperationen. Arch. Psychiat. Nervenkr. **199**, 377–385 (1959).

RIECHERT, T., MUNDINGER, F.: Stereotaktische Geräte (Stereotaxic instruments). In: Introduction to Stereotaxis with an Atlas of Human Brain. G. Schaltenbrand, P. Bailey (eds.), pp. 437–471. Stuttgart: Thieme 1959.

RIECHERT, T., MUNDINGER, F., GABRIEL, E.: Hochfrequenz-Koagulation und lokalisierte Vereisung bei stereotaktischen Hirnoperationen. Selecta, Ausgabe A, **6**, 1351–1354 (1964).

RIECHERT, T., RICHTER, D.: Operative Behandlung des Tremors der Multiplen Sklerose und des essentiellen Tremors. Münch. med. Wschr. **114**, 2015–2028 (1972).

RIECHERT, T., RICHTER, D.: Stereotaktische Operationen zur Behandlung des Tremors der Multiplen Sklerose. Schweiz. Arch. Neurol. Neurochir. Psychiat. **111**, 411–416 (1972).

RIECHERT, T., SCHWARZ, R.: Erfahrungen mit kortikalen und intrazerebralen Ableitungen der Hirnströme. Dtsch. med. Wschr. **77**, 1075–1077 (1952).

RIECHERT, T., WOLFF, M.: Zielgerät zur intracraniellen elektrischen Ableitung und Ausschaltung mit besonderer Berücksichtigung der Eingriffe am Trigeminus. Vortrag Deutsche Neurochirurgen-Tagung Bonn 13.9.1950. Nervenarzt **22**, 437 (1951).

RIECHERT, T., WOLFF, M.: Über ein neues Zielgerät zur intracraniellen elektrischen Ableitung und Ausschaltung. Arch. Psychiat. Nervenkr. **186**, 225–230 (1951a).

RIECHERT, T., WOLFF, M.: Die Entwicklung und klinische Bedeutung der gezielten Hirnoperationen. Med. Klin. **46**, 609–611 (1951b).

RIECHERT, T., WOLFF, M.: Die technische Durchführung von gezielten Hirnoperationen. Arch. Psychiat. Nervenkr. **190**, 297–316 (1953).

RIESE, W.: Beiträge zur Faseranatomie der Stammganglien. J. Psychol. Neurol. (Lpz.) **31**, 81–122 (1925).

RINVIK, E.: The cortico-nigral projection in the cat. An experimental study with silver impregnation methods. J. comp. Neurol. **126**, 241–254 (1966).

RINVIK, E.: The corticothalamic projection from the pericruciate and coronal gyri in the cat. An experimental study with silver-impregnation methods. Brain Res. **10**, 79–119 (1968).

RINVIK, E.: Demonstration of nigrothalamic connection in the cat by retrograde axonal transport of horseradish peroxidase. Brain Res. **90**, 313–318 (1975).

RINVIK, E., GROFOVA, I.: Observations on the fine structure of the substantia nigra in the cat. Exp. Brain Res. **11**, 229–238 (1970).

RINVIK, E., GROFOVA, I., OTTERSEN, P.: Demonstration of nigrotectal projections in the cat by axonal transport of proteins. Brain Res. **112**, 388–394 (1976).

RONDOT, P.: Contraction du muscle lors de son raccourcissement passif dans certains états pathologiques. C.R. Soc. Biol. (Paris) **159**, 1524–1527 (1965).

ROSEGAY, H.: An experimental investigation of the connections between the corpus striatum and substantia nigra in the cat. J. comp. Neurol. **80**, 293–321 (1944).

ROSENSCHON, G., WECHSLER, W.: Zur Chemopallidektomie und der Ausschaltung des N. reticulatus thalami in der stereotaktischen Behandlung des Parkinsonismus. Arch. Psychiat. Nervenkr. **205**, 100–115 (1964).

ROUSSY, G., MOSINGER, M.: Etude anatomique et physiologique de d'Hypothalamus. Rev. neurol. **1**, 848–887 (1934).

von Sántha, K.: Zur Klinik und Anatomie des Hemiballismus. Arch. Psychiat. **84**, 664 (1928).

von Sántha, K.: Hemiballismus und Corpus Luysi: Anatomische und pathophysiologische Beiträge zur Frage des Hemiballismus nebst Versuch einer somatotopischen Lokalisation im Corpus Luysi. Z. Neurol. **141**, 321 (1932).

Sasaki, K., Prelević, S.: Excitatory and inhibitory influences of thalamic stimulation on pyramidal tract neurons. Exp. Neurol. **36**, 319–335 (1972).

Sasaki, K., Staunton, H.P., Dieckmann, G.: Characteristic features of augmenting and recruiting responses in the cerebral cortex. Exp. Neurol. **26**, 369–397 (1970).

Schachter, J.M., Bravo, G., Cooper, I.S.: Involuntary movement disorders following basal ganglia surgery in man. J. Neuropath. (Baltimore) **19**, 228–237 (1960).

Schaerer, J.P.: Stereoencephalotomy, observations in Parkinson's cases. Confin. neurol. (Basel) **22**, 351–355 (1962).

Schaltenbrand, G.: Thalamus und Schlaf. Allg. Z. Psychiat. **125**, 48–62 (1949).

Schaltenbrand, G., Bailey, P.: Einführung in die stereotaktischen Operationen, mit einem Atlas des menschlichen Gehirns, Vol. 3. Stuttgart: Thieme 1959.

Schiff, J.M.: Lehrbuch der Physiologie des Menschen. I. Muskel- und Nervenphysiologie. Lahr: Schauenburg 1858.

Schmidt, K.: Zur blutdrucksenkenden Wirkung der Stereoencephalotomie in den Stammganglien beim Parkinsonsyndrom. Dtsch. med. Forsch. **1**, 115–118 (1963).

Schmidt, K.: Hirndurchblutung und Hirnsauerstoffaufnahme beim Parkinsonsyndrom, ihre Beziehung zum Funktionszustand des vegetativen Nervensystems, zur motorischen und psychischen Symptomatik und zum Elektroenzephalogramm. Neurochirurgia (Stuttg.) **8**, 142–157 (1965).

Schmidt, K.: Kreislauf und Atmung beim Parkinsonsyndrom und deren Beeinflussung durch umschriebene und akute Ausschaltung im Stammganglienbereich. Fortschr. Med. **84**, 805–808 (1966).

Schmidt, K., Kaniak, G.: Die Atemfunktionsstörungen beim Parkinsonsyndrom. Neurochirurgia (Stuttg.) **3**, 182–193 (1960).

Schmidt, K., Umbach, W.: Simultanuntersuchungen der Hämodynamik und des Elektroencephalogramms zur fortlaufenden Kontrolle der Hirnfunktion während neurochirurgischer Eingriffe. Acta neurochir., Suppl. **VII**, 510–513 (1961).

Segundo, J.P., Machne, X.: Unitary responses to afferent volleys in lenticular nucleus and claustrum. J. Neurophysiol. **19**, 325–339 (1956).

Selby, G.: Stereotactic surgery for the relief of Parkinson's disease. 2. An analysis of the results in a series of 303 patients (413 operations). J. Neurol. Sci. **5**, 343–375 (1967).

Selby, G.: Long-term treatment of Parkinson's disease with L-dopa: A clinical study of 148 patients. In: Advances in Parkinsonism. W. Brinkmayer, O. Hornykiewicz (eds.), pp. 473–482. Basel: Editiones "Roche" 1976.

Sheer, D.E.: Brain and behavior: The background of interdisciplinary research. In: Electrical Stimulation of the Brain. D.E. Sheer (ed.), pp. 3–21. Austin: Univ. of Texas Press 1961.

Sherrington, C.S.: The integrative action of the nervous system. Cambridge: Cambridge University Press 1947.

Shimazu, H., Hongo, T., Kubota, K., Narabayashi, N.: Rigidity and spasticity in man. Electromyographic analysis with reference to the role of the globus pallidus. Arch. Neurol. (Chic.) **6**, 10–17 (1962).

Siegfried, J.: Die Parkinsonsche Krankheit und ihre Behandlung. Wien u. New York: Springer 1968.

Sloan, N., Jasper, H.H.: Studies on the regulatory functions of the limbic cortex. Electroenceph. clin. Neurophysiol. **2**, 317–327 (1950).

Smith, M.C.: Stereotactic operations for Parkinson's disease. Anatomical observations. Mod. Trends Neurol. **4**, 21–52 (1967).

Smith, M.C.: Eight cases of hemiballismus: historical findings. J. Neurol. (Brux.) **32**, 66–67 (1969).

Sontag, K.-H., Wand, P.: Decrease of muscle rigidity by dimethylaminoadamantan (DMMA) in intercollicularly decerebrated cats. Arzneimittel-Forsch. **23**, 1737–1739 (1973).

Sourkes, T.L., Poirier, L.J.: Neurochemical bases of tremor and other disorders of movement. Canad. med. Ass. J. **94**, 53–60 (1966).

Spatz, H.: Über Stoffwechseleigentümlichkeiten der Stammganglien. Z. ges. Neurol. Psychiat. **78**, 641–648 (1922).

Spatz, H.: Physiologie und Pathologie der Stammganglien. In: Handbuch der normalen und pathologischen Physiologie, Vol. X, pp. 318–417. Bethe u. Bergmann (eds.) 1927.

Spatz, H.: Ergebnis der anatomischen Untersuchung von 70 Fällen von Encephalitis epidemica. Zbl. Neurol. **56**, 435–437 (1930).

Spatz, H.: Anatomie des Mittelhirns. In: Handbuch der Neurol., Vol. I, pp. 474–540. Bumke u. Foerster (eds.). Berlin: Springer 1935

Spatz, H.: „Die systematischen Atrophien." Eine wohlgekennzeichnete Gruppe der Erbkrankheiten des Zentralnervensystems. Arch. Psychiat. Nervenkr. **108**, 1–18 (1938).

Spiegel, E.A.: Die Bedeutung des Forelschen Feldes für die Neurochirurgie und Neurophysiologie. Acta neurochir. **13**, 292–304 (1965).

Spiegel, E.A., Wycis, H.T.: Physiological and psychological results of thalamotomy. Proc. roy. Soc. Med. **42**, 84–92 (1949).

Spiegel, E.A., Wycis, H.T.: Thalamic recordings in man with special reference to seizure discharges. Electroenceph. clin. Neurophysiol. **2**, 23–27 (1950).

Spiegel, E.A., Wycis, H.T.: Stereoencephalotomy (Thalamotomy and Related Procedures). New York: Grune and Stratton 1952.

Spiegel, E.A., Wycis, H.T.: Stimulation of the brain stem and basal ganglia in man. In: Electrical Stimulation of the Brain. D.E. Sheer (ed.), pp. 487–497. Austin: Univ. of Texas Press 1961a.

Spiegel, E.A., Wycis, H.T.: Chronic implantation of intracerebral electrodes in humans. In: Electrical Stimulation of the Brain. D.E. Sheer (ed.), pp. 37–44. Austin: Univ. of Texas Press 1961b.

SPIEGEL, E.A., WYCIS, H.T.: Stereoencephalotomy. Part II. Clinical and Physiological Application. New York: Grune and Stratton 1962.
SPIEGEL, E.A., WYCIS, H.T., BAIRD, H.W., SZEKELY, E.G.: Functional state of basal ganglia in extrapyramidal and convulsive disorders. Arch. Neurol. (Chic.) **75**, 167–174 (1956).
SPIEGEL, E.A., WYCIS, H.T., FREED, M., LEE, A.L.: Stereoencephalotomy. Proc. Soc. exper. Biol. **69**, 175–177 (1948).
SPIEGEL, E.A., WYCIS, H.T., MARKS, M., LEE, A.L.: Stereotaxic apparatus for operations on the human brain. Science **106**, 349–350 (1947).
SPIEGEL, E.A., WYCIS, H.T., SHAY, H., CONGER, K.B., FISCHER, H.K.: Effects of lesion of human thalamus in region of dorsomedial nuclei. Fed. Proc. **9**, 119–120 (1950).
SPIEGEL, E.A., WYCIS, H.T., SZEKELY, E.G., ADAMS, J., FLANAGAN, M., BAIRD, H.W., III: Campotomy in various extrapyramidal disorders. J. Neurosurg. **20**, 871–884 (1963).
SPIEGEL, E.A., WYCIS, H.T., SZEKELY, E.G., BAIRD, H.W., III, ADAMS, J., FLANAGAN, M.: Campotomy. Trans. Amer. neurol. Ass. **87**, 240–242 (1962).
SPULER, H., SZEKELY, E.G., SPIEGEL, E.A.: Stimulation of the ventrolateral region of the thalamus. Arch. Neurol. **6**, 208–219 (1962).
STARLINGER, J.: Die Durchschneidung beider Pyramiden beim Hunde. Jber. Psychiat. Neurol. **15**, 1–42 (1897).
STARZL, T.E., TAYLOR, C.W., MAGOUN, H.W.: Ascending conduction in the reticular activating system with special reference to the diencephalon. J. Neurophysiol. **14**, 461–478 (1951).
STAUFFER, H.M., SNOW, L.B., ADAMS, A.B.: Roentgenologic recognition of habenular calcification as distinct from calcification in the pineal body. Amer. J. Roentgenol. **70**, 83–89 (1953).
STEG, G.: Efferent muscle innervation and rigidity. Acta physiol. scand. **61**, Suppl. 225, 1–53 (1964).
STEG, G.: Biochemical aspects of rigidity. In: Parkinson's Disease. Rigidity, Akinesia, Behavior. J. Siegfried (ed.), Vol. 1, pp. 47–63. Bern: H. Huber 1972.
STEPHAN, H.: Vergleichend anatomische Untersuchungen an Insektivorengehirnen. II. Oberflächenmessungen am Allocortex im Hinblick auf funktionelle und phylogenetische Probleme. Morph. Jb. **97**, 123–142 (1956).
STRUPPLER, A., LÜCKING, C.H., ERBEL, F.: Neurophysiological findings during stereotactic operations in thalamus and subthalamus. Confin. neurol. (Basel) **34**, 70–73 (1972).
STRUPPLER, A., STRUPPLER, E.: Veränderungen der Motoneuronaktivität auf elektrischen Reiz eines Thalamuskerns (V.o.a) während stereotaktischer Parkinson-Operationen. Arch. Psychiat. Nervenkr. **203**, 483–499 (1962).
STUMPF, C.: The fast component in the electrical activity of rabbit's hippocampus. Electroenceph. clin. Neurophysiol. **18**, 477–486 (1965).
SUGITA, K., MUTSUGA, N., TAKAOKA, Y., DOI, T.: Results of stereotaxic thalamotomy for pain. Confin. neurol. (Basel) **34**, 265–274 (1972a).
SUGITA, K., TAKAOKA, Y., MUTSUGA, N., TAKEDA, A., HIROTA, T.: Correlation between anatomically calculated target points and physiologically determined points in stereotaxic surgery. Confin. neurol. (Basel) **34**, 84–93 (1972b).
SZABO, J.: Topical distribution of the striatal efferents in the monkey. Exp. Neurol. **5**, 21–36 (1962).
SZABO, J.: The efferent projections of the putamen in the monkey. Exp. Neurol. **19**, 463–476 (1967).
SZABO, J.: A silver impregnation study of nigrostriate projections in the cat. Anat. Rec. **169**, 441 (1971).
SZABO, J.: Strionigral and nigrostrial connections. In: Basal Ganglia – Cellular and functional aspects. R. Hassler, J.F. Christ (eds.). Appl. Neurophysiol. **42**, 9–12 (1979).

TALAIRACH, J., DE, AJURIAGUERRA, J., DAVID, M.: Etudes stéréotaxiques des structures encéphaliques profondes chez l'homme. Presse méd. **28**, 605–609 (1952).
TALAIRACH, J., DAVID, M., TOURNOUX, P., CORREDOR, H., KVASINA, T.: Atlas d'anatomie stéréotaxique. Paris: Massone & Cie. 1957.
TALAIRACH, J., HÉCAEN, H., DAVID, M., MONNIER, M., DE AJURIAGUERRA, J.: Recherches sur la coagulation thérapeutique des structures sous-corticales chez l'homme. Rev. neurol. **81**, 4–24 (1949).
TALAIRACH, J., TOURNOUX, P.: Apparail de stéréotaxie hypophysaire pour voie d'abord nasale. Neuro-chirurgie **1**, 127–131 (1955).
TASKER, R.R., RICHARDSON, P., REWCASTLE, B., EMMERS, R.: Anatomical correlation on detailed sensory mapping of the human thalamus. Confin. neurol. (Basel) **34**, 184–196 (1972).
TITECA, L., VAN BOGAERT, L.: Herido-degenerative hemiballismus. A contribution to the question of primary atrophy of the corpus Luysii. Brain **69**, 251–263 (1946).
TOKIZANE, T., SHIMAZU, H.: Functional differentiation of human skeletal muscle. Tokyo: University of Tokyo Press 1964.
TRAVIS, A.M.: Neurological deficiencies after ablation of the precentral motor area in Macaca mulatta. Brain **78**, 155–173 (1955a).
TRAVIS, A.M.: Neurological deficiencies following supplementary motor area lesions in Macaca mulatta. Brain **78**, 174–198 (1955b).
TRÉTIAKOFF, C.: Contribution à l'étude de l'anatomie pathologique du locus niger de Soemmering avec quelques déductions rélatives à la pathogénie des troubles du tonus musculaire de la maladie de Parkinson. Thèse de Paris 1919.

UMBACH, W.: Tiefen- und Cortexableitungen während stereotaktischer Operationen am Menschen. Acta med. belg. **1957**, 161–170.
UMBACH, W.: Electrophysiological and clinical observations in 1280 stereotactic operations in man. Excerpta med., Int. Congr. Ser. No. 36, 152–153 (1961).
UMBACH, W.: Elektrophysiologische und vegetative Effekte bei stereotaktischer Reizung und Ausschaltung im menschlichen Hirn. In: Lectures on the Diencephalon. W. Bargmann, J.P. Schadé (eds.), pp. 46–55. Amsterdam: Elsevier 1964.
UMBACH, W.: Elektrophysiologische und vegetative Phänomene bei stereotaktischen Hirnoperationen. Berlin-Heidelberg-New York: Springer 1966a.

UMBACH, W.: Vegetative und emotionale Reaktionen auf umschriebene intracerebrale Reizung und Ausschaltung beim Menschen. In: Abhdlg. Dtsch. Akademie d. Wissensch., Nr. 2, pp. 389–397. Berlin: Akademie-Verlag 1966b.

UMBACH, W., EHRHARDT, K.J.: Ableitungen mit Mikroelektroden in den Stammganglien des Menschen. Arch. Psychiat. Nervenkr. **207**, 106–113 (1965a).

UMBACH, W., EHRHARDT, K.J.: Micro-electrode recording in the basal ganglia during stereotaxic operations. Confin. neurol. (Basel) **26**, 315–317 (1965b).

UMBACH, W., RIECHERT, T.: Bewußtseinsstörungen unter dem Bild akinetisch-mutistischer Verhaltensweisen nach stereotaktischen Ausschaltungen in den Stammganglien. Arch. Psychiatr. **204**, 96–112 (1963).

UMBACH, W., RIECHERT, T.: Elektrophysiologische und klinische Ergebnisse stereotaktischer Eingriffe im limbischen System bei temporaler Epilepsie. Nervenarzt **35**, 482–488 (1964).

USUNOFF, K., DIMOV, G., DIMOV, S.: Ascending efferent nigral fibers to diencephalon and telencephalon. C.R. Acad. bulg. Sci. **25**, 549–552 (1972).

USUNOFF, K.G., HASSLER, R., ROMANSKY, K., USUNOVA, P., WAGNER, A.: The nigrostriatal projection in the cat. Part 1. Silver Impregnation Study. J. Neurol. Sci., **28**, 265–288 (1976).

USUNOFF, K.G., HASSLER, R., WAGNER, A., BAK, I.J.: The efferent connections of the head of the caudate nucleus in the cat: an experimental morphological study with special reference to a projection to the raphe nuclei. Brain Res. **74**, 143–148 (1974).

VAN BOGAERT, L.: Aspect clinique et pathologique des atrophies pallidales et pallidoluysienes progressives. J. Neurol. Neurosurg. **9**, 128–134 (1946).

VAN BUREN, J.M., BORKE, R.C.: Variations and Connections of the Human Thalamus. Vols. 1, 2. Berlin-Heidelberg-New York: Springer 1972.

VAN BUREN, J.M., MACCUBBIN, D.A.: A outline atlas of the human basal ganglia with estimation of anatomical variants. J. Neurosurg. **19**, 811–839 (1962).

VAN MANEN, J.: Stereotactic methods and their applications in disorders of the motor system. In: Van Corcum and Comp. N.V. Dr. H.J. Prakke and H.M.G. Prakke 1967, pp. 1–230.

VOGT, C.: La myéloarchitecture du thalamus du cercopithèque. J. Psychol. Neurol. (Lpz.) **12**, 285–324 (1909).

VOGT, C.: Quelques considérations générales à propos du syndrome du corps strié. J. Psychol. Neurol. (Lpz.) **18**, 479–488 (1911).

VOGT, C., VOGT, O.: Zur Kenntnis der elektrisch erregbaren Hirnrindengebiete bei den Säugetieren. J. Psychol. Neurol. (Lpz.) **8**, 277–456 (1907).

VOGT, C., VOGT, O.: Erster Versuch einer pathologisch-anatomischen Einteilung striärer Motilitätsstörungen nebst Bemerkungen über seine allgemeine wissenschaftliche Bedeutung. J. Psychol. Neurol. (Lpz.) **24**, 1–19 (1919).

VOGT, C., VOGT, O.: Zur Lehre der Erkrankungen des striären Systems. J. Psychol. Neurol. (Lpz.) **25**, Erg.-H. 631–846 (1920).

VOGT, C., VOGT, O.: Sitz und Wesen der Krankheiten im Lichte der topistischen Hirnforschung und des Variierens der Tiere. I. J. Psychol. Neurol. (Lpz.) **47**, 237–457 (1937).

VOGT, C., VOGT, O.: Thalamusstudien, I–III. J. Psychol. Neurol. (Lpz.) **50**, 32–154 (1941).

VOGT, C., VOGT, O.: Morphologische Gestaltungen unter normalen und pathologischen Bedingungen. J. Psychol. Neurol. (Lpz.) **50**, 161–524 (1942).

WAGNER, A.: Veränderungen der Gamma-Aktivität durch Reizungen im Zwischen- und Mittelhirn bei der Katze. In: Progress in Brain Research. W. Bargmann, J.P. Schadé (eds.), Vol. 5, pp. 67–73. Amsterdam: Elsevier 1964.

WAGNER, A.: Über die häufig fehlende Korrelation zwischen der Aktivität der Muskelspindeln und derjenigen der Fusimotoren unter cerebraler Bahnung und Hemmung. Arch. Psychiat. Nervenkr. **206**, 525–536 (1965).

WAGNER, A., DUPELJ, M., HASSLER, R.: Activity of gamma and alpha motoneurons after reserpine and L-dopa administration. Proc. 25th Internat. Congr. Physiol. Sci., Vol. IX, Nr. 2763, München 1971.

WAGNER, A., HASSLER, R., KIM, J.S.: Striatal cholinergic enzyme activities following discrete centromedian nucleus lesions in cat thalamus. Abstr. Nr. 59, 5. Int. Meet. Int. Soc. Neurochem., p. 116, Barcelona 1975.

WAGNER, A., KALMRING, K.: The dynamic and static sensibility of the Ia affarents during electrical stimulation of the substantia nigra. Brain Res. **10**, 277–280 (1968).

WALKER, A.E.: An experimental study of the thalamocortical projection of the macaque monkey. J. comp. Neurol. **64**, 1–41 (1936).

WALKER, A.E.: The primate thalamus. Chicago: Chicago Univ. Press 1938.

WALKER, A.E.: Cerebral pedunculotomy for the relief of involuntary movements. Acta psychiat. scand. **24**, 723–726 (1949a).

WALKER, A.E.: Thalamocortical relationships. Electroenceph. clin. Neurophysiol. **1**, 451–454 (1949b).

WALKER, A.E., GREEN, H.D.: Electrical excitability of the motor face area; a comparative study in Primates. J. Neurophysiol. **1**, 152–165 (1938).

WALLER, W.H.: Progression movements elicited by subthalamic stimulation. J. Neurophysiol. **3**, 300–307 (1940).

WALSHE, F.M.R.: Observations on the nature of the muscular rigidity of paralysis agitans and on its relationship to tremor. Brain **47**, 159–177 (1924).

WARD, A.A., JR., MCCULLOCH, W.S., MAGOUN, H.W.: Production of an alterating tremor at rest in monkeys. J. Neurophysiol. **11**, 317–330 (1948).

WARD, A.A., JR., STERN, J.: Thalamic inhibition of the myotatic reflex in man. J. Neurosurg. **20**, 1033–1039 (1963).

WEISSCHEDEL, E.: Die zentrale Haubenbahn und ihre Bedeutung für das extrapyramidal-motorische System. Arch. Psychiat. **107**, 443–579 (1937).

WHITE, R.J., MACCARTY, C.S., BAHN, R.C.: Neuropathological review of brain lesions and inherent dangers in chemopallidectomy. Arch. Neurol. (Chic.) **2**, 12–18 (1960).

WHITTIER, J.R.: Ballism and the subthalamic nucleus (Nucleus hypothalamicus; Corpus Luysi). Arch. Neurol. (Chic.) **58**, 672–692 (1947).

WHITTIER, J.R., METTLER, F.A.: Studies on the subthalamus of the rhesus monkey. II. Hyperkinesia and other physiologic effects of subthalamic lesions, with special reference to the subthalamic nucleus of Luys. J. comp. Neurol. **90**, 319–370 (1949).

WILSON, S.A.K.: Progressive lenticular degeneration: A familial nervous disease associated with cirrhosis of the liver. Brain **34**, 295–509 (1912).

WILSON, S.A.K.: An experimental research into the anatomy and physiology of the corpus striatum. Brain **36**, 427–492 (1914).

WINKELMÜLLER, W.: Wirkung von Reizeffekten und Ausschaltungen der Substantia nigra auf das motorische Verhalten der freibeweglichen Katze. Acta neurochir. **24**, 269–303 (1972).

WISNIEWSKI, H., TERRY, R.D., HIRANO, A.: Neurofibrillary pathology. J. Neuropath. exp. Neurol. **29**, 163–176 (1970).

WYCIS, H.T., SPIEGEL, E.A.: Ansotomy in Paralysis agitans. Confin. Neurol. (Basel) **12**, 245–246 (1952).

WYSS, O.A.M.: Ein Hochfrequenzkoagulationsgerät zur reizlosen Ausschaltung. Helv. physiol. pharmacol. Acta **3**, 437–443 (1945).

YASARGIL, M.G., WYSS, O.A.M., KRAYENBÜHL, H.: Beitrag zur Behandlung extrapyramidaler Erkrankungen mittels gezielter Hirnoperationen. Schweiz. med. Wschr. **89**, 143–150 (1959).

YORK, D.H.: Alteration in spinal monosynaptic reflex produced by stimulation of the substantia nigra. In: Corticothalamic Projections and Sensorymotor Activities. T. Frigyesi, E. Rinvik, M.D. Yahr (eds.), pp. 445–447. New York: Raven Press 1972.

YORK, D.H.: Motor responses induced by stimulation of the substantia nigra. Exp. Neurol. **41**, 323–330 (1973).

YOSHIDA, M., PRECHT, W.: Monosynaptic inhibition of neurons of the substantia nigra by caudato-nigral fibers. Brain Res. **32**, 225–228 (1971).

ZIMMERMANN, H.M.: Cerebral apoplexy: Mechanism and differential diagnosis. N.Y.St.J. Med. **49**, 2153– (1949).

Subject Index

Abbreviations of Thalamic and Hypothalamic Structures

Abbreviations of Thalamic and Hypothalamic Structures

A.d	Nc. anterodorsalis
A.le	Ansa lenticularis
A.m	Nc. anteromedialis
A.pd	Ansa peduncularis
A.pr	Nc. anterior principalis
Amy	Amygdala
Aq	Aquaeductus cerebri
B	Nc. basalis
Br	Fasciculus diagonalis (BROCA)
Br.cj	Brachium conjunctivum
Br.c(o).if	Brachium colliculi inferioris
Br.co.s	Brachium colliculi superioris
CA	Cornu Ammonis
Ca.i	Capsula interna
Cb	comb system of the peduncle
Cd	Caudatum; Nc. caudatus
Ce	centre médian
Ce.mc	centre médian magnocellularis
Ce.pc	centre médian parvocellularis
Co	Nc. commissuralis
Co.a (C.a)	Commissura anterior
Co.hb	Commissura habenularum
Co.hyp.p	Commissura hypothalamica posterior
Co.p (C.p)	Commissura posterior
Co.s	Colliculus superior
Co.s.II.d	Commissura supraoptica dorsalis (MEYNERT)
Cu	Nc. cucullaris
D.im	Nc. dorsalis intermedius
D.im.e	Nc. dorsalis intermedius externus
D.im.i	Nc. dorsalis intermedius internus
D.o	Territorium dorso-orale
D.o.e	Nc. dorsalis oralis externus
D.o.i	Nc. dorsalis oralis internus
D.sf	Nc. dorsalis superficialis
Dec.Br.cj	Decussatio brachiorum conjunctivorum
F.Fo	Fasciculi tegmentales FORELI
F.r	Fasciculus retroflexus (MEYNERT)
Fa	Nc. fasciculosus (ex pedunculo inferiori thalami)
Fi	Fimbria
Fo.Mo (F.Mo)	Foramen interventriculare MONROI
Fx	Fornix
G.l	Geniculatum laterale
G.m	Geniculatum mediale
G.m.fa	Nc. geniculatus medialis fasciculosus
G.m.fi	Nc. geniculatus medialis fibrosus
G.m.li	Nc. geniculatus medialis limitans
G.m.mc	Nc. geniculatus medialis magnocellularis
Gl.pi	Glandula pinealis
H	(tegmental) field H (FOREL)
H_1	Fasciculus thalamicus H_1 (FOREL)
H_2	Fasciculus lenticularis H_2 (FOREL)
H.l.c	**Nc. hypothalamicus lateralis caudalis**
H.l.o	**Nc. hypothalamicus lateralis oralis**
Hb	Ganglion habenulae
Hb.ist	Nc. habenularis interstitialis
Hb.m	Nc. habenularis medialis (parvocellularis)
Hb.mc	Nc. habenularis magnocellularis
Hb.mix	Nc. habenularis mixtocellularis
i.La	Nc. intralaminaris
i.La.c	Nc. intralaminaris caudalis
i.La.d	Nc. intralaminaris dorsalis
i.La.if	Nc. intralaminaris inferior
i.La.im	Nc. intralaminaris intermedius
i.La.z	Nc. intralaminaris zentralis
Ist	Nc. interstitialis (CAJAL)
L.po	Territorium lateropolare
L.po.b	Nc. latero-polaris basalis
L.po.e	Nc. latero-polaris externus
L.po.i	Nc. latero-polaris internus
L.po.mc (mc)	Nc. latero-polaris magnocellularis
La.im	Lamella intermedia
La.l	Lamella lateralis
La.pa.i	Lamella pallidi interna
Li.m	Nc. limitans medialis
Li.opt.	Nc. limitans opticus
Li.pt	Nc. limitans portae
M	Territorium mediale
M.b	Nc. medialis basalis
M.c.e	Nc. medialis caudalis externus
M.c.i	Nc. medialis caudalis internus
m.F.B.	medial forebrain bundle
M.fa.a	Nc. medialis fasciculosus anterior
M.fa.p	Nc. medialis fasciculosus posterior
M.fi.a	Nc. medialis fibrosus anterior
M.fi.p	Nc. medialis fibrosus posterior
M.pL	Nc. medialis paralaminaris
Ma	Corpus mamillare
ma-th	Tractus mamillo-thalamicus
n.III	Radices nervi oculomotorii
Nc.III	Nc. oculomotorius
Ni	(substantia) nigra
p.Pd	Nc. peripeduncularis
Pa.e	Pallidum externum

Abbreviations of Thalamic and Hypothalamic Structures

Pa.i	Pallidum internum
pa.-s.th	Tractus pallido-subthalamicus
Pc.dm	Nc. parvocellularis dorsomedialis (hypothalami)
Pc.o	Nc. parvocellularis oralis (hypothalami)
Pd	Pedunculus cerebri
Pd.if	Pedunculus thalami inferior
Pf	**Nc. parafascicularis**
Pf.m	Nc. parafascicularis medialis
Pfx	Nc. parafornicalis (hypothalami)
Pfx.s	Nc. parafornicalis superior (hypothalami)
Pfx.v	Nc. parafornicalis ventralis (hypothalami)
Pi (pi)	Glandula pinealis
Pl.ch	Plexus chorioideus
Pm.o	Nc. paramedianus oralis
pr.co	Nc. praecommissuralis
pr.G	Nc. praegeniculatus
Pr.st	Nc. praestitialis
Pt	Nc. parataenialis
Pu	Pulvinar
Pu.ig	Pulvinar intergeniculatum
Pu.ig.fa	Nc. pulvinaris intergeniculatus fasciculosus
Pu.ig.g	Nc. pulvinaris intergeniculatus griseus
Pu.l	Pulvinar laterale
Pu.l.if	Nc. pulvinaris lateralis inferior
Pu.l.s	Nc. pulvinaris lateralis superior
Pu.m	Pulvinar mediale
Pu.m.d	Nc. pulvinaris medialis dorsalis
Pu.m.i	Nc. pulvinaris medialis internus
Pu.m.v	Nc. pulvinaris medialis ventralis
Pu.m.z	Nc. pulvinaris medialis zentralis
Pu.o.l	Nc. pulvinaris oralis lateralis
Pu.o.m	Nc. pulvinaris oralis medialis
Pu.o.v	Nc. pulvinaris oralis ventralis
Pu.sf	Pulvinar superficiale
Put	Putamen
Pv	Nc. paraventricularis (hypothalami)
Pv.c	Nc. paraventricularis caudalis (hypothalami)
Q	bundle Q (pallido-mesencephalic) of SANO
R	bundle R of SANO
R.pr.Lm (R.pL)	Radiatio praelemniscalis
Rt.c.v	Nc. reticulatus caudalis ventralis
Rt.c.z	Nc. reticulatus caudalis zentralis
Rt.gl	Nc. reticulatus geniculatus
Rt.im.d	Nc. reticulatus intermedius dorsalis
Rt.im.v	Nc. reticulatus intermedius ventralis
Rt.im.z	Nc. reticulatus intermedius zentralis
Rt.o.d	Nc. reticulatus oralis dorsalis
Rt.o.v	Nc. reticulatus oralis ventralis
Rt.o.z	Nc. reticulatus oralis zentralis
Rt.po	Nc. reticulatus polaris
Rt.po.d	Nc. reticulatus polaris dorsalis
Rt.po.v	Nc. reticulatus polaris ventralis
Rt.pu.d	Nc. reticulatus pulvinaris dorsalis
Rt.pu.v	Nc. reticulatus pulvinaris ventralis
Rt.pu.z	Nc. reticulatus pulvinaris zentralis
Ru	Nc. ruber
s.II	Nc. supraopticus
s.Hb	Nc. subhabenularis
S.th	**Nc. subthalamicus**
spi.-th (spi)	**Tractus spinothalamicus**
Spl	Splenium corporis callosi
St.m (St.md)	Stria medullaris thalami
st.-ni	Tractus strio-nigralis
St.t	Stria terminalis
T.tg.c	Tractus tegmenti centralis
Tb.l	Nc. tuberis lateralis (hypothalami)
Tb.M (TM)	Tubero-mamillar-complex
Tu	Tuber cinereum
Tu.l	Nc. tuberis lateralis (hypothalami)
V.c	Nc. ventralis caudalis
V.c.a.e	Nc. ventrocaudalis anterior externus
V.c.a.i	Nc. ventrocaudalis anterior internus
V.c.e	Nc. ventrocaudalis externus
V.c.i	Nc. ventrocaudalis internus
V.c.p.e	Nc. ventrocaudalis posterior externus
V.c.p.i	Nc. ventrocaudalis posterior internus
V.c.p.i.b	Nc. ventrocaudalis internus basalis
V.c.pc.e	Nc. ventrocaudalis parvocellularis externus
V.c.pc.i	Nc. ventrocaudalis parvocellularis internus
V.c.por	Nc. ventrocaudalis portae
V.t (V.ter)	Vena terminalis
Ve.l	Ventriculus lateralis
V.im	Nc. ventralis intermedius
V.im.e	Nc. ventro-intermedius externus
V.im.i	Nc. ventro-intermedius internus
V.o.a	Nc. ventro-oralis anterior
V.o.i	Nc. ventro-oralis internus
V.o.m	Nc. ventro-oralis medialis
V.o.p	Nc. ventro-oralis posterior
X	Chiasma
X	pallido-hypothalamic bundle X (FOREL)
Z.c.	Nc. zentralis caudalis
Z.i	Zona incerta
Z.i.c	Zona incerta caudalis
Z.i.v	Zona incerta ventralis
Z.im	Nc. zentralis intermedius
Z.im.e	Nc. zentro-intermedius externus
Z.im.i	Nc. zentro-intermedius internus
Z.o	Nc. zentralis oralis
II	Tractus opticus
III	Ventriculus tertius
IV	Ventriculus quartus

Printed in Great Britain
by Amazon